W9-APV-938

The FUTURE of the ARMY PROFESSION

with a Foreword by
General (Retired) Frederick M. Franks, USA

project directors
Don M. Snider
Gayle L. Watkins

edited by
Lloyd J. Matthews

United States Military Academy

Boston Burr Ridge, IL Dubuque, IA Madison, WI New York San Francisco St. Louis
Bangkok Bogotá Caracas Lisbon London Madrid
Mexico City Milan New Delhi Seoul Singapore Sydney Taipei Toronto

McGraw-Hill Higher Education

A Division of The ***McGraw-Hill*** *Companies*

THE FUTURE OF THE ARMY PROFESSION

Copyright © 2002 by The McGraw-Hill Companies, Inc. All rights reserved. Printed in the United States of America. Except as permitted under the United States Copyright Act of 1976, no part of this publication may be reproduced or distributed in any form or by any means, or stored in a data base retrieval system, without prior written permission of the publisher.

McGraw-Hill's Primis Custom Publishing consists of products that are produced from camera-ready copy. Peer review, class testing, and accuracy are primarily the responsibility of the author(s).

1 2 3 4 5 6 7 8 9 0 DOC DOC 0 9 8 7 6 5 4 3 2

ISBN 0-07-255268-9

Editor: Margaret Hollander/Ann Jenson
Cover Design: Fairfax Hutter
Printer/Binder: R. R. Donnelley & Sons Company
Cover photos courtesy United States Army
Top: U. S. Army soldier with Bosnian child, near Tusla, Bosnia by SFC Larry Lane
Bottom: U. S. Army photo by Sarah Underhill, Fort Benning, Georgia

Contents

Acknowledgements

This research project's roots go back to the 1991-96 tenure of Lt. Gen. Howard Graves, the 54th Superintendent of the Military Academy, as he began to consolidate a staff and faculty team to assist him over the five challenging years that lay ahead. Among these, he chose Gayle Watkins in 1991 to be a faculty member in the Eisenhower Program, which provided graduate-level education to the men and women serving as tactical officers within the U.S. Corps of Cadets. En route to West Point, Gayle completed a doctorate in sociology at Stanford University, choosing to study the social organization of work and professions—a serendipitous choice, given the present project. Lieutenant General Graves then chose Don Snider in 1995 to be the Olin Distinguished Professor of National Security Studies, a position he occupied for three years before joining the civilian professoriat of the Academy's Department of Social Sciences. Thus we first acknowledge our personal debt to Howard Graves, a superb Army officer, intellectual, and leader whose tenure at the Academy will forever reflect his own immense character and wisdom.

Secondly, we are indebted to those whose support made possible this anthology and the several conferences leading up to its publication. Primary among these are the project's executive director, Lt. Col. Kevin Dopf, and conference organizer, Ms. Jo May. Their boundless energy and enthusiasm sustained not only us but also the editor and the numerous other participants. We drew intellectual and personal inspiration from Lt. Gen. Walter Ulmer, USA Ret., one of the authors of the landmark 1970 *Study on Army Professionalism* and keynote speaker at the 2001 Senior Conference. Walt's timely and insightful guidance kept us on track throughout this endeavor. We also owe an enormous debt to the Smith Richardson Foundation and the Robert McCormick Tribune Foundation, which generously provided the financial support needed for this expansive project. The leaders and members of the West Point Department of Social Sciences provided a positive, supportive environment for our work, proving that scholarly research could flourish alongside extraordinary classroom teaching.

We also acknowledge the individual authors whose works appear in this volume. In most cases they have been close colleagues and friends over the lives of public service we have each pursued; in all cases they are American patriots and personally dedicated to this vitally important topic. To work in collaboration with them, from the first framework discussions to the final research report, has been a warm and genuine delight in professional camaraderie. From friends such as these, we can and shall always seek to learn. We must make special mention of the editor of this volume, Lloyd Matthews, who truly went above and beyond the call of duty to fashion a finished and integrated product from the works of so many independent authors; and of his wife Phyllis, who spent countless after-duty hours entering corrections and proofing text.

We also affirm our debt to Col. Kerry Pierce, USMA 1974, a Rhodes Scholar, School of Advanced Military Studies graduate, Professional Engineer, fellow Army officer, and close personal friend whose intellectual and organizational skills brought about a renewal of the study of officership at West Point in the late 1990s. Without our work with him in that effort, we would not have seen so clearly the needed redefinitions encompassed in the role of the Army officer in the 21st century and in the future of the Army profession; nor could we have sensed the urgent necessity to undertake this research project when we did.

And, of course, others have paid a perhaps unfair price because we chose to invest so much time in this project—beloved spouses, family, and friends. For their gracious forbearance we are simply grateful.

Don M. Snider
Gayle L. Watkins
West Point, New York

List of Illustrations

List of Tables

Foreword

Today as I write this foreword, the U.S. Army is doing its duty in the front lines of America's war against global terrorism in Afghanistan and the Philippines, as well as being on the front lines of freedom in other operations in the Balkans, the Sinai, Kuwait, and South Korea. In all, over 173,000 soldiers are deployed outside the United States, to include 24,000 mobilized reservists. Soldiers and leaders of the Army profession are performing these duties in the highest traditions of service to our nation.

President George W. Bush paid tribute to this performance of duty on 28 January 2002 in his State of the Union address to the Congress and the American people:

> When I called our troops into action, I did so with complete confidence in their courage and skill. And tonight, thanks to them, we are winning the war on terror. The men and women of our armed forces have delivered a message now clear to every enemy of the United States: even 7,000 miles away, across oceans and continents, on mountaintops and in caves, you will not escape the justice of this nation. . . . Our war on terror is well begun, but it is only begun. . . . History has called America and our allies to action, and it is both our responsibility and our privilege to fight freedom's fight.

Yet, at the same time the Army profession is executing these duties to defend our freedoms, it is also in the midst of a vital transformation. Even as our military takes the fight to the terrorists in this war, there is also a concurrent need to think about changing our defenses in deeply fundamental ways. The Army Chief of Staff, Gen. Eric Shinseki, in the fall of 1999 disseminated a far-reaching *Vision for the U.S. Army* to transform it into a more strategically responsive and dominant force at every point on the spectrum of military operations. He sees the Army's duty today as not only to win decisively this current war but also to accelerate Army transformation so as to win the next one. As Secretary of Defense Donald Rumsfeld recently told us, the transformation of the Army is a part of Defense Transformation, and "the war on terrorism is a transformational event."

Some early actions already indicate the clear path forward. Early in 2000, General Shinseki commissioned a series of studies of Army training and leader development. They began first with officers, then expanded to the noncommissioned officer study in 2001, and will finish in 2002 with warrant officers and civilians. These studies are demonstrating, once again, that at the very core of transforming the Army are, and must continue to be, its professionals. Vital to transforming the Army in this early decade of the 21st century are their dedication, commitment, and renewed focus on what it means to be an Army professional. General Shinseki confirmed as much in late November 2001:

> People are central to everything else we do in the Army. Institutions don't transform, people do. Platforms and organizations don't defend this nation, people do. And finally, units don't train, they don't stay ready, they don't grow and develop leadership; they don't sacrifice; and they don't take risks on behalf of the nation; people do.

The Army Chief of staff has established from this series of studies that there is now an imbalance between what the Army profession says it is and what it is proving to be in actual practice. The studies reported, after extensive interviews with members of the profession, that Army culture is "out of balance"—the profession's doctrine and ethos, its beliefs, are not consistent with many current operational practices:

> The soldiers interviewed in the field transmitted their thoughts in clear text and with passion. They communicated the same passion and dedication for selfless service to the nation and the Army as any generation before them. Pride in the Army, service to the nation, camaraderie, and Army values continue to strongly influence the decisions of officers and their spouses to make the Army a career. However, they see Army practices as being out of balance with Army beliefs.... the Army's service ethic and concepts of officership are neither well-understood nor clearly defined. They are also not adequately reinforced throughout an officer's career.

Upon receipt of these conclusions and recommendations in early 2001, General Shinseki directed a series of actions to reform Army practices in training and leader development. For example, one specific action now being implemented will "define and teach an Army Service Ethic and Officership throughout OES (Officer Education System) from Officer Basic Course through the War College."

So, with the path to transformation becoming increasingly more clear, how are Army professionals to think about their profession in such times as these? That is what this very timely anthology is all about, filled as it is with the results of extensive research on the future of the Army profession, research conducted for the first time within a new framework of "competitive professions." Clearly, it is the duty of all professionals to reinterpret continually their profession and their individual roles within it, particularly as the external world and our nation's role in that world change so radically. Such continuous examination is a vital strength of our Army profession and marks the very core of its members' ability to best serve our nation.

My own involvement in the Army profession, like that of so many others, has been one of a lifelong calling. I knew I made the right decision in life a long time ago. In that choice I was influenced by the World War II generation in my own family and in my community as I grew up in Pennsylvania. I wanted to earn the right to lead such great Americans if it ever again became necessary to defend our freedoms. At West Point, I was influenced by many positive role models who inspired in me a desire to lead the kind of life they had led up to that point. I trusted what they had told me about the profession.

After graduation with the West Point Class of 1959, I took my place as a young Army officer among the professional officers and NCOs in the 11th Armored Cavalry Regiment along the Iron Curtain in Germany. They reinforced what I had learned and seen at West Point. And later, when my mandatory service was completed, I chose to continue to serve, to stay on, as did so many others through Vietnam and through the Army renaissance of the 1970s and early 1980s. I chose to stay because of the insights and inspiration I had earlier received from my professional associates and because of a "hot blue flame" of commitment lit in me at Valley Forge General Hospital in the early 1970s.

What a life it has been! It was not always easy and not always fair, as the old song goes, but given the opportunity to make those choices all over again, I would not do anything other than choose to be a soldier and a member of the profession of arms in the United States Army. It is a profession I have been privileged to be a part of almost my entire life from the young age of my cadet days in 1955 until I retired from active duty on 1 December 1994. I am forever grateful to our country and to the American people for entrusting me with high responsibilities and with the lives of their sons and daughters in two wars and countless other duties. And I am forever grateful to the splendid men and women soldiers with whom it has been my privilege to serve in peace and war and for the steadfast friends my wife and I have come to know and admire and respect over those years.

In May 1999 the Superintendent of the United States Military Academy invited me to become the "visiting scholar" in the newly formed Simon Center for the Professional Military Ethic (SCPME). It was an honor to be asked, and without hesitation I accepted those duties. In a sense it was an opportunity to continue that commitment made a generation ago to our Army profession and to the nation we serve. The vision for SCPME is not only to develop in each cadet a self-concept of what it means to be an officer in the United States Army prior to graduation, but also to reach out to the Army and foster continuing education within the profession itself. Fulfilling that mission is one reason I have prepared this foreword.

My own sensing is that there has not been any lessening of the desire of Army professionals for professionalism. Thus, it is both timely and relevant to the Army's transformation that this text be made available to current and future Army professionals, as they dedicate themselves to renewing, and indeed transforming, the very essence of the profession.

I am aware that many might interpret such a momentous initiative as a sign that something is drastically "wrong." Nothing could be further from reality. One of the great strengths of the U.S. Army has been, and is, its ability to learn, renewing its considerable strengths and transforming where necessary to continue its role in the vanguard of honorable and effective service to our nation. If anything is "wrong" now, it is that the prevailing rate of change and pace of operations has, until very recently, simply outstripped the opportunities for needed reflection and re-examination.

To help correct that, Professors Don Snider and Gayle Watkins have done our Army profession a splendid service by initiating this research project, drawing on the work of many military and civilian researchers, and then compiling this anthology. The ideas, analysis, and conclusions are far-reaching, just what are needed to provoke candid reflection and to build renewed consensus regarding a more effective and relevant, but transformed, Army profession. Even more important is the new framework they provide for analysis of the Army as one of several professions, and nonprofessions, competing for the same jurisdiction and legitimacy before the American people. This is the reality the Army now faces, and to which it must adapt.

I believe we are at the end of one period of professionalism begun in the U.S. Army late in the 19th century, and are now, in the first decade of the 21st century, beginning another period. We are examining how the Army profession should adapt itself to the new set of strategic realities that confront us today. As before, the challenge is to adapt but also to retain the many continuities of our profession we know are essential to success on the battlefield. We must continue, indeed strengthen, our values and ethos from the past, especially the moral requirement for expertise in land warfare, remaining always able to fight and win decisively in close quarters against enemies of every sort. We must also adapt the profession to the new strategic conditions for the use of force, more integrally combined with other elements of national power and necessitating new inter-relationships with civilian agencies of our government in this information age. Such adaptations will include evolving civil-military relationships and forms of joint warfare unforeseen at the time of the Goldwater-Nichols Act in 1986. This anthology will give readers issues galore to think about, and more.

I would not expect every professional who studies this text to agree with every idea here; I take exception with some of those ideas myself. Yet I do agree with many of them, and provide this foreword to make that point. This text offers an accessible and most useful place for individual professionals, and for the profession's strategic leaders, to join together in dialogue and mutual reflection. I recall that in 1962, British Gen. Sir John Hackett delivered a splendid series of lectures at Trinity College in Cambridge. For a time the U.S. Army reprinted these lectures in a Department of the Army pamphlet titled "The Profession of Arms." About the U.S. Army, Hackett noted: "The years between 1860 and the World War saw the emergence of a distinctive American professional military ethic, with the American officer regarding himself as a member of a learned profession whose students are students for life."

Hackett was an unusually insightful soldier-scholar. To follow that military ethic today, as we must, demands broad professional re-invigoration in the context of the ongoing transformations both at the Defense level and within the Army. I firmly believe that only through such a concerted, introspective process will the profession complete—as was done earlier in the 1970s and 1980s—the successful transformation that is needed to fight effectively the wars we'll confront in this new century.

As we were making our final preparations in VII Corps before attacking into Iraq in February 1991, I was visiting one of our divisions. As I was explaining our attack plan to a group of soldiers, one of them stopped me and said, "Don't worry, general, we trust you." I was humbled by that statement and vowed to do everything I could as corps commander to continue to earn that trust in battle. I also realized that in an instant that soldier had captured the essence of what we do as professionals: we gain and maintain our soldiers' trust by executing our duties to the highest levels of professionalism. We owe nothing less to our soldiers, to our nation, to ourselves, and to our profession.

Frederick M. Franks, Jr.
General, U.S. Army Ret.
West Point, N.Y.
February 2002

Introduction

The Army today faces broad challenges. During the Cold War era, the Army prepared to fight and defeat a well-defined enemy with a similar force structure. The dramatic results of Operation Desert Storm during the Gulf War validated these efforts, but also served notice to other nations that the United States could not be symmetrically opposed. In the ensuing decade, faced with asymmetric challenges across the full spectrum of operations, the Army has forged peace in the Balkans, restored democracy in Haiti, and remained globally engaged through participation in numerous other missions. Current operations now include counterterrorism worldwide and homeland defense. The call for transformation of the force to respond better to today's threats is clear. This transformation requires not only a change in equipment, but also a change in the mindset of Army leaders concerning how we view our profession.

As Superintendent of the United States Military Academy, it is a pleasure to introduce this text, drawn from an independent research project conceived and directed by members of the West Point faculty. As an undergraduate institution of higher education and the nation's premier leader development institution, West Point educates, trains, and inspires our graduates for service as commissioned leaders of character in the Army. The research results deployed in the present book will assist us in accomplishing that mission. Within the Army profession and beyond, this work demonstrates our commitment to provide the nation we serve with a better understanding of its Army and its evolution into the 21st century.

This book, an anthology by many hands, contains the results of research conducted by dedicated Americans both in and out of uniform. They represent multiple disciplinary perspectives on the military in general, and on the Army in particular, lending needed richness and diversity to such an investigation. But in spite of their different perspectives, they share an abiding respect for the Army, recognizing this venerable institution as being absolutely essential in providing for the common defense of the republic. They gave selflessly of their time and talents over the past year to bring this project to fruition. And for that devotion to the Army profession, we are grateful.

In the aggregate, these authors make two significant contributions to the literature of the Army profession that I want to notice specifically. First, they capitalize upon a common analytical framework for their various research contributions. This framework applies a fundamentally new and valuable view of professions to the American military profession. Second, they contribute to the ongoing dialogues within the Army about its professional future, including questions of what its expert knowledge should include for the 21st century and how that unique expertise should be imparted to the members of the profession. In these areas, I doubt that all readers will agree with all that they find here. The far more important point, however, is that these issues be surfaced, studied, and debated to ensure a better future for America's Army as a profession.

This year, the United States Military Academy celebrates its bicentennial, marking two hundred years of service to the nation. Throughout these two centuries, the Academy has provided graduates with a solid education in the military art and its professional practice so that they are prepared to be leaders in the U.S. Army. Thus, we welcome this contribution toward fulfilling our mission as West Point begins its third century of service to the American republic.

William J. Lennox, Jr.
Lieutenant General, U.S. Army
Superintendent

I

Setting the Stage

1 Introduction[1]

Don M. Snider and Gayle L. Watkins

> Over two years ago now, we released the Army Vision, and it talked about three things . . . it talked about people; it talked about readiness; and then it talked about transformation. And though today we'll talk about transformation . . . this thing called the objective force—I think it's important for me to point out that in that vision statement it begins talking about people and it ends talking about people. And we should not miss the importance of this construct. People are central to everything else we do in the Army. Institutions don't transform, people do. Platforms and organizations don't defend this nation, people do. And finally, units don't train, they don't stay ready, they don't grow and develop leadership, they don't sacrifice, and they don't take risks on behalf of the nation, people do. What we say about people in the vision inextricably links the other two parts—transformation and readiness; without people in the equation, transformation and readiness are little more than academic exercises.
>
> GEN. ERIC SHINSEKI, ARMY CHIEF OF STAFF, 8 NOVEMBER 2001[2]

The Army that won the battles of the Gulf War in 1991 was one of the most professional ever fielded by America. As Gen. Norman Schwarzkopf commented, "We could have traded equipment with the Iraqis and still won."[3] The extraordinary success met by the Army in the Gulf was based on the quality and training of its people.

A decade later, as the Army continues to transform from a Cold War force to one appropriate for the 21st century, it must grapple with many issues. In this book, we propose that the most critical challenge the Army now faces in its planned transition is to reinforce the professional nature of the institution and to provide the opportunity for its soldiers to be members of a profession—the Army profession. The Army is neither a public-sector bureaucracy manned by civil servants nor is it a business with employees. It has been and must continue to be a profession, one in which military professionals serve with deep pride and immense personal satisfaction.

How Militaries Transform—Various Schools of Thought

History shows that many armies do not adapt well in peacetime to changing environments; some do not adapt at all and no longer exist as armies. And even for those that are able to innovate and adapt in order to remain effective militarily and relevant to the societies they defend, the process is often long and difficult. Such processes take at least a decade or longer and often are not resolved when the next war starts, thus requiring the even more difficult process of wartime innovation and adaptation.

Since modern armies reflect the national culture from which they are drawn and its state of technology, and since they each have unique relationships with the civilian society and the nation defended, it is not surprising that there is little consensus in the scholarly literature on just how military transformations occur in peacetime—the specific causal relationships. In fact, there are three rather distinct schools of thought that address the phenomena of military transformations, each with its own historical literatures and each recently renewed and invigorated since the end of the Cold War.

The most recent and popular school of thought on military transformations—populated by historians, defense and policy analysts, technologists, and futurists—tends to view military institutions in relation to the changing conduct of warfare and how militaries fight—particularly the influence of technology.[4] This school claims that the military services of the United States are now undergoing a true "revolution in military affairs" (RMA).[5] Such claims started during the last decade of the Cold War based on the conventional arms race then occurring between the Soviets and NATO (largely American), and gained momentum as a result of the remarkably successful one-hundred-hour Gulf War, in which the high-tech American-led forces overwhelmed the Iraqi army.[6]

Then in 1996, all three U. S services adopted a vision of future warfare—*Joint Vision 2010*[7]—that required a peacetime transformation in all of their conventional capabilities. Both the joint vision and the subsequent service visions focused on gaining for their forces significant increases in military effectiveness, increases to be brought about by creating a "system of systems" integrating ubiquitous sensors and strike systems in real-time decision cycles more compact than those of any enemy. Situational awareness within the battle space, sometimes described as "dominant battlefield knowledge," became the term to describe the uniquely American high-technology advantage in future conventional warfare.

But whether such a transformation is actually occurring today, or even should, is contested both within the three services themselves, by the larger community of defense policy-makers in recent administrations, and within the powerful defense industries.[8] Thus the service perspectives on transformation have often been, not surprisingly, more related to the endless political game of near zero-sum resource allocations within the Department of Defense than to future war-fighting realities.[9] Currently, within this school there simply is no consensus on an RMA other than the general agreement that, if affordable, one is desirable. Moreover, what role current conventional force structures—the so-called "legacy" forces left over from the Cold War—will play in such transformations, and at what opportunity costs, remain undecided. The war in Afghanistan, following the "first modern war"—Kosovo—by less than two and a half years, is drawing to a close as this chapter is written.[10] It will doubtless heighten within this school the relevance of asymmetric responses to an RMA-led transformation of U.S. defense capabilities. All of which can only leave the future of Army transformation even more uncertain, but no less urgent.

The second school of thought—populated by political scientists and organizational theorists—views armies as large, standing institutions whose behaviors in peacetime can be best understood in strategic context by the application of evolving organizational theories originally found in business, economics, and public administration literatures.[11] According to this school, major influences on an army's decisions issue from its own intrinsic character and its relationships with the external environment, particularly its structural relationship with the other governmental institutions (executive and legislature) with authority over it.

Of particular interest regarding this school is its emphasis on the culture of the military institution, the relationships between culture and strategic doctrines, and the influence such organizational culture has in interpreting the external environments in which the army is operating, both domestic and international. In this school one of the roles of strategic leaders is to influence organizational cultures in such a manner as to make profound change more acceptable and likely, such as abetting "learning" within the organization.[12] A recent Army Chief of Staff, Gen. Gordon Sullivan, adopted such an approach to Army modernization during his tenure.[13] In our view, this school places more emphasis on the human resources of the Army than does the technology-oriented RMA school.

The third school has focused on civil-military relations, those observable relationships in the post-Westphalian western world between military institutions and the states they serve. The scholars within this school—historians, political scientists, sociologists, and anthropologists—have long focused on the central issue of a state's ability to control a military institution sufficiently strong to defend that state.[14] The challenge has been to maintain equilibrium between the two imperatives introduced by Samuel Huntington forty-five years ago, that is, to balance the *military* imperative of a strong, effectively coercive military institution of immense power, with the *social* imperative for that same institution to remain legitimate before the society served and subordinate to its elected and appointed civilian officials.[15] How different and how separate the military should be from civil society, and to what degree it has the need for a different culture and ethos, have long been issues of concern to this school.

More recently, as western societies have entered the post-industrial era and states have abandoned mass conscripted armies in the tremulous hope that major land warfare is obsolete, this school has widened its interests. Now, this school's salient themes are "post-modern militaries," the influence on them of "other-than-war" missions,[16] and the potential of new cultural gaps growing between such militaries and the societies they serve.[17] It is fair to say that scholars of the civil-military school have focused on human resources more than the other two, both in terms of individual soldiers and leaders and collectively as social organizations. Today, however, their focus has largely narrowed to a sociologically identifiable set of Army soldiers—women, gays, single-parents—and the equity and ethics involved in their military service, or the lack thereof.[18]

Key to Military Transformations—"People"

As can be inferred from our review of the second and third schools—organizational and civil-military—they share certain characteristics. One such is the proposition advanced by General Shinseki in the introductory quotation: *people count*. In his view they count more than technology and readiness when it comes to transforming the U.S. Army. Unfortunately, the currently popular RMA school is almost devoid of considerations of human resources, beyond the notion that they constitute "cultural impediments" to change.[19]

But it is not clear in General Shinseki's vision and its implementing White Paper who those "people" are.[20] Are they the Army's soldiers and their leaders who count in military transformations, i.e., just the uniformed people? Or do civilian contractors and Department of the Army civilians count also as "people"? And what of retired soldiers, now increasingly contracted back into the Army? Further, General Shinseki has not told us *how* these people count. What identities, roles, attitudes, and behaviors will they need to effect such a transition? Do they now have them? How are the before-and-after roles different, especially for the Army officer corps? And, most importantly, what relationship must these "people" have with the Army, and the Army with them, to be effective at transformations?

We suggest in the present book—in fact, it is the major theme—that General Shinseki is entirely correct in his judgment that "people count" most in the transformation of the U.S. Army. But we go even further by suggesting that in recent decades none of the three schools of thought about military transformations has emphasized sufficiently, *nor in policy relevant ways*, in what sense "people count." And until that is done, the necessary roles that "people" must play in order for any transformation effort to be successful within the U.S. Army, and the policies necessary to enable them to fulfill those roles, cannot be determined, thus rendering futile such transformation efforts.

We believe the reason the various schools of thought have neglected the roles and policies of people is that *they have failed to view these "people" as "professionals" and the U.S. Army as a "profession," or calling*. They have not introduced the analytical frameworks and necessary understandings of how modern professions prepare and compete for the opportunity to apply their expertise that would facilitate, indeed enable, such a people-oriented perspective.

Accordingly, we present in this book the results of a collective research effort, undertaken by over a score of military and civilian scholars, to renew study of the U.S. Army as a profession, its "people" as members of a profession, and its transformation as one that, if it is to occur, will have to occur within a highly competitive system of military professions of which the Army profession is only one. As will become clear, our thesis is that competitive professions are quite different from other producing organizations in America, and the three military professions even more so. Thus they must be studied, understood, and led in light of this difference. Simply stated, the transformation of a military profession is not the same as that of a business or other type of organization, and cannot be approached in the same way.

The U.S Army as a Profession

America has three military professions: army, maritime, and aerospace. These professions are generally (but not wholly) contained within the Departments of the

Army, Navy, and Air Force and surrounded by the immense bureaucracies of the Department of Defense, the Congress, and the Executive. Therefore, the military services are simultaneously professions and government bureaucracies. For example, a newly promoted major of infantry stationed in Korea has developed professionally to the same level of expertise as a newly promoted major of infantry stationed in Germany. By promoting them to the rank of major, the Army as a profession has certified their competence at the field grade level to the client, American society. But the Army as a bureaucracy uses its personnel management system when making the two majors' future assignments, considering them merely as "faces" to be matched with a particular manpower "space." This dual nature—profession and bureaucracy—creates a challenging but healthy tension within all such military institutions.

Before considering how the Army can better deal with this tension, let us review briefly the nature of military professions. First, expertise and the knowledge underlying it are the coins of the professional realm. More so than other occupations and organizations, professions focus on generating expert knowledge and the ability of its members to apply that expertise to new situations.[21] Medical professionals perfect medical techniques to apply to patients, lawyers apply legal expertise in trying cases, and the military develops new technologies, capabilities, and strategies to "provide for the common defense," most often in places and under circumstances that cannot be foreseen. Such professional expertise is ultimately validated by the client (or professions would not exist!), and forms the basis for the trust between the profession and the society served. Given such trust, professions are granted limited autonomy to establish and enforce their own professional ethics, the maintenance of which further enhances such trust. Furthermore, success in professional practice stems from effective and ethical application of the expertise—the patient is cured, the case is won, conflict is deterred, or if fought, settled on terms favorable to the United States. Thus measures of efficiency, while important, rank behind effectiveness as measures of success for professions.

Although they may appear static, modern professions also have a hidden, dynamic, and occasionally indecorous side. They are continuously engaged in fierce competitions for control over the jurisdictions in which they apply their expertise.[22] A well-publicized current example of these competitions is occurring in the medical arena where physicians and HMOs battle over the right to make patient-care decisions. Other professions face similar challenges as they seek to gain legitimacy in new fields while retaining legitimacy in their traditional areas.

Since the end of the Cold War, the Army has been embroiled in such competitions in a variety of settings. These include counter-drug operations on our southern border, operations other than war in the Balkans, and now homeland security and counter-terrorism. The Army's competitors within these jurisdictions range from other professions (the other military services and foreign armies) to other governmental agencies, private contractors, and nongovernmental organizations, both American and international. Simultaneously the Army has continued to compete in its traditional jurisdiction, conventional war-fighting, where its ability to do so has been compromised by a legacy force that is now too strategically immobile to be fully relevant to the new global situations.

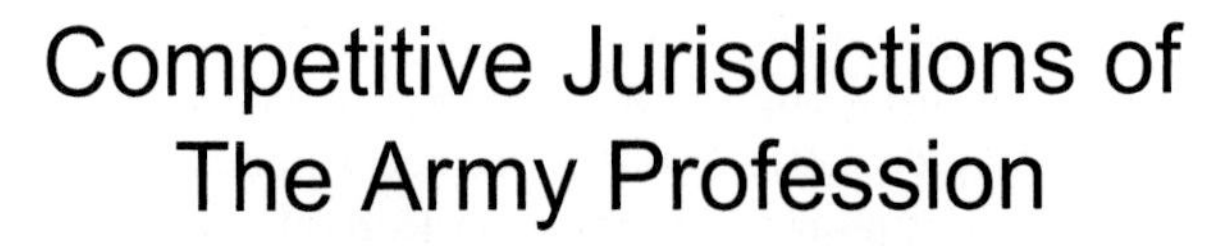

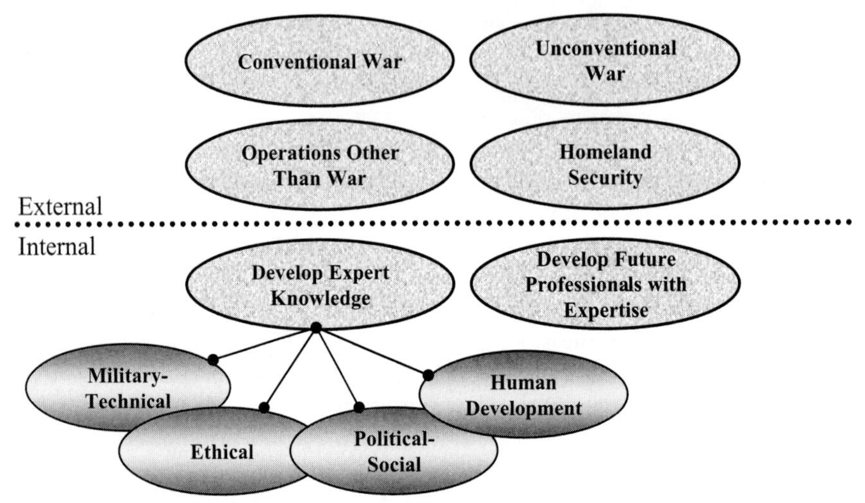

Figure 1-1. *Competitive Jurisdictions of the Army Profession.*

Figure 1-1 presents a schematic of the four external jurisdictions in which the Army profession competes. Also displayed are the internal jurisdictions in which the Army as a profession develops and maintains its expert knowledge and teaches it to the members of the profession and their units. Since the end of the Cold War even these internal jurisdictions have become more competitive as the Army increasingly "contracts out" such professional services as doctrinal studies, the training of foreign armies, staff functions within the Army General Staff, and the leadership of ROTC detachments.

The U.S. Army as a Bureaucracy

In contrast to the more traditional professions, which have not normally been lodged within a controlling organization, the Army is also a huge hierarchical bureaucracy, only one of many within the U.S. government. Unlike professions, bureaucracies focus on application of knowledge embedded in organizational routine and process rather than in their employees. The extreme of this is represented by government bureaucracies—think of a state department of motor vehicles in one of the 50 states. Bureaucracies implant such knowledge in their structures through standing operating procedures, operating routines, personnel policy handbooks, regulations, pay structures and bonuses, equipment, and machinery more so than in the development of their employees and their functional competence.[23] Efficiency and survival are normally the dominant organizational goals within bureaucracies.

Herein lies the natural tension within the Army's dual character. On the one hand, the Army is a profession focused on effectively developing military expertise and its members' ability to collectively apply their expertise to new situations. On the other hand, it is a hierarchical bureaucracy focused on efficiency, "doing more with less." This dual nature is unavoidable and healthy for the nation, although it may be cause for considerable tension both for the individual professional and for the institution as a whole.

Maintaining an appropriate balance between the Army's two natures is elusive; at any point in time, one is likely to predominate over the other. The result is an Army whose leaders, self-concepts, decisions, and organizational climate for soldiers reflect either a high degree of military professionalism and effectiveness or a high degree of bureaucracy and efficiency. In the former, the self-concept of the Army's members is likely to be one of "professional," while in the latter their self-concept is one of "employee." The Army of Desert Storm was a case of the former. Historically, militaries that do not resolve this tension in favor of their professional nature can experience the "death" of their professional character. As their bureaucratic nature comes to dominate, they cease to be a profession and become little more than an obedient military bureaucracy.[24]

What Is Dominant Today within the Army—Profession or Bureaucracy?

For a number of reasons, most having to do with the past decade of downsizing and budget cuts, the Army of the late 1990s has behaved more like a bureaucracy than a highly effective military profession. We have argued this point elsewhere, with our rationale resting on indicators of reduced Army effectiveness in applying its expertise to tasks at hand (e.g., the 1999 deployment of Task Force Hawk to Albania where the force was never subjected to the hazards of combat; reports from the Army's combat training centers of declining proficiency among tested battalions and task forces; declining readiness ratings among Army divisions; and a study by the RAND Arroyo Center documenting declining tactical proficiency among combat arms captains).[25] Our rationale also rested on indicators of low morale caused by an excessively bureaucratic personnel management system (e.g., the exodus of captains from the Army; the Army's own reports of declining soldier morale; and increased declinations by lieutenant colonels and colonels selected for command).

Also weighing heavily in our conclusion was an independent study of American military culture by retired senior officers and defense policy-makers who documented within the active Army a serious gap in trust between junior and senior Army officers. The gap is worsened by the rise of a "zero-defects" mentality too often displayed by commanding officers at all levels, just the opposite of what is needed in military organizations if the developmental processes of professions are to flourish.[26]

Now, however, such conclusions about the state of the Army profession need not rest primarily on external analyses. Early in 2000, in response to several of the issue studies just cited, the Army Chief of Staff directed an internal panel of general officers to study officer perceptions and understandings of the Army's training and leader development systems to determine the scope of current problems and to recommend corrective actions. Not surprisingly, after extensive field work the panel thoroughly documented by the fall of 2000 many of the same deficiencies earlier indicated by the external studies, as well as several not previously aired, particularly in Army training systems.[27]

With respect to the Army's culture, the panel concluded that the gap between the institution's professed ideals and its actual practices in the areas of training and leader development had spread outside the Army officer corps' "band of tolerance," within which there is no loss of trust between officers and the institution. Seven specific conclusions supported this overall finding:

1. "Within the Army Officer Corps, the service ethic and concepts of officership are neither well-understood nor clearly defined;
2. An excessive operational pace is a major source of the degradation. [It is] detrimental to readiness, leader development, and officer job satisfaction; leads to micromanagement; and is a major reason for attrition among all cohorts;
3. Retention is a significant issue across [all commissioned grades below flag rank], a result of a perceived lack of commitment from the Army, limitations on spouse employment and a perceived imbalance between Army expectations and the family, the lack of work predictability, and only limited control over assignments;
4. Micromanagement has become part of the Army culture [, producing] a growing perception that lack of trust stems from the leader's desire to be invulnerable to criticism and blocks the opportunity for subordinates to learn through leadership experiences;
5. The officer efficiency report (OER) is a source of mistrust and anxiety [and] is not yet meeting officer expectations as a leader development tool . . . aspects of the OER are seldom used, and senior raters seldom counsel subordinates;
6. Assignment requirements, instead of individual leader development needs, drive officer personnel management; and
7. Officers believe mentoring is important for both personal and professional development, yet a majority of officers report not having mentors."[28]

While the panel did not study the Army as a profession nor use the language of professions in their report, there can be no doubt that they are describing quite accurately an Army that is behaving far more bureaucratically than professionally. There are, of course, many causes of the bureaucratic state of the Army. The panel noted "excessive operational tempo"; the Center for Strategic International Studies study earlier noted several other causes, both internal and external to the Army. The primary research presented in Chapter 5 invokes the words of Army officers themselves to describe this phenomenon. One of our most striking findings is that the Army has bureaucratized its professional knowledge into checklists and forms to such a degree that it has reduced the expertise of its officers by reducing their opportunities to exercise it, frustrated them in their attempts to apply their knowledge and experience, and decreased the effectiveness of professional decisions. All of these sources clearly indicate that *the Army profession is seriously compromised by excessive bureaucratization of major leadership and management systems and is so perceived by the individual members of the Army officer corps.*

Profession or Bureaucracy—What Difference Does It Make?

One might argue that for purposes of efficiency or even to make the planned transformation "easier," the Army should simply continue to deprofessionalize, becoming an obedient but nonprofessional military bureaucracy. One need look no further than Europe to see Western democratic societies readily accepting this outcome. But America has a different role to play in the world. If deprofessionalization happened here, American society would lose two key benefits of military professionalism: (1) the development and adaptation of expert knowledge and the expertise of military professionals; and (2) social control of individuals within an institution capable of terrible destruction.

One reason for the nearly unanimous public support for the current war on terrorism in Afghanistan and elsewhere is that the leaders commissioned to be responsible for the most powerful army in the world are considered to be experts at their art and unquestionably under the control of elected and appointed civilian officials. Professions excel where bureaucracies do not in the creation, adaptation, and application of abstract expert knowledge to new situations. Therefore, if the Army is to remain on the cutting edge of military strategy, applied technology, and operations, its professional nature must be renewed, adapted, and preserved. Other nations may be willing to renounce the leadership role in military theory and practice, but America cannot.

With regard to social control, by nurturing the profession's ethic within its members, a profession offers a better means of shaping human behavior in situations of chaotic violence, stress, and ambiguity than bureaucratic management can ever hope to achieve. In other words, within the culture of a profession is embedded an ethos that strongly informs the actions of the individual professionals, even in the direst of circumstances. A good example appears in a letter then Maj. Gen. George Patton wrote from North Africa in 1942 to classmates in the states after his division had fared poorly against the Germans: "We did not turn and run, because we were more afraid of our consciences than we were of the enemy."[29] Such social control within the Army profession—by development of professional ethics—has been under stress during the past decade owing to political guidance to deployed forces to avoid any casualties, particularly in unpopular peacekeeping operations.[30] Such risk-averse behavior is natural and fitting for a bureaucracy. In contrast, restoring the discretion of the Army officer corps to do its duty in combat as best fits the needs of mission accomplishment, always the first priority, is vital to renewing the ethos of the Army profession.

Since the continual development of military expertise and effective control of an Army operationally engaged on behalf of American society are both essential to the nation's future security, a nonprofessional Army is certainly not in America's best interest. Equally important, neither is it in the Army's best interest. To remain relevant to the nation it serves and to effect the transformation it plans, the Army must renew its essence as a profession.

From Bureaucracy to Profession—the Missing Component of Army Transformation

The research project serving as the basis of this book was designed to study the Army as a profession, to assess the state of the profession, and to make recommendations for the redefinition and renewal of the Army's professional character as an essential part of its planned transformation. Surprisingly, of all the studies done both by the Army and for the Army in recent years, this is the first time since 1970 that the Army has been studied *as a profession*. However, it was no surprise to discover that many of the ideas offered the last time the Army undertook to renew its professionalism—after Vietnam and the end of the draft in the early 1970s—were quite responsive to what is needed today.[31]

To coordinate the efforts of such a diverse group of researchers in the conduct of a substantive review of the state of the Army profession, we applied a common framework for the individual research as reflected in the chapters of this book (see Figure 1-2).

This framework, which incorporates both the traditional view of military professions as well as the newer view of professions as competitive vocations, allows visualization of several factors vital to the Army officer corps' successful introspection and dialogue regarding the planned transformation.

First, it becomes clear that each broad area of expert knowledge of the profession—the military-technical, the moral-ethical, the political-social, and human development—should be analyzed and understood from three different perspectives: the client's (society); the professional institution's (the Army); and that of the individual professional (officer, noncommissioned officer, Army civilian). Second, owing to the significant differences between these perspectives, the two horizontal boundaries dividing them are loci of recurrent tensions (civil-military relations and the gap in trust between the Army and its junior professionals) about which our researchers could develop policy-relevant conclusions and recommendations.

To this framework we added the new understanding of the Army profession as operating within a competitive systems of professions.[32] As discussed earlier, modern professions are competitors—for members, resources, and, most important, jurisdiction—within a "system of professions." This system includes other professions, occupations seeking professional status, and other producing organizations, each of which vies for jurisdiction and a legitimated claim to apply their expertise to specific situations. It is in the context of such jurisdictional competitions that strategic leaders of the profession must develop the detailed requirements for professional systems—education, ethics, oversight, and credentialing, to name but a few.

	Expert Knowledge of the Army Profession			
Level of Analysis	**Military-Technical**	**Moral-Ethical**	**Political-Social**	**Human Development**
Societal National and global context in which Army exists	National and international uses of military forces	National and international values and beliefs	National and international political and societal systems	
		Civil-Military Tensions		
Institutional Internal context and systems	Internal Army systems supporting military-technical capabilities	Internal Army systems that establish, communicate, and maintain the profession's norms and values	Internal Army systems focused on political and societal actions	Internal Army systems focused on developing, educating, training, and managing the Army's human resources
		Army-Soldier Tensions		
Individual People who constitute the Army	Individual knowledge and skills needed to be successful in the Army	Individual moral-ethical values	Political and social knowledge andskills held by or necessary for Army members	Individual commitment, motivation, and expectations of the Army

Figure 1-2. *Framework for Analysis of the Army Profession.*

The Scope of the Book: Selected Analyses of the Army Profession

This book, actually an anthology by many hands, is divided into seven sections each of multiple chapters and contains the most relevant of the research efforts. Some chapters were researched and prepared individually, others by teams; but each was done by scholar(s), both military and civilian, well qualified to contribute to this specific project.

Section I is introductory, while the last section contains the project's broad conclusions and recommendations for transforming the Army from bureaucracy to profession. From various perspectives and levels of analysis, the intervening five sections report, first, on the state of the Army profession within each of the four broad areas of its expert knowledge and expertise (military-technical, moral-ethical, political-social, and human development/leadership); and, second, on two areas we have called areas of "internal expertise" (Figure 1-1), areas where the profession's expert knowledge, such as direct and strategic leadership, is applied to its own preparations for subsequent competition in external jurisdictions. Each of these five sections can be studied alone, but all are complemented by the chapters of the first and last sections of the book.

In Section I we set the stage for the research project with chapters designed to bring the reader up to date on the Army profession in order that he or she may gain a fuller appreciation of the subsequent sections. To aid in understanding, James Burk in Chapter 2 incorporates the new "competitive" views of professions with the traditional views of military professions by Samuel Huntington and Morris Janowitz, suggesting a synthesis useful for looking to the future. Chapters 3 and 5 present analyses of current aspects of the Army that are of deep concern: the well-documented gap of trust between junior and senior members of the profession; and the alarming fact that the Army officer corps does not now have below the rank of lieutenant colonel a common understanding of, nor language to communicate about, the Army profession. Chapter 4 looks historically at the character of service the Army profession has rendered to the American people in the past.

Section II of the book explores on several levels of analysis four aspects of the profession's military-technical expert knowledge. In Chapter 6, James Blackwell reviews the Army's war-fighting doctrines and analyzes the external competitions in which the Army is currently negotiating for legitimacy (Figure 1-1). Chapter 7 by Elizabeth Stanley-Mitchell does the same for the Army's efforts at digitization, expressing useful insights into effects on both the profession and its junior leaders. The influences of operations other than war are analyzed by Thomas McNaugher in Chapter 8, while Deborah Avant opens up a whole new area for discussion with her analysis of the Army's often uncritical approach to "contracting out," particularly in military training (Chapter 9). She identifies some of the potential pitfalls not yet experienced from this practice. For example, concerning the status of contractors, are they members of the Army profession? If so, who has certified to the client their military-technical competency? While these four chapter topics certainly do not map the entirety of military-technical expert knowledge of the Army profession, they do provide incisive ways to think about it and in the process reveal some startling (and unsettling) results.

Section III of the book deals with four aspects of the profession's political-social expertise, those practices by which the profession's strategic leaders relate the profession to the nation's political leaders and to the profession's client, the American people. Separate chapters cover how, and how well, this was done in the past (Chapter 13), and how it is currently being done within two contentious areas: the joint arena (Chapter 11) and use-of-force decision-making at the national level (Chapter 10). In Chapter 12 of this section, Marybeth Ulrich presents a new view of the Army officer corps' responsibilities in leading a "democratic" army, suggesting a specific set of professional norms to govern their behavior in such a role. Without doubt that new set of norms will generate renewed debate, as is intended.

The ethics according to which the profession applies its expertise is the focus of Section IV. The three chapters each shed light on different aspects of that ethics. In Chapter 14 Mark Mattox presents a precise functional derivation of the moral obligations of an Army officer. Those same obligations, sharply focused on the case of a notional colonel commanding an Army brigade in operations other than war, are dramatized in James Toner's imaginative epistolary argument in Chapter 15. In

Chapter 16, the final chapter in this section, Martin Cook raises the thorny question of how Army professionals may react to missions of which they or the American public question the legitimacy.

Perhaps the most important area of expert knowledge for Army professionals is leadership and human development. Leaders must understand their subordinates if they are to help them develop and to lead them successfully in combat. As Gen. Eric Shinseki stated in the opening quotation of this chapter, "Institutions don't transform, people do." He is referring, of course, to human development. In Section V, four aspects of human development are analyzed: the definition and measurement of psychological maturity of officers (Chapter 17); whether "principles of officership" can be useful in officer development programs (Chapter 18); the spiritual needs of soldiers and leaders, particularly in preparation for combat (Chapter 19); and the necessary roles for strategic leaders of a profession, roles quite at odds with what is currently contained in Army leadership doctrine (Chapter 20).

In contrast to the earlier sections, Section VI turns inward, focusing on the Army's key internal jurisdictions (see Figure 1-1). Since professions instill their expert knowledge in people rather than machines or organizational systems, the heart of many of the Army's internal jurisdictions is the selection, education, development, and career management of its members. The chapters in this section address the importance of continually advancing the abstract knowledge in these jurisdictions. For example, David Segal and Meyer Kestnbaum challenge the Army to reconsider some of its personnel management policies that are based on out-of-date research (Chapter 21). Their chapter highlights the importance to the Army of understanding people, showing that human resource research must enjoy a priority equal to that of technological research endeavors.

This section also presents case studies on three aspects of the Army's internal jurisdictions: soldier well-being, physical readiness, and officer training and development (Chapter 22). These case studies highlight the importance of abstract knowledge—its creation, development, and maintenance—to a profession's success. Furthermore, they show the power of tying that abstract knowledge to organizational outcomes and goals. Case Study No. 1 discusses the Army's recent efforts to use abstract knowledge to regain control over a critical jurisdiction, the care of its members, now known as "well-being." This case shows the vital role abstract frameworks play in jurisdictional competition. In Case Study No. 2, Maureen LeBoeuf and Whitfield East argue that the present state of the Army physical readiness program indicates a loss of important expertise in this area and disconnection from the Army's goals and missions. Finally, Case Study No. 3 by Joe LeBoeuf traces the conclusions of the Army Training and Leader Development Panel in seeking to recover ground for the Army in the mission-essential areas of training and leader development.

In Section VII, we present the major conclusions and recommendations (Chapter 25) aggregated from the overall research project along with one chapter addressing the way ahead. In Chapter 24, Andrew Abbott, whose conception of "competitive" professions was applied within this project for the first time ever to the Army profession, discusses the necessity of redefining the concept of "career" if the Army hopes to accommodate successfully its two natures of bureaucracy and profession.

Finis: In Army Transformation, People Count but Professionals Count Even More

There are many fine professionals in the Army, those who go to work each day seeking no more personally than the opportunity to serve their country by improving their expertise and applying it to the tasks at hand. To us, these professionals include anyone who has made and is living a resolute commitment to a lifetime of ethical service in the Army profession—officer, noncommissioned officer, or Department of Army civilian. Many others aspire to such status, and will doubtless someday attain it.

But among these professionals one group must be singled out for special responsibility in the planned transformation. From the members of the Army officer corps, as the commissioned agents of the American people responsible for the continued stewardship of the profession and for the development of the sons and daughters of America who serve in it, more is expected, legally and morally. It is in behalf of these officers' professional education and preparation for the role of Army steward, and equally in behalf of reinforcing professionalization for the superb noncommissioned officers with whom they serve, that this book has been prepared.

We have no doubt that, under their leadership in the years ahead, the Army profession will be redefined and renewed as the fundamental precondition for a successful Army transformation.

Notes

1. The views and opinions expressed in this chapter are those of the authors and are not necessarily those of the Department of the Army or any other U.S. government entity.
2. Speech by Gen. Eric Shinseki, Army Chief of Staff, AUSA Symposium, 8 November 2001, accessed on 10 November at *http://www.army.mil/leaders/csa/speeches/20011108CSAREMARKSAUSA.doc*.
3. Stephen Biddle, "Victory Misunderstood: What the Gulf War Tells Us about the Future of Conflict," *International Security* 21 (Fall 1996): 139-79.
4. For examples see Adm. William Owens, *Lifting the Fog of War* (Baltimore, MD: Johns Hopkins University Press, 2001); Michael O'Hanlon, *Technological Change and the Future of Warfare* (Washington, DC: The Brookings Institution, 2000); Stuart J.D. Swartzstein, *The Information Revolution and National Security* (Washington, DC: Center for Strategic and International Studies, 1996); and Williamson Murray and Allen R. Millett, eds., *Military Innovation in the Interwar Period* (Cambridge: Cambridge University Press, 1996).
5. See Richard O. Hundley, *Past Revolutions, Future Transformations* (Washington, DC: RAND, National Defense Research Institute, 1999).
6. For the Army's version of the war, see Robert H. Scales, Jr., *Certain Victory* (Washington, DC: Office of the Chief of staff, U.S. Army, 1993).
7. See Gen. John Shalikashvili, *Joint Vision 2010* (Washington, DC: Office of the Chairman, JCS, 1996). There is now an updated version, *Joint Vision 2020*, but the emphasis remains the same.
8. For the best summary of this debate, see O'Hanlon, Chapter 2.

9. See George C. Wilson, *This War Really Matters* (Washington, DC: CQ Press, 2000); and William Greider, *Fortress America* (New York: PublicAffairs Press, 1998).
10. For the first "modern" war, see Chapter 1 of Wesley K. Clark, *Waging Modern War* (New York: Public Affairs Press, 2001).
11. See Deborah D. Avant, *Political Institutions and Military Change* (Ithaca, NY: Cornell University Press, 1994); Elizabeth Kier, *Imagining War: French and British Military Doctrines between the Wars* (Princeton, NJ: University Press, 1999); Stephen Peter Rosen, *Innovation and the Modern Military: Winning the Next War* (Ithaca, NY: Cornell University Press, 1991); Theo Ferrell, "Figuring Out Fighting Organizations," *Journal of Strategic Studies* 19, No. 1 (March 1996): 122-135; and Michael James Meese, "Defense Decision Making Under Budget Stringency: Explaining Downsizing in the United States Army," (Ph.D. diss., Princeton University, May 2000).
12. See Peter Senge, *The Fifth Discipline* (New York, NY: Doubleday, 1990); and Emily O. Goldman, "The U.S. Military in Uncertain Times: Organizations, Ambiguity, and Strategic Adjustment," *Journal of Strategic Studies* 20, No. 2 (June 1997): 41-74.
13. See Gordon R. Sullivan and Michael V. Harper, *Hope is Not a Method* (New York: Random House, Times Business, 1996).
14. Peter D. Feaver, "The Civil-Military Problematique: Huntington, Janowitz, and the Problem of Civilian Control," *Armed Forces and Society* 23, No.2 (Winter 1996): 149-78.
15. The classics of the civil-military school include: Alfred Vagts, *A History of Militarism* (rev. ed.; New York: Free Press, 1959); Samuel P. Huntington, *The Soldier and The State* (Cambridge, MA: Harvard University Press, 1959); Morris Janowitz, *The Professional Soldier: A Social and Political Portrait* (Glencoe, NY: The Free Press, 1960); Bengt Abrahamsson, *Military Professionalism and Political Power* (Beverly Hills, CA: Sage Publishing, 1972). For more recent literature see also Samuel Sarkesian and Robert E. Connor, Jr., *The U.S. Military Profession into the Twenty-First Century* (London: Frank Cass Publishers, 2000); Michael C. Desch, *Civilian Control of the Military* (Baltimore, MD: Johns Hopkins University Press, 1999); and Don M. Snider and Miranda Carleton-Carew, eds. *U.S. Civil-Military Relations: In Crisis or Transition* (Washington DC: Center for Strategic and International Studies, 1995).
16. See James Burk, ed., *The Adaptive Military: Armed Forces in a Turbulent World* (London: Transaction Publishers, 1998); Charles C. Moskos, "Toward a Post-Modern Military: The United States as Paradigm," in *The PostModern Military*, ed. Charles C. Moskos, John Allen Williams, and David R. Segal (Oxford, Eng.: Oxford Univ. Press, 2000): 14-31; and Don M. Snider, "America's Post-Modern Military," *World Policy Journal* 18, No. 1 (Spring 2000): 47-56.
17. For the widening gap in political identity, see Oli R. Holsti, "A Widening Gap Between the US Military and Civilian Society? Some Evidence, 1976-1996," *International Security* 23 (Winter 1998/99): 8; and commentary and reply by Joseph J. Collins and Oli Holsti, *International Security* 24 (Fall 1999): 199-207. For other gaps, see Peter D. Feaver and Richard D. Kohn, eds., *Soldiers and Civilians* (Cambridge, MA: MIT Press, 2001).
18. For example, Mary Katzenstein and Judith Reppy, *Beyond Zero Tolerance: Discrimination in Military Culture* (New York: Rowman and Littlefield Publishers, 1999). For an alternative view see Don M. Snider, "The Uninformed Debate on Military Culture," *Orbis* (Winter 1999): 11-26.
19. See discussion between the Defense press corps and newly appointed Director of Force Transformation, Rear Adm. (Ret.) Arthur K. Cebrowski, 27 November 2001 accessed at *http://www.defenselink.mil/news/Nov2001/t11272001_t1127ceb.html.*
20. See "U.S.Army White Paper: Concepts for the Objective Force," 8 November 2001 accessed at *http://www.army.mil/features/WhitePaper/default.html.*
21. See A. P. Carr-Saunders and P. A. Wilson, *The Professions* (Oxford, Eng.: Oxford University Press, 1933) for initial case studies on professions. H.L. Wilensky's "The Professionalization

of Everyone?" *American Journal of Sociology* 70 (1964), 137-58, presents the professionalization sequence for American professions. For more recent literature, see Eliot Freidson, *Professional Powers: A Study of the Institutionalization of Formal Knowledge* (Chicago, IL: University of Chicago Press, 1986).

22. This discussion is drawn primarily from Andrew Abbott's, *The System of Professions: An Essay on the Division of Expert Labor* (Chicago, IL: University of Chicago Press, 1988). Additional sources include Elliott Krause, *Death of the Guilds: Professions, States, and the Advance of Capitalism, 1930 to Present* (New Haven, CT: Yale University Press, 1996); Eliot Freidson, *Professionalism Reborn: Theory, Prophecy and Policy* (Chicago, IL: University of Chicago Press, 1994).
23. For bureaucratic behavior within military institutions, see David C. Kozak and James M. Keagle, eds., *Bureaucratic Politics and National Security: Theory and Practice* (Boulder, CO: Lynne Reiner Publishers, 1988).
24. One need only look to the armies of the post-Cold War Western Europe nations for examples of this phenomenon.
25. See Don M. Snider and Gayle L. Watkins, "The Future of Army Professionalism: A Need for Renewal and Redefinition," *Parameters* 30 (Autumn 2000): 5-20.
26. See Edwin Dorn and Howard D. Graves, *American Military Culture in the 21st Century* (Washington, DC: Center for Strategic and International Studies, 2000).
27. See the Executive Summary to *The ATLDP Study Report to the Army*, accessed at *http://www.army.mil/features/ATLD/ATLD.htm.*
28. ATLDP, Executive Summary, OS-10 and OS-11.
29. The quotation is from an unpublished manuscript, "The Inadvertent Demise of the Traditional Academy, 1945-1995," dated November 1995, by Professor Roger H. Nye, Department of History, USMA.
30. See Don M. Snider, John Nagl, and Tony Pfaff, "*Army Professionalism, the Military Ethic and Officership in the 21st Century*" (Carlisle Barracks, PA: Army War College, Strategic Studies Institute, 2000).
31. See *Study on Military Professionalism* (Carlisle Barracks, PA: U.S. Army War College, 1970). For the official history of this period, see Robert K. Griffith, *Today's Army Wants to Join You: The US Army's Transition from the Draft to an All-Volunteer Force* (Washington, DC: Center of Military History, 1995); and Anne W. Chapman, *The Army's Training Revolution, 1973-1990* (Ft. Monroe, VA: TRADOC, 1990). For a very readable current history of the same events, see James Kitfield, *Prodigal Soldiers* (New York: Simon and Schuster, 1995).
32. As noted earlier, this understanding is drawn primarily from Andrew Abbott, *The System of Professions.*

2 Expertise, Jurisdiction, and Legitimacy of the Military Profession

James Burk

This chapter critically reviews the concept of the "professional soldier" as it applies to the American military from the end of the nineteenth century to the present. There are many reasons why such a review is needed. First, the end of the Cold War closed a long period in which preparation to fight and win world wars was the primary mission of the military, thus defining the occupation of the professional soldier. It is natural to wonder whether (or how) the professional soldier's occupation has changed as a result of this historical event. Second, the two classic and still influential studies of the professional military—Samuel Huntington's *The Soldier and the State* and Morris Janowitz's *The Professional Soldier*—were written over forty years ago, in the shadow of the world wars and during the Cold War.[1] Neither is historically up to date, although that (in principle) is relatively easy to fix. The fundamental question about them is whether recent experience reveals flaws in the logic of their arguments or suggests that their influence on our thinking about the professional military should be qualified or limited. Third, since Huntington and Janowitz wrote, much new scholarly work has been done on the professions in general. This work has implications for contemporary studies of the military profession that have not yet been fully explored.

We should not pretend that a critical review of this issue is value-free or disinterested. To call an occupation a "profession" is usually to make a positive normative judgment about the work being done, and, since we think that professional work is a social good, whatever we call professional work also reveals something about what we believe is required for the well-being of society.[2] When we inquire about the state of the military profession, we do not want a simple description of what the military *does*; we want a description compared to particular standards that prescribe how military activities *ought* to be done if they are good. That comparison is essential. It lets us know whether the military is measuring up to (or falling short of) our normative expectations. As should be obvious, the quality of the comparison (and of the normative judgment reached) depends on the validity and currency of the prescriptive statements used to characterize professional work. Yet how do we know whether these statements are valid and current?

Our answer to that question depends largely on the theoretical position we take about what a profession is and how professions develop. We might hold that the concept of a profession is widely accepted, clear, and unchanging. If so, then the prescriptive statements are universal claims that define the essence of professional activity. Once they are stated, they are recognized as sufficiently authoritative to rely on and use without further question. The task of evaluating professional activity is limited to getting the description right and fairly comparing what is done with what ought to be done. If we suppose, however, that the profession is a contested concept whose definition is historically opaque and changing, then the authority and universality of prescriptive standards are questionable and require defense. This enormously complicates the task of evaluating professional activity because, in addition to description, we have to identify the normative standards appropriate to a particular period. Nevertheless, the most recent studies of professional development argue that this more complicated task is the one we face.[3] As the historian Samuel Haber tells us, statements defining professional activity are "social artifacts fashioned by public events and usage."[4] Their validity and currency are not given once for all time, but are socially constructed and reconstructed by the choices and chances affecting particular generations. In this case, substantive prescriptions about what the professions ought to be evolve historically, influencing and influenced by what the professions are.

Adopting the second position to guide this chapter, I argue that the military profession's role has expanded over the course of the last century, widening from the management of violence early in the century to encompass the management of defense following the Second World War and the management of peace after the Cold War. This role expansion has resulted largely from the changing nature of war, war being the object of professional expertise, and brought the military into closer contact and competition with other professions when it performs its tasks. No less important, when we examine historical fluctuations in the prescriptive factors that define what a profession ought to be and do, we find that changing public views of what constitutes legitimate expert knowledge pose a challenge to the military's claim to professional status. In short, compared to their counterparts early in twentieth century, military professionals today must work harder to define and defend the domain within which they work and to overcome public skepticism about the value of their expertise.

My argument is developed in four parts. The first examines the idea of the professional in general. Here I propose a definition of the professions that depends on three prescriptive factors. These factors can be used to guide and characterize historical changes in professional activity. The second part considers how debates about military professionalism are connected to these prevailing ideas about the professions. Here I consider in particular (though not exclusively) the classic models of the military professional proposed by Huntington and Janowitz. I argue that neither model provides a wholly reliable guide for thinking about the military profession today. The third part tells how the military profession has changed over time. The description is not a simple narrative, but is organized around the three prescriptive factors identified in the first part that characterize professional activity. This analytic narrative tests the usefulness of these prescriptive factors for interpreting long-term trends in professional development. The fourth and last part draws out the practical implications of this study for creating and maintaining an effective military profession at the present time.

The Idea of a Profession

Social scientists and historians agree that a profession is a distinct kind of work, but disagree about what actually distinguishes it from other kinds of work. Eliot Freidson suggests that we should not become preoccupied with this disagreement, as if it were a matter that might someday be resolved. A resolution would be possible only if the idea of a profession were a "generic" concept, the one true meaning of which we might someday finally discover. In fact, Freidson argues, a profession is a "historic concept."[5] What it means changes with the particular period of history we examine. The standing of medicine, law, and the clergy as professions meant one thing in medieval Europe, another thing in eighteenth-century British society, and something else again in present-day American society. The different meanings are telling. Beyond what they say about the progress of the division of labor, they connote something about the evaluative and often conflictual processes that make professions occupations of higher rather than lower social status. In short, disagreement about what a profession is stems partly from the historical development of occupational structures and partly from social conflicts about the allocation of occupational prestige.

Embracing this historical view of the idea of the professions does not excuse us from offering a definition of the profession. Some working definition is needed, if only to clarify what a historical approach to the study of professions should consider. In this chapter, I rely on the following definition: *A profession is a relatively "high status" occupation whose members apply abstract knowledge to solve problems in a particular field of endeavor.* The definition identifies three elements as critical to the idea of a profession: high status, which I will link to a notion of legitimacy; applied abstract knowledge, the source of expertise; and a field of endeavor or jurisdiction for problem-solving. These three elements are commonly, though not always or exclusively, referred to by sociologists writing on this subject.[6] It will be worthwhile briefly to consider what these elements mean and how they are interrelated.

The British sociologist, T. H. Marshall, identified the importance of high status to the professions in an early essay on the subject, first published in 1939.[7] The professions, he wrote, were "occupations suitable for a gentleman," which in Britain meant a person of high status. We should not suppose that he meant doing professional labor necessarily gave one high status. In eighteenth-century British society it was rather the other way around. An occupation qualified as a profession if it entailed work that a gentleman would do, that is, if it were compatible with "the good life." The source of prestige, on this account, was the high status of the person who performed the activity. As commerce was then a disreputable activity, work could not be professional if it entailed striving for money (although professionals had to make a lot of money to meet "the needs of a gentlemanly life"). By the twentieth century, a profession's high status derived more from the work done than from the social standing of the worker. What counted were the effects of professional labor on the lives of the clients. "The idea of service," Marshall wrote, "became more important than the idea of freedom."[8] To ensure the quality of service, thus protecting their prosperity and respect, professionals organized into associations. Professional associations guaranteed the technical competence of their members by controlling their training and testing their ability; they imposed a code of ethics that put the needs of clients in first place and limited intra-professional competition by barring customary commercial practices (haggling over fees, advertising, etc.); and they protected the domain of professional practice from encroachment by non-members, not to preserve the revenue for association members, but in order (they said) to protect clients from unqualified encroachers.

According to Marshall, then, the contemporary professions enjoyed high status because they were "functional" occupations organized to meet important social needs. His was an idea widely shared in the mid-twentieth century.[9] By the 1970s, as sociologists became more skeptical of functional explanations, they began to locate the source of professional status in monopolies maintained by the operation of power, income, and education.[10] The main idea was that control over these resources could be used to secure and defend an advantaged place in the social structure. Without denying the relevance of these resources, Andrew Abbott has shown that, by themselves, they provide no satisfactory account of professional status.[11] Writing in 1981, he was prepared to revive a functionalist account, agreeing with Edward Shils that professionals possess esoteric knowledge that allows them to confront disorders—disease, crime, sin, war, etc.—and often to impose order on them or bring them to heel.[12]

By the late 1980s, however, Abbott had reformulated this idea to de-emphasize the functionalist assumption that professions met needs found naturally in society. He came to emphasize rather a constructionist account, arguing that an occupation's identification as a profession and its standing within society were outcomes of social competition within a system of professions for control over abstract knowledge applied within particular jurisdictions.[13] In keeping with this account, he defined professions quite loosely as "exclusive occupational groups applying somewhat abstract knowledge to particular cases."[14] In contrast with my own definition, his focuses on just two elements, professional expertise and jurisdiction, ignoring social status (though he seems to acknowledge the importance of legitimacy, he never incorporates it in his definitional construct).[15] I think that is a mistake because it wrongly assumes that an occupation's struggle for professional standing rests entirely on possessing and applying abstract knowledge.

It is certainly true that professionals apply abstract knowledge to solve social problems. All professionals have been instructed in and mastered a body of knowledge; their entry into professional practice is predicated on receiving some form of higher education. But the *form* abstract knowledge takes varies by profession and within professions over time. This variation in the form of knowledge is important because it affects the social standing of a profession among other professions and nonprofessional occupations. The historian Bruce Kimball has shown for example that in mid-eighteenth century America, the highest-ranked profession was the clergy, but by the early nineteenth century the highest-ranked profession was the law, and by the early twentieth century it was medicine. While each of these professions masters an abstract body of knowledge, the *form* of the knowledge of the highest-ranked professions has moved from the logic of theology to the logic of jurisprudence to the logic of natural science.[16] This historical succession indicates that the prestige of professions does not depend only on the mastery of knowledge. Professional prestige also depends on the legitimacy society accords to the form that knowledge takes.

It may be tempting to argue, as sociologists used to do, that the historical succession from theology through jurisprudence to science represents the growth and triumph of reason over superstition or at least over claims that cannot be tested empirically or verified. Talcott Parsons argued this way; although the professions originally descended from a "religious matrix," he contended they were now committed to the primacy of "cognitive rationality."[17] If we succumbed to Parsons's reasoning, we would say that professional expertise (i.e., the mastery of abstract knowledge) was proven by the ability actually to solve problems faced by individuals and society. With apologies to Christian Science, we would then say that medicine is ranked higher than the clergy among professions in the twentieth century because the application of medical science to the problems of disease and public health is demonstrably more effective than the application, say, of prayer. Indeed, that was Parsons's view. By the twentieth century, he said, occupational groups anchored in religious commitments were not professions; they only approximated professions.[18]

Nevertheless, this is a temptation we should avoid. It is true that professional standing requires control over a domain of social life—a jurisdiction—within which members of the profession try to solve particular problems by applying the special knowledge at their command. Because control over a jurisdiction is usually contested, as Abbott contends, professions are involved in competition to secure their place in society. They wage this competition by various means, from redefining the nature of the professional task to be performed to fostering legislation that bars competitors from practicing in the field. Still, the most important factor for gaining and maintaining control over a jurisdiction is demonstration that the professional activity succeeds, that it solves the problems it confronts.

Try as they might, however, professions do not perfectly control either the definition of problems to be solved or what counts as a solution. They cannot answer the question on their own. Consider the ascent of modern medicine among the professions. The rising social status of medicine rested in large part on its successful application of germ theory to the diagnosis and treatment of diseases. But the development of germ theory to approach the problem of disease reflected—it did not precede—the growing belief in society that science (not theology or jurisprudence) was *the* legitimate form of abstract knowledge, superior in practice to any other. In the future, should belief in the practical efficacy of scientific knowledge decline, as it has done over the last half-century, then professions whose abstract knowledge is scientific in form would lose status and legitimacy, even if they were demonstrably solving more problems than before.

In sum, my definition identifies three prescriptive factors that, when found together, mark an occupation as a profession. One is mastery of abstract knowledge, which occurs through a system of higher education. Another is control—almost always contested—over a jurisdiction within which expert knowledge is applied. Finally is the match between the form of professional knowledge and the prevailing cultural belief or bias about the legitimacy of that form compared to others, which is the source of professional status. We can refer to these three simply as expertise, jurisdiction, and legitimacy.

A final comment on my definition is warranted. Applying knowledge to solve social problems was thought by earlier students of the professions to entail a public service both to the client and to society. The idea of public service was an important element in their definition of the professions, because its demands supposedly overrode the dictates of the marketplace and self-interest. Relations with clients were based on the principle of trust (not caveat emptor), and professionals were expected to perform their duties even if that required an element of self-sacrifice.[19] Recent studies of the professions, however, have defined the professions more sparely as occupations based on expert knowledge, with the connection of service to the public welfare dropped by the wayside. As Steven Brint argues, the idea that professionals were "social trustees," applying knowledge to serve the public good gave way in the 1960s to the idea that professionals were knowledge "experts" hired to serve organizational authorities or market forces without much regard for the public good.[20] This change in ideas, he says, reflects a change in reality. It is at least an important shift in our understanding of the professions and one to which we will return.

The Profession of Military Service

Social science studies of the military as a profession began in the mid-twentieth century most notably with the work of Huntington and Janowitz. It is not difficult to say why military professionalism became an object of study at this time. The Cold War required that the United States maintain a large professional military during "peacetime." The country had never done this before. For a liberal political culture, traditionally ambivalent about war and standing armies, maintaining a large force in being posed questions about the quality of civil-military relations. Many thought that a high level of military professionalism was needed to ensure a technically competent military establishment and to ensure that the military establishment would act only to serve appropriate civilian authority. This thought was based on the idea that professionals were, to use Brint's term again, "social trustees" who were organized to serve the public good and not their own self-interest. But was this a reasonable idea? Was military service a profession, and if so what made military professionals effective?

The question of whether military service is a profession received different answers depending on who was asked for an answer and when. Scholars studying the professions in the first half of the twentieth century did not usually include military service in their field of inquiry. In their seminal survey of the professions, published in 1933, A. M. Carr-Saunders and P. A. Wilson explicitly refused to count the Army as a profession. They were concerned with the professions only in relation to the ordinary business of life. "The Army is omitted," they wrote, "because the service which soldiers are trained to render is one which it is hoped they will never be called on to perform."[21] Unless the military's expertise is applied, it does not count. This is a novel and, some might say, utopian ground for exclusion that tells us more about social thought between the world wars than it does about the professions. Other scholars pointed out that military service was not professional because the military was controlled by the state; military officers lacked professional autonomy, which the scholars thought was a critical trait of the professions.[22] But the ideal of autonomy as a characteristic of the professions has lost influence as professional work is increasingly embedded within corporate hierarchies. More typically, sociologists just ignored the military when they studied professions, failing to say why. It is likely that, for sociologists in this period, the prototype of a legitimate profession was science-based medicine; the military seemed unlikely to fit the mold. If the military was a profession, it was so only in the eighteenth-century sense that T. H. Marshall discussed. That is, it was a profession only because being a military officer was an occupation fit for the life of a gentleman. In the twentieth century, this idea of the profession was no longer relevant.

The public held a more inclusive view of the professions, one that had room for the military. They did not count it among the higher-ranked professions. In 1955, Janowitz notes, the public placed the prestige of the military officer "below that of the physician, scientist, college professor, and minister"—and also below the public school teacher.[23] Still, they thought the military was a profession. Indeed, public consensus on this point, though unnoticed by social scientists studying the professions, was not recent but had already formed by World War I.[24] Early in the twentieth century, it was plain that a cadre of professional military officers was necessary to train military forces in peacetime and to direct them during war.

What Huntington and Janowitz did was connect this public view of the professional military with then current social science theories about what a profession was. For them, it was largely rhetorical to ask whether military service was a profession. Neither had any doubt that it was. Huntington made a more explicit defense of the proposition than Janowitz, but both were engaged in a task of conceptual clarification, to place the military within the bounds of the professions. At the very beginning of *The Soldier and the State*, Huntington states: "The modern officer corps is a professional body and the modern military officer a professional man."[25] He then argued that expertise, responsibility (or service to society), and corporateness (or group unity) are the marks of a profession and that the military possesses all three of these, just as do the prototypical professions, medicine and law. In *The Professional Soldier*, Janowitz simply took for granted that the military was a profession. He too identified the officer corps as a professional group like medicine or law, marked by expert skill acquired over time, the rendering of a specialized service, and a sense of group identity, which entails a system of internal administration and a professional ethic.[26] In short, they agreed that the military was a profession and about what made it so.

Where they disagreed was in their answer to the second question, what made the military profession effective? Huntington believed that military professionals were effective under conditions of objective control; by objective control he meant that civilians would dictate the military security policy but leave the military free to determine what military operations would secure the policy objectives. There was, in essence, a swap of military loyalty to civilian authority in return for professional autonomy to control one's work. The alternative, he thought, was subjective control in which civilians would meddle in military affairs, trying to make the military more like society; the military would by necessity become politicized, with the result being a decline in military effectiveness. Janowitz, in contrast, believed that no clear line could be drawn between the military and civilian realms. He thought that effective civil-military relations in a liberal democracy required that the military identify with and be representative of the society it served. Isolation of the military from the larger society would reduce professional effectiveness. He also thought that the development of weapons of mass destruction fundamentally changed the nature of war in ways that required professional soldiers explicitly to consider the political consequences of military operations. Effective military professionals redefined their understanding of successful problem-solving. They downplayed their role as "heroic warriors" in pursuit of absolute military victory to emphasize their role as pragmatic managers of the minimum use of force needed to achieve political settlements and maintain viable international relations. To achieve this latter ideal, Janowitz advocated the development of what he called a constabulary force.

Important as both these models are, neither stands without criticism today. Huntington argues that the military professional is to master the functional requirements of war, to organize and train the military to meet them, and to lead the military to fight when ordered by political authorities to do so. But the presumption that there is a clearly delineated military sphere defined by war-fighting that is independent of the social and political sphere is dubious at best, especially in an era when we possess weapons of mass destruction. There is no real distinction between the ends and means of war. What ends are possible to think about depend to a large extent on the means with which they are to be pursued. As a practical matter, this means that the distinction between the political sphere, where ends are decided on, and the military sphere, where means are deployed in pursuit of ends, is highly misleading. If so, then Huntington's theory of objective civilian control does not really tell us how military professionalism can be strengthened.

Also, Huntington's confidence in the value of military professionalism is uncritical. He ignores the possibility that professionals (like all social groups) may act in ways contrary to the interests of the larger group. But sociologists have identified a number of ways in which professionals may act in their own interest against the public good: cultivating an incapacity to grasp the insights from any perspective but one's own; treating the means of professional practice as if they were ends in themselves, forgetting the original purpose they were intended to fulfill; acting to protect the station and privileges of one's group—service or branch—above any other, simply because it is one's own; shirking those tasks one is assigned to undermine the assignment; or acting opportunistically to do what is required to advance one's career.[27] Taking these into account, we have to conclude that the degree of military professionalism varies for reasons other than the institution of objective or subjective control.

Janowitz's model is more helpful because it anticipates change in the character of the military profession, something Huntington fails to do. The change is driven by the need to adapt professional practice to the changing nature of war: that affects what counts as a solution to problems of military security. But he was perhaps naïve in expecting that the profession could alter its self-conception away from the heroic warrior to the pragmatic constabulary model without intensifying competition (from within the military and without) for control over its professional jurisdiction. And while he recognized that tensions existed among officers of differing specialties—e.g., work oriented primarily to management, technology, or combat—he did not thoroughly explore the difficulties encountered between bureaucratic and professional cultures.

Even more objectionable, Janowitz failed to resolve the problems his theory poses for military professionalism. If the changing nature of war blurred the boundaries between the military and political spheres, as Janowitz said, then how was military professionalism to be reformed to ensure that its expertise was applied in the service of the state? The question is a serious one. In a constabulary force, military officers were inevitably "politicized" as they prepared for their new roles to deliver strategic deterrence and fight limited wars. But a politicized military at least implicitly challenges civilian supremacy.[28] How was the challenge to be met? In *The Professional Soldier*, Janowitz's discussion of mechanisms of civilian control to ensure professional responsibility is extremely limited, being confined to control mechanisms that are easily worked around. The problem Janowitz faced was an old one in the theory of the professions, namely, how to ensure that professionals are acting in the best interests of society when those they serve lack the expertise to define the service they need, or even to know whether their needs have been adequately addressed. His answer was to strengthen the commitment of professional soldiers to the system of civilian control through programs of political education that would help them connect their professional training to national and transnational purposes.[29] But the thrust of this argument is circular: responsible military professionalism is made to depend on the strength of military professionalism.

When Huntington and Janowitz devised their pioneering models, they were both confident that strong military professionalism was essential to ensure the military security of American democracy. Their confidence rested in part on the logic of the functionalist theory of the professions that prevailed at mid-century and on which they both relied. According to that theory, professionals were thought to be social trustees, acting in a fiduciary capacity to ensure the public good. They possessed systematic knowledge acquired through a long period of training, and, as required by their occupational ethic, they applied it competently, objectively, and impartially to meet the needs of the client and improve public welfare.

A decade after Huntington's and Janowitz's classic works appeared, sociologists abandoned this simple functional theory. In its place, they conceived a more conflict-oriented, class-based model that saw professions as group "projects" to obtain social mobility. According to this conflict approach, the professions did not necessarily serve social needs found in the natural order of things. They rather imposed their own definitions of needs on society and created self-serving means for meeting them.[30] It is mere speculation to wonder how Huntington or Janowitz might have concluded differently about the military profession had they written from the perspective of these later models. But the effects of these more cynical models of the professions were incorporated (though not explicitly) in the work of Charles Moskos and his associates.[31] Beginning in the late 1970s, Moskos argued that the military profession was moving away from its traditional institutional principles of social organization to embrace occupational or market-based principles of organization. It has proven difficult to trace development of the military profession along this single theoretical dimension. Over the years, those working on this model have multiplied the number of dimensions taken into account. Yet, despite refinements, in the end, there is consensus that a significant change has taken place. The military profession in its current state is, we must conclude, more like Brint's "expert" profession simply doing its job for an organized authority than like a "social trustee" profession rendering self-sacrificial service to the country.

Trends in Military Professionalism

My approach to the study of change in the military profession is guided by our definition of the factors that mark an occupation as a profession: expertise, jurisdiction, and legitimacy. My hypothesis is that when we examine military professionalism from the late nineteenth century to the present, we see first a convergence of these three factors to strengthen military professionalism, and then an unraveling of these factors, especially in the last half-century, that subjects the military profession to strain. The reasons for this pattern of change are interrelated. As we shall see, they are not factors always within the power of the military profession to control.

To develop these ideas, I concentrate on three pivotal periods in the history of the Army professionalism: the period between Reconstruction and World War I, when on most accounts the Army officer corps turned itself into a modern profession; the period just following World War II, when the Army faced a professional crisis centered largely on the problems of defining and defending its jurisdiction at a time when military roles and missions were hotly contested; and finally the period since the end of the Cold War, when the Army profession has been challenged by lost resources, increased deployments, and increased competition for control of its jurisdiction. The task is to identify the main themes that connect these three periods along the analytic dimensions of expertise, jurisdiction, and legitimacy.

Expertise

The period following Reconstruction was one in which the Army began seriously to cultivate the study of war as an applied science which officers must master if they were to be professional specialists in the management of violence. The emphasis on the *science* (not the art) of war, while not entirely new, was new in its scope. It underwrote the development of a new professional education system and the founding of professional associations and journals.[32] This movement within the Army had its parallels in the Navy and in the larger society. This was a period in which the "culture of professionalism"—a term used by Burton Bledstein—was widespread.[33] Many middle-class occupational groups were organizing themselves to stake a claim to professional status. The success of their claims often hinged on their being able to demonstrate expertise over an abstract body of knowledge that took the form of science. Here the themes of expertise and legitimacy are intertwined. By the late nineteenth century, science was believed to be *the* source of reliable knowledge about the world; claims to expertise were more readily recognized if they were obviously grounded in science. Even the ancient professions of theology and law were affected by this development—they began to recast the form of their specialized knowledge in a scientific mold.[34]

The scientific turn in the Army professional's claim to expertise was accepted without significant challenge until the end of World War II. Before then there was a close connection between the use of military force and the possibility of military success or victory. If superior resources did not prevail, it was thought to be a failure to apply the science of war, not a challenge to the idea that greater military strength yields greater military security. But the development of weapons of mass destruction in the wake of the world wars made plain (sooner for some than for others) that sheer military might was no longer a guarantee of security if the use of that might was so destructive that its use was self-defeating. This shift in outlook spurred development of a new and broader science of national security and strategic studies, and the experts in this new science were not confined to those in the uniformed services.[35] Nevertheless, the Army professional did not abandon the claim to expertise in military science, but if anything pursued the claim with increasing vigor, extending its system of professional military education and augmenting it by dispatching officers to pursue graduate degrees in fields ancillary to or supportive of the military sciences. At the end of the twentieth century, the Army's claim to expert knowledge in the management of violence, certified by science, was a foundation for its professional identity.[36] That is not to say, however, that the empirical challenge to that expertise posed by weapons of mass destruction had been successfully resolved.

Jurisdiction

A professional jurisdiction is defined by the boundaries of the domain within which expert knowledge is applied. It is sometimes an actual place, like a hospital, court room, or battlefield, and sometimes a slice or aspect of life. So, while an Army commander focuses on directing the course of battle, the chaplain is focused on the soldiers' souls and the medic on the soldiers' wounds. In broad strokes, the Army's jurisdiction narrowed over the late nineteenth century, as it shed domestic functions and focused on protecting the country's borders and fighting land wars abroad. Competitors who would ply their knowledge in this jurisdiction at the expense of the professional Army, notably volunteers and state militias, were effectively curbed early in the twentieth century. State militias were upgraded, but subordinated to professional Army control by the Militia Act of 1903 and the National Defense Act of 1916. By World War I, well-meaning amateurs like Theodore Roosevelt were no longer allowed direct entry into the officer corps. The Army was following an old practice of the professions, protecting its jurisdiction by resort to the law. The Navy was a rival for resources, but not a competitor to fight land battles.

This neat arrangement was disturbed by the advancing technology of war over the course of the twentieth century. The disturbance was severe at the end of World War II as evinced by the conflict over the creation of a "unified" Department of Defense and of an Air Force separated from the Army.[37] There was severe competition among the various services over what their roles and missions—their jurisdiction—should be. This contest arose because the technological development of war over the course of the twentieth century had the effect of homogenizing the theater of war, undermining the clear connection of the three major services with a particular environment of warfare—land, sea, and air—and so creating an impetus for "joint" military operations. One could say that there is just one military profession, but the services resist this more abstract identity, in part because (as Kenneth Allard has shown) their dominant operating environments continue to support different kinds of expertise and modes of intra-professional cooperation, even to this day.[38] The interservice competition over jurisdiction has blown hot and cold ever since, and it has certainly not gone away.

However, the more important competition over the Army's professional jurisdiction has come from external sources. In effect, I argue that the jurisdictional claims of the military in general, including the Army, have expanded since the end of the world wars. We can reduce that to slogan form by saying the Army has moved from the management of war early in the century to the management of defense (a wider concern) during the Cold War to the management of peace (an even wider concern) since the end of the Cold War. With each broadening of the Army's jurisdictional claims, competition over the Army's jurisdiction has increased. As already suggested, the growth of national security studies after the world wars was propelled by the threat of weapons of mass destruction; war narrowly conceived could no longer be seen as a tool in the kit of foreign policy. The outbreak of general war had to be averted, and civilian scientists in relevant disciplines began to develop expertise outside the military profession to determine how this might be done. Within the Department of Defense, new techniques of rational planning were introduced to determine the most efficient use of scarce resources. This eroded claims of the Army and other service professionals that such judgments should not be based on the more general, quantitative language of cost-benefit analysis but rather on particular service expertise. Since the end of the Cold War, encroachment of civilian experts on the Army's jurisdiction increased in large part because military operations were no longer confined to actual war but had spread to include peacekeeping missions, broadly defined. Unlike actions in a theater of war, peacekeeping operations are not an endeavor of "sole practice" by the Army professional; they occur within what Abbott called a "multi-professional workplace."[39] Successful peacekeeping requires cooperative action between the Army and a wide array of other civilian service providers who answer to no single authority.[40]

This new competition over the Army's jurisdiction could be limited if the boundaries of its jurisdiction were narrowed, as they were at the turn to the twentieth century. But this approach, preferred by some, is unlikely given the changing nature of war, the object of the Army's professional expertise. As Janowitz argued long ago, it is the changing nature of war that is driving the expansion of the Army's definition of its professional jurisdiction, thus inviting the conflicts just observed.

Legitimacy

We have noted that the Army's claim to be a modern profession a century ago was bolstered by its ability to develop its expertise in the form of military science. This new science encompassed the study of battle, logistics, weapons development, and even the mobilization, assignment, and leadership of military personnel. Science at the time, enjoying a high point of its prestige, was considered the most reliable form of practical knowledge. It was not only the Army that advanced its claim to professional status by showing that its expertise was scientific. A new source of strain for the Army professional—as for many professionals—has been society's increasing skepticism of the legitimacy of expert knowledge based on science and the ideal of objectivity.

In 1962, Thomas Kuhn, a historian of science, famously cast doubt on the claims of science to provide objective and reliable knowledge with his analysis titled *The Structure of Scientific Revolutions*.[41] His argument, however, was only one of a number of studies to show that scientific knowledge often rested on hidden non-rational social factors. In any case the stock of scientific knowledge was always tentative and subject to revision. Sometimes the revisions required were disturbing, because the science once relied on was shown now to threaten human well-being. Rachel Carson documented the harmful unintended consequences of applying modern science to industry in her study, *Silent Spring*, which created a foundation for the ecology movement.[42] In the case of military science, even reliable knowledge was a threat to human welfare, symbolized powerfully by the threat of thermonuclear weapons and literally brought home to many whose milk contained strontium 90, a radioactive particle released in fallout from above-ground nuclear testing.

There are two important results of this increased skepticism about the legitimacy of science for the professions in general and for Army professionalism in particular. First, increased skepticism about science has encouraged nonexperts to question the exercise of professional authority more boldly than they had done in the past. Before World War II, students of the professions could write confidently, for example, of doctors giving orders that only foolhardy clients would ignore. By the end of the twentieth century, resort to alternative medicine was more common, and medical doctors found it more difficult to silence competing claims from the practioners, for example, of chiropractics, homeopathy, and acupuncture. More generally, professional pleas for deference had a hollow ring, resting, as they were accused of doing, on an ideological footing to secure the privileges associated with their practice. This is true for the military as well. A good illustration of the point can be found in debates over whether and how military technology and performance standards should be revised to accommodate greater integration of women.[43] Those skeptical of the wisdom of such integration regard the redesign, say, of airplane cockpits or basic training regimes as evidence that professionalism is on the decline, as expertise and technical competence are diluted for reasons of "political correctness." In contrast, those skeptical of professional expertise argue that the established standards are not universal, objective truths, but rest on social constructions of gender that can be modified without compromising military effectiveness.[44] Without a general belief in what constitutes a legitimate form of reliable knowledge, such debates are impossible to resolve except by the exercise of politics.

Second, when we lack common belief about what constitutes legitimate, truthful, and objective knowledge, it is difficult to sustain a professional ethic of service. Unquestioned trust between the professional and the client, which is needed to sustain the ethic, no longer exists. When professionals claim to be offering impartial, disinterested, even self-sacrificial service, their claims are not accepted at face value, but are examined cynically by clients and other interested observers who look for the self-interested "angle" in such claims, and find claims of self-interest more believable than claims of altruistic service. Uncertainty about "objective truth" also makes it more difficult for professionals to engage in self-sacrificing service with confidence that their self-sacrifice is justified, that the service is rendered in pursuit of what is unambiguously good. In the Army, such doubts show up at the end of the Cold War, with rising doubts about the value of losing life when engaged in peacekeeping operations. It is not enough to assert that self-sacrificing service is central to professional practice.[45] The norm of self-sacrifice must be grounded in a general belief that the application of expert knowledge within a particular jurisdiction is legitimate.

Implications for Military Professionalism

Professional practice is strong when the conditions of all three prescriptive factors are met, that is, when the application of expertise within a particular jurisdiction is uncontested and thought in general to be legitimate. Otherwise the quality of professional practice is cast into doubt. Applying this standard of evaluation to our historical analysis, it is clear that the military profession was strongest at the end of the nineteenth century through the end of World War II, but has weakened since then as a result of heightened jurisdictional competition and declining public confidence in science as a form of legitimate expertise.

Increased jurisdictional competition stems largely from the changing nature of war. The enormous destructive power of weapons of mass destruction has made clear the self-defeating character of unrestrained armed conflict; military security was no longer (if it ever really was) a simple function of military strength. The practical effect of this development, in the aftermath of World War II, did not diminish, but expanded the military's role. Still centrally concerned with the management of violence between armies at war, the military profession came also to manage defense, the aim of which was to avoid the outbreak of general war through a strategy of deterrence; and after the Cold War, the military began to manage the peace—not alone, but with others—with the aim of limiting armed conflict to the maximum extent and using the least force possible to promote a political settlement of differences, maintain respect for human rights, and provide humanitarian relief. The difficulty here is that the military does not possess the same level of abstract knowledge about how to conduct peacekeeping missions (or how to deter global war) that it has for waging war. Even if it did, the military does not and cannot pretend to possess a monopoly of expert knowledge relevant for mission success in its expanded jurisdictions. To a greater extent than when fighting war, the management of defense and peace requires the military profession to cooperate and compete with other professions as it goes about its work.

To strengthen military professionalism under these circumstances, one might consider the obvious possibility of contracting the military's jurisdiction. The Army could attempt to restrict its professional role by confining itself to problems of war, traditionally understood as conventional warfare between uniformed armies of nation-states and fought for the control of land—and forsake all other concerns. When threats stem less from the prospects of interstate war than from irregular religious and ethnic conflicts and attacks by terrorists, this possibility seems unrealistic. The need to prepare for general war cannot be ignored, but the Army could not justify a return to its world war jurisdiction while retaining its contemporary relevance.[46] Alternatively, one might deny the problem and emphasize the military's evident success in defending its expanded jurisdiction. The absence of a general war over the last half-century, the military could assert, demonstrates its professional competence in managing defense. Yet it is difficult to say with certainty whether or to what degree the absence of general war in this period resulted from the application of military expertise to the problem of deterrence. In fact, many wonder whether deterrence based on a nuclear standoff is a viable strategy. Consideration of an antimissile defense system suggests there are doubts about that even within the defense community.

A third approach considers how the Army professional can operate most effectively in its multi-professional workplace. Within the broad jurisdiction of the management of peace, mission success, say in Bosnia or Kosovo, cannot be achieved by relying on the expertise of the Army professional alone. Success requires the combined expertise of all the relevant professionals involved. The analogy is imperfect, but one can think for example about the combined expertise required in a hospital operating room before any complex surgery can succeed. It cannot be assumed, however, that professionals engaged in a peacekeeping arena will wish to cooperate for a common goal. Even if they did, at the moment there is no well-established institutional authority with the expertise to identify, much less integrate, the various professional activities that are needed to maintain peace and substitute nonviolent for violent means of conflict resolution. Nevertheless, it is important that the various professional organizations on the mission employ their particular expertise in a way that contributes to the overall objective of building peace in the region.

It is becoming clear with experience that the military's role in building peace is highly circumscribed. David Last, a major in the Canadian military, observes that while "military forces are effective at guaranteeing military security against organized military opposition . . . they are impotent in the face of bricks through windows or threatening telephone calls in the night."[47] He also notes that other agencies are needed to police, govern, provide relief, and move toward reconciliation if the conflict is to be resolved. He does not deny that military expertise is required in peacekeeping operations, especially to stop and prevent organized physical violence. But suppressing violence is only one step in the process, providing temporary relief.[48] Jurisdictional competition is limited and professionalism enhanced if the military's role within peacekeeping and humanitarian interventions is limited to fit its peculiar expertise. The same is true for other agencies involved. Achieving this outcome requires deliberate coordination across professional boundaries in which respect for the capacity of the others is acknowledged. There is no benefit in this setting if the military professional maintains a false sense of its own autonomy or attempts to do everything it possibly can. [49]

The public's declining confidence in science as a reliable form of practical knowledge is another troubling trend for military professionalism. This decline is a problem for all science-based professions, not just the military, because it makes it more difficult for all to establish the legitimacy of their authority. While science-based expertise continues to be used, and will be used far into the future, its use is no longer inevitably considered to be self-justifying. To shore up their authority, professions spend time building "sustainable partnerships" with those outside the profession whose expertise is other than their own and yet affects what they do.[50] That requires entering into a dialogue across professions and with the public (including the news media). The purpose is to increase the public's understanding of the reasons for a particular course of professional action and the profession's understanding of the reasons why that course of action might be resisted. There is no reason to expect that dialogue alone will lead to consensus. But without such a dialogue, it is hard to gain the public's trust on which professional legitimacy depends or to create the professional self-confidence required to embrace an ethic of self-sacrificing service.

This course, however advisable, also has a utopian ring to it. It seems to presume the persistence of the professions as "social trustees," committed to serving the public good. It seems to ignore the rise of "expert" professionals who serve only the interests of organizational authorities and market forces. But, as Brint has shown, the presumption of social trusteeship and the ignoring of self-advancing experts are not warranted; social supports needed for social trustee professionalism to thrive have eroded. This erosion has helped to elevate the authority of the market over moral elements in the life of the professions.[51] Brint's argument suggests that the new competitor to science as the form of objective and reliable practical knowledge is the market itself: what is true, objective, and useful knowledge is what sells in the marketplace; anything else is fancy. However crude that sounds when stated baldly, it is not far away from ordinary experience. We are used to people asking about the "bottom line" of our arguments and having them conclude that they "buy it" or not.[52] Making the market the new form by which the legitimacy of professional knowledge is judged does not unseat the science-based professions. But it poses a challenge to the customary pecking order of the professions, with those that can wrap their expertise in marketable form rising to the top (as economists and clinical psychologists have risen in the profession of social science). It also means that pressures will build within professions to formulate their expertise in the vocabulary of the marketplace. Such a change is what Moskos and others observed in the military profession beginning in the 1970s. The seeds of change were planted perhaps with Robert McNamara's introduction of cost-benefit analysis to evaluate the military budget and procurement programs in the early 1960s.

The triumph of the market cannot be ignored because it arises from and is part of society itself. Nevertheless, the triumph is not complete and so the choice is not between legitimating expertise based *either* on the market *or* on science (or other more traditional social forms). The problem is to recognize when market legitimation is an appropriate basis of action and when it is not. In principle, this problem is not difficult to solve. Markets provide effective and legitimate diagnoses and prescriptions about the worth and allocation of social goods when there is real uncertainty about what goods to pursue at what cost. Historically, the rise of market-based legitimation in modern liberal democracies is connected with the growing heterogeneity of their populations and the diversity of their cultures. With the prevailing commitment to the principle of toleration as a foundation for society, there is reluctance to impose definitions of what is good or which goods are to be preferred over others. Instead, these definitions and valuations are established by free exchange whether in the marketplace of commerce or in the marketplace of ideas. Whenever the military profession confronts such uncertainty—as it may do for instance in decisions about the design of future weapons systems or in calculations about how much money soldiers should be paid—there is no reason why the profession should not employ the language of the market. Doing so is not completely foreign to military experience. The battlefield, after all, is full of uncertainty, and battles are decided by the exchange of fire. A great deal of military science is devoted to the end of gaining and sustaining the advantage in this deadly competition.

Many times of course it is not clear whether expertise rooted in the logic of the market or expertise rooted in some other moral or scientific logic should prevail. That is true for instance with respect to privatizing activities that once were handled within the military. Market efficiencies may justify privatizing some health-care or finance functions or closing some bases. But other matters are more difficult to judge. For instance, Deborah Avant, in her chapter on privatizing military training in the present book (Chapter 9), questions whether market efficiencies justify privatizing initial training of cadets who will become Army officers. The efficiencies are tempting to a downsized force that must stretch resources. Still, when the military relies on private contractors for instructors, it may subtly teach new entrants into the profession that, despite the rhetoric of self-sacrificing leadership, market logic trumps other considerations. As a result, young officers may leave the profession, abandoning command for the pursuit of economic opportunity.[53]

But when there is relative certainty about the goods to be sought or about comparative worth, then the language of the marketplace may be superfluous and the legitimacy of expert knowledge may rest on other grounds. The higher the degree of certainty, the more likely the public will regard such justifications as legitimate. Examples here include scientific questions about how to build thermonuclear weapons or to treat the psychological consequences for individuals subjected to prolonged stress in combat. Once the decision is made to do either, then scientific expertise is required. Professional diagnoses and prescriptions based on this knowledge should be defended against the logic of economic efficiency and "market tests." The same is true with respect to certain moral issues that the military professional must face. Relying on forms of discourse borrowed from the marketplace, one cannot explain or justify self-sacrifice for the public good that military (and other professional) service often requires. This was a lesson that Robert McNamara had yet to learn as he tried to monitor the "progress" and "efficiency" of the war in Vietnam by relying in part on body counts and ratios of enemy to American losses.[54]

In short, when possible, the military should wrap its expertise in the language of the market to enhance its professional legitimacy. But when that is counter-productive or not possible, the military should patiently explain both to itself and to the public why that language is sometimes inappropriate and why we might sometimes wish to act on alternative forms of knowledge to pursue a social good. Such explanations are especially important when military missions of war or peace put lives at risk, whether our lives or the lives of others.

Notes

1. Samuel P. Huntington, *The Soldier and the State* (Cambridge, MA: Harvard University Press, 1957); Morris Janowitz, *The Professional Soldier* (Glencoe, IL: Free Press, 1960).
2. I say usually because we sometimes use the adjective "professional" in a derisive way, as we do when we speak about "professional politicians" or "professional wrestlers." In these cases,

we expect the word professional to denote a full-time occupation that is somehow dishonorable—the politician because politicians are often required to compromise on principles in order to rule and the wrestlers because we know that the outcome of the sporting events in which they compete is scripted. The use of the term professional in these cases is ironical; we use it to cast doubt on their integrity and on their contribution to the well-being of society.

3. See, for example, Edward M. Coffman, "The Long Shadow of *The Soldier and the State*," *Journal of Military History* 55 (January 1991): 69-82; and Matthew Moten, *The Delafield Commission and the American Military Profession* (College Station: Texas A&M University Press, 2000), 6-15.
4. Samuel Haber, *The Quest for Authority and Honor in the American Professions, 1750-1900* (Chicago, IL: University of Chicago Press, 1991), x.
5. Eliot Freidson, *Professionalism Reborn: Theory, Prophecy, and Policy* (Chicago, IL: University of Chicago Press, 1994), 16. Freidson is not the only one to argue so. A brief but important summary of the historical development and changing meaning of the professions is found in Talcott Parsons, "Professions," *International Encyclopedia of the Social Sciences*, 12: 536-47. More comprehensive treatments can be found in Haber, *Quest for Authority*; and Bruce A. Kimball, *The "True Professional Ideal" in America* (Lanham, MD: Rowman & Littlefield, 1995).
6. For a review of definitions of the professions, see Geoffrey Millerson, *The Qualifying Associations: A Study in Professionalisation* (London: Routledge & Kegan Paul, 1964); and Freidson, *Professionalism Reborn.*
7. T. H. Marshall, "The Recent History of Professionalism in Relation to Social Structure and Social Policy," in *Class, Citizenship, & Social Development* (Chicago, IL: University of Chicago Press, 1977).
8. .Ibid., 160.
9. So, for instance, Talcott Parsons argued that the professions were an "institutional framework in which many of our most important social functions are carried on." Talcott Parsons, "The Professions and Social Structure," in *Essays in Sociological Theory*, Rev. Ed. (New York: Free Press, 1954), 48.
10. See Eliot Freidson, *Professional Dominance* (New York: Atherton Press, 1970); Jeffrey Berlant, *Profession and Monopoly* (Berkeley and Los Angeles: University of California Press, 1975); and Magalí Sarfatti Larson, *The Rise of Professionalism* (Berkeley and Los Angeles: University of California Press, 1977).
11. Andrew Abbott, "Status and Status Strain in the Professions," *American Journal of Sociology* 86 (January 1981), 828-29.
12. Ibid., 829. See also Edward Shils, "Charisma, Order, and Status," in *Center and Periphery* (Chicago, IL: University of Chicago Press, 1975), 267-71.
13. Andrew Abbott, *The System of Professions: An Essay on the Division of Expert Labor* (Chicago, IL: University of Chicago Press, 1988).
14. Ibid., 8.
15. See, ibid., 184-195. Abbott's discussion of legitimacy underscores his belief that legitimacy is peripheral—not central—to what constitutes the profession. He acknowledges that professional legitimacy rests on a fit between professional work and values in the larger culture. He notes that "changes in values can recast the meaning of a profession's legitimation arguments without any change occurring in the arguments themselves," as meaning changes when put in a different context (p. 187, original italics). And he shows how changing values have affected the professions. Nevertheless, he contends that value shifts over the last two centuries have had "surprisingly small effects on the actual history of [professional] jurisdiction" (p. 195)—a contention that ignores effects they may have had on the form of professional expertise.
16. Kimball, *The "True Professional Ideal" in America.*
17. Parsons, "Professions," 537-38.
18. Ibid., 539.

19. These ideas are expressed, for instance, in the writings of Marshall, Parsons, and Wilensky.
20. Steven Brint, *In an Age of Experts: The Changing Role of Professionals in Politics and Public Life* (Princeton, NJ: Princeton University Press, 1994).
21. A. M. Carr-Saunders and P. A. Wilson, *The Professions*, Repr. Ed. (London: Frank Cass, 1964).
22. See Arthur Larson, "Military Professionalism and Civil Control," *Journal of Political and Military Sociology* 2 (Spring 1972): 57-72; and Bernard J. Piecznski, "Problems in U.S. Military Professionalism, 1945-1950" (Ph.D. diss., State University of New York at Buffalo, 1985), 34.
23. Janowitz, *Professional Soldier*, 4. More current occupational prestige scores, part of the General Social Survey, do not include military officers as an occupational category.
24. Allan R. Millett, *Military Professionalism and Officership in America* (Columbus, OH: Mershon Center, 1977).
25. Huntington, *Soldier and the State*, 7.
26. Janowitz, *Professional Soldier*, 5-7.
27. These professional pathologies that might affect military effectiveness are taken mainly from Larson, "Military Professionalism," 65-66. The idea that the military professional is an agent who might shirk his responsibilities comes from Peter Feaver, "Crisis as Shirking: An Agency Theory Explanation of the Souring of American Civil-Military Relations," *Armed Forces & Society* 24 (Spring 1998): 407-34.
28. Peter D. Feaver, "The Civil-Military Problematique: Huntington, Janowitz, and the Question of Civilian Control," *Armed Forces & Society* 23 (Winter 1996): 149-178.
29. Morris Janowitz, "Civic Consciousness and Military Performance," in *The Political Education of Soldiers*, eds. Morris Janowitz and Stephen D. Wesbrook (Beverly Hills, CA: Sage, 1983), 74-76.
30. Abbott, *The System of Professions*, 3-6.
31. See Charles C. Moskos and Frank R. Wood, eds., *The Military: More Than Just a Job?* (Washington, DC: Pergamon-Brassey's, 1988); and Charles C. Moskos, John Allen Williams, and David R. Segal, eds., *The Postmodern Military: Armed Forces After the Cold War* (New York: Oxford University Press, 2000).
32. In 1877, Maj. Gen. John M. Schofield told an audience of Army officers: "It is the Science of War in the broadest sense, not simply the *Art* of War, that we are to study." Quoted in Carol Reardon, *Soldiers and Scholars: The U.S. Army and the Uses of Military History* (Lawrence: University Press of Kansas, 1990), 9. For overviews of military professionalization in this period, see in addition to Reardon, for example, Huntington, *Soldier and the State*, 222-88; Millett, *Military Professionalism*; Russell F. Weigley, *History of the United States Army* (New York: Macmillan, 1967), 313-54; and John M. Gates, "The 'New' Military Professionalism," *Armed Forces & Society* 11 (Spring 1985): 427-36. Historians have recently explored the seeds of American military professionalism in the first half of the nineteenth century. See, e.g., William B. Skelton, *An American Profession of Arms* (Lawrence: University Press of Kansas, 1992); Moten, passim; and, more briefly, Allan R. Millett and Peter Maslowski, *For the Common Defense* (New York: Free Press, 1984), 126-130.
33. Burton J. Bledstein, *The Culture of Professionalism* (New York: Norton, 1976).
34. Kimball, *The "True Professional Ideal" in America.*
35. For a broader discussion of these trends, see the chapters by Christopher Dandeker and James N. Rosenau in *The Adaptive Military*, ed. James Burk (New Brunswick, NJ: Transaction, 1998).
36. I am indebted to James S. Powell, a history graduate student at my university, for his content analysis of articles appearing in *Army* in 1980, 1990, and 2000. His analysis confirms the centrality of applied scientific expertise for professional identity within the Army.
37. "Piecznski, "Problems in U.S. Military Professionalism"; Townsend Hoopes and Douglas Brinkley, *Driven Patriot: The Life and Times of James Forrestal* (New York: Knopf, 1992).
38. Kenneth C. Allard, *Command, Control, and the Common Defense* (New Haven, CT: Yale University Press, 1990).

39. Abbott, *System of Professions*, 69-79, 151, 153.
40. See Ian Wing, *Refocusing Concepts of Security: The Convergence of Military and Non-military Tasks* (Duntroon ACT: Land Warfare Studies Centre, 2000).
41. Thomas S. Kuhn, *The Structure of Scientific Revolutions* (Chicago, IL: University of Chicago Press, 1962).
42. Rachel Carson, *Silent Spring* (Boston, MA: Houghton Mifflin, 1962).
43. James Burk, "Three Views on Women in the Military," *Society* (forthcoming).
44. See, e.g., Rachel N. Weber, "Manufacturing Gender in Military Cockpit Design," in *The Social Shaping of Technology*, second ed., eds., Donald MacKenzie and Judy Wajcman (Philadelphia, PA: Open University Press, 1999), 372-81.
45. For an interesting discussion of the current doubts about self-sacrificing service within the Army, see Don M. Snider, John A. Nagl, and Tony Pfaff, *Army Professionalism, the Military Ethic, and Officership in the 21st Century* (Carlisle Barracks, PA: U.S. Army War College, Strategic Studies Institute, 1999).
46. This means that the slogan—the Army's purpose is to fight and win the nation's wars—provides little guidance to any serious discussion of military professionalism.
47. David Last, "Twisting Arms or Shaking Hands? How to Put Peacekeepers Out of Business" (paper presented at conference, Taking Stock of Civil Military Relations, sponsored by the Centre for European Security Studies [The Netherlands], the Centre for Security and Defence Studies [Canada], and the Geneva Centre for the Democratic Control of Armed Forces [Switzerland], held at The Hague, 9-11 May 2001), 4.
48. For a fuller discussion of these issues, see Thomas G. Weiss, *Military-Civilian Interactions: Intervening in Humanitarian Crises* (Lanham, MD: Rowman & Littlefield, 1999).
49. An official with the International Committee of the Red Cross (ICRC) observes that "it is somehow paradoxical that in several tragic situations over the past decade, when the need for military force to restore peace was critical . . . the military (implementing mandates assigned by their political masters) were more involved in activities of an essentially humanitarian nature. The most prominent example is the U.N. Protection Force in Bosnia (UNPROFOR)." Meinrad Studer, "The ICRC and Civil-Military Relations in Armed Conflict" (paper presented at conference, Taking Stock of Civil Military Relations), 5. See note 47 for complete citation.
50. The phrase is used in *The United Nations and Civil Society: The Role of NGOs,* Proceedings of the 30th United National Issues Conference (Muscatine, IA: The Stanley Foundation, 1999), 2, as quoted and discussed in Wing, *Refocusing Concepts of Security*, 99.
51. Brint, *In an Age of Experts*, 203-05.
52. On the importance of simple metaphors for the construction of practical knowledge, see George Lakoff and Mark Johnson, *Metaphors We Live By* (Chicago, IL: University of Chicago Press, 1980).
53. Thomas E. Ricks, "Younger Officers Quit Army at Fast Clip," *Washington Post*, 17 April 2000, A1 (available from Lexis-Nexis, academic edition; accessed 27 June 2001). For data on the rates of loss of officers from lieutenant through colonel, see the briefing by Lt. Gen. Timothy J. Maude, Deputy Chief of Staff for Personnel, U.S. Army, presented at the Commanders' Conference, 19 October 2000. Available from: *http://www.d-n-i.net/DCSPER_10_00/index.htm*; Internet; accessed 27 June 2001.
54. Robert S. McNamara, *In Retrospect: The Tragedy and Lessons of Vietnam* (New York: Times Books, 1995), 48. In his own words: "I always pressed our commanders very hard for estimates of progress—or lack of it. The monitoring of progress—which I still consider a bedrock principle of good management [a market-based principle]—was very poorly handled in Vietnam. . . . Uncertain how to evaluate the results in a war without battle lines, the military tried to gauge its progress with quantitative measurements such as enemy casualties (which became infamous as body counts)."

3 Trust in the Profession of Arms

Joseph J. Collins and T. O. Jacobs

Trust is both simple and complex. It is simple because everyone has a sense of its essence and its importance. Trust is also complex because, while it can be examined as a distinct strand, it is also woven into many other fabrics, such as ethics, leadership, cohesion, and professionalism. In respect to professionalism, trust is so critical that it is hard to imagine a healthy profession of arms without widespread trust. If nothing else, the demands of the battlefield dictate that the military profession advocate and achieve genuine trust, both horizontally (i.e., within and among units), and vertically (i.e., up the chain of command, ideally from the lowest private to our most senior general officers).

The trust problem in today's armed forces is evident beyond a shadow of a doubt. In the Military Climate/Culture Survey (MCCS)—the cornerstone of the Center for Strategic and International Studies report, *American Military Culture in the 21st Century*—analysts found that only about a third of armed forces officer and enlisted respondents clearly agreed with the statements that "when my service's senior leaders say something, you can believe it is true," and "in my service an atmosphere of trust exists between leaders and their subordinates."[1]

Focus group comments and survey results from all services also found significant dissatisfaction with senior officer and unit-level leadership. A recent article based on interviews of nearly 700 Navy junior officers described serious leadership problems and a malaise among lower-ranking naval officers, many of whom did not aspire to command.[2] Trust-related issues were frequently cited as obstacles to good leadership and an effective organizational climate in all of these studies.

In the spring of 2001, the Army released the results of a study conducted by the Army Leadership and Training Development Panel (ATLDP). After surveying or interviewing nearly 14,000 soldiers and officers, the study authors noted that "there was some mistrust of senior leadership," and that "micromanagement has become part of the Army culture." It concluded: "There is a growing perception that lack of trust stems from the leader's desire to be invulnerable to criticism and blocks the opportunity for subordinates to learn through leadership experience."[3]

In this chapter, we shall define trust, especially as it relates to the military profession; assess the state of trust in the Army today; analyze factors that underlie trust problems; and present individual and institutional recommendations for improving trust today and in the future.

Trust and the Profession of Arms

Trust can be defined as total confidence in the integrity, ability, and good character of another.[4] While trust is a revered quality, it is not a highly prized guide for action. It is somehow both virtuous and inherently unsafe. In much of the popular literature, we are warned against a reliance on trust, except perhaps for trust in God or a strong sword. In international relations in particular, trust is fleeting, except where there exist reciprocal mechanisms for righting violations of that trust. President Ronald Reagan, for example, told Soviet President Mikhail Gorbachev, over and over, that the appropriate attitude toward arms control was "trust but verify." In other words, although we must coexist, I cannot trust you unless I have the ability to keep tabs on your behavior.

While trust is not highly regarded in commerce or politics, a high level of trust is critical to military life. On the human level, trust is the cornerstone of interpersonal relations and a sine qua non for successful leadership. Dr Jonathan Shay, a Harvard-trained psychiatrist, writes that cohesion, leadership, and training are force multipliers and "create and embody well-founded trust and well-founded confidence." Moreover, trust is locked in a circular relationship with integrity and honesty. Commanders and command policies must make it *safe* to tell the truth. . . . Whatever causes risk aversion, career fear, courtier skills, or a climate of fear, lack of truthfulness is like a steady blood loss for *everyone* in the organization. Trust makes consistent truthfulness possible; consistent trustfulness makes well-founded trust possible. There is no way out of this circularity.[5]

> Moving beyond the issue of interpersonal trust, one must ask: What is trust in a professional organization? Why is trust important? How does the military profession differ from others in its dependence on trust, if indeed it does? Before we tackle these questions directly, however, we need to glance briefly at the military's professional context.

All professions worthy of the name have an explicit or implicit ethical code, a code of conduct that governs members in relation to the profession's client system. Many professions have standing bodies empowered to review compliance with their ethical codes. The reason for this concern about ethical practices is straightforward. All true professions provide services to a set of clients, but their members have a level of special skill, knowledge, and understanding that goes beyond that possessed by the client system. Unscrupulous practitioners may therefore with relative ease exploit their clients.

For the body as a whole to remain respected and privileged as a profession, the body must police itself to ensure that client exploitation does not occur. In this way, the individual members are constrained to conform to standards of ethical conduct that protect the body as a whole. Members monitor service to their clients to ensure they are truly being served. Thus the military profession has a set of ethical, legal, and professional standards. Violate these standards, and there may be legally enforceable penalties or other types of sanctions. In this way, the military is much like other professions.

However, there are some important differences that stem from the extreme demands the military profession sometimes places on its members. A strong military norm is the expectation of unquestioned obedience to lawful orders. (Of course, the word "lawful" introduces more complexity. The implication is that the recipient of the order must make a judgment about its lawfulness, which means that obedience is of necessity provisional.) The rule of obedience runs directly counter to an important societal norm in this country, i.e., that citizens have the right to individual self-determination, assuming they do not in the exercise of that right do harm to or otherwise impinge on the rights of others. In essence, the military ethic demands that its members surrender that right. If a soldier is ordered to move forward into harm's way, he or she is expected to do just that. The freedom to decide against engaging would jeopardize the lives of those who have engaged. So the profession must demand that its members surrender their right to self-determination when circumstances require. Trust—upward toward leaders and laterally toward comrades—is the essential dynamic that makes it possible for full engagement with the enemy to happen.

When ordered into harm's way, the individual cannot avoid asking, "Is this really necessary?" If obedience is expected, then the member must have trust that the order is both necessary and prudent. He or she must also have trust that others will obey as well. "Necessary" means that the objective of the action warrants the potential sacrifice envisioned. "Prudent" means that the potential sacrifice is the minimum that can be engineered into the situation, and that the risk involved is fully justified and understood.

If a member is ordered to do something that is either unnecessary or imprudent, then that member has been exploited. Surrender of the right to self-determination has not been reciprocated by prudent care and concern on the part of the organization. That is a violation of the equity norm; no other norm is more central to human relationships and to the mutual trust between an organization and its members. If the member cannot trust the organization ("I believe someone up there really cares whether I live or die"), then the member will not surrender his/her right to self-determination, with consequent risk that orders will not be obeyed. If the leadership does not look out for the members' welfare, then the members will look out for themselves.

Thus the link between trust and selfless service is very strong. Trust is key in the military because without it one cannot have confidence that all members will obey lawful orders in the service of the body as a whole. They may instead decide to act in their own self-interest. Any erosion of trust therefore will erode the core ethic of the military profession. Viewed this way, trust is the essential foundation for cohesion and commitment. Every soldier who stands ready to sacrifice for the service has reason to expect other soldiers to be willing to make the same sacrifice.

This makes apparent a second important equity concern, equity in the sacrifice expected. One of the most powerful norms in the Army is that leaders be willing to expose themselves to the same sacrifice demanded of other members. Those who command their front-line troops from the rear are despised. The heroes do not always command from the front, but they are always willing to be at the front if that is where they best serve the needs of the body as a whole. The catch is that "the same sacrifice" has a different meaning at each level of the organization. At the bottom, the soldier who risks life and limb to accomplish a mission has every reason to expect comrades to undertake the same risk. But privates do not really expect generals to do exactly the same thing. So how does the soldier frame his/her expectations about the top-level leader's willingness to accept risk? In most of the data we have seen, there are two concerns, and these concerns parallel the two equity considerations mentioned above.

The first concern is that the Army does not fairly reciprocate the soldier's loss of self-determination. For example, soldiers may find that commitment to a service career compromises their ability to care for their families, while at the same time service-provided benefits are allowed to erode. What soldiers counted on—institutional care and concern for their families—has diminished but the institution has not diminished its expectations for soldier commitment. Soldiers expect leaders to do what it takes to maintain that care and concern.

A second concern is that the top leadership is not equally committed to the institution. For example, the soldier may be expected to risk his/her life to accomplish a mission. The corresponding expectation on the part of a soldier, especially those soldiers who at any one time are on the cutting edge, might be that the top-level leaders be prepared, if necessary, to risk position and rank in defense of the institution.

While our focus here is on trust within the profession, few soldiers maintain so lofty a perspective. For most, trust is both highly personal—about their immediate leaders and comrades—and institutional—about their senior officers and service leaders. Problems real or perceived in the quality of leadership, integrity, or devotion to the institution can have negative effects on the levels of trust in an institution that relies on it the way the human body relies on blood.

General Evidence of Trust Problems in the Active Army

There is a mountain of evidence that the Army along with the other services has significant leadership problems, many of which adversely impact on trust inside units and between and among seniors and subordinates at various levels. The proximate causes of these problems—discussed in detail below—range from human nature to resource management, to careerism, to inflexible command selection processes, to the environmental pressures on today's force, perhaps the most overworked peacetime force in American history. Below are verbatim examples of what soldiers have been saying about leadership and trust issues in today's Army.

> LIEUTENANT: "The largest problem affecting retention of junior officers is the perception that senior leadership (LTC+) is completely out of touch with soldiers and their needs. Ticket-punching and looking good are the priority." [6]

> NCO: "Civilian and senior military leadership are so out of touch with reality. They are more concerned with votes or their next star or civilian job than that of the welfare of their soldiers."[7]

JUNIOR OFFICER: "The more senior the leader is, the more out of touch he is with reality. It has been a long time since I have seen a leader with the intestinal fortitude to tell his boss he's wrong." [8]

Other problems concern micromanagement and a zero-defects mentality. Not having the freedom to fail has led many officers to adopt a new organizational ethos: management-by-minutiae. This micromanagement style fosters untruthfulness and attitudes that stress form over substance, appearances over reality. A major at Ft. Leavenworth said to a focus group:

> The Army has evolved into a zero-defects culture. Given intense competition for limited command and promotion opportunities, officers can ill afford to fail. Commanders who cannot afford failure tend to put primary importance on appearing to excel in all things. This tendency also results in a lack of tolerance of mistakes by subordinates. Couple this with the lack of proper counseling, [and it] results in hesitant and underdeveloped junior officers. This will eventually permeate all grades.[9]

In addressing the problem of a significant increase in attrition among captains, the Army's Vice Chief of Staff, Gen. John Keane, cut to the heart of the matter:

> Why are they getting out? The issue is more complex than the lure of a strong economy. . . . The fourfold increase in the number of deployments, coupled with a 300,000 soldier reduction, is causing officers to have to do more with less. . . . They are discouraged by long hours spent in garrison . . . and there is no end in sight; they witness field grade officers working even longer hours. . . .They are frustrated by what they perceive as a "zero-defect" mentality and a resulting culture of micromanagement. They came into the Army to lead soldiers and to willingly shoulder the immense responsibility that goes with command; however, they tell us that this responsibility has been taken away from them by leaders more concerned with making sure that nothing goes wrong on their watch. And most importantly, they are frustrated because they feel we, as senior leaders, are either unwilling or unable to understand and address their concerns.[10]

The problems outlined by General Keane are not only discussed openly in the Army, they are also evident to civilian observers as well. For example, the Defense Advisory Committee on Women in the Service, a civilian group that studies issues related to women in the military, found leadership problems to be one of the three most pressing issues cited by focus groups in the Committee's latest travels throughout the military. Their report released in the spring of 2001 indicated a widespread perception that senior leaders don't know how to lead: "Our troops at every level expressed a lack of faith in the leadership." Among the comments that men and women made were: "We are being micromanaged . . . there are too many layers. . . . Generals [are] trying to be platoon leaders . . . senior officers and NCOs are afraid to discipline because they are not backed up by the chain of command," etc.[11]

From 1998 to 2000, the Center for Strategic and International Studies (CSIS), a Washington think tank, conducted a study of military culture and organizational climate in the military. CSIS surveyed 12,500 men and women in uniform from the Army, Marine Corps, Coast Guard, and various service or joint headquarters. It also conducted 6-8 person focus groups with a total of over 750 people in uniform. The study found that force reductions, high operations tempo, and resource constraints in units have exacted a toll on the people in a mostly married (56 percent) force. The study concluded that, while apparently not as damaged as it was in the post-Vietnam era, the military in 1998-99 was far less ready and less satisfied than it was a decade ago. Overall, the biggest force management problems noted by military personnel in the CSIS survey were:

- Pay/quality of life shortfalls;
- Excessive operations tempo, too much time away from home, and a resulting work-family imbalance;
- Shortage of material and human resources in units affecting both readiness and morale; and
- Questionable quality and training of new personnel just entering the service.

In survey and focus groups, there were numerous complaints of leadership problems: micromanagement, a zero-defects mentality, and continually having to do more with less. Participants noted problems of trust between seniors and subordinates, both in the field and between the field and Washington. These problems have been magnified by global telecommunications and access to the Internet for people in the field. People may not understand more, but they hear and see more. Overall, the study also found that the quality of officer leadership in units was not uniformly excellent and was in need of improvement. This complex set of problems has taken a toll. Only a fourth of the survey respondents considered morale high. Recruiting and retention have become problematical. Readiness and quality-of-life problems have severely impacted the very people who have enabled the force to do more with less.

Specific Evidence of Trust Problems

An examination of active Army data from the Military Climate/Culture Study (MCCS) used in the CSIS study reveals specific evidence of trust problems in units in the field and between people in the field and the senior leaders of the service.[12] This chapter will examine selected questions that concern honesty, a cornerstone of trust; perceptions of the selflessness and unit focus of officers and NCOs; and the demonstrated level of trust existing in units and the service.

The MCCS surveys were conducted in deployable units in 1998 and 1999. Thus, junior NCO and enlisted personnel are disproportionately represented in the results. For example, of the 7,808 active Army members surveyed, 5,470 were junior enlisted E1 to E4, 1619 were NCOs (E5 to E6), 287 were senior NCOs (E7 to E9), and only 318 were company grade officers (O1 to O3). In the questions noted below, the analysis will *not* record the responses of the very few (79 of 7,808) senior officers in the troop unit portion of the database, who always registered the highest, i.e., least critical, scores. Throughout the analysis, emphasis will be on those who definitely agreed or disagreed with a particular statement. This will reveal the degree to which there is unambiguous support for these trust-sensitive statements. The average score for each rank group will also be shown. In all, a score of 4.0 marks the beginning of the "slightly agree range" and was felt by the survey administrators to be the minimum for a good score. In total, including enlisted and senior officer respondents in the active Army database, roughly 40 of the 99 survey responses hit the 4.0 mark.[13]

Honesty

As noted above, honesty and trust have a circular relationship. A climate of trust makes honesty possible, and honesty is a prerequisite for the development of trust. The Army's own field manual on leadership says it best: "Subordinates learn how to trust their leaders if the leaders know how to do their jobs and act consistently—if they say what they mean and mean what they say."[14]

Item 41.You can tell it like it is in our unit, we don't hide bad news.

	E1-4	E5-6	E7-9	O1-3
% Agree + % Strongly Agree	31	30	48	54

Item 79. When my Service's senior leaders say something, you can believe it is true.

	E1-4	E5-6	E7-9	O1-3
% Agree + % Strongly Agree	28	31	47	44

An analysis of the active Army data in MCCS suggests serious honesty problems at the unit level and between personnel in the field and the senior service leadership. While respondents often scored unit-relevant responses higher than service-oriented questions, there were clearly many unsatisfactory unit ratings as well.

In Items 41 and 79, the first concerning the unit, the second concerning the service's senior leaders—only a third to about a half of our NCO and officer unit leaders showed a definite belief in the prevalence of honesty in the unit or among the most senior leaders. Junior enlisted E1 to E4 scores were even lower. In a pattern typical of nearly all questions, respondents showed more faith in the prevalence of honesty in their units than in a system or their service. Most startling here is that there is marginally more (but not statistically significant) skepticism about senior service leaders among junior officers than among senior NCOs. The trust felt by officers and NCOs toward their most senior leaders is marginal to say the least, although it is stronger than that shown by the soldiers and the lower-ranking NCOs. Trust among Army professionals is based on honesty, and Army MCCS respondents clearly did not perceive that there was enough honesty in their units or even among the senior leaders in their service. Responses to the next few items suggest that selected aspects of leadership—selflessness and focus on the unit before self—were also weighed in the balance and found wanting.

Taking Care of Troops

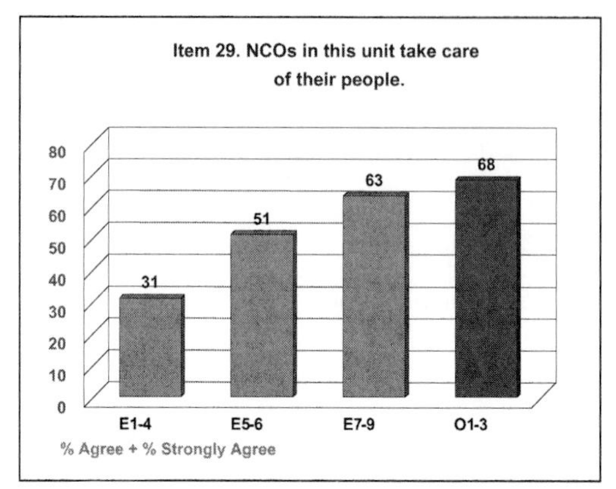

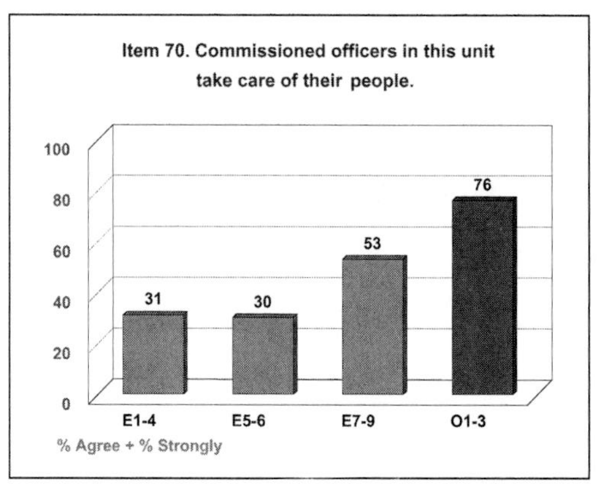

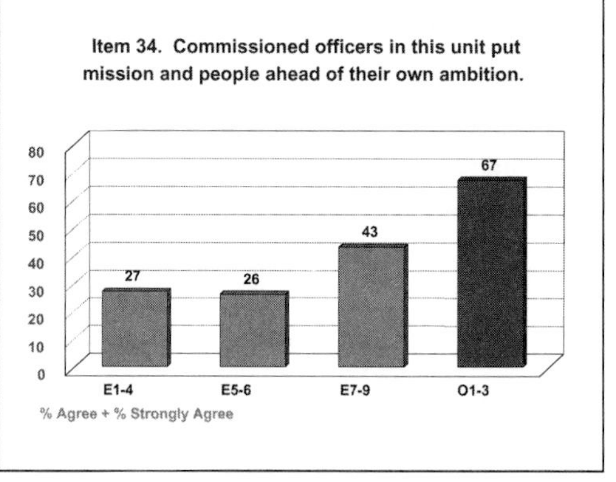

The first duty of a leader is to accomplish the mission, and the second is to further the welfare of his subordinates. The implied contract between soldiers and their leaders demands that officers and NCOs behave selflessly toward their people and their unit. Soldiers must trust their leaders to do their job because they cannot strike, quit, or otherwise compel their leaders to behave according to the appropriate standard.

Problems of trust in military units can impinge on combat readiness and performance. Military units are especially trust-sensitive since they must have the cohesion to hang together in harm's way and to perform tasks that go beyond those anticipated by any compensation system. Life in the military is both a job and an adventure, and sometimes it is a dangerous one. Leaders who don't take care of their soldiers will never develop cohesive units. Their example of looking out for self before unit is likely to create a trend that will be dysfunctional for combat effectiveness.

The three survey items about taking care of troops suggest that:

- NCOs are generally seen by lower-ranking soldiers as doing better than officers in taking care of the troops;
- NCOs generally perceive officers as unsatisfactory in selflessness; and that officers tend to put their career ahead of the well-being of their unit;
- Junior officers are dramatically less dubious of officer selflessness and taking care of troops than NCOs are. The officer-NCO/enlisted gap on this issue is vast, suggesting either NCO and enlisted aspirations beyond reason, or officer perceptions that are unrealistic, to put it mildly.

- Enlisted (E1 to E4) scores indicate that they are less satisfied with NCO care for them than NCOs are. At the same time, the junior enlisted clearly agree with the junior NCOs that commissioned officers are failing by a long shot to meet standards on this important issue.

The next three questions go to the issue of perceived levels of trusts in units and in the active Army as a whole.

Perceived Levels of Trust in Units

Army doctrine has long held that "effective leaders strive to create an environment of trust and understanding that encourages their subordinates to seize the initiative and act."[15] Countless battles have been won because young officers and sergeants felt empowered to act to achieve the higher goal in the absence of detailed orders. Americans are well aware of the many telling vignettes from our country's frequent experience with war. On a fluid battlefield, junior leaders with initiative may well make the difference between defeat and victory.

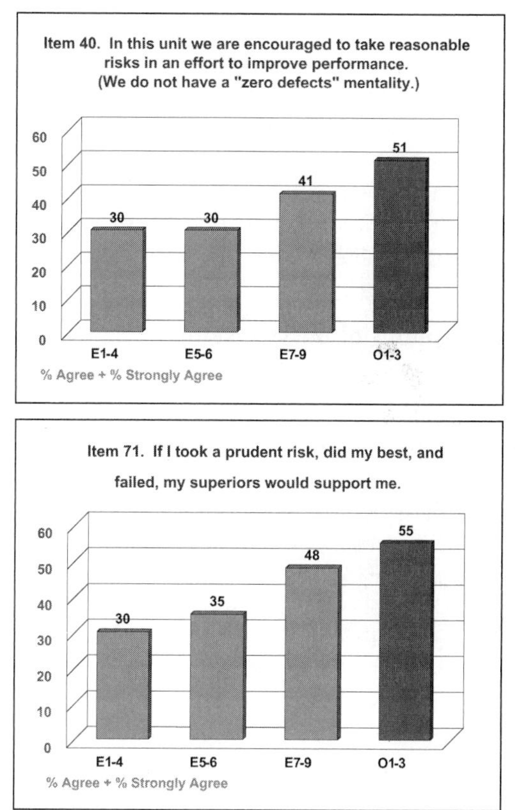

Success breeds success. Army leadership doctrine says, "Trust also springs from the collective competence of the team. As the team becomes more experienced and enjoys more success, it becomes more cohesive."[16] However, units short on trust may not build a record of success. Alternatively, those units that succeed through excessive micromanagement may find that success is a bitter pill and that cohesion may even decline with every "successful" act of mission accomplishment.

In Items 40, 71, 74, some large differences are revealed between the scores of junior enlisted and junior NCOs, on one hand, and the more closely aligned scores of senior NCOs and junior officers on the other. All agree, however, that the biggest problem seems to be risk avoidance. NCOs and junior officers, judging especially by Item 40, feel that a zero-defects climate pervades their units. While a zero-defects level of performance is often required in some high-stress tasks such as the assembly of special weapons or air traffic control, we are not dealing with such tasks here.

The proximate cause of this zero-defects climate is micromanagement. One pernicious effect of a no-mistakes, zero-defects climate is to encourage—indeed to insist in many cases—that after being micromanaged themselves, junior officers and NCOs in turn micromanage their subordinates. General officers who micromanage thus create micromanaging brigade and battalion commanders, who in turn create micromanaging company commanders, and so on ad nauseam. There is no easy exit from the vicious cycle created by micromanagement and the zero-defects climate. The admission by the Army leadership that "micromanagement is pervasive in the force" is a good first step.[17] Judging by these three items, there does indeed seem to be a significant problem in trust in Army units. The next two items concern whether this is a systemic problem.

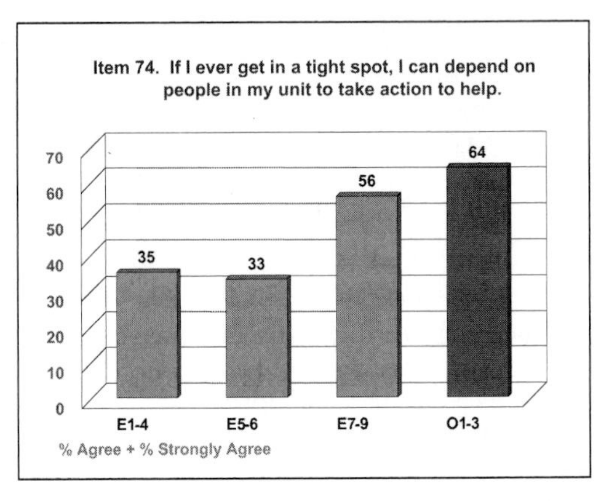

Perceptions of Service-Wide Trust

Item 82 shows a perception of a low level of trust within the active Army, with the overall response indicating that there is a service-wide trust problem. Scoring even lower, Item 88 was, along with Item 82, a bottom third response on the survey. The response to Item 40 concerning zero defects in the unit and the even more negative response to Item 88 suggest a widespread perception that one mistake could ruin an entire career. They demonstrate a clear belief that one cannot make mistakes in today's Army.

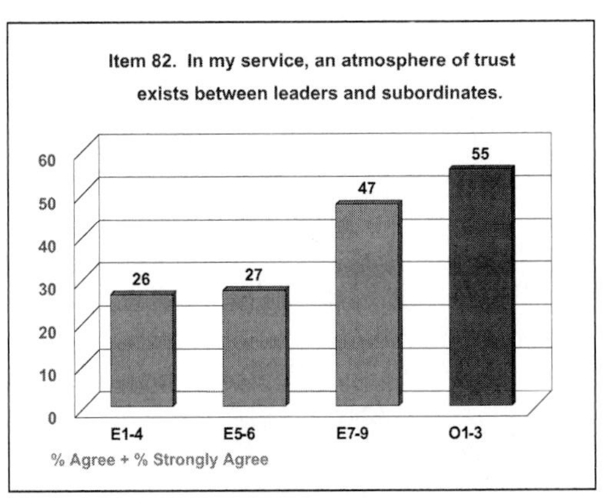

On one level, this can lead to over-supervision of one's subordinates to make sure that the leader doesn't get "tubed" by an inattentive, lazy, or ineffective subordinate. And as noted above, overly close supervision tends to compound itself. On another level, a zero-

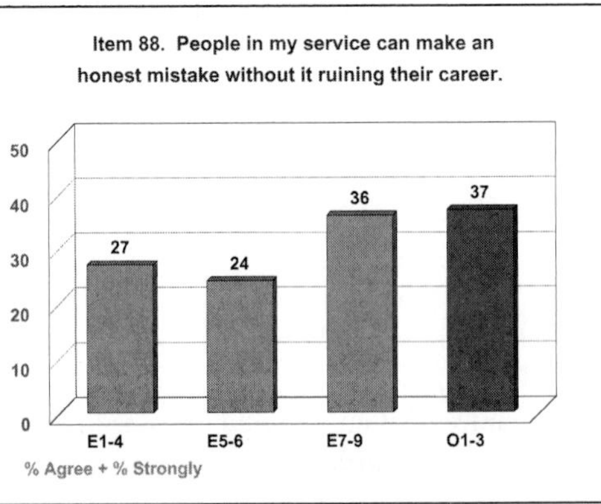

defects climate is likely to have a negative effect on honesty and selflessness. Little trust, less truth. Less truth, even less trust. Our analysis now shifts to an examination of hypotheses on why trust has become problematical in the military today.

Origins of the Trust Problem

A common response when problems of leadership or trust are exposed is that "it has always been that way." Others are likely to add that griping is part of the "Army way," or even that "a bitching trooper is a happy trooper." Despite these pearls of wisdom, it is important to ask whether the apparent erosion of trust in the military is growing, constant, or on the decline.

The situation is clearly better than it was in the post-Vietnam era. In 1970—toward the end of the Vietnam War and the start of the all-volunteer force—then-Col. Walt Ulmer and a few dozen colleagues at the U.S. Army War College conducted a landmark study on Army leadership and professionalism. This effort found significant perceived performance shortfalls in Army leadership behaviors as measured against the well-accepted principles of leadership.[18] There are some interesting parallels between the findings of that study and today's situation. For example, the USAWC study found that good local leadership was critical to command climate, and that internal conditions tended to outweigh external pressures as causes of problems in units. Integrity was an issue, then and now, but in the 1971 study there did not seem to be a problem between the field and the senior leaders of the service.[19]

Despite the similarities, however, it is clear to most veterans that the armed forces are far better off today than they were at the end of the Vietnam War, if for no other reason than the presence of so many well-trained NCOs and other personnel in the modern military. No stress on today's force could have anywhere near the impact that the Vietnam War had on the Army. Along with higher-quality people, it is also clear that the armed forces have achieved—with Panama, Desert Storm, and Kosovo as markers—a quantum increase in operational effectiveness. As far as public standing is concerned, since 1988 the American people have had more confidence in the military than any other institution. In June 2000, the military's confidence rating was 17 percent higher than that of the Supreme Court, 22 percent higher than that of President Clinton, and 40 percent higher than that of the U.S. Congress. [20]

Despite this sustained popularity, it is not at all clear that the Army—in readiness, morale, or training—is as well off today as it was in the early 1990s. In the Army's Sample Survey of Military Personnel (SSMP), according to data extending through 1998, officers put unit morale that year at its lowest point in the decade. Likewise, the percentage of officers who said they planned to stay in the Army until retirement declined from 70 percent in the fall of 1991 to 60 percent in the fall of 1998. This was the lowest point in the decade, and a full 6 percent below the average for the 8-year period. Those officers who were satisfied or very satisfied with the overall quality of Army life declined from 76 percent in the fall of 1992 to 58 percent in the fall of 1998.

Enlisted percentages showed comparable trends.[21] The recent findings of the Army leadership panel suggest that the conditions responsible for such trends have not abated.

While the military is probably among the least mercenary of the professions, the erosion of benefits and services is a common theme, and one closely connected with the sense of violated trust. One serving captain noted in the spring 1999 SSMP:

> The only reason I would depart the Army is my family. Army programs in support of dependents have continually eroded over my 10 years of service. Dependent medical care is under-staffed and in many cases incompatible [with my family's needs]. Dependent dental care is nonexistent. Youth Services is a shadow of its former self. At some point my family will no longer tolerate the combination of substandard housing and poor family/dependent services, and I will depart the service. [22]

One might logically conclude from such comments that the climate of trust is worse than in the early 1990s.

How did trust become an issue for today's Army? The short answer is inequities, frustrated expectations, and indifference to that wise old rule: "Take care of the troops and they'll take care of the mission." The mission continues to be accomplished, but somehow the "payback" has not kept pace. For some reason, pay, benefits, and even readiness have ebbed, and it appears to many that senior officers have not spoken out for the force. In all, there is a widespread sense of grievance on the part of our dedicated, patriotic men and women in uniform. It is difficult to speak concisely of "causes" of this mixed set of trust problems. For some, violated trust is a matter of specific cases and incidents. For others, the trust problem is much different, a mélange of causes that bring about effects that generate other causes.

In looking for the sources of the trust problem, five factors—environmental pressure, leadership problems, internal Army pressures, communication problems, and generation attitudinal change—range from catalyst to direct cause. Let us examine each in turn.

Environmental Pressure

Trust problems are in part a function of pressures that emanate from the operational environment. To fix this element, we will have to make strategic and operational changes to lessen operations tempo, decrease other situational sources of stress on the force, and restore work-family balance in the force. We have a small, mostly married force that has too many commitments and too much OPTEMPO for its own good. Many men and women in the Army complain that they are unable to balance work and family life (see Chapter 5, by Gayle Watkins and Randi Cohen in the present volume). The frantic pace of operations also creates problems for an increasingly aging set of equipment, as well as for the overworked soldiers who have to maintain it. In all, this pressure on the force contributes to poor leadership and to problems in the climate of trust.

We have been bombarded with statistics on deployments and stress on the force: "After the Cold War, the United States reduced the size of its military by 36 percent and cut military spending by 40 percent from its Cold War peak. . . . While the

United States has never been more secure, its armed forces have never been busier."[23] Most experts agree that the armed forces have experienced a threefold increase in the number of deployments. Many members of the armed forces exceed a 150 days per year on deployments or otherwise away from home station. Repetitive tours in places like Bosnia, Kosovo, Saudi Arabia, or Kuwait are not popular either.

While there has been a marked increase in the number of deployments for major operations, it is only when these deployments are added to regular training, joint exercises, forward stationing, and engagement activities that one gets a full picture of stress on the force. These stress-producing activities, singly or in combination, contribute to long hours, family separation, and personnel turbulence. In turn, this turbulence is magnified by a personnel management system that requires moving personnel frequently and having them rapidly change jobs to "punch tickets" to meet career milestones. The urge to micromanage subordinates may also increase in new, sometimes dangerous, and often unique activities. Pressure to maintain readiness in units with aging equipment may also both increase stress and make it harder to tell the truth under such demanding conditions.

In the CSIS study, survey and focus group subjects noted that the inability to balance work and family life was second in magnitude only to problems concerning pay and allowances. For officers, it is likely that work-family issues are the greatest problem. The Department of Defense found that O-3 (Navy lieutenants, captains in the other services) retention across the board in the late 1990s was the worst since 1973, and that the Army was the most problematical service in that regard. The Army has said that OPTEMPO and the frequency of individual reassignments and relocations were the top reasons for captains leaving the service.[24] The number who have left have caused the Army's selection rate for major—historically 55-75 percent—to top the 90 percent mark.

Perhaps the greatest single indicator of stress on the force and associated trust problems is command declination. Officers arrange their careers to get commands at the 0-5 and 0-6 levels, the keys to becoming competitive to be a general officer. All of the services have had problems in this area, and the Army is no exception. Beginning in 1996, command declinations have risen from almost zero to 6 percent for colonels, and as high as 5 percent for lieutenant colonels. Even allowing for the increased number and types of command billets, these numbers, while apparently low, are quite alarming.[25]

Such factors may put pressure on the force, but military units are meant to withstand great pressure. Units or individuals who cannot cope with demanding conditions away from garrison are unlikely to be good in combat. However, when excessive OPTEMPO issues interact with questionable or inexperienced leadership, it can have a devastating impact on the climate of trust within military units.

Leadership Problems

Many officers told the CSIS team that good leadership isn't everything, it's the only thing. At the same time, many reported that the supply of good leadership in local units or in their service as a whole did not come close to meeting the demand for it. This presents us with an interpretive mystery. Do such responses indicate periodic, widespread outbreaks of bad leader behaviors, such as those documented in the Army War College leadership study and the CSIS study? Or are leadership problems chronic in the Army and the other services?

While complaints of leadership deficiencies are constant, there are clearly times when bad or dysfunctional leadership is more noticeable and apparently more salient. Moreover, a volunteer-based system requires more creative leadership than a system where an endless supply of conscripts allows poor leaders to treat soldiers as nearly free goods. Three culprits appear to be at work here.

- First, as described above, is a high degree of pressure on the force, such as a war or a decade of high OPTEMPO. This stress weakens the "body" of the Army and appears to strengthen leadership "viruses" such as micromanagement and the associated zero-defects mentality.
- Second are structural problems attendant to a personnel system that rotates personnel frequently, moves people into and out of command too quickly, and then creates an efficiency report system that leads to frustration, perceived high levels of competition, or both. Rather than reinforcing good leadership, the personnel management bureaucracy may be retarding it.
- Finally, there are in fact many good officers who are not great commanders. Our iterative school, assignment, and command selection systems do a spotty job of identifying and advancing those officers who will be the best commanders.

Still, it is important to note that some leaders are doing better than others. CSIS analysts frequently found battalion-sized units that were doing much better than their peer units in relatively similar environments. In all cases, good officer leadership seemed to yield great advantages, creating better units and partially immunizing them against the stress of high OPTEMPO or even combat. Conversely, bad officer leaders can create trust problems of their own and even retard the efforts of good NCO leaders. It is clear that to improve trust, we will have to improve leadership, and that to improve leadership, we will have to work hard to improve leader selection and development.

Internal Army Pressures

The third factor associated with trust problems centers on intra-systemic stress on leaders as a possible causative factor. In essence, it is closely related to the leadership issue. Trust problems in today's Army are partly the effect of a hyper-competitive career environment in a smaller force with more limited promotion and command selection opportunities. This helps to create a zero-defects mentality and the need to appear, if not perfect, at least "mistake-less" to one's superior officers.

On first look, this factor seems to be a weak one. By all accounts, officers today have as good if not better opportunities for promotion and command selection than their Cold War forbears did. While it is possible that the post-Desert Storm drawdown—now a decade past—left the Army with a greater supply of hard-charging perfectionists than in the "good old days," there is no reliable proof of that. Indeed, in our experience the reductions of the early 1990s took a significant chunk of the most talented officers, especially those with highly marketable skills. As with pilots and technical specialists in other services, Army officers with the most marketable skills left the service in great numbers.

The new Officer Evaluation Report (OER) is problematical and might be contributing to the perception of hyper-competitiveness or lack of opportunities. The recent Army Training and Leader Development Panel Report echoed the many complaints from the field when it noted that "the Officer Evaluation Report is unfair both in its system and application."[26]

The technical details are important here: senior raters are limited in the number of top ratings they can give and are forced to give many center-of-mass (COM) ratings, viewed by many—perhaps incorrectly—as the career kiss of death. Chief among complaints from the field appears to be that officers are pooled into large groups so that their peers in command positions—widely viewed as the most important jobs—can receive the highest ratings without violating limitations on the number of "top-block" ratings. Others, of course, must receive COM ratings to fill out the senior rating officer's profile. This problem impacts directly on the climate of fairness and trust within units, because an artifact of the system is seen as preventing accurate assessment of performance. Ironically, many hard-charging lieutenants see being assigned to a captain's billet on battalion staff, once a significant honor, as a career blow.

Although there is huge dissatisfaction with the current OER, some personnel managers will defend it as an excellent instrument. The question is, excellent for whom? In an up-or-out system, personnel managers need tools for eliminating those who should be out. Euphemistically, the system is a meritocracy, looking for the best to be promoted. In practice, promotion boards often must process hundreds or thousands of files at a pace that allows perhaps only a few minutes to study an individual file. When there are more files than promotions, the board member may initially just look for ways to eliminate files.

How does this relate to equity? In order to get a spread of quality in the files, personnel managers require the senior rater to have a distribution of ratings. To obtain that distribution, the senior rater may be forced to give a COM rating to a highly competent officer, especially if that officer is not in a position critical to branch qualification. In effect, for that officer the Army is no longer a meritocracy. He will have received an evaluation that almost by design has not reflected his performance.

In the larger scheme of things, the current OER is the product of a personnel management system founded on conscription and the potential need for mobilization in an era that can support the need for neither. This is a significant violation of equity and a major contributor to trust problems throughout the Army.

In sum, while there is little or no proof that the officer pool is more competitive, there is a strong sense of scarcity when it comes to promotion, command, and schooling opportunities. This sense of scarcity is magnified by problems with the current OER and the negative perception—understandable in a zero-defects environment—that COM ratings offer little protection to those who receive them. It is also likely that

OER fairness problems are connected to perceptions of honesty and integrity, both central to the development of a climate of trust.

Communications Problems

Trust problems in the Army and the armed forces as a whole are not caused by but are strongly affected by improved telecommunications and the widespread availability of the internet and other sources of information. Senior officer behaviors are more transparent, and personnel at the lowest levels can monitor—through news media and web sites—the activities of their most senior officers without the benefit of interpretation by intervening layers of management.

Although there is no doubt that this is true, its effects are incalculable and may well vary with individuals. In MCCS focus groups, CSIS analysts frequently heard that senior service leaders were lying, ill-informed, or co-opted by politicians. To others, their senior leaders were simply out of touch with the turbulence and problems in their units. In these focus groups, time after time people in field units would evince a relatively intimate knowledge of senior officer public behaviors, such as in congressional testimony. Four-star officers were especially visible. General officer scandals are not hushed rumors in small circles of officers in the know; instead, they are stories on CNN.com, repetitively pasted to emails that circle the globe and pervade the Army in a day. MSNBC and electronic national newspapers (for example, *www.Washingtonpost.com*) were also cited as common, well-known sources of information.

Modern telecommunications and the internet spread concern about the condition and readiness of the armed forces. CSIS focus groups found a sense among officers and senior NCOs that senior officers were portraying readiness higher in their reports and testimony than actual conditions in the field or fleet warranted. Two field grade officers at a high-level joint headquarters, for example, said in a focus group that when their four-star forwarded to the Pentagon the readiness report they had prepared, "it bore little resemblance to what they had submitted" to him through their bosses.[27] Command information materials on Department of Defense web sites were occasionally cited as the source of the conclusion that senior leaders were out of touch, misinformed, or just plain lying about readiness.

None of this is meant to accord these negative opinions the title of "facts." Rather, it is more likely that there are honest differences in perspectives on issues at different levels in the organization, and that there may well be some degree of information overload going on. It is similarly true that many personnel at the unit/ship level do not understand the environment at the highest levels of command where politics, policy, and military affairs come together. Perception here, however, has a reality all its own. Clear and direct corporate communications are more necessary than ever before.

In a similar vein, many focus group participants noted the pervasive and negative effects of the overuse of email, power point graphic presentations, and video teleconferences. "Command by email" was an oft-heard complaint about bosses who commanded their units electronically. All of these phenomena contributed to over-supervision and magnified the power of senior leaders to manage details, thus further demonstrating their lack of trust in their subordinates. One respondent noted that even unmanned aerial vehicles had been used in the field as agents of over-supervision, again perpetuating a lack of trust.

Generation Attitudinal Changes

In a ground-breaking study, Dr. Leonard Wong of the Army War College's Strategic Studies Institute, comparing younger officers (Generation X) to their Baby Boom generation senior officers, noted that:

> Generation X officers are more confident in their abilities, perceive loyalty differently, want more balance between work and family, and are not intimidated by rank. Additionally, while pay is more important to Generation X officers, it alone will not keep junior officers from leaving. [28]

The statistical evidence to back up this description is substantial. Adding to the generational differences is a cultural difference in the eras in which the two groups came into the Army. Today's colonels entered an Army that was dominated by the Reagan buildup and Desert Storm, a high-water mark for the U.S. Army. Today's captains came in during the drawdown and are more familiar with peace operations than they are with combat. The colonels at one time had resources aplenty; the captains have lived their whole career with growing shortages. To bridge this gap, Dr. Wong recommends more family time, a better overall quality of life, more mentoring and personal relationships, less reliance on hierarchical leadership, elimination of the zero-defects mentality, and highlighting technology, a focal point for Xers.

In essence, trappings aside, the Xers appear to have a devotion to plain speaking, a more fluid view of careers, and a higher demand for creative leadership than the system is currently supplying. Their different attitudes do not portend less devotion to Army values than their predecessors. Indeed, the CSIS study documented that junior officers today have a strong dedication to Army values.[29] Thus, for the issue of trust, the generational divide is not a terminal illness, only a complicating condition that calls for greater doses of good leadership and common sense.

Summarization

We list below the main points developed in this chapter:

- Trust is important in every profession, but because of the nature of combat it is critical to the military.
- There is at present a significant trust problem in the Army and the other services. It is difficult to state its magnitude and trend with scientific precision, but it is highly likely that this problem increased in magnitude and severity throughout the 1990s.
- The trust problem exists inside of units, and within the service as a whole.
- There is a separate and pernicious trust problem between people in the field and their most senior uniformed leaders.
- In essence, trust problems are about equity and the concern that the Army does not fairly reciprocate the soldier's loss of self-determination, and that leaders at

all levels may not be as committed to the institution, and as willing to take risks for it, as soldiers are.

- The trust problem in units is evident in perceptions of quality of life, integrity, leader selflessness, fairness in performance ratings, adequacy of rewards, the ability to balance work and family time, the zero-defects mentality, and the level of over-supervision, commonly known as micromanagement.
- The trust problem between people in the field and their most senior uniformed leaders revolves around issues of plain speaking and perceived honesty or accuracy on the part of senior leaders. Divergent perceptions of readiness (e.g., conditions in one's unit versus what was reported to higher headquarters or Congress) are particularly sensitive.
- Trust problems are exacerbated by a range of factors from environmental pressure to information overload. A comprehensive strategy to fix trust problems must include both near-term actions to reduce these factors, and long-range actions to address the more fundamental causes of perceived inequity.

Conclusion

All five of the trust-related factors discussed earlier remain connected to problems of trust in today's Army. Leadership in units is problematical and made all the more so by pressure on those units from excessive, but perhaps mission-essential, demands on the time of leaders and subordinates alike. While the Army was not found to be hyper-competitive, the problems associated with the new OER were a major equity issue, and thus a source of distrust. Trust problems were not caused by but only magnified by institutional communications problems as well as the misuse at all levels of modern devices such as email and power point. It is very clear that any attempt to improve trust in the Army will have to attack each of these areas of concern.

The focal point of "fixes" to the trust problem must be to restore a sense of equity by reciprocating the soldier's surrender of his/her right of self-determination and by demonstrating that the senior leadership is prepared to make appropriate sacrifices for the force. It will also be necessary to maintain clear institutional and interpersonal communications within the force. To improve trust in units, across the Army, and in the profession as a whole, the biggest payoffs will come by improving leadership and the understanding that Army leaders have of the environment around them; and by altering the personnel management and leader development systems.

Overall, to restore or improve trust within the force, the 21st century Army leader—to adopt some well-worn language—must reestablish and refine the informal contract with the soldier, both today and in the future. As noted in the preface to the 30-year-old Army War College leadership study:

> In this informal contract, the Army expects a professional and disciplined response from the soldier. The soldier, on the other hand, expects fairness, worthwhile work, and sufficient pay from the Army. If each party to this informal contract meets the

expectation of the other, a mutually satisfactory relationship will exist—a relationship that will create the loyalty and dedication which are the cornerstone of true discipline and professionalism.[30]

Notes

1. Walter F. Ulmer, Jr., Joseph J. Collins, and T.O. Jacobs, *American Military Culture in the Twenty-first Century: A Report of the CSIS International Security Program* (Washington, DC: Center for Strategic and International Studies, 2000). Dr. Edwin Dorn and Lt. Gen. (USA Ret.) Howard Graves chaired the project. Lt. Gen. (USA Ret.) Walter F. Ulmer, Jr., chaired the working group that did the survey and managed the preparation of the report. Joseph J. Collins was the CSIS project director and an active participant in survey administration and focus groups, and Owen Jacobs coordinated the preparation of survey results and advised all concerned on methodology. This report will henceforth be referred to as the CSIS Report. The published results of the multi-service, multi-component Military Climate/Culture Survey (MCCS) are on pages 83-108 of the CSIS Report. For a discussion of senior officers and trust, see pages 68-73 of the report. Much of this chapter will use hitherto unpublished data from the active Army database of the study. Active Army surveys were done in 1998 and 1999.
2. For a study that looks at leadership problems in the Navy's officer corps, see Rear Adm. John T. Natter (USNR) et al., "Listen to the JOs [Junior Officers]: Why Retention is a Problem," *U.S. Naval Institute Proceedings* (October 1998).
3. Department of the Army, *The Executive Summary of the Army Training and Leader Development Panel,* April 2001, ES2, ES9. This report will hereafter be cited as ATLDP.
4. See *Webster's New Riverside University Dictionary* (Riverside, CA: Riverside Publishing, 1994), under "trust."
5. Jonathan Shay, M.D., " Trust: Touchstone for a Practical Military Ethos," in *Spirit, Blood, and Treasure: The American Cost of Battle in the 21st Century*, ed. Donald Vandergriff (Novato, CA: Presidio Press, 2001), 4, 13. Emphasis in the original.
6. C.E. Elczuk, *Content Analysis of Written Comments [on the] Spring 1999 Sample Survey of Military Personnel* (an unpublished research product of Swan Research for USARI, November 1999), 4-5. This document will hereafter be cited as SSMP Spring 1999.
7. SSMP Spring 1999, 4-5.
8. A hitherto unpublished written comment on the MCCS, a basis for the CSIS Report.
9. An unpublished, anonymous comment recorded in a Command and General Staff College Focus Group, Spring 2000.
10. Gen. Jack Keane, "What is the Army Doing About Captains Getting Out," at *http://www.perscom.army.mil/Opcptret/attrit.htm*, accessed 16 October 2000.
11. Unpublished report of field visits by Defense Advisory Committee on Women in the Service, April 2001, 4.
12. For a full explanation of the MCCS and a copy of the survey, see the CSIS Report, 36-46, 99-109.
13. The Army active component database is not in the CSIS report and has never been published.
14. Field Manual 22-100, *Army Leadership* (Washington, DC: Headquarters Department of the Army, 1999), 5-19. This document is hereafter cited as FM 22-100.
15. FM 22-100, 1-14.
16. FM 22-100, 5-19.
17. ATLDP, ES2.
18. U.S. Army War College, *Leadership for the 1970s: USAWC Study of Leadership for the Professional Soldier*, October 1971.

19. Personal communication (15 July 2000) to the authors from Lt. Gen. Walter F. Ulmer, Jr., (USA Ret.), who, as a colonel, was a major participant in formulation and execution of the USAWC Study.
20. For the full report, see: *http://www.gallup.com/poll/indicators/inconfidence.asp* , 3-4, accessed 5 June 2001.
21. Unpublished analysis of the SSMP from Spring 1991 to Spring 1998 forwarded to the authors by Lt. Gen. David Ohle (USA), then the Army's top uniformed personnel official.
22. SSMP Spring 1999.
23. CSIS Report, 18.
24. Jim Garamone, "DoD Examines Captain/Lieutenant Retention," Armed Forces Press Service, 17 May 2000. This article is built around an interview of Vice Adm. Pat Tracey, DoD's top uniformed personnel officer.
25. Personal communication in spring 2001 with a former member of the Army staff who is an expert on senior officer leadership.
26. ATLDP, ES2. See also Mike Galloucis, "Is It Time for a 360-Degree Officer Evaluation System?" *Army*, November 2001, 47-55.
27. The focus group comments that supplemented the survey data in the CSIS Report remain unpublished.
28. Leonard Wong, *Generations Apart: Xers and Boomers in the Officer Corps* (Carlisle Barracks, PA: U.S. Army War College, Strategic Studies Institute, October 2000), v.
29. CSIS Report, 63-66. See also Jody L. Bradshaw, "Motivating Future Leaders," in *Building and Maintaining Healthy Organizations: The Key to Future Success*, ed. Lloyd J. Matthews (Carlisle Barracks, PA: U.S. Army War College, Department of Command, Leadership, and Management, January 2001, 79-81.
30. Maj. Gen. Franklin M. Davis, Jr., in his preface to *Leadership for the 1970s: USAWC Study of Leadership for the Professional Soldier*, October 1971, ii.

4 Serving the American People: A Historical View of the Army Profession[1]

Leonard Wong and Douglas V. Johnson II

In one of his final acts as Secretary of the Army, Louis Caldera authorized the Army Superior Unit Award for the U.S. Army War College commending "the great work that has been done on the transformation of the Army for the period 1 October 1999 to 1 October 2000." In addition to authorizing ribbons and lapel pins for soldiers and civilians assigned to the War College during that period, the award also authorized a streamer for the War College flag. Although the streamer will be a source of pride to the War College, it will doubtless be overshadowed by the 173 campaign streamers hanging on the Army flag, streamers with names like Yorktown, Meuse-Argonne, Normandy, Tet Counteroffensive, and Liberation and Defense of Kuwait. Battle streamers, not peacetime streamers, properly signify the Army's defining essence. Nevertheless, the contrast between the two types of streamers illustrates the breadth of the actual professional identity of the Army. The Army has always been more closely associated with its war-fighting function despite the fact that it has operated in a peacetime role for most of its existence.

The heavy attention given to war-fighting rather than peacetime duties is particularly interesting in view of Army doctrine. According to the Army's capstone doctrinal manual, FM 1, *The Army* (2001) , the Army's purpose is "to serve the American people, protect our enduring national interests, and fulfill our national military responsibilities. The Army, with the other Services, deters conflict, reassures allies, defeats enemies, and supports civil authorities."[2] This foundational manual also identifies, for the first time as a formal doctrinal statement, the Army's Core Competencies. Among six subheadings, the first and the last deal with matters other than war. Here, the breadth of the Army's jurisdiction is fully and properly spelled out. War-fighting is still, of course, the central feature of the Army's overall role—*to serve the American people.*

This chapter examines how the Army as a profession has served the American people over its lifetime, even as the very definition of a profession has evolved and matured and the demands placed by the American people on its Army have changed. We specifically analyze what "service to the American people" has entailed and how the boundaries of the profession have expanded or contracted over time. We take a systems approach to professions, using the work of sociologist Andrew Abbott.[3] A systems approach suggests that there is a universe of tasks or work performed by professions. The link between a profession and its work is called its jurisdiction. According to Abbott, to analyze a profession "is to analyze how this link is created in work, how it is anchored by formal and informal social structure, and how the interplay of jurisdictional links between professions determines the history of the individual professions themselves."[4] We take a historical look at the Army as a profession, examining the jurisdiction of the Army, its interaction with other professions, and its response to the demands of the American people expressed through the Executive and the Congress.

A systems approach to professions, as with many sociology-based methodologies, is concerned with conflict—in this case, conflict between professions. Viewing the Army in an ecology of professions implies that only the fittest will survive as competing professions jockey for jurisdiction. The Army is unique in that while it is a profession, decision-makers not in the profession often dictate its jurisdiction. Thus, while uniformed soldiers comprise the profession, civilian appointees—often people with limited military backgrounds—can decide to shift the Army's jurisdiction to include new tasks or exclude old ones. Examples of recent taskings include provision of environmental protection by creating artificial offshore reefs out of old tank hulls and of providing exemplary day care on Army posts. Other professions such as medicine, law, or engineering do not have members outside the profession making jurisdictional decisions. That the Army's jurisdictional boundaries may change not only due to competition from other professions but also because of civil control of the military becomes an important point when considering the authority of senior Army officers, who in fact may not be the key decision-makers determining the Army's jurisdiction.

Before we explore the historical development of the Army as a profession, it will be useful to review some of Abbott's key concepts. According to Abbott, professional work determines the vulnerability of professional tasks to competitor interference. Objective and subjective characteristics of the tasks determine the vulnerability of the tasks to poaching by other professions. This vulnerability is affected by how strong the link is between the profession and its tasks.

Objective characteristics are empirical qualities that resist any redefinition of work. For example, in the treatment of alcoholism, an objective fact is that consumption of alcohol always produces central nervous system depression. Different professions may put alcoholism treatment in their jurisdiction, but the effects of alcohol on the body is an objective fact that sets limits to how far professions will adjust their jurisdiction. Objective qualities can include technological, organizational, natural, and cultural characteristics.

Subjective qualities of tasks originate from the redefinition of the problem bounded by objective properties. Thus, alcoholism, while maintaining its objective characteristics, could be redefined as a disease, mental disorder, or sin. A redefinition of the work may allow a different profession to put alcoholism in its jurisdiction. Subjective characteristics include diagnosis, treatment, and inference. *Diagnosis* takes information into the professional knowledge system by seeking the right professional category for a problem, but also removing the problem's extraneous qualities. For example, an architect would want to focus on design, not on costs. *Treatment* parallels diagnosis, but instead of taking information about the problem, treatment gives results. The more specialized a treatment, the more a profession can retain control of it. Laser eye surgeons, for example, enjoyed a protected jurisdiction for a period of time, but are now experiencing vulnerability as new technologies emerge. Finally, *inference* is undertaken when the connection between diagnosis and treatment is obscure. Inference is the application of professional knowledge to a particular problem. If the inference process is too routinized or too esoteric, the profession can be degraded. Legal advice on the preparation of wills via the internet is an example of the former, while fortune-telling is an example of the latter.

Using Abbott's concepts and a systems approach to professions, we analyze the Army's history in four eras. In each era, we discuss the tasks and jurisdiction of the Army at that time and then examine the competition for jurisdiction using the objective and subjective characteristics of the tasks.

1781-1901

Following the Revolutionary War, the task granted to the Army at its inception was consistent with its size—80 men, 55 of whom were detailed to secure military stores at West Point.[5] The day following the reduction of the standing Army to 80 men, Congress called for 700 militia to be mobilized to secure the frontier, opening the door to expansion of the force and the Army's new jurisdiction.[6] This expansion was not accompanied by an outpouring of enthusiasm for service on the remote frontier, and the Army experienced great difficulty in raising the 700 militia. Further, as the duties assigned this body offered little in the way of incentives, desertions and avoidance of service became common.

With the exception of the War of 1812 (really no more than the concluding act of the Revolutionary War), the Mexican War, the Civil War, and the Spanish-American War, most of the Army, in the decades that followed, was involved in the business of doing whatever the nation, through the Executive and the Congress, directed it to do. The initial tasks set before the Army were principally concerned with security of the frontier—protecting the treaty rights of the Indians from aggressive and unscrupulous settlers, and in turn protecting the legitimate settlers from Indian depredations.[7] But it was the Army that began the systematic exploration and development of the frontier region, later extending its work westward with the "march of civilization." Accompanying these tasks were those incident to sustaining life on a remote frontier—surveying, road building, local subsistence farming, peacemaking or peacekeeping, and some law enforcement.[8]

The southern frontier became the focus of attention during the 1830s with what was probably the most unpopular war before Vietnam—the Seminole War. This too was a guerrilla war that lasted some six years and came close to ruining the Army. The war was partly motivated by aggressive land-grabbing which was so transparent to the soldiers involved that there was considerable sympathy for the Seminoles combined with a great reluctance to serve in that hot, inhospitable climate.[9] Officers used every imaginable device to avoid duty in Florida including ways we today would view as chargeable offenses.[10]

The Army's role in the life of the nation in these early years was essentially to fill in the gaps where no civilian institutions existed, or to perform the functions of existing institutions unwilling to operate in the less secure regions in which the

Army tended to be stationed. Always at the forefront was the requirement to "Fight and Win the Nation's Wars," but there was no special threat that required a force capable of doing more than suppressing an occasional outbreak of Indian violence. Essentially, the Army as a profession emerged to embrace any tasks levied by the American people that necessitated the deployment of trained, disciplined, manpower under austere conditions on behalf of the nation. During the early part of this period, continuing hostilities with Great Britain gave substance to an external threat that put some pressure on the development of coastal fortifications, but even the aggressive behavior of the British along the Canadian border area was insufficient to significantly increase the size of the Army.[11] Natural and cultural objective factors such as the absence of a threat due to a landmass bounded by oceans, the absence of a functioning fiscal system for the first decades of the republic, and the attendant shortage of entrepreneurs coupled with a relatively low population density all served to secure the jurisdiction of the Army from any serious competition.

Coast Artillery was the first branch to have a professional school, thus confirming the unique requirements of those charged with the land portion of coastal defense.[12] It was not so much that a viable threat existed, but the memory of recent disasters and the relative weakness of the Navy combined to cement this portion of the Army's jurisdiction into place until well into the missile age.[13]

From 1821 onward, West Point was looked to as an incubator of Army professionalism, going somewhat beyond a mere engineering school.[14] The influence of West Point came from the simple fact that it was the only engineering school in the entire country before 1824, and its graduates were in high demand throughout the nation.[15] These engineers were the only portion of the Army in close continuing contact with the civil population. Their function, aside from the relatively minor involvement in coastal defense, was infrastructure development. Roads and canal systems had obvious dual-use merits. Beginning in 1833, military periodicals approximating professional journals began to appear, and in 1846 Lt. Henry Halleck published the first systematic study of military theory by an American seeking to establish "that principles of military art and science constitute a body of knowledge to be understood and mastered only through a professional education."[16] With the profession focusing on tasks demanding disciplined, trained personnel capable of being deployed to austere conditions, West Point implicitly put nation-building within the purview of the military profession.

The remarkable combat performance of the Army during the Mexican War demonstrated military skills not normally associated with the Army of this period, but also provided a precursor to a new jurisdiction—civil administration. Immediately following Winfield Scott's conquest of Mexico, a new civil administration was inaugurated in and around Mexico City that served as a model for years to come. Until its withdrawal from Mexico, the occupying forces of the U.S. Army were the civil administration of the Mexican capital region.[17] That experience gave some impetus to the extension of Army jurisdiction that was further developed following the Spanish-American War with the administration of Cuba, briefly of Puerto Rico, and the suppression of the Philippine Insurrection. This jurisdiction was again exercised during the brief American occupation of the Rhineland following World War I, and returned full force with the occupations of Germany and Japan following World War II. It was again resurrected when the U.S. Army moved into Somalia, Bosnia, and Kosovo, and for a brief period following the liberation of Kuwait.

A unique feature of early Army life was that quartermaster operations were largely contracted out. For instance, the Mexican War offered the first instance in which the Quartermaster Department was solely responsible for clothing the Army.[18] As warfare became more sophisticated and the forces involved grew larger, the responsibilities and militarization of quartermaster operations increased. The debacle of quartermaster support during the Spanish-American War set expansionary measures in motion that increased further during World War I. By the time World War II became a reality for America, quartermaster operations had grown to rival major civilian industries. Interestingly, the Army is now in the process of reversing that trend as the perceived need for military control over its supply and support system rapidly diminishes. As long as the threat remains low and the apparent ability of civilian industry to meet military demands in a timely fashion is not seriously questioned, the military jurisdiction over its own support structure will probably continue to diminish.[19]

From a system of professions viewpoint, the first era in the history of the Army saw the profession relatively unchallenged by other professions. The nation needed a force capable of expanding its frontiers and developing the infrastructure necessary for growth. As a young nation, the United States had only the Army to turn to for numerous functions. Fighting the nation's wars was the central task, but frontier protection and infrastructure-building dominated the early years of the Army's existence.

1902–1939

Events in 1902 represented a sea change in Army activities. Suddenly and without design, the United States found itself a colonial power. While the Army had administered the former rebel states in what were the darkest days of its existence, long-term administration of conquered foreign territories was something entirely new and it changed the nature of Army business. The Army found itself administering distant possessions of largely unknown cultures, and with the now familiar lack of guidance. Although the regulars and volunteers had sufficed to handle the war, it had been essential to call new volunteers to the colors to aid in suppressing the Philippine insurgency. In this regard, the viability of the non-regular/reserve component of the Army regained a reputation it had lost during the Revolutionary War where its performance when it was good, was very, very good, but when it was bad it was deadly.[20] Since that time, the voice of the reserve components and particularly of the Army National Guard has been loud in proclaiming alternately that it is the nation's strategic reserve (which it will be if ever fully equipped and trained) and that it is the military' essential link to the American people (which Desert Storm demonstrated fully). In an unusual twist in the jurisdictional reach of the profession, a subcomponent periodically emerges from the civilian population to assume regular duties and sometimes to outnumber the regular members of the profession.

Repetitive tours in foreign stations exposed the Army and the nation to distant events formerly of little concern. It was as if a new nerve system had grown up and with it new sensitivities. U.S Navy ships on distant stations could act largely on the initiative of their individual captains until the arrival of telegraph cables constrained their independence. The Army was dispatched after telegraph connections had been established. The severing of telegraph cables from Cuba in 1898 was one of the early acts of what we recognize as information warfare today. But telegraph lines ran to major cities only and not into the remote Philippine, Cuban, Nicaraguan, or Puerto Rican countryside. Thus, soldiers like Lt. Conrad Babcock, U.S. Cavalry, found themselves acting as mayors of small Philippine towns, responding to their captain or major, the Provincial Governors. Like their Navy counterparts before them, their own good judgment, tempered by training and personal discipline, was their guide with little else to rely on.[21]

In the continuing rebalancing of its jurisdictional limits, the Army's courts-martial system underwent several major changes. The reforms of 1880 did away with physical punishments that today would be considered torture.[22] Other reforms followed including the establishment of the Uniform Code of Military Justice in 1950, which "extended civilian substantive and procedural legal principles to the armed forces."[23] The U.S. Supreme Court finally went the last mile in 1969, when Justice William O. Douglas in *O'Callahan* v. *Parker* dissolved the Army's jurisdiction over offenses committed by servicemen outside a military installation. This was not reversed until 1987, but the ebb and flow demonstrated here sharply defines just how dramatically the Army's jurisdiction can shift.[24]

In 1905, San Francisco was shaken and almost ruined by an earthquake. The Army quickly helped establish a functional infrastructure in and around the Bay City, received the citizens' thanks, and then went back to work policing the National Parks and trying out all sorts of new technologies. It was a period of intense technological experimentation, as basic research had matured to produce numerous marketable devices. The Europeans majored on military equipment, the Americans on other things.

When Pancho Villa raided the southern Texas border, a "division" was formed to police the border and eventually to launch a suppression operation.[25] The greatest fear was that Mexican political disorders would cross the border. Banditry and such disorders seemed to be synonymous.[26] Brig. Gen. John J. Pershing was blessed with public and political attention throughout his abortive counter-Villa campaign. His thoroughly apolitical conduct offered a refreshing contrast to past senior Army officer behavior. It set a widely noted example that lasted well into the waning years of the century. In terms of jurisdiction, the Army acted as the U.S. Government's border police agent. Then, in 1914, it acted with a heavy hand to land a military force at Vera Cruz to maintain American "rights."[27] But beyond those essentially military events, the Army gradually slipped into a quiet existence in the continental United States and the Philippine Islands.

American involvement in World War I exposed the problems attendant upon having an Army manned and equipped at a bare-bones level. Policing the Caribbean, the Philippines, and the Mexican border did not require large numbers of soldiers, and other duties had dwindled away. The national leadership's inability even to conceive of participation in a major European war kept force structure on a level commensurate solely with immediate needs. While the war revealed for all to see the Army's insular thinking, at the same time it drove the Army into broader jurisdictional realms including issues touching national industrial mobilization hitherto the purview of civilians. The 1930 Industrial Mobilization Plan grew out of the war materiel manufacturing chaos of World War I.[28]

The sudden need to expand to a multimillion-man force was a surprise to the nation, but at least the Plattsburgh summer camp programs had kept the idea of leadership training and honorable service to the nation alive and relatively healthy. Although initially created and administered by civilians, later assisted by Regular Army officers, the Plattsburgh program turned out to be the first in an expanding series of leadership training initiatives. Accompanying the upsurge in hostilities against the Mexicans, the Plattsburgh movement gained increasing momentum such that the Congress paid for the 1916 camps rather than the attendees. The Army thus found itself in the business of leadership training under the guise of providing basic training to everyone interested enough to attend the camps, which came to 16,000 in 1916.[29] Ironically, the ideas and initial execution of both leadership training and rifle marksmanship came from outside the Army.[30]

During this era in the Army's professional history, imparting discipline became an Army jurisdictional claim. At first viewed by the freeborn men of a free land as the imposition of tyrannical rules of dubious value with the intent of inducing mindlessness, it gradually came to be seen as the panacea for salvaging damaged youth. The presence of regular troops has almost always been sufficient to quiet even the worst civil disturbances. The riots in the mining camps in the west, the Pullman strike (1894), and more recently the racial integration of the University of Alabama were all attended to by regulars who won the grudging admiration of those whose presence they were sent to control.[31] It has not been the case uniformly to be sure, but when civil law and order collapse, the appearance of the Regular Army is seen as a signal that nonsense will no longer be tolerated and that evenhanded treatment will be accorded to all. This is an uncomfortable continued projection of Army jurisdiction into civilian government and is closely regulated, but it remains a core function. Even today, as the trait of self-discipline is more absent from a broadly self-indulgent society, military discipline has achieved new value. Undisciplined youth get into trouble, and in desperation many civil entities have turned to the military for assistance in creating "Boot Camps" whose purpose is to inculcate the discipline society no longer provides.[32]

Along this same line, with the onset of the Depression a similar occasion arose. Over its vigorous objections, the Army was charged with operation of the Civilian Conservation Corps.[33] Whatever else the CCC accomplished in terms of conservation, it did build character and taught leadership, even if not overtly. Here the Army's strength was once again demonstrated and recognized by outsiders. Only later would the Army itself awaken to this strength, but it had gradually established a unique jurisdiction that was being alternately reinforced or challenged by developments in business practices. Alert progressive officers who admired scientific management practices being developed by businessmen noted the new management practices with interest. But while civilian business management practices developed along scientific management lines, the civilian element never embraced leadership development to the extent the Army did.

When the Army was founded, its resources for producing its own equipment, though slight, were very nearly adequate and remained so for most of its early history with the exception of the Civil War period. The Spanish-American War sent some signals that Army management practices were not up to the mark, and the arrival of World War I caught the nation and the Army still only partially prepared. The scale of war caught everyone by surprise and opened the door to such extensive Army involvement in the civilian industrial base that eventually President Eisenhower was prompted to caution against the "military industrial complex."[34] The interaction between industry and the Army has recently matured to such an extent that General Electric claims to have benefited significantly from a series of seminars with Army officers at the U.S. Army War College.[35] The intermingling of industry and military has developed to the point that those relationships are tightly circumscribed by congressionally mandated rules of conduct to protect the government against irregularities stemming from too cozy a relationship.

The second era in the Army's professional history was marked by an expansion in several jurisdictions. The Army continued to develop infrastructure, but now technological advances enabled this to be done across the oceans. Administration of foreign lands reinforced the already well-established jurisdiction of frontier administration. Likewise, war continued to be the central jurisdiction, but this too was taken to distant lands. Jurisdictional competition continued to be minimized, as the young nation had no other force to turn to for its needs.

1945-1991

The world had shifted by the time World War II ended, and this shift affected the U.S. Army profoundly. No longer just a small constabulary force administering some Pacific or Caribbean island, the Army found itself essentially governing two major nations and operating dozens of smaller entities from Trieste to Saipan. Standing most of the Army down failed to reduce the burdens or the extent of dispersion. In some places the native citizens were able to reconstruct their societies and take charge of their own lives in relatively quick order. Italy did not require an occupation force although it might have been better for its economy had it had one. Soldiers acted in all kinds of civil capacities other than as warriors for the first few years after the war and returned to the combat role only after the "Iron Curtain" was emplaced. The Army thereupon began to resume some of its focus on arms and war-fighting as the threat of "monolithic Communism" grew and the Berlin airlift and Korea gave weight to the fears quietly developing.

For the first time since the Revolution, America faced a major direct threat to its existence and it reacted accordingly. What had emerged from World War II was a new jurisdiction. The traditional functions had expanded enormously and were suddenly resident over the entire globe, not just in a few remote places. Further, the missions of preserving the world's peace and protecting our own vital national interests had expanded as well. The Army found itself doing something it had never seriously contemplated and for which, at the time, there was no substitute—serving as the world's policeman. The United Nations was too young and too weak to make any serious impact, and only the American armed forces had any capability to perform the role.

What is important to realize, however, is that what became the Cold War was an aberration in American military history. This kind of war demanded a new description of the Army profession at least in terms of the variety of duties in which one expected soldiers to be competent. The singular focus on a clear threat—even if its manifestations showed remarkable variety—drove a unique condition that persisted until The Wall came down. The Army profession, at least in the minds of its uniformed leaders, shifted its jurisdiction from infrastructure-building and administration to a clear combat capability to win the "First Battle." Never before had that imperative existed with quite the same force; never before had it seemed to matter that much.

Behind all of this lay the advent of nuclear weapons and intercontinental delivery systems. To some extent, and for one specific period, it looked as if the Army had become irrelevant. Quickly adapting to the nuclear situation, the Army attempted to recapture the jurisdiction lost to the Air Force's Strategic Air Command. The Davy Crockett was a first feeble step, and that eventually matured into the Atomic Annie, the Army's first authentic field nuclear delivery system. From that point on experiments continued as the Army struggled to maintain its place on the nuclear battlefield. Although the struggle was never fully resolved, several generations of soldiers grew up in nuclear-capable artillery units and the associated planning staffs, knowing all the while that their efforts were next to suicidal and that the intercontinental ballistic missiles were the real insurance policy. Mutually assured destruction gave some credibility to battlefield nuclear weapons and the Army forces associated with them, in a fond and fortunately realized hope that neither would ever have to be used.

Then came Vietnam and the frustration of another guerrilla war. Like the Seminole War, the goal was generally clear, but its legitimacy was not. Protecting Western Europe from the rapacious Soviet hordes was one thing, but protecting people who seemed to hate or were mostly indifferent to their own government put the required sacrifices in a questionable light. Not everyone saw the war that way, but the frustrations inherent in guerrilla warfare and the ambiguities at the strategic level—the failure to apply full power against known enemy refuges or lines of communication—brought serious questions into plain view. The result was catastrophic. The Seminole War could never have ended quite the same way as Vietnam since it was homeland or land immediately adjacent thereto. An intervening ocean made Vietnam very different. In its latter days, the Vietnam experience provided little assurance that the Army was "serving the American people" as it had done in the past. However, the profoundly compromised Army that returned from that conflict underwent a rapid transformation through what came to be known as the "training revolution."

The training revolution inspired a new sense of purpose in an Army that once again found itself focused on the protection of Western civilization. Whether it would have had the same effect had the war in Vietnam still been in progress is questionable. By taking their jobs apart and analyzing then in minute detail, then assembling the details into trainable bits, the Army recovered its balance and its self-confidence—it knew what its jobs were and it knew whether it was proficient or not. From that time on, the organization has majored in training and is widely proclaimed to be a model worthy of emulation by many segments of the civil population. This jurisdictional expansion paralleled that of leadership and instilling discipline.

During the Cold War era, the Army's jurisdictional claims continued to be uncontested. Several objective characteristics helped reduce the vulnerability of the jurisdiction. First, technological factors such as the advent of nuclear weapons brought the prospect of war to the homeland. National survival became a tangible goal and the Army became a key element in that pursuit. Second, the aftermath of World War II combined with the immaturity of national governance structures in many areas left the Army administering significant parts of the world. Only the Army had the professional knowledge and ability to continue administration in foreign lands. Finally, the repercussions of the Vietnam War, along with memories of Desert One (the abortive embassy hostage rescue mission in Iran in 1980), changed the diagnostic criteria for the Army's relevance from situations requiring the deployment of trained, disciplined manpower in austere conditions to situations where the nation's vital interests as well as the political and military objectives are clear and there is public support of a commitment to victory.

1992–Today

The late 1980s and early 1990s were marked by the confluence of several factors that impacted the Army profession. First, the tremendous successes in Operations Just Cause in Panama and Desert Shield/Storm in Southwest Asia reinforced the Army's emphasis on winning the first battle with overwhelming force after political and public support is gained. The focus on training, the realistic exercises at the Combat Training Centers, the reform of officer and noncommissioned officer personnel management systems, and the modernization of the Army succeeded in finally exorcizing the demons of Vietnam. While the battlefield successes buttressed the Army's war-fighting jurisdiction, the fall of the Berlin Wall eroded the nation's urgency in winning the next first battle. With the United States emerging from the Cold War as the world's only superpower, the Army found itself at the peak of its warrior preparation with no apparent opponent.

As the nation rushed to cash its peace dividend, the Army underwent a 40 percent downsizing that, while much more compassionate in execution than the post-Vietnam drawdown, still had debilitating effects on the profession. As competition within the officer ranks spiraled upward, morale began to plummet. Internally, the force struggled with its professional identity as the warrior ethos it had nurtured during the Cold War was questioned. Madeline Morris, a Duke law professor hired as a consultant by the Secretary of the Army, attacked the aggressive, "masculinist attitudes" of the military culture.[36] Much of the push to open the military to homosexuals and combat positions to women focused on the apparently diminishing role of war in forming the Army's identity.

With the absence of a clear threat to the nation, the meaning of "serving the American people" began to revert to providing disciplined, trained manpower capable of deploying to a possibly dangerous environment. Soldiers found themselves once again patrolling the borders—this time in counterdrug operations. The Army deployed again to the national parks—this time to fight forest fires. Soldiers once again took postings to far away lands—this time under the military-to-military engagement policy.[37] Support to civil authorities continued as in the past with troops assisting with the Olympics and the presidential inauguration. The nation, through Congress, recalled the Army's earlier role and deployed it to the horrendous scene of Hurricane Andrew in a domestic humanitarian effort. Then it was deployed to Haiti, Somalia, Rwanda, Bosnia, and Kosovo. In each of these latter deployments, the general mission could have been stated as "Stop the Killing." In several ways this was a unique sort of employment. Haiti was actually more of an effort to install a government pledged to reestablish law and order, in the process bringing a halt to the long-standing practice of killing anyone who might threaten the regime. The other deployments were efforts arising indirectly out of the Nuremburg trials, implicitly giving the United Nations authority to intervene *within* a nation's borders to halt genocide.[38] Constabulary forces have been left behind in Bosnia and Kosovo with no end date in sight.

Resistance to the shift in jurisdiction was heard as some complained about the detrimental effects of peacekeeping or operations other than war on the Army. Senior officers called for calm, assuring members of the profession that these missions were nothing new. Yet, while the missions were replays of history, something had changed. The Army was still carrying the self-image of warrior. Peacekeeping, humanitarian assistance, disaster relief, and civil support were all worthy missions, but the Army still retained an equal priority on winning the next first battle. The psychological strain of sustaining two major jurisdictions continues to burden the force today.[39] Furthermore, some of the peacekeeping missions quickly devolved to police activities with too little attention being paid to the bald fact that military police of any description cannot cope with thugs who have tank and artillery arsenals. For these, heavily armed soldiers are required.

Additionally, recent deployments, although paralleling those to the Philippines (1898-1902), Dominican Republic (1916-1924), and Cuba (1898-1902), are qualitatively different. In past deployments, troops were deployed and remained on duty as constabulary forces (to be completely accurate, only Cuba made it out of the tutelage stage in a timely manner). Their mission was to remain on station until the adversaries were subdued and legitimate civil government—preferably local—could be established. Recent deployments follow the same pattern, but do not confront the element of a clear-cut enemy. What is worse, although having the establishment of the rule of law under local government as an ultimate aim, these deployments are given no real impetus in that direction. Hence, operations in Haiti, Somalia, Bosnia, and Kosovo began as if a clearly identifiable enemy existed, e.g., with the 10th Mountain Division launching an air assault, or the 1st Armored Division crossing the Sava River, but then evolved into operations in which political factors prevented the "bad guys" from being eliminated. The establishment of local government operating under the rule of law and order was not on the mission list. Failing that, a date of withdrawal remains uncertain. The result is uneasiness in the profession as this new implication of deployment without a mandate to accomplish the implicit mission is added to the Army's jurisdiction.

This internal professional angst has arisen against a backdrop of intense jurisdictional competition. The world has become a much safer place in a macro sense. Consequently the capability of being deployed to dangerous environments that helped secure the Army's jurisdiction has been in less demand. With that change, the deployment of disciplined, trained manpower *other than the Army* has become more feasible. Contractors, nongovernmental organizations, and other U.S. governmental entities all compete for jurisdiction once the environment is deemed secure. It is routine to see civilians working side by side with soldiers on deployments—a development that gnaws at the profession's self-identity.

Although the United States is the world's sole superpower, it is not the world's only peacekeeper. Thus, armies from the North Atlantic Treaty Organization, often under the United Nations flag, compete for the peacekeeping jurisdiction. The Russian race to the Pristina airport in Kosovo graphically illustrated the competition for the peacekeeping jurisdiction. Although the presence of U.S. peacekeepers unquestionably ups the international ante, it is interesting that the U.S. Army is getting involved in a jurisdiction traditionally held by countries as diverse as Fiji and Norway.

While the Army scrambles to find its professional identity in peacekeeping, its warfighting jurisdiction is also being subjected to unprecedented competition. The Marines have tried to establish themselves as the urban warfare experts—fighting the future "three block war."[40] The Air Force has repeatedly claimed that precision engagement makes land warfare moot since "it will be possible to find, fix, or track and target anything that moves on the surface of the earth."[41] Meanwhile, the failure to secure the Pentagon's approval for committing Task Force Hawk to the fight in Kosovo, along with memories of the 82nd Airborne Division potentially serving as no more than "speed bumps" before the advance of Saddam Hussein's armored divisions, has given the Army renewed impetus to transform itself into a more deployable multi-use force.

The jurisdiction for developing leadership and discipline within society has experienced some interesting developments. The country still looks to the Army as a source of moral discipline, and the recent "Army of One" marketing campaign did bring serious debate about how that aspect of the profession was being manipulated to appeal to today's youth. Yet under the perhaps disingenuous individualistic veneer of an "Army of One" lies the same discipline-conscious, team-oriented Army that has served the nation for centuries. A new trend that seems to be gaining momentum has been the hiring of retired officers as school administrators to bring the needed leadership and vision to troubled school systems to include Seattle and Washington, DC.

Leadership development, however, is starting to show signs of a unique type of jurisdictional shifting with the introduction of contracted retirees. Quasi-military organizations, that is, service organizations manned almost entirely of retired military personnel, are already at work performing military-to-military functions, providing engineering assistance to remote, barely pacified or even hostile areas, and administering overt training to foreign militaries.

The Army has always informally and formally welcomed the wisdom of retired officers. Recently, however, Army alumni through contracts have cemented their role as logisticians, ROTC professors of military science, and trainers of the Croatian, Macedonian, and Colombian armies.[42] The rationale for these contracts has usually been personnel shortages in the active force. The practical effect, however, of using contracted retirees is to extend the membership of the profession to include retirees. Retired officers are contracted as "senior mentors" to battle command training programs and other exercises as well as to battalion and brigade commanders themselves. Other professions also use alumni to pass on the precepts of the profession, but future research might well explore the effects on the profession of more questionable activities such as political involvement of retired general officers.[43]

Army futurists have considered the wisdom of divorcing the military from functions derived from the frontier garrison traditions. These include commissaries, chaplains, finance, and a host of logistics activities. The added capabilities of the internet, which provides instant access and greater visibility to the market, have given even greater impetus to out-sourcing in a fashion reminiscent of the early quartermaster functions that were largely civilian contract operations. In the absence of war or the imminent prospect of a major war, efficiency tends to trump effectiveness. The arguments that chaplain's functions should be contracted out are typical. Any clergy in the surrounding community should be able to meet the soldier's spiritual needs. Moreover, chaplains are more or less contracted out from their denominations anyway—denominational representation is proportionate to that within the service and chaplains serve only with the endorsement of their host denominations. Additionally, chaplains have comprised an officer corps only since World War I. Oddly enough, however, the present demands for chaplain support to the deployed force are as high or higher than in a combat situation since there are more difficult issues in the panoply of peacekeeping situations. Civilian contract chaplains for these forces would require intensive training before they could possibly be relevant and would, most likely, be unable to make a rapid transition to deal with combat should it break out—would it even be written into their contracts?

If Paychex administers almost the entire payroll operations of large and small businesses, why shouldn't the military payrolls be contracted out to it? Is there a good reason in peacetime not to replace the Quartermaster Corps with Wal-Mart, Exxon, Mobil, Giant Foods, and a clutch of other civilian commercial service providers? Despite the current open-ended war on terrorism, the efficiency experts can continue to carry on such arguments and demonstrate the reasonableness of their position. The surface credibility of the civilianization threat to the Army's jurisdiction is troublesome to many within the profession, for it clearly suggests, "Your days are numbered," at least in certain areas long regarded as sacrosanct. But for those who heed the arguments based on effectiveness, the nostrums of efficiency gurus can be easily dismissed as destructive of the force whose essential core competency is to fight wars.

In the same fashion that many of the military logistics and support functions are susceptible to being contracted out, so too one might argue that military communications ought to be contracted out to AT&T or some comparable organization. Ultimately, everything except the killing systems could be contracted out and even some of them could be for smaller operations. Sandline and the now-defunct Executive Outcomes were the first mercenary forces established after the end of the Cold War in the West. They are likely to remain afoul of United Nations organizations and others wary of modern freebooters for whom the niceties of the Geneva Convention are an impediment to action. But the mercenary principle, representing a theoretical approach attractive to efficiency experts, would be deadly to national security in the long run. Certainly soldier pay could be managed from afar, but it is not clear what premium must be paid to coax a FedEx delivery vehicle to enter a combat zone with that just-in-time replacement part.

The disciplined use of force against enemies without such discipline must be the measure of merit against which jurisdiction is tested. Almost everything else can be done by others, albeit at increased cost and with increased risk. However, so long as an existing military force retains the ability to perform the deadly force mission, it will likely be less expensive—in the terms currently employed to make such measurements—to also employ that force on operations other than war if no other serious threat looms near. But the employment of such forces in too many lesser missions will, at some point, dilute their capability for combat. Conversely, the too-frequent use of other organizations in roles requiring force, but of a more police-like nature and scale, could gradually militarize the larger society, a trend already underway in the proliferation of SWAT teams and their clones. We see today as in the past that involvement of the military in nonmilitary tasks lowers the incentive of potential recruits to serve in it, thus suggesting that successful recruitment emanates from clarity and purity of mission.

Ultimately the Army exists to fight wars. In light of the 11 September 2001 terrorist attacks, those now serving will confront a new kind of war on a different scale, involving different techniques and occurring in different settings than those formerly thought normal. These will co-exist with a parallel routine of nonwar activities. Current training practices will need to be refocused to prepare soldiers for both missions. As long as the prospect of war was distant and the violence of our age tended to be principally a police matter, the jurisdiction of the Army could in theory have shrunk safely. All that changed on 11 September, with the Army jurisdiction's sudden expansion to encompass domestic functions. Because this present condition is not war in the strictest legal sense, the Army is visible simultaneously in both its Active and Reserve guises, doing for the nation what the nation requires. The peacekeeping missions will remain and the unique training requirements associated with them. The realities of this new "war against terrorism" will, despite the fondest hopes and best promises of technology, require trained, disciplined soldiers to close with and destroy the enemy. This function cannot be contracted out, nor can this jurisdiction be passed off to any other organization; this must be retained with all the robustness required of an activity in which the margin for victory is often not much over two percent.

So what is to be made of all this? The prolonged tendency toward inculcating a "Hooah" attitude in the Army has finally come up against the dual necessity to do more than just all that noise tends to suggest. Teaching the history of battles and campaigns that end with the military defeat of the enemy's forces is no longer adequate. The simple truth is that the Army has spent most of its existence doing things other than war. It is time the instruction in Army history be honestly presented to include all the other seemingly mundane "stuff." And that is also "stuff" in which the Army should take equal pride of performance. It is the "stuff" that will in no small measure determine the long-term success of our present involvement in Afghanistan. Today, planning for post-conflict operations is, as we have seen recently, a key part of our national strategy. It is also part of Army doctrine. These need to appear in the Army history curriculum, recognizing that repetitive episodes of military government are facts of Army life. Additionally, an explanation of the conditions under which support to civil authorities is legal and appropriate needs to be included, as does a thorough explanation of the reason soldiers are prohibited from aiding civil officials in enforcing the law or suppressing civil disorders unless expressly ordered to do so by the President (Posse Comitatus Act of 1878). These are all matters of established doctrine and now are highlighted in the newly updated field manual, FM 1, *The Army*. If the glory of combat operations is all that officers are taught and acculturated to, their understanding of their jurisdiction will suffer, they will misunderstand the breadth of their commitment, and some will leave the profession feeling betrayed.

This extended period of Army transformation presents a unique opportunity to initiate change in public dialogue describing the Army and its roles as well as reorienting its educational program. "Hooah" should mean a job well done, including pulling survivors from collapsed buildings and then setting up and running refugee camps, delivering babies, providing clean water and adequate sanitation. It should also mean that the menacing mob in future Bosnias and Kosovos will be faced down and dispersed without casualties because the key agitators are identified through human intelligence sources and publicly taken into custody. It also means eradicating terrorists and their support systems in campaigns that more closely resemble in concept the Indian Wars of the Western frontier. Everything the Army does in this new era is shaped through the development of "campaign plans." Campaign plans now address a great deal of nonhostile activity—as does the Transformation Campaign Plan itself—and appropriately signal the connectivity between the use of a disciplined force for other than violent activities as well as for combat. While the institution is already in the process of changing its lexicon to reflect the dualities with which it is dealing, a more formal approach is necessary to inculcate the fullest understanding of the profession's evolving jurisdictions.

We acknowledge that these conclusions rest on the frail promise that education can actually make a difference. We acknowledge further that acting on these conclusions will require a revision of existing instructional materials and will likely aggravate the competition for curricular time to present it, but we have implicit faith in the ability of the Army, when seized by an idea, to bring it to fruition more thoroughly than any other institution.

Notes

1. The views and opinions expressed in this chapter are those of the authors and not necessarily those of the Department of the Army or any other U.S. government entity.
2. Headquarters, Department of the Army, Field Manual 1, *The Army* (Washington, DC: U.S. Government Printing Office, 2001), 25.
3. Andrew Abbott, *The System of Professions: An Essay on the Division of Expert Labor* (Chicago: University of Chicago Press, 1988).
4. Ibid., 20.
5. Russell F. Weigley. *Towards an American Army: Military Thought from Washington to Marshall* (NewYork: Columbia University Press, 1962), 81.
6. Allan R. Millett and Peter Maslowski, eds., *For The Common Defense: A Military History of the United States of America,* Rev. and enl. (New York: The Free Press, 1984), 91.
7. Russell F. Weigley, *History of the United States Army* (New York: The Macmillan Company, 1967), 83; and Robert M.Utley, *Frontiersmen In Blue: The United States Army and the Indian, 1848-1865* (Lincoln: University of Nebraska Press, 1967), xii - xiii, 2, 4-5.
8. Utley, xii.
9. Weigley, *History of the United States Army,* 162.
10. Samuel J. Watson, "'Daily More Dissatisfied': Dissent Among U.S. Army Officers During the Second Seminole War" (Paper presented to Houston Area Southern Historians, Houston, TX, 20 September 1994).
11. Weigley, *History of the United States Army,* 98.
12. Ibid.,153.
13. Ibid., 163. The coastal defense gun forts were essentially replaced by bombers in the late 1930s and by NORAD interceptor and NIKE air defenses in the 1950s and 1960s.
14. Weigley, *History of the United States Army*, 143.
15. Ibid., 164.
16. Ibid., 151.
17. Ibid., 188-89.
18. Ibid., 180.
19. On the flip side of this reduction in direct control of the means of production of military supplies is the increase in the management of the development and procurement processes personified in the Program Manager's program and now the Acquisition Corps. These developments have, in some measure, extended the Army's jurisdiction into areas previously ceded to civilianization of traditionally quartermaster functions. The argument seems to be that if the Army can no longer afford to control the metal-bending industries, it should at least have significant influence over how such industries execute Army contracts.
20. Weigley, *History of the United States Army,* 1-9.
21. Draft manuscript titled "In the Beginning," Papers of Conrad Stanton Babcock, Jr., Hoover Institution for War and Peace, Stanford, CA, 108-227; also Edward M. Coffman, *The War to End All Wars: The American Military Experience in World War I* (New York: Oxford University Press, 1968), 18-19; and Allan R. Millett, *The General: Robert L. Bullard and Officership in the United States Army, 1881-1925,* Contributions in Military History, Number 10 (Westport, CT: Greenwood Press, 1975), 168-83, 194-98.
22. Donna Marie Elean Thomas, "Army Reform in America: The Crucial Years, 1876-1881" (Ph.D. diss., University of Florida, 1980, MMI #8115686), 136-137.
23. Weigley, *History of the United States* Army, 503; Millett and Maslowski, 505.

24. O' Callahan v. Parker, 65 Supreme Court 1969; 395 US 258, 89 SCt. 1683, 23 L.Ed. 2d 29, (1969).
25. Millett and Maslowski, 337.
26. Coffman, 13-14.
27. Weigley, *History of the United States* Army, 347.
28. Millett and Maslowski, 398-99.
29. Coffman, 14-15.
30. Donald N. Bigelow, *William Conant Church: Army and Navy Journal* (New York: Columbia University Press, 1952), 184-86.
31. Jerry M. Cooper, "The Army's Search for a Mission, 1865-1890," in *Against All Enemies: Interpretations of American Military History from Colonial Times to the Present*, eds. Kenneth J. Hagan and William R. Roberts, Contributions in Military Studies Number 51 (Westport, CT: Greenwood Press, 1986), 189.
32. By 1997, thirty-four states had instituted boot camps. See for example Corey Kilgannon, "Changing Youth Attitudes About Police," *New York Times*, 13 August 2000, A7. Interestingly, however, the effectiveness of boot camps is being questioned, as participants are not becoming model citizens upon their return to society. See Jayson Blair, "Boot Camps: An Idea Whose Time Came and Went," *New York Times*, 2 January 2000, A4.
33. John W. Killegrew, "The Impact of the Great Depression on the Army, 1929-1936" (Ph.D. diss., Indiana University, 1960), xii, 1-22.
34. Dwight D. Eisenhower, "Farewell Radio and Television Address to the American People," 17 January 1961, in *The American Military: Readings in the History of the Military in American Society*, ed. Russell F. Weigley (Reading, MA: Addison-Wesley Publishing Company, Inc., 1969), 153.
35. "Why Welch Did an About-Face on GE's Strategy," *USA Today*, 28 March 2001, 2A.
36. Madeline Morris, "By Force of Arms: Rape, War, and Military Culture," *Duke Law Journal* 45 (1996): 651.
37. See for example the series by Dana Priest, "The Proconsuls: America's Soldier-Diplomats," *Washington Post*, 28-30 September 2000, discussing the role of the military in diplomacy.
38. William G. Eckhardt, "Nuremberg—Fifty Years: Accountability and Responsibility," *University of Missouri Kansas City Law Review* 65 (1996): 5-7.
39. See Volker C. Franke, "Warriors for Peace: The Next Generation of U.S. Military Leaders," *Armed Forces and Society* 24 (Fall 1997): 33, for a discussion of conflicting war-fighting and peacekeeping attitudes being developed in pre-commissioned officers.
40. See Joel Garreau, "Point Men for a Revolution: Can the Marines Survive a Shift from Platoons to Networks?" *Washington Post*, 6 March 1999, A1.
41. Air Force Competencies, U.S. Air Force Force Protection Battlelab, *http://afsf.lackland.af.mil/Organization/AFFPB/org_affpb_precisionengagement%20.htm.*
42. See David Adams and Paul de la Garza, "Contract's End Hints of Colombia Trouble," *St. Petersburg Times*, 13 May 2001, A1, for claims that contractors were not successful in Colombia.
43. William Becker, *Post-Retirement Political Action*, Strategy Research Paper, AY 2001, U.S. Army War College, Carlisle Barracks, PA, 2001. Becker found that general officers expect to regain most of their civil rights willingly laid aside for the good order and discipline of the profession once they retire. Most expect to be able to take an active part in politics even though they recognize that they will be most effective when acting as individuals rather than as a bloc.

5 In Their Own Words: Army Officers Discuss Their Profession[1]

Gayle L. Watkins and Randi C. Cohen

In their article titled "The Future of Army Professionalism: A Need for Renewal and Redefinition," published in *Parameters* in 2000, Don Snider and Gayle Watkins posit that the U.S. Army is at risk of transitioning from a professional organization to an obedient bureaucracy, an organizational form that while not an overt threat to the nation will not provide the level of national defense demanded by the American people. Furthermore, they claim that this shift is deleterious for both the nation and the Army itself. By its nature, the professional mode has been and continues to be best able to address the Army's unique challenges because of its two key strengths: knowledge development and individual control.

The Army's first challenge is the development of military knowledge in advance of our adversaries. The coin of the professional realm is knowledge; professions are built on abstract knowledge and its application. This knowledge or expertise resides in professional members, rather than in organizational systems or technology; professions are positioned to create, expand, and develop knowledge. Military knowledge—doctrine, strategy, tactics, and technologies—is constantly evolving as adversaries seek advantage over one another. Many nations depend upon the advances made by others, preferring to be consumers of military knowledge rather than creators. However, the United States has chosen to take a leadership role in the development of military knowledge. Therefore, although other nations' armies can become bureaucratized and still succeed, America's Army must maintain its professional status if it is to meet the country's expectations and needs in the 21st century. It must remain a professional institution capable of continued evolution and revolution in military affairs.

The second challenge the Army faces is controlling service members as they engage in the most lethal of social activities during security operations marked by chaos and complexity. This need to control behavior is established by the military's unique place among professions and organizations as the only American institution sanctioned to kill in the pursuit of national interests and objectives. Because of the potential threat inherent in this distinctive role, societal norms and laws govern how this professional task will be carried out. Not only do we expect our soldiers to fight with courage and skill but we also demand that they do so honorably, showing compassion for noncombatants and captured enemies, applying a sense of fair play even in combat, and always maintaining an awareness of the political requirements and repercussions of their actions.[1]

The environment in which the Army's professional tasks are undertaken further complicates the achievement of this degree of behavioral control. Much has been written about the uncertainty and chaos of military operations, both combat and noncombat. In fact, few organizational endeavors can match their inherent complexity, pace, risk, and uncertainty. Soldiers in these situations face the competing and often conflicting demands of self-preservation and honorable service.

Internalized values, traditions, trust, and member commitment provide the most powerful means of achieving a high degree of control over those engaged in this most lethal of societal services under arduous conditions that cannot be easily monitored. These control systems, known to organizational theorists as "clan control," are most often developed in small, horizontal organizations and professions.[2] With its hundreds of thousand members and its hierarchical structure, the Army is unable to develop the type of clan control present in small organizations across its depth and breadth. Instead, it is the internalized values and unwritten norms arising from the Army's professional nature that enable its members to fight honorably, to risk and even give their lives for the nation, and to support national and international conventions on the moral and ethical conduct of military operations.

Given that knowledge development and member control are critical to its success, sustaining the Army as a profession is paramount. We therefore set out to study the status of the profession, specifically from the perspective of some key professional members, the Army officer corps. To organize our efforts, we used the theoretical framework on professions presented by Andrew Abbott in his 1988 book, *The System of Professions*. Abbott's framework presents professions as exclusive occupations embedded in a competitive system of professions and organizations, vying for control over where and when their knowledge can be applied. Within this system, professions are organized around the abstract knowledge that they develop, apply, and protect. Their workplace tasks consist of applying this knowledge under specific controls that enable the profession to gain and maintain social legitimacy. Battles for jurisdiction, resources, and legitimacy occur among professions and organizations through claims on abstract knowledge and the definitions and expansions of tasks.

This framework has been used recently to study a variety of professional systems, from physicians to funeral directors. A number of studies have been done on the development, expansion, and defense of professional jurisdictions, including studies of physicians, nurses, lawyers, social workers, vision care providers, and funeral directors.[3] Researchers have also focused on professional expertise, particularly on how professions determine the tasks of their professional work such as R.C. Cohen's study of financial planners (1997) and Y.M. Johnson's investigation of indirect tasks in social work (1999); and on the relation between expertise and professional education.[4] Following in the footsteps of these researchers, we applied Abbott's framework to the Army profession in order to understand how officers reflect the Army's professionalism in their thoughts about their work.

Methodology

Our chapter seeks to understand the Army profession from the perspective of a key segment of its membership, the officer corps. Since there have been no qualitative studies of the profession for the past thirty years, our work is neither longitudinal nor comparative; we could not determine whether there has been a change in Army officer concepts of professionalism. However, we do provide a current and accurate portrait, a rich and detailed description of how Army officers think about their profession's knowledge, tasks, control, and jurisdiction. This information will enable us to make some recommendations to strengthen the profession.[5]

Data Gathering

As we designed our research, we were well cognizant of problems confronted during the last study of the Army profession, the U.S. Army War College's *Study on Military Professionalism* (1970).[6] We were particularly concerned that officers would find it difficult to express their feelings about their profession because of the intimate and abstract nature of the topic. Earlier Army research had found that officers were either reluctant or unable to discuss their profession from an individual perspective:

> Without exception, in group discussions or interviews, respondents, irrespective of grade level or experience, avoided coming to grips with the problem of definition. This finding coincides with the empirical research and theoretical studies of values. Values and value systems defy verbalization because they are abstract feelings and sentiments, and because they remain largely a personal matter.7

In addition to the intimate nature of the topic, the researchers found that its abstract nature made it difficult for officers to discuss it:

> Values and value systems defy verbalization, not only because they represent ideological *feelings*, but because they are general and not linked to specific objects. For example, it is exceedingly difficult to translate accurately a value such as "Duty" into operative guidelines for behavior.[8]

Furthermore, selection of the optimal data collection method was critical since researchers have found that focus groups often impede accurate data collection: the responses from one person often bias what another person says; an outspoken person can dominate the discussion, thus preventing the ideas of others from being expressed; and there may not be sufficient time to explore each participant's views in depth.[9]

To resolve these problems, we turned to recent advances in marketing research, featuring methods that enable researchers to investigate concepts that people have trouble verbalizing.[10] These techniques are designed to be consonant with how humans think and communicate, thereby tapping into respondents' personally held opinions and feelings. The methods use descriptive imagery—stories, pictures, objects—to enable individuals to access their feelings indirectly by linking abstractions to more concrete images. Using such methods for data elicitation is best done in individual interviews rather than in focus groups. Therefore, we designed a multiple-method interviewing technique that incorporated the following steps:

- We trained interviewers to ask unbiased and non-leading questions. While it may seem obvious that interviewers would ask neutral questions, it actually requires expertise to conduct interviews without assuming the meaning of words the respondents offer. Additionally, when Army officers interview, they need to take special care not to slip into their well-trained role of counselor.
- During the interviews, we focused on and leveraged the descriptive imagery that the respondents used. Stories, images, objects, metaphors, etc. enable individuals to better express their feelings and opinions. The interviewers therefore asked questions about the descriptive imagery to get in-depth responses.
- We also used laddering techniques common in marketing to connect concrete examples or attributes to more abstract values and consequences. By asking the respondents to explain why certain issues matter to them, we uncover important associations among ideas.

Sample

We selected six Army posts with deployable or deployed units, both in the United States and overseas, that would provide a well-rounded sample of branches and experiences. Since the environment heavily influences work identities, such as members' feelings about their profession, we selected officers who are in the field Army rather than in school or the Pentagon.[11] In this type of research, sample size does not need to reach that of statistical significance; rather, it need only be large enough to impart a consistent set of responses for any relevant subgroup. We asked the staff of each participating post to provide participants of certain ranks, but placed no restrictions in terms of gender, race, or career intentions. Our sample consisted of 80 company- and field-grade officers from combat arms, combat support, and combat service support branches who were presently serving in division or corps positions.[12] The distribution of ranks, branches, race, and gender is shown in Table 5-1. Additionally, in our sample, about twice as many officers indicated that they intended to stay in the Army as those indicating they were going to leave.

		Number	Percentage
Rank	Lieutenant	18	22.50
	Captain	33	41.25
	Major	24	30.00
	Lieutenant Colonel	5	6.25
Race	White	74	92.50
	African-American	6	7.50
Branch	Combat Arms	32	40.00
	Combat Support	16	20.00
	Combat Service Support	32	40.00
Gender	Male	73	91.25
	Female	7	8.75

Table 5-1. *Distribution of Rank, Branch, Race, and Gender in the Sample.*

All interviewers were trained in the multiple-method interviewing technique. We used male and female, military and civilian interviewers to reduce potential bias in the interviewer-respondent interaction. The participants seemed comfortable with the interviewers and, in general, were quite engaged in the process. Many officers expressed appreciation that someone was listening to their opinions, and that they had the opportunity during the interview to reflect on the "bigger issues." Some officers also became emotional during the interview, and we often had to stop the discussion before the officers were ready to end due to time constraints.

Interview

In preparation for the interviews, we provided all the participants with an assignment letter, asking the respondents to think in advance about what it means to be a professional officer so that they would be primed for the interview.[13] Additionally, we asked the participants to bring three items to the interview that symbolized their thoughts on the topic.[14] Such pictures or objects served as a springboard for discussion, facilitating access to deeper insights and feelings. For example, one officer brought a washer that signified technical expertise while another brought a green beret that symbolized his commitment to his team.

The one-on-one interviews lasted approximately 45 minutes each, were tape-recorded, and transcribed into hard copy. All the interviewers began the discussion with the same introduction, emphasizing the confidentiality of the interview and requesting full and honest responses.

During the interview, the participants explained why they selected the items they brought with them and what those items and their associated ideas meant in terms of being a professional officer. The interviewer based follow-up questions on the words and ideas the participating officers offered, asking for definitions, feelings, and consequences associated with key ideas. As the final step in the interview, the interviewers asked the respondents to select someone (living or dead, real or fictional, military or civilian) who best epitomized to them what it means to be a professional officer.

Data Analysis

We followed widely accepted standards for the analysis of qualitative data, starting our inductive analysis by using a representative subset of transcripts to build a coding schema.[15] Multiple analysts looked at these same transcripts and developed a common list of variables that we then applied systematically to all of the transcripts. Using *Atlas.ti* software, we were able to link quotations to codes and then generate reports on these variables.[16] We iterated between the reports and hypotheses, developing themes and identifying issues.

We also approached the data in a deductive mode, looking for the answers to theory-driven questions. We specifically sought out information pertaining to professional knowledge, tasks, control, and jurisdictional issues. Additionally, we categorized the responses to the question about who best epitomizes what it means to be a professional, thereby gaining insight into the "ideal type" of the professional officer, which further clarified what matters in the profession.

The findings that we present in the results section emerged in multiple interviews. We took care in our analysis to weigh the frequency and intensity of responses. We generally indicate these using descriptive terms. Terms we use to describe the number of officers discussing a theme include: *most*, which indicates a substantial majority; *many*, which indicates a majority; *some*, which indicates a substantial minority; *several* and *few*. We also use very specific adjectives to indicate the respondents' level of intensity, ranging from *anguished* to *indifferent*.

Although this sample was not intended to be statistically significant, we were sensitive to potential patterns associated with the demographic profiles of the participants. Unless explicitly stated, we found no discernible differences by location, branch, rank, gender, race, or career intention.

Results

As noted in the methodology section, we used a two-pronged approach to analyze the data: a deductive assessment of the key components of a profession, including knowledge, task, control, and jurisdictional issues; and an inductive inquiry that, much like the data-gathering, enabled the officers' voices to emerge unfiltered.

Do Army Officers Consider the Army a Profession?

During these latter efforts, while letting the data speak for themselves, one dominant issue surfaced that was quite basic: officers do not necessarily consider themselves to be part of a profession. Although a majority of officers spoke positively about their work as a profession, a sizable minority revealed serious concerns. Some company-grade officers were surprised by the question we posed since thinking about the Army as a profession was not something they had considered previously. As one captain put it,

> I know very few Army officers that consider ourselves under the term "professional" in the same form that you consider lawyers and doctors. . . . I think professional—I've never heard that term used before. I've never heard senior officers use that . . . term.[17]

Expressing indifference, disappointment, or even sorrow, other officers referred to actions and policies of senior leaders and the institution as a whole that undermine the military as a profession. These officers' thoughts about the profession focused on its diminishment. For some officers, a profession requires a level of expertise or competence that is not currently present in the Army:

> People aren't confident in their unit's ability to do the wartime mission because people aren't technically competent. If the thing is a profession then it is a profession, I don't go to a doctor that doesn't know what he is doing 'cause the doctor works at the hospital XYZ then that must mean he is good. [If] the guy doesn't know anything, it doesn't make any sense. If the thing is a profession then the guy should be

> an expert in what he does. People [in the Army] aren't experts because they are way too heavily involved in the technical or the interpersonal but they don't have the right mix.

Finally, some of this group argued that the profession is being degraded by inducements to susceptible officers who stay for reasons other than the profession, such as educational benefits (italicized questions embedded in the quotations are those of the interviewer):

> I read in the *Army Times* where they are thinking of offering $100,000. . . . They are offering $100,000 bonuses to lieutenants if they sign something saying "Hey, I will spend twenty years in." *What does that mean in terms of what it means to be professional in the Army?* I think the Army is corrupting to what it means to be a professional. They have thrown the idealism of being a professional in the garbage to get numbers and the numbers they are getting are not the kind of numbers they need to sustain the profession.
>
> Officers have been given a choice of whether or not to consider it a profession. You can look at it as a stepping stone or something merely to pay back college or you can look back to its history. Like the anachronistic qualities of the pride or the glory and accomplishments and take that to embody a profession and then continue to live that and somewhat ignore the reality of what is going around you as far as how other people view it. And so well people are just using it to pay back college and they are not really leading soldiers and they are not really giving it their 100%. They are not really being an Army officer. They are being someone who wears the uniform and punches a clock.

What Does It Mean To Be A Professional Officer?

Beyond this initial, unexpected finding that a significant number of officers did not consider their vocation to be a profession, we further examined the officers' descriptions in terms of the professional characteristics outlined by Abbott: knowledge, tasks, control, and jurisdiction. Each of these attributes was developed by at least some of the respondents but the number varied widely. While offering their thoughts on the problems facing the Army, as well as on ways the profession could be strengthened, most of the officers articulated a strong sense of pride and satisfaction in their service.

Officers' discussions of their professional knowledge centered on the Army's unique abstract knowledge and expertise. They often distinguished the general expertise required of all officers from the specific expertise required by their branch or specialty. During these conversations, they also spoke of the importance of continuing education and self-study to their development as Army officers.

Abstract Knowledge. Some officers made the connection between the strength of the profession and the abstract knowledge associated with it. For example, one officer said that Gen. Omar Bradley epitomized the army professional because of the way he used his intelligence to understand soldiering and accomplish the mission. Several of the interviewed officers emphasized the importance of lifetime study of military strategy and the warrior ethos when they brought in texts—some historical, others technical—as their objects. Others emphasized the uniqueness of their knowledge, such as the officer who described how a map his unit made while in Kosovo was representative of the unique knowledge held by the Army profession:

> Another part of it is kind of a unique item as opposed to another profession. A doctor as part of his profession doesn't make maps. Some other professions, it's not a unique artifact because obviously civil engineers and other types of people like that do it, but it is fairly specific to a profession to take a map and add control measures onto it, add land marks, TRPs, boundaries, whatever you have. And that kind of defines a more unique aspect of the military profession how you'd use something like that. . . .
>
> I would say that every profession has a fundamental base of actual knowledge that is different from everybody else. Engineers know about concrete, doctors know about anatomy, military officers are supposed to know about leadership, about capabilities of their equipment, about how to direct multiple elements, how to make a plan and communicate it to subordinates, and those are kind of the elements to acknowledge.

Expertise. For many of the participants, professional expertise is the hallmark of being an officer. The participants articulated two key aspects of this expertise: the general expertise of leadership and the more specific expertise of technical competence. The answer "leaders" dominated responses to the question, "Who best epitomizes what it means to be a professional officer?" Most officers cited individuals who set a standard for leadership, primarily past and present Army commanders, but also people such as Dr. Martin Luther King, Jr. Even when citing Army examples, admirable leadership qualities often had less to do with military prowess and more to do with personal accomplishment and bearing, e.g., Gen. Colin Powell. During the interviews, officers explicitly and implicitly discussed the importance of leadership expertise as part of the profession. Most felt that there was something unique to being a leader in the Army:

> If you looked up in the dictionary what a professional is it is pretty much a life's career, I would think. It's more than just a skill. It incorporates many skills. A lot of them cerebral if you have that power, and officership is a profession I think because it [includes] the task of leading soldiers into combat. A lot of people have lost sight of what our main job is. To lead soldiers into combat, uphold and defend the Constitution of the United States, and at the same time provide the tools, the equipment, the lifestyle for the Army soldiers.

Several officers, however, claimed that there was no special expertise to Army leadership, that it was no different from the basic management tasks undertaken by supervisors in other organizations, as this young officer describes:

> I thought about that, "Do I consider the Army to be a profession?" and I am not so sure I do or that I even care because I got a buddy that works for Sprint and he goes to a training class for four months which is [Officers Basic Course] for me. I majored in management but I could have majored in French or Chemistry and I go to my training class for four months and now I am an officer and he did the same thing. He sells whatever, long distance to businesses, and he majored in whatever he majored in and he goes to four months of training and then he sells long distance, so I don't know that the Army is any different than what he is doing. . . . What we

> are doing is just general management training. I majored in management in college and I know more just from my undergraduate studies than I learned in actual Army training.

Many officers argued that being able to apply their expertise was essential to being a professional officer. They had acquired specific expertise through self-education and schooling, and that technical competence provided a sense of accomplishment and purpose, pervading their professional tasks and identity. They also admired, for example, the technical competence of golfer Tiger Woods. However, many of these officers felt that they were unable to act on their expertise because of Army management systems that limited their options. Responses to these limitations varied by rank. Company-grade officers were disturbed primarily by what they saw as hypocrisy:[18]

> So if I'm supposed to lock in my training calendar and that's the way it is, then how is it O.K. for me to receive an e-mail that tells me, "You will do this tomorrow?" Well, why didn't you, as the leader who's supposed to follow your own rules, plan this more than 24 hours in advance? What kind of example is that? So it's O.K. for you to tell me I can't do that to my soldiers, but you as a senior officer can now tell me that you're going to violate your own rule. I don't take kindly to that, and that happens quite frequently.

Other young officers expressed frustration with what they saw as a lack of trust in their abilities:

> When you're a staff member or whatever member of an organization you should be able to . . . work with your peers, work with other agencies. Go out and achieve things and make them happen and for me to go to the G4 or work within the DISCOM or do something, I shouldn't have to funnel it through and ask permission. Or I shouldn't have to provide information overload with a lot of details. I should be able to achieve it and work those missions, work those issues externally without being forced by the Commander to constantly provide updates and constantly provide reports. It's almost like a lack of trust.

This lack of trust is evident not only to the officers we interviewed but to others outside the Army, as related by this field-grade officer:

> And the Swedish officer on my staff group turned to me and said, "You know, they treat you like twelve-year-olds." So here's an officer also in the military profession but outside of our world looking at what we do and maybe not taking away what we hoped he would've taken away. I told him "Well, sir, in a certain respect you are right. We are being treated like twelve-year-olds."

It was the lieutenant colonels who expressed what could best be called anguish over their inability to apply their professional expertise with appropriate autonomy. The depth of these feelings was quite profound, sometimes overwhelming the officers, and quite apparent to the interviewers:

> What we've essentially done is we've replaced a fundamental belief in the professionalism of our officer corps and a notion that we will train our officers and equip our officers to apply their judgment in running their units. We've replaced that with a series of centralized programs and checklists, and it fundamentally undermines professionalism because it removes from us an individual responsibility that is

> beyond that of a technician, that is in fact a calling . . . because in the Army you're asked to make decisions of life and death.

With cross-sectional data such as these in our study, we cannot know whether officers feel differently about this topic at different stages of their careers. However, it is worth noting that it is the captains who express uncertainty or disparagement regarding professional expertise issues compared with the profound anguish of the lieutenant colonels. One hypothesis is that having a sense of the uniqueness of expertise, a sense of being a professional and not a manager, develops with one's career. One officer described his personal recognition of this development in this way:

> You know you fill out the little form when you buy a new appliance and stuff and they always ask you what kind of job you have, and I tell you I check the professional block. When I was a platoon leader I think I always checked the middle management block and then one day it dawned on me, "You know I'm not really a middle manager." *Why aren't you a middle manager, why do you now check professional?* I don't know, it's just one of those things where I guess it's just the evolution as you grow a little older and you actually realize what you are doing and the impact that you have.

The Profession's Tasks

One element of being a professional officer that is truly unique involves the management, including the infliction, of violence. Some but not a majority of officers recognized that this expertise was central to the profession—a distinct feature they lay claim to on the ground of their associated knowledge.

> I mean a lot of people manage, and we manage money and we manage people. We manage equipment, maintenance fleets. As far as civilian-type managerial skills. But on the extreme end we manage violence or the applied use of violence and the graduations thereof. To me that is a very specialized skill that not everybody can do. My brother-in-law is a professional musician, and he actually supports himself and his family doing it, he's very good. And that's a skill I couldn't hope to replicate. On the other hand, doing what I do, he couldn't hope to replicate what I do either. So how we apply our managerial skills is one main difference. One application of that makes us different.

A number of officers linked this expertise to the importance of professional values and also to how the diminishment of the profession affects the specific application of violence.

> But ultimately we are a warrior profession. We are supposed to engage and kill other people so you have to rely on people having a certain level of self-reliance if you will. And if you are going to turn to your squad leader and say, "Go set up a position over there and defend it." You have got to depend on him and you've got to instill in him, "Here are your standards and you are held responsible for achieving those standards." And if we have a safety briefing to you majors every time we have a long weekend, we are kind of taking that responsibility away. We are kind of taking that responsibility from you as a professional to meet those standards of the profession.

Officers applauded frank discussions of the warrior facet of their profession. For example, one officer described how much he appreciated candor in a professional development session on "killing."

> This book *On Killing* was given by our commanding general to us in Special Forces when he took command. And the reason why I think that's an important part of professionalism is a lot of people when they talk about coming in the Army they advocate all the positive sides that those in civilian life would think is important. You're going to get these education benefits if you come in the Army. You're going to get these medical benefits if you come in the Army. If you stay for twenty years, you'll get these great financial benefits if you come in the Army. Very few people address up front what Special Forces soldiers have to do. And that is to kill other people, which is a very difficult thing to do. Which is the thesis of this book that a retired Army lieutenant colonel wrote and our commanding general not only got a copy of this book for all the leadership in all the Special Forces units. But he also had an [Officer Professional Development and Noncommissioned Officer Professional Development class] where this guy was the guest speaker and talked about that.

While the claim on abstract knowledge for the Army profession may lie with managing violence or even leadership, another fundamental task the officers consistently and emotionally discussed involved neither; there was a remarkable emphasis during the interviews about the work surrounding taking care of soldiers. A few officers clearly related the importance of taking care of soldiers to the accomplishment of their mission, as this officer did:

> If you're in a leadership position, serving soldiers is serving the nation. And I don't want to try to make it as codling soldiers and serving soldiers is the same thing. But the idea that soldiers can look at that leader and know that that leader's perspective is to train them, to prepare them but also to—it's not for an evaluation report—it's for their ability to meet their mission and to save their lives but also to set them up for success whether it is in their civilian world if they're moving on or if it's family life. I'm mentoring them in different problems and taking extra time to know the soldiers instead of just being there to be in charge.

One field-grade officer describes the moment when he truly understood the relationship between the mission and taking care of soldiers as his unit prepared to parachute into a combat zone:

> It just instilled in me my duty and responsibility as a leader to prepare them because once I got them on the plane, the door opens and they exit the aircraft, I am no longer in control. So, before that I have to prepare them to do their job. As a leader I am responsible for getting down to that level and insuring that they are capable of doing that without me around, that that squad leader and that soldier can move out and do what is necessary.

Another related the breadth of his responsibility to his soldiers, regardless of time and place:

> I am always an officer. I am an officer on Sunday morning, I am an officer Saturday night. I am always accountable for the obligations and the responsibilities that I have assumed. If something comes up or I see a soldier downtown, I see something happening on post it is my responsibility to take any action that is necessary. The job comes first. *And what does that have to do with calling?* Well, it is not like you work

> for IBM where you are done. You don't act like you have a total separate life. My life is integrated with my job. I think to some extent a civilian's life it too, but a civilian can leave, can just say, "I quit." He does not have to worry about what is happening to the soldier and his subordinates on the weekend. That is our responsibility.

However, for the majority of officers, their range of tasks extended far beyond accomplishing the mission or caring for soldiers in the field. A more accurate label for the tasks that the officers described as dominating their time and energy would be that of social work. The officers cited example after example of the work they do on behalf of their soldiers, including buying groceries for families or even more:

> I had an E7 that got a DUI. I went to the court with him. I went to all his meetings with the company commander. I stood there to look out for his best interests because if he's not being taken care of, then how can I expect him to do his job?

The Profession's Control Mechanisms

Discussions of control mechanisms, the means by which the profession controls its members' behavior, dominated many interviews. Although the Army also uses bureaucratic control systems (promotion, bonuses, etc.) like other organizations, its primary means of control is clan, based on internalized values, traditions, trust, and commitment. The officers clearly recognized the importance of these mechanisms when they discussed two categories of control: professional values and lifestyle.

Professional Values. Officers clearly understood the relative importance of values since they were the professional element most commonly identified by those we interviewed. The officers saw either the specific values or their Army application as unique. Some clearly recognized the importance of these values in gaining the confidence of the American public: "We are trusted by the public to perform a duty regardless of our personal feelings. When we are called on to do the job, and we report to do the job, we do it."

Officers often emphasized these values by identifying the profession as a calling. Sacrifice and patriotism were repeatedly mentioned as hallmarks of the Army that were rarely found in other jobs. A number of officers emphasized the extraordinary responsibility the Army is given when citizens allow their children to enlist. One officer described how values underlie the Army profession in this way:

> Well, basically I think what mainly sets apart being an Army officer as a profession is more in its ideals. And in the past, especially in the past being an officer was a calling and it was more of a, well, if you start off being an officer that unless something came up either physically or financially then you were going to make it a career. And a profession is something you begin that is a career not just a stepping stone.

The more prevalent professional value mentioned in the interviews was moral courage, often described as "doing the right thing." Examples of this ranged widely, from an officer speaking his mind even when he knew his boss did not want to hear it, to defending subordinates from unjust actions, to having the courage to tell a lieutenant he should not be in the Army.

Moral courage was frequently linked with communication skills, woven together in the officers' descriptive imagery. Some officers valued communication skills in isolation as "the ability to communicate effectively and clearly to a wide range of people," up and down the chain of command. But most often communication was tied to candor, a willingness to stand up for what was right regardless of whom the officer was speaking to. People who were seen to epitomize Army professionalism included those who were willing to tell the truth even when it was unpopular, or confront their superiors over important issues, or stand up for their subordinates. For example, one officer identified his former battalion commander as his exemplar of the profession because he had pulled their unit from a Grafenwoehr rotation when it was not properly prepared.

Although some officers mentioned the importance of "doing what's right even when no one is looking," most connected moral courage with being a role model for junior officers and soldiers:

> Because here you have folks that are coming in off the street and in many cases nobody has ever talked to them about the basic values that they should lead their life by. Their first examples of what it means to be a professional soldier, and naturally officers are soldiers but also professional officers, are consistent with what they've seen in basic training or what they've heard in basic training. And that becomes internalized to them and they understand that it's not just rhetoric and also they aspire to be like the best example of those core values, which in most cases should be the officers. We really are the standard-bearers of the values of the Army.

Officers also described physical courage as an important professional value. In these discussions, they often mentioned people they admired who were willing to risk their safety and lives for a cause they believe in. These exemplars of courage included historical figures such as King David, contemporary figures such as Lance Armstrong, the great cyclist, as well as people the officers had known personally who had risked themselves for others.

> There was a battery commander that I went to the desert with who put a vehicle in front of me and ran over a mine and . . . it didn't cause a lot of damage to the vehicle, but he was injured and . . . he didn't worry about himself. He ran over a mine and he knew that this had happened and he immediately got out of the vehicle and went around to the driver's side of the vehicle to see if his driver was all right without looking at himself, considering himself, so that was something that was admirable.

The concept of duty was another frequently mentioned professional value, often described as "getting the job done" without hesitation or comment regardless of what it takes to do it. Officers identified this quality through their objects and exemplars as a central facet of Army professionalism. One officer brought in a book and described its personal significance in this way:

> *A Message to Garcia* is a very short story that was published in a magazine in 1899, I believe. And it's the story of Lt. Andrew Rowan, who during the Spanish-American War, was told to take this message to Garcia. And he took the message and he got the message there. And it's talking about duty and accomplishing your duty and not asking, "Well, where is he at?" "Where is he at?" the famous quote within the book.

> You know he didn't stop and say, "Where is he at? How do I get there? What's the message say? Why does he have to get the message? When does he have to get it?" It was just get the message to Garcia. And it's a very controversial kind of essay but it's something that I always think about. I don't always need to know where I'm going with that message or what's in the message. But I need to be confident that I'm taking the message for a good reason. *And how do you get that confidence?* By having confidence in the leader who gave me the message. And unfortunately I've had a lot of confidence in my leaders, good and bad, and that confidence is eroding, it really is. I'm starting to ask why I'm taking the message.

More than duty, commitment to the Army and the profession were often mentioned as important values. It was during these discussions on this topic that officers pointed out the inconsistencies they saw between the Army's recruiting programs and its base values.

> And a profession is something you begin that is a career not just a stepping stone. And one of my thoughts that I find amusing . . . is that the Army nowadays attempts to bring people in [with mercenary appeals] . . . "To get paid for four years of college you only owe four years." So people are doing the four years of college and giving them the four years as good as they can, going all out and then saying "Okay, I am ready to go step on to my next point of life." Because the Army sold this bill of goods as "Hey, four years of college, four years of leadership experience, the civilian world is going to love you." And then so you go "Okay, I am going to go and see if the civilian world loves me" and the Army acts hurt.

Discussions of values differed across ranks. Young officers often described it as follows: "I do the right thing. I show up. I mean I am in the right place in the right uniform." More experienced officers however described duty as a more difficult task, e.g., giving up weekend family time for unexpected work requirements. One of these officers described this process of professional maturation in this way:

> In my mind the only professionals, true professionals, you are going to have in the military are the masters. I kind of look at lieutenant colonel, colonel level as about where you get to that. Everything below that may be a journeyman at best, certainly at other levels in the enlisted . . . they are apprentices. . . . The apprentices and the journeymen are not the masters, and I think it's the Brigade Command level where the master is really. . . . That's where we truly understand the implication of our profession and then can implement that into policy whether it be on the battlefield or shaking the organization to face changes in the world. . . .
>
> Not that everybody can't be imbued with the attributes of professionalism that we want them to. An apprentice by serving time is going to develop and strive to attain those attributes and professionalism and loyalty and duty and selfless service, personal integrity. So they are going to be growing those things that become part of their person. They now have a grasp and are living out the professional ethic but they aren't a professional yet because they just don't have the education and experience to be a master.

Lifestyle. Many officers pointed to the uniqueness of the Army lifestyle as evidence of their profession, particularly its all-encompassing nature. Although not always specifically mentioned, lifestyle elements were observable in the officers' indications of physical characteristics that set Army officers apart from the rest of society, from berets and Stetsons to physical standards. Others focused on the seamlessness between Army work and personal life—living on post, socializing with other officers, being "on duty" twenty-four hours a day, seven days a week, as this officer viewed it:

> It is unique because you are in your own society. When you . . . change jobs or you cease to be a doctor or you cease to be a lawyer, you don't "get out." When you leave the military, you "get out." That is a common phrase. "I am getting out." And it is its own society and I think that is what truly makes it a unique-type job because you are influencing the lives of soldiers and trying to develop them professionally in a job that has the utmost magnitude of importance which most of us have never experienced which is part of the problem. Because it is hard for us to imagine a mission that we have never done. I think that is what is most unique. You are in your own society of leading soldiers and this is all they have. They live on post. They eat and breathe Army. There are plenty of professions out there where people give as much time to their jobs, definitely so for the most part. But setting the example and never wavering, never letting soldiers see you get really flustered and frustrated, I think it is different. I don't know if it is more stressful or deeper than being a doctor or whatever.

The Profession's Jurisdictions

Prior to this study, we did not consider the family to be a competitor within any Army jurisdiction; however, we do today. The officers that we interviewed clearly identified their spouses and families as powerful agents who, although outside the control of the Army, affect critical career decisions and choices. This internal jurisdiction, the management of the Army's human resources, is an essential one for the Army.[19] Family members often seek to control their military spouses, sometimes to the point of forcing resignation or retirement. In the past, when relatively few wives held jobs outside the home, these families did not have much power in a jurisdictional conflict with the Army. They do now; the changing role and power of women in our society coupled with a new recognition of the importance of personal relationships with children have turned the tables on the Army. The needs of the Army no longer trump the needs of families.

Most officers, both single and married, repeatedly and consistently mentioned their family and (potential) spouse as being one of the most powerful forces bearing on their continuance in the profession. Many of the symbolic objects that the officers brought to the interviews reflected this substantial role of the family in the professional equation, for example, calendars demarcating time spent away from loved ones and photos of children. A large number of officers described the personal conflict they felt as they were pulled between the extraordinary demands of their profession and their own desire to be an active part of their families' lives. Numerous officers echoed the comment, "I never want to look back on my life and say that my career was more important than my family." One officer described his deployment to the Balkans this way:

> But the opposite side of the coin was that as much as it was professionally rewarding there was all this time flying by that I wasn't . . . and all these pictures my kids were doing and sending me and part of their life that I was missing. Just at the same time that I was having more fun than I could have imagined applying myself in this profes-

> sional regard, everything was on hold in Heidelberg. . . . [My wife] and the kids and, you know, kids losing teeth, kids finding out about sex for the first time, really it was explained to them in my absence and that's just when it came up, when they started asking those questions. And I mean that's just broken bones and school concerts and kind of all captured by these pictures. Just personal life happening without me. My family's life going on without me. So that's the sacrifice of being in this profession.

Many officers expressed greater frustration about this Army-family conflict when it arose from non-deployment-related requirements, ranging simply from the administrative demands of their jobs to post-deployment National Training Center rotation schedules. As one recently married officer said:

> I've been married for 24 months and I've been gone for 14. So the time that I have with my family is crucial. But unfortunately I find myself getting there early and meeting other people's timelines that aren't there early until 1700, 1800; and as an average I never get home till 1930, 2000 back in garrison. So unfortunately I don't feel like I have much control over that at all. And it definitely, for my wife, it's really affected her desire for me to stay in the military. And she seriously wanted me to consider whether or not my kids are going to know me or know of me 20 years from now. And I do see a lot of, unfortunately, a lot of the senior leaders who I think are really good military officers who don't have very good families at all.

Some officers indicated their priorities had changed over their career, shifting from an Army focus to placing their families first. One officer described the effect of focusing too little on his wife when he was a young officer and how he has changed:

> *How does keeping your family first balance with your profession?* I got to tell you I didn't know how to do it when I was younger and that is why I lost my first wife but now it works out really well. You make some calls. You make some decisions. I am not going to stay here and work on this report tonight. I'm going home to see my wife. I can work on the report tomorrow or I can . . . pass it off, give it to somebody else. You just realign your calls. . . . When we went to Saudi I was afraid of changing my outlook. *Changing your outlook from?* The Army first, family second. But it's been the best thing I've ever done simply because I feel now it is the right thing to do. If I'd have been that mature as a 2nd lieutenant, I probably wouldn't have lost my first wife.

The officers' interest in and devotion to the profession often reflected, or even hinged upon, spousal support. The involvement of their spouse in their professional decisions throughout their careers was significant. For example, a Medical Service Corps officer described his fiancée's role in his branch selection:

> Oh the one other thing I was going to tell you about was the Ranger Challenge team. That was the third thing. *That dominated your life back then?* Yes, sir. My wife now was my girlfriend back then and she hated it. *She never saw you?* She hated it and like I got into the Army and told her, "I want to go Special Forces." I'm like "I've got to go Special Forces." I'm like Special Forces and Delta Force, seems to be like the last safe haven for warriors and people who don't get screwed with picking weeds out of the volley ball court anymore. But then she started talking to some of my buddies' wives who are in the Group and they are like "Oh, I've been single the five years he's been in the Group." And she's like "You are not going Special Forces." A good challenge for life? She is like "No, you will not. It will be even worse than the Ranger Challenge Team." She hated it. We didn't have a social life. I mean here we are in col-

> lege in the prime of our life and on the weekends I would just be like "Oh, if I was there at all." So yeah she wasn't too wide about it. *What does she think about it now?* She is honestly counting the days down until I get out.

Additional jurisdictional competition emerged as an issue during the officers' discussions about the various kinds of missions the Army is engaged in. While some officers noted that they did not care for the broadening range of missions being undertaken in recent years, most indicated that they did not mind or had even benefited personally and professionally. At one end of the continuum, officers were receptive to diverse missions being part of the profession:

> I'd say something like the military is a profession. We're professional soldiers. And what we do is, we're here to make sure that we're ready to fight our nation's wars, and we also do other things such as fight wild fires and do other public good, hurricane relief, things like that. It's all part—if we weren't professionals then we wouldn't be able to provide that service and we'd probably be challenged by some other country if we weren't professional in a military way.

At the other end of the continuum, they described fighting and winning wars as the sole focus of the profession:

> I think that has a definite impact on the hurting of the ideal of professionalism in the Army. You have Bosnia. I am a compassionate person. The whole Bosnian Serb and the Croats and the Muslim factions over there and all the genocide. Yes, on a personal level that affects me but if you look at what the Army is for, I feel that we have no business over there. . . . I mean it hurts when you join the Army to uphold and defend the Constitution of the United States and then you are sent for six months away from your family for a no-conflict because those people are smart. They are not going to do anything while we are over there. The day after we finally say "Oh, everything is okay" and leave, that is when they will re-begin the atrocities and everything like that. But they are not going to just do it while we're there and they haven't.

But most saw humanitarian missions, including peacekeeping, to be in line with what it means to be a professional officer:[20]

> This is a camel-riding crop. A nomad, camel herder, in Somalia made this for me. Every morning . . . I used to toss this guy a little water every now and then. . . . And you wouldn't see him the rest of the day. But I would see him every morning about daybreak and so one morning I threw him over a quart of water and he threw this stick to me. I think he carved it because it is pretty uniform. . . . I have heard people say that you know we are not the humanitarian Green Peace of the world. But I'll tell you, if you go on one of those missions and you see the people. And Somalia was an ugly thing. But anyway, when you see people like that and little kids starving, it was worth going over there to me and doing that and being part of that.

Conclusion

Our intention in this chapter was to examine the Army profession from the perspective of a key segment of its membership, the officer corps. We have sought to organize this investigation using Andrew Abbott's theory of professions, which emphasizes a system of professions in which each profession controls a unique body of abstract knowledge, conducts tasks based on that knowledge, exerts behavioral control over members, and jostles for jurisdiction with other occupational groups. We gathered data to speak to the profession using a proven methodology for tapping into the authentic thoughts and feelings of the Army profession's members.

This methodology as used in this research enabled us to acquire an in-depth and thoughtful view into the minds of Army officers. The approach allowed officers to dictate which issues got raised and define which issues were important regarding what it means to be a professional officer. Therefore, rather than responding to pre-set questions or other participants' opinions, officers were able to indicate their individual and genuine priorities. The analysis of these data illuminated consistencies across accounts that inform our understanding of Army officers and the study of professions in general.

Our analysis led us to six major conclusions: (1) the officer corps does not have a shared emphasis on or understanding of the Army's unique abstract knowledge, which is managing violence; (2) this lack arises from a failure to focus on the Army's unique abstract knowledge and an over-emphasis on another profession's abstract knowledge, that of social work; (3) the Army is bureaucratizing its professional expertise into checklists and forms, thereby reducing the expertise of its officers, frustrating them in their attempts to apply their knowledge and experience, and quite likely decreasing the effectiveness of professional decisions; (4) social or behavioral controls over officer's conduct are essential yet weaker than Army officers consider ideal; (5) most officers are open to broadening the Army's jurisdiction across the spectrum of conflict; and (6) the Army and officers' families are in competition over the officers' professional commitment: this remains an unresolved jurisdictional conflict. The foregoing conclusions can be aggregated into the following fundamental issue areas: the Army profession's knowledge and tasks; professional expertise; social or behavioral control; and the Army's jurisdictions. Let us discuss each in turn.

The Profession's Knowledge and Tasks

Key to understanding what it means to be a professional is the connection to and application of abstract knowledge. Although the respondents spoke to this issue in compelling ways indicating they felt strongly about the content of the Army's abstract knowledge and their professional responsibilities in its application, there was no agreement on the Army's *unique* abstract knowledge. Rather, officers focused heavily on a domain of abstract knowledge, taking care of people, that is dominated by another profession, namely social work. We found it striking that only a small number of officers focused their discussion on the Army profession's unique area of expertise, managing violence or killing, indicating it is not broadly viewed as the central, unique task of this profession.

Instead, most officers emphasized taking care of soldiers as the key element of their professional knowledge, either in terms of the amount of time they spent on the task and/or in terms of the priority it has in their minds. These data indicate that taking care of soldiers has become an end in and of itself for most officers, disconnected from the mission and unrelated to the profession's formal training in killing and managing violence. This disjunction leaves the Army's abstract knowledge base, managing violence, vulnerable to direct competition from other occupational groups and places the Army in direct competition with social work and other occupations over the subordinate knowledge regarding taking care of soldiers and families.

Our recommendations regarding the Army's professional knowledge emphasize making officers more aware of the Army's unique abstract knowledge through existing socialization mechanisms and educational programs. Specifically, senior leaders need to keep the unique task of managing violence central to officers at all levels and determine the emphasis officers should place on taking care of soldiers relative to other key professional tasks and activities. If managing violence is to rise in importance, then officers should be encouraged to talk about this knowledge in the context of what it means to be a profession through officer professional development programs, counseling sessions, and informal mentor opportunities. If taking care of soldiers is to be reduced in relative importance for officers, then senior leaders must determine who does it, under what circumstances, and to what extent. If, on the other hand, it is to be maintained or increased in emphasis, senior leaders should exploit it by formally incorporating it into the profession's expert knowledge and making the officer corps thoroughly aware of this change.

Professional Expertise

The details surrounding the application of a profession's abstract knowledge are as important as its content. Since professions differ from other occupational forms largely because they embed expertise in human beings rather than machines or organizational systems, ensuring that professionals are free to develop and then apply their professional expertise is essential for a profession's success. If instead it is organizational systems or machines that determine how, when, and where expertise is applied, then the occupation is not considered a profession and its members are not professionals. Furthermore, if the tasks truly are professional in nature, then the limitations of organizational systems and machines will result in less effective decision-making.

Our data indicate a clear encroachment of organizational management systems or bureaucracy into the domain of the professional. The issue of the application of professional expertise elicited the strongest emotional response from officers, indicating an understanding on their part of the importance of professional discretion. They felt thwarted by centralized organizational systems that kept them from exercising their core sense of being a professional, their expertise. As one officer stated:

> This is the way one soldier put it to me, the thing I don't like about the Army, he said, was the Army says if one guy steps on the nail we're going to give everybody in the Army a tetanus shot. And I thought there was a lot of insight in that statement, because instead of holding commanders responsible for the command climate they have, instead of court martialing those who have failed, instead of removing them from those things, instead we set up a whole bunch of centralized programs to make sure that no one will fail. And then it's really not for the purpose of making sure no one will fail. It's for the purpose of making sure that we look good in the end.

Moreover, these officers described the very outcome that arises from the bureaucratization of professional tasks—decreased effectiveness—when they worried that the lives of their soldiers and others would be risked because of their inability to exercise their expertise.

From our perspective, it is essential that the Army's senior leaders, through their own actions and organizational systems, establish trust in Army officers' ability to exercise the professional expertise appropriate to their rank. This will require that senior leaders scrutinize and vastly reduce the centralized systems and programs that hinder officers in these efforts. If Army professionals are to be able to act as professionals, the Army's organizational systems—operations, personnel management, training, research and development, etc.—must be conformed to support this.

Social (Behavioral) Control

In response to the discretion they are given by society, all professions enact some degree of behavioral control over their members to ensure that professional expertise is applied ethically. They use a variety of means to control their members' behavior including codes of ethics, espoused values, legal mechanisms, and regulated lifestyle. In an operational environment, they enforce the Geneva Conventions and the civilian control of the military, insist upon compliance with rules of engagement, indoctrinate members to protect innocent citizens, etc. These controls are essential for the military professions, which are endowed with a society's violent power, its military machine designed to kill human beings and conduct wars. Not only is social control important for the military but it is also one of the greatest challenges since it seeks to ensure ethical behavior from soldiers in times of chaos, uncertainty, danger, and fear. In fact, being ethical may conflict with a soldier's innate instinct for survival and self-defense. Therefore, to accomplish its mission, the Army must have very powerful and consistent social control mechanisms.

Our data indicate that Army officers clearly understand the importance of social control mechanisms in their profession. However, their interviews also indicated that they did not feel the Army's present mechanisms were adequate. Moral courage dominated the officers' discussions of values and their selection of professional exemplars. Bluntly, they desire candor. They want the freedom to speak and act without penalty, and they want honesty from their superiors, both immediate and the most senior. But many of them felt the Army's senior officers were not speaking candidly, neither to the profession's members nor to those important constituencies outside the profession.

Furthermore, unique Army lifestyle elements such as post housing, officers' clubs, and uniforms clearly mattered to officers in terms of their profession. These items serve both to publicly reinforce that the Army is a unique profession and to facilitate the sharing of values, thereby strengthening the profession.

Thus our recommendations focus on making senior leaders more cognizant of messages they are sending, directly and indirectly, to the profession's members. Modern media, such as C-SPAN and the Internet, enable officers at all levels to see and hear the words of senior leaders in all venues, including testifying before

Congress. Inconsistencies between what these officers experience in their daily lives and the messages they hear from senior leaders weaken their confidence in the profession's values, especially moral courage. Recruiting commercials are one of the most critical means of communicating professional values *to the profession*, as well as to the American public and potential recruits. If the messages in these commercials contradict the profession's values (and the reasons are unexplained), then the profession's social control mechanism and status as a profession will be weakened.

Senior leaders should also consider establishing or reestablishing more lifestyle programs, such as officer and NCO clubs. When faced with lifestyle program decisions, these leaders must consider the impact on the profession.[21] Such programs are very important to officers, giving them access to each other so that they might share their values and serving as a touchstone, a public display of the profession's uniqueness.

The Army's Jurisdiction: Focus and Family Tension

Professions compete for control over jurisdictions where they are free to apply their abstract knowledge. Over the last decade, the Army has engaged in an extensive debate regarding the boundaries of its jurisdiction and the appropriateness of its involvement throughout the spectrum of violence. Some have argued that war is the only suitable jurisdiction for the Army; others have supported a broader jurisdiction that includes activities across the spectrum of conflict. From a professional perspective, the debate deals with changing the size and contours of the Army's jurisdiction—whether it should be narrowed, perhaps risking irrelevancy to the nation, whether it should be broadened, perhaps encroaching on other occupations, or whether it should be resized and reshaped into a configuration falling somewhere in between. Resolution of this debate should take place in negotiation between the profession's senior leaders and the nation's elected officials. However, the perspective of the profession's members should certainly be taken into account.

Our respondents indicated that most officers are comfortable undertaking a variety of missions involving potential or actual destruction, damage, injury, or death. The profession's key tasks—managing violence (be it from human or natural causes) and caring for others (be they soldiers or civilians)—enable them to apply their professional expertise to a range of missions. Therefore, for most officers humanitarian missions are consistent with the skills and knowledge developed within the profession and contribute to the values essential for professional control. From this perspective, the jurisdictional boundaries of the profession are in flux, with the members' sentiment being to expand.

Finally, it is quite interesting that although the topic was the military profession, most officers at some point in their interviews broached their families, or lack thereof, as being integral to the discussion. What does it mean that so much control over the commitment of individual professionals resides outside of the organization? Clearly, the profession no longer dominates its members' commitment and their associated career decisions. In essence, there is a jurisdictional battle between the family and the profession.

In regard to jurisdiction, we recommend that the Army's senior leaders think strategically about the leadership of the profession, its jurisdiction, and competitors. They must decide and then negotiate the profession's boundaries, cognizant that most officers believe humanitarian and peace-keeping missions are within the profession's domain. Furthermore, they must enable members of the profession to take care of their families, i.e., reduce workloads to give them time to do this, rather than establishing centralized programs that do it for them. It will also be important to find ways to directly involve spouses in career decisions and discussions since these are joint decisions with both partners essentially having veto power.

Future Research

As a rich mosaic of officers' views of what it means to be a professional Army officer circa 2000, this study provides a reliable starting point for research on the profession as experienced by individual officers. Future research could effectively build on this study in three main ways:

- Focus on understanding the tasks of the profession better. How much time, energy, and feeling do officers expend on given tasks? Are the tasks central to the abstract knowledge of the profession? For a time study of tasks, methodologies geared to structured self-reports or that employing ethnographic techniques would provide the most accurate data.
- Make the research longitudinal. Do officers' views of themselves as professionals change over time? In what ways? As changes in mission, the economy, and public support occur over time, it would be worthwhile to conduct the same type of study as reported here—two, five, or even ten years out. This longitudinal study of officers' views would, in effect, be a study of peacetime versus wartime/post-wartime understandings. It would be an invaluable investigation of the role of abstract knowledge in an evolving profession.
- The research could be broadened. It would be beneficial to compare the beliefs and feelings of U.S. Army officers to others in the military profession. Studying officers across services and internationally would help to discern which issues are fundamental to the profession and which ones are of particular relevance to the American Army.

Studying what the Army profession means to officers provides key insights into how to strengthen the profession in terms of its knowledge, tasks, control systems, and jurisdiction, as well as reinforcing the reality that the Army is indeed a profession. It is worth underscoring that these leverage points involve expanding the connection officers have to their profession, through underscoring its core abstract knowledge and emphasizing the relevance of certain tasks, rather than expending money on better technology or higher pay. The education of the profession in this sense requires foresight, vision, leadership, and diligence, not necessarily a larger budget.

Notes

1. Field Manual 100-5, *Operations* (Washington, DC: Headquarters, Department of the Army, 14 June 1993).
2. William G. Ouchi, "Markets, Bureaucracies, and Clans," *Administrative Science Quarterly* 25 (1980): 129-141; and "A Conceptual Framework for the Design of Organizational Control Mechanisms," *Management Science* 25 (1979): 833-48.
3. For physicians, see T.J. Hoff, "The Social Organization of Physician-managers in a Changing HMO," *Work and Occupations* 26 (1999): 324-51; nurses, see L.H. Aiken and D.M. Sloane, "Effects of Specialization and Client Differentiation on the Status of Nurses: The Case of Aids," *Journal of Health and Social Behavior* 38 (1997): 203-22; lawyers, see J.W. Stempel, "Theralaw and the Law-Business Paradigm Debate," *Psychology, Public Policy, and Law* 5 (1999): 849-908; social workers, see R. Fisher, "Speaking for the Contribution of History: Context and the Origins of the Social Welfare History Group," *Social Service Review* 73 (1999):191-217, M.S. Twui and R.K.H. Chan, "The Future of Social Work: A Revision and a Vision," *Indian Journal of Social Work* 60 (1999):87-98, B.S. Vourlekis, G. Edinburg, and R. Knee, "The Rise of Social Work in Public Mental Health through Aftercare of People with Serious Mental Illness," *Social Work* 43 (1998):567-75, and Andrew Abbott, "Boundaries of Social Work or Social Work of Boundaries?" *Social Service Review* 69 (1995):545-62; vision care providers, see F.F. Stevens et al., "The Division of Labour in Vision Care: Professional Competence in a System of Professions," *Sociology of Health and Illness* 22 (2000): 431-52; and funeral directors, see S.E. Cahill, "The Boundaries of Professionalization: The Case of North American Funeral Direction," *Symbolic Interaction* 22 (1999): 105-119.
4. For financial planners, see R.C. Cohen, "Who's Planning for Your Future? Jurisdictional Competition Among Organizations and Occupations in the Personal Financial Planning Industry" (unpublished diss., 1997); for tasks in social work, see Y.M. Johnson, "Indirect Work: Social Work's Uncelebrated Strength," *Social Work* 44 (1999):323-34; and on expertise and professional education, see Fisher, "Speaking for the Contribution of History," and D.M. Austin, "The Institutional Development of Social Work Education: The First 100 Years and Beyond," *Journal of Social Work Education* 33 (1997):599-612.
5. This study is purely qualitative in nature; our data arise solely from the words that officers used to express their feelings and thoughts about their profession. For readers more accustomed to quantitative research, our findings may appear vague and imprecise because they are not presented as percentages or numbers with significance levels. Instead, we will support our conclusions with statements such as "many officers" or "a few respondents." However, we want to assure our readers that these data have been rigorously collected and thoroughly analyzed using state-of-the-art qualitative software, in this case *Atlas ti*, and methods, such as those presented by Matthew B. Miles and A. Michael Huberman in *Qualitative Data Analysis: An Expanded Sourcebook* (Thousand Oaks, CA: Sage Publications, 1994). As a result, this study reflects the strengths of qualitative research—it provides a rich description of reality in today's Army officer corps and the meaning that officers make of their profession. This description provides an opportunity to draw conclusions and develop hypotheses unfettered by preconceived notions and expectations.
6. U.S. Army War College, *Study on Military Professionalism* (Carlisle Barracks, PA: U.S. Army War College, 1970).
7. Ibid.
8. Ibid.
9. D. Morgan, "Focus Groups," *Annual Review of Sociology* 22(1996): 129-152.
10. G. Zaltman, "Rethinking Market Research: Putting People Back In," *Journal of Marketing Research* (November 1997): 424-37.

11. B.E. Ashforth and F. Mael, "Social Identity Theory and Organization," *Academy of Management Review* 14 (1989): 20-39.
12. We focused this initial study on Active Component officers because of their traditional role in the leadership of the Army profession. However, the other elements of the Army profession—Reserve Components and noncommissioned officers—are also important and worthy of future study.
13. For detailed information concerning the interview protocol, including copies of the assignment letter, contact the authors.
14. If the participants did not bring in any items, the interviewers based the discussion on three issues the participant suggested regarding the profession.
15. Miles and Huberman, *Qualitative Data Analysis.*
16. T. Muhr, *Atlas.ti, version 4.1 for Windows 95 and Windows NT*, Scientific Software Development, Berlin.
17. This response and all those following are generally verbatim renderings of the respondents' oral remarks. In a few cases, responses have been emended to a very slight degree on behalf of clarity.
18. This finding is quite similar to those found in the *Army Training and Leadership Development Panel Report (Officers): Final Report* (Ft. Leavenworth, KS: Combined Arms Center, 2001).
19. See a more extensive discussion of internal jurisdictions in Chapter 1.
20. For additional research on officer feelings about non-traditional missions, see Deborah Avant and James Lebovic, "U.S. Military Attitudes toward Post-Cold War Missions," *Armed Forces & Society* 27 (2000):37-56.
21. Many of these lifestyle programs now fall under the Army Well-Being Program. See Chapter 22, Case Study No. 1, for more about this program.

II

The Profession's Military-Technical Expert Knowledge[1]

Before we can focus on the future of the Army profession, we must understand its current state. As we discussed in Chapter 1, professions are all about expert knowledge and the creation of human expertise to apply that knowledge to new situations. The next five sections of this book (II through VI) establish a benchmark for the Army profession's current expert knowledge. We do not presume that these five sections provide an exhaustive map of this expert knowledge, but they do highlight important areas of knowledge and, in so doing, raise serious questions for consideration by all members of the Army profession.

We begin this section with four chapters about the Army's military-technical knowledge—how the profession prepares for and conducts land operations combining Army soldiers with organizations, doctrine, and technology. Several themes arise from these chapters, as recapitulated in the paragraphs below.

First, although all authors initially focused on the Army's expert knowledge, they were quickly led toward work jurisdictions and the Army's ability to compete within them. This approach necessitated viewing the Army as existing within a system of professions.[2] For the purposes of this project, the Army's expert knowledge should thus be viewed only relative to jurisdictional competitions rather than in absolute terms. In other words, we are not asking the simplistic question: "Can the Army effectively win wars with its present state of knowledge?" Rather, we are asking a far different question: "Within the same jurisdiction, is the Army's expert knowledge more applicable to defense-related problems than that of its professional competitors?" Other agencies and organizations are competing for legitimacy within the Army's jurisdictions, so if it is to maintain control within these arenas, it must both establish and defend the superiority of Army expertise in these jurisdictions.

The second theme highlights the effect of two current trends—jointness and privatization—on these jurisdictional competitions. These forces are strongly influencing the Army's claim on its professional knowledge, its professional practices, and the boundary distinguishing members of the profession from others. All of the authors in this section, based on extensive research, expressed concern that the Army had neither recognized the power of these forces nor critically and adequately thought through its own responses.

Lastly, a very sobering theme centered on the questions of whether Army officers were cleaving into two types, perceived as either "thinkers" or "doers," and, if so, whether the professional future for each was equally bright. Since knowledge is the foundation of professions—expansion of that knowledge being fundamental to a profession's evolutionary success—it is essential to have valued members whose role is to create and develop expert knowledge *in addition* to those who apply professional expertise. If the Army is to flourish as a profession, both types of Army professionals need to be equally esteemed, and to have equally bright futures. Unfortunately, this is not the case today, nor without deep cultural change is it likely to be so in the future.

All three of these themes reinforce the necessity for the profession's strategic leaders to renew the profession's identity and to communicate this identity in multiple and enduring ways to its members during the Army's transformation.

Notes

1. The authors express their appreciation to Maj. Edmund M. Ackerman of the USMA Department of Social Sciences, who served as Panel Rapporteur for this section's topic during the USMA Senior Conference, June 2001.
2. See Andrew Abbott's *The System of Professions: An Essay on the Division of Expert Labor* (Chicago, IL: University of Chicago Press, 1988).

6 Professionalism and Army Doctrine: A Losing Battle?

James A. Blackwell

Army doctrine, and the Army's doctrinal process, have served the war-fighter well. "Doctrine," according to *Webster's Third International Dictionary*, is the body of principles in any branch of knowledge. For the Army it is the way the Army fights.[1] Presently, 634 publications define Army doctrine authoritatively.[2] The Army's doctrine development process is maintained rigorously by the U.S. Army Training and Doctrine Command (TRADOC), with doctrine centers at Fort Monroe, Virginia, and at each of the relevant "school-houses."[3] There are over 100 soldiers and 47 Department of the Army Civilians[4] assigned as full-time doctrine developers and writers throughout the Army, and they are working on the largest doctrinal publication list the Army has ever developed. The workload is such that other soldiers provide doctrine writing support in a matrix management process, and contractors are frequently hired to supplement the work force.

The Army's methodology for doctrinal review and revision is more rigorous than ever before. As Army units conduct training exercises at the Combat Training Centers (CTCs), Observer-Controller teams assess not only unit training proficiency in doctrinal aspects of operations, they also conduct periodic reviews of the adequacy of doctrine as demonstrated by those units. The CTCs prepare assessments of Army doctrine every eighteen months or so and make recommendations for doctrinal changes needed.[5] This includes thorough review of doctrine for non-war-fighting operations.

Training and Doctrine Command has recently completed a thoroughgoing revision of its doctrine development process and now has a Five-Year Doctrine Literature Master Plan and a new implementing regulation. The new, strengthened process includes provisions for the integration of results from experimentation efforts, joint doctrine, and future concepts, and provides for the development of doctrine for the Interim and Objective Transformation Forces.

The Army's doctrine and doctrine process have never served the war-fighter better. It does an outstanding job of teaching soldiers *how to fight*. But professional doctrine must do more than that. It must also educate soldiers *on how to think about how to fight*. In this, Army doctrine falls short.

When soldiers think of themselves as members of a *profession*, they tend to think of themselves in terms of the classic organizational approach to professionalism. Typified in Samuel Huntington's book *The Soldier and the State* (1957), Army professionals focus on organizational patterns—the idea that there is a common process of development that cuts across such otherwise disparate callings as medicine, law, accounting, religion, and the military. A profession, in this view, is an organized body of experts who apply esoteric knowledge to particular cases. They have elaborate systems of instruction and training, together with entry examinations and other formal prerequisites, and they normally possess and enforce a code of ethics or behavior.[6] The focus of understanding the nature of a particular profession is, in this view, the structure of the organization and how closely a specific group comes to reaching the ideal.

Sociologist Andrew Abbott, however, says that there is more to it than that. He maintains that a profession is an occupational group that controls the acquisition and application of various types of knowledge. His theory goes beyond the identification of the ideal to suggest that the defining quality of a profession is how it does in the competition for dominance over knowledge: "Jurisdictional boundaries are perpetually in dispute, both in local practice and in national claims. It is the history of jurisdictional disputes that is the real, determining history of the professions."[7]

In Abbott's scheme, regardless of how well or poorly a group fits the ideal model of a profession, an organization must have firm control over its knowledge base in order to compete successfully with all the other organizations that contend for the same jurisdiction. He calls that knowledge base the profession's *abstraction*, that is, how it thinks about what it does. He argues that professions are in a constant state of struggle for jurisdiction over that knowledge base, a state that he calls *the ecology of professions.*

It is in this sense that we should consider whether the Army's doctrine is serving the Army well as a fundamental element of its institutional professionalism. If Abbott's theory is right in the distinctions he chooses to emphasize—and there is every indication that he is—then Army doctrine must be more than "the concise expression of how Army forces contribute to unified action in campaigns, major operations, battles, and engagements."[8] In the context of the ecology of professions, doctrine is an occupational group's codification of the abstractions it employs to control the acquisition and applications of the various kinds of knowledge over which it asserts jurisdiction.[9] The Army claims primacy over the use of lethal force in land warfare.[10] It is losing that claim in the current competition for jurisdiction over land warfare largely because its doctrine and doctrine process do not provide sufficient cognitive power in ongoing jurisdictional disputes with rival professions.

In the general system of professions, abstraction is an essential function that any profession must master if it is to survive in the struggle over occupational control.[11] It provides the basis for a profession's inferences—its link between diagnosis and treatment. Only a knowledge system governed by abstraction can redefine problems and tasks to defend them from interlopers and seize new problem areas. It provides a profession with the strongest form of control over its jurisdiction by controlling its knowledge domain. A profession challenged by objective change in technology must have a system of abstraction to survive. Some professions employ abstraction as a strategy or tactic in the professional ecology, though the Army's institutional value structure tends to shun such stratagems. For example, in the early 1990s a consensus emerged among military analysts that a Revolution in Military Affairs (RMA) was emerging. The Air Force, Navy, and Marine Corps argued that their then-current plans, programs, and budgets in fact already embodied that revolution. They made their case in glossy vugraph presentations such as "Global Reach, Global Power," "From the Sea," and

"Operational Maneuver from the Sea." The Army, in contrast, treated the notion of a Revolution in Military Affairs as a hypothesis. It never argued that its existing programs were *the* RMA, as did its sister services; instead it made the case that the RMA was a concept yet to be demonstrated. The Army did not link its then-current plans, programs, and budgets to the RMA, but it embarked on a series of conceptual exploratory inquiries, including "Louisiana Maneuvers," "Advanced Warfighting Experiments," "Force XXI," and "The Army After Next." The Army did not engage the other services at that level of abstraction until nearly a decade later, after it had made major changes in several of its principal programs (for example the complete restructuring of the Future Combat System Program, and the design of the Objective Force). Only then did Army Chief of Staff Gen. Eric Shinseki pronounce at the October 2000 Association of the United States Army Convention the Army's abstraction under the "Transformation" label.

An effective system of abstraction must be strong enough to compete, although this does not require reaching some absolute standard of abstraction across all professions. At a minimum the system of abstraction should provide rational consistency to the system of inference and clarify the definitions of the boundaries of the profession as well as their rationale. The body of doctrine must legitimize the work of the profession by clarifying its foundations and tracing them to major cultural values. It should of course provide for research and instruction among the members of the profession. But it must also provide for the generation of new diagnoses, treatments, and inference methods (to use Abbott's medical analogy.)

It is against this measure of effectiveness that Army doctrine needs to be evaluated in order to judge the role of doctrine in Army professionalism. The effectiveness of Army doctrine as a pedagogical instrument is indisputedly best-in-class. The object of this chapter, however, is to reach a judgment on how well Army doctrine serves the profession in the competition for jurisdiction.

The outcomes of jurisdictional competition form six types:[12]

- Full and final jurisdiction, one profession winning at the expense of all others;
- Subordination of one profession to another, either cognitively or practically;
- Split jurisdiction into two interdependent parts;
- Shared jurisdiction without a division of labor;
- A losing profession is allowed an advisory role vis-à-vis the winning one; or
- Division of jurisdiction by client type.

To the extent that the Army is able to assert full and final jurisdiction as technology and organizations change, it maintains and even enhances its position in the ecology of professions. But if the Army increasingly subordinates, splits, shares, advises, or divides jurisdiction, its claims to legitimacy and monopoly over the use of lethal force in land combat erodes. Doctrine, as the systematization of the Army's abstraction about its occupational authority and control, is the essential measure of effectiveness on how the Army is competing in this milieu.

The Army finds itself today in three elemental professional competitions (see Table 6-1). Each competition forces the Army to adapt professionally as the nature of its core competence changes, and as other professions challenge the Army's traditional occupational exclusivity. The first of these is technological in nature, brought on by the emergence of the Revolution in Military Affairs.[13] The RMA represents a dramatic change in the nature of the conduct of warfare, resulting from a complex interaction among new operational concepts, innovative organizational designs, and emerging technological capabilities. The currently emerging RMA consists of new warfare areas in Long Range Precision Strike, Information Warfare, Dominating Maneuver, and Space Operations. The Army terms its engagement in the RMA as *Army Transformation.*

The second set of competitions that Army doctrine must engage in is the evolving joint character of military operations. The Goldwater-Nichols Act of 1986 created a fundamental shift in the nature of the professional competition for jurisdiction, among other changes, by legally and institutionally legitimizing the contribution of joint functions, organizations, and people to U.S. war-fighting. That the Army has found it necessary to engage in this competition was dramatically revealed in a survey of Army general officers conducted by TRADOC in 1998.[14] Respondents uniformly concluded that TRADOC, rather than publish a revision of Field Manual 100-5, *Operations,* should instead defer that revision and immediately begin to align Army doctrinal publications, including their numbers, titles, and content, with the growing body of joint doctrinal publications.

The third area of engagement is in non-war-fighting competencies. While the Army has always had a role in such operations, the decade of the 1990s, for a variety of reasons, heightened the role of the Army in them. Most Army operations after Operation Desert Storm have been of this variety.

Competitions	Competitors
Revolution in Military Affairs	Other Services Office of the Secretary of Defense Joint Forces Command
Jointness	Joint Staff Commanders-in-Chief
Nonwar	Federal Agencies Multinationals Non-Governmental Organizations

Table 6-1. *The Army's Competitions for Occupational Jurisdiction.*[15]

Army doctrine provides an objective indicator of how effectively the Army is able to assert its claims of jurisdiction in these competitions among rival professions. Trends in Army doctrinal change can be observed at the institutional level in official documents, both in terms of the titles and in the content of key writings. The impact of doctrinal change can also be observed at the individual level, although in a more subjective manner, by means of a survey of published professional writings and in the results of recent surveys of soldiers. In both approaches, the doctrinal writings themselves can be evaluated as to the emergent settlements of the three current jurisdictional competitions.

Levels of Analysis

In this chapter, I examine Army doctrine at the institutional level through two methodologies. First, the Army reveals its professional cognitive map over time by way of its Index of Doctrinal Publications. The titles of the manuals themselves and the subject areas into which they are classified change as the Army's definition of its occupational jurisdiction changes. It is a straightforward process to identify the deletions and additions as a quantitative measure of change in Army doctrinal coverage, then to subjectively classify the implications of those changes along the dimensions of the three jurisdictional competitions. In this analysis I have examined this index in decennial increments from 1940 to 2000.

The second measure of Army abstractional adaptation is in the content of the doctrinal statement contained in Field Manual 100-5, *Operations*.[16] This is the "heartland of work over which it has complete, legally established control, legitimated by the authority of its knowledge."[17] The published versions of FM 100-5 over time form a data set from which a classic content analysis can reveal trends in terms of the three jurisdictional competitions. In this chapter I have limited the content analysis to the 1982, 1986, and 1993 versions of FM 100-5 and the October 2000 edition of United States Army Command and General Staff College Special Text 3-0 (ST 3-0). Because FM 100-5 was in formal revision by the Army, ST 3-0 was the only authoritative source of current Army doctrine on *Operations* available publicly during the conduct of this research. ST 3-0 therefore served as the most recent document for the purposes of developing this trend analysis.

At the individual level of analysis, two data sources provide some insight into the effectiveness of Army doctrine in the ecology of professions. First, the Center for Strategic and International Studies commissioned the Center for Creative Leadership in Greensboro, North Carolina, to conduct a survey of attitudes among military professionals on a number of cultural issues, some of which relate to the jurisdictional competitions considered in this analysis. Secondly, there is a robust professional Army literature available, primarily in the publications *Military Review* and *Parameters*, which, though journals of ideas as opposed to outlets for prescribed policy, nonetheless reveal which way the doctrinal winds are blowing.

Institutional Level: The Army's Cognitive Map

In 1940, before the outbreak of World War II, the Army had perhaps forty-five identifiable field manuals.[18] Numbering was not consistent, with separate publications variously identified by title, volume, and chapter as well as by number. This system was the culminating point of the Army's doctrinal transformation after World War I. That process resolved ongoing jurisdictional disputes within the Army among its infantry, cavalry, and field artillery branches (infantry emerged dominant at that time), producing the first comprehensive codification of the Army's jurisdictional abstraction in the Field Service Regulations of 1923, the

Old Series Number	Subject Area	1945	1970	2000	New Series Number	New Subject Area
1	Aviation	15	8	15	3-04/3-xx	Aviation
2	Cavalry	6	0	0		
3	Chemical Warfare	8	2	14	3-11/3-xx	Chemical (NBC)
4	Coastal Artillery	59	0	0		
5	Engineer	16	18	40	3-34/3-xx/4-04	Engineer
6	Field Artillery	24	129	19	3-09/3-xx	Field Artillery
7	Infantry	9	3	13	3-21/3-xx	Infantry
8	Medical	7	6	26	4-02/4-xx	Medical
9	Ordnance	6	4	6	4-xx	Ordnance
10	Quartermaster	3	5	86	4-xx	Quartermaster
11	Signal	6	16	9	6-x/6-xx	Signal
12	Adjutant General	1	2	2	1-x/1-xx	Adjutant General
14	Finance	0	0	1	1-06	Finance
16	Chaplain	0	3	1	1-05	Chaplain
17	Armor	26	7	9	3-20/3-xx	Armor
18	Tank Destroyer	9	0	0	3-xx	Management Information Systems
19	Military Police	2	11	9	3-19/3-xx	Military Police
20	Miscellaneous	1	10	3	3-xx/7-xx	General
21	Individual Soldier	19	20	14	3-xx	Individual Soldier
22	Infantry Drill	1	3	4	3-xx/7-xx	Leadership, Courtesy, and Drill
23	Basic Weapons	21	31	15	3-xx/7-xx	Weapons
24	Communications Procedures	14	7	12	6-xx	Communication Techniques
25	Transportation	4	0	5	7-x/7-xx	General Management
26	Interior Guard Duty	1	0	0	7-xx	Organizational Effectiveness
27	Military Law	3	2	4	1-04/1.xx	Judge Advocate/Military Law
28	Welfare, Recreation and Morale	2	0	0		
29	Combat Service Support	0	15	0	7-xx	Composite Units and Activities
30	Military Intelligence	13	17	0	2-x/2-xx/3-xx	Military Intelligence
31	Special Operations	7	29	9	3-05/3-xx	Special Forces
32	NOT USED				7-xx	Security
33	Psychological Operations	7	29	9	3-53	Psychological Operations
34	Intelligence	0	0	20	2-xx/3-xx	Combat Electronic Warfare and Intelligence
35	Women's Army Corps	1	1	0		
36	NOT USED				3-xx/4-xx	Environmental Operations
38	Logistics	0	19	2	4-xx	Logistics Management
39	NOT USED				3-xx/7-xx	Special Weapons Operations
40	NOT USED				3-14	Space
41	Civil Affairs	0	2	1	3-57	Civil Affairs
42	Quartermaster	0	0	2	4-xx	Supply
43	NOT USED				4-xx	Maintenance
44	Anti-Aircraft/Air Defense	10	30	10	3-01/3-xx	Air Defense Artillery
45	Censorship	0	2	0		
46	Public Affairs Operations	0	0	1	3-61	Public Information
50	NOT USED				7-xx	Common Items of Nonexpendable Material
51	NOT USED				3-xx.x	Army
52	NOT USED				3-xx.x	Corps
54	Higher Echelons	0	8	2	4-xx	Logistics Organizations and Operations
55	Transportation	6	17	17	4-01/4-xx	Transportation
57	Airmobile/Pathfinder	0	3	2	3-xx	Airborne
60	Amphibious	1	1	0	3-xx	Explosive Ordnance Disposal Procedures
61	Divisions	0	2	0	3-xx.x	Division
63	Combat Service Support	0	0	9	4-x/4-xx	Combat Service Support
67	NOT USED				3-xx	Airmobile
70	Mountain and Winter	2	0	0	7-xx	Research, Development, and Acquisition
71	Division Operations	0	0	7	3-xx.x	Combined Arms
72	Jungle	0	1	0		
74	NOT USED				7-xx	Military Missions
75	NOT USED				7-xx	Military Advisory Groups
77	NOT USED				3-xx	Separate Light Infantry
90	Operations	0	0	17	3-xx/3-xx.x	Combat Operations
97	NOT USED				7-xx	Division (Training)
100	Operations	4	5	37	3-x/3-xx.x	General Operational Doctrine
101	Staff	4	9	3	5-x/5-xx	Planning/Staff Officers
105	Umpire	2	3	0	3-xx	Maneuver Control
145	NOT USED				7-xx	Reserve Officers' Training Corps
300	NOT USED				7-xx	TOE Consolidate Change Tables
J-Series	Joint	0	0	25		

Table 6-2. *The Army Cognitive Map.*[23]

progenitor of FM 100-5.[19] During and after World War II the Army's doctrinal system became more routinized.

Eric Heginbotham has attributed the effectiveness of the U.S. Army in executing combined arms warfare, compared to the British experience, in large part to doctrinal processes just before and during World War II. He argues that doctrine became a common language for all American Army officers to employ in discussions about employment of combined arms, thus becoming the base upon which improvements were made and guidelines for operations were created. Doctrine was the mechanism for producing rapid adaptation within a dense network of channels among Army professionals for communication within the force.[20]

This networking among Army professionals produced, by 1945, the first consistently codified cognitive map of the Army's abstraction of its occupational domain. For example, *Basic Field Manual List of Training Publications*, the 1945 version of FM 21-6, was the first to apply integrated regular groupings of subject matter and to assign consistent numerical designations for Army field manuals. The robustness and rigor of the U.S. Army's doctrinal process continued to grow throughout the post-World War II period.[21] The broad changes that have occurred are observable in the indexes that report the titles and subject matter covered by the Army's field manuals.[22]

The 1950s emphasized traditional combined arms as tactical nuclear weapons doctrine emerged. Manuals of the 1960s reflect the epitome of tactical nuclear weapons doctrine, especially with the emergence of Army missile systems. The 1970s manuals were characterized by a resurgence of combined arms warfare, as applied to conventional war-fighting in Europe and by a growth in coverage of subjects required for fighting in Vietnam. The dominant changes in the 1980s manuals reflect the training revolution that occurred within the Army beginning in the late 1970s. Books on individual occupational specialties, covered earlier (and since) in technical or training manuals, were elevated to field manual status in the 1980s. By the 1990s this began to change, with the training revolution being gradually displaced by the operational revolution characterized by AirLand Battle doctrine and the Persian Gulf War. The latest index of Army doctrinal publications reveals a growth in coverage of logistics matters and jointness of Army operations.

There are some interesting continuities demonstrated in the Army's cognitive map (Table 6-2). Some weapons are apparently timeless, with the M2 caliber 50 machine gun serving as a familiar example in FM 23-65. Some subjects change name and number but haven't really gone away since ancient history; the 1940 index lists FM 25-5, *Animal Transport*, while the 2000 index shows FM 31-27, *Pack Animals in Support of Army Special Forces Operations*. This database also reveals some insight into the effectiveness of the Army's system of abstraction in its current jurisdictional competitions.

While there are no immediately observable references in the latest index to the Revolution in Military Affairs, there is an implicit gradual movement in subject matter coverage in the direction of the dramatically new ways of waging warfare contained in RMA concepts. There is now a separate manual devoted exclusively to information warfare, FM 100-6. Likewise, the Army recognizes the importance of one of the other emerging new RMA warfare areas in the publication of FM 100-18, *Space Support to Army Operations*, although the promise of the RMA is that space operations will themselves become a new warfare area. The precision strike warfare area of the RMA is covered, implicitly, in Army doctrinal coverage of new fire support and communications techniques, revealed in such new manuals as FM 6-20-10, *Tactics, Techniques and Procedures for the Targeting Process*; FM 6-24.8, *TADIL-J, Introduction to Tactical Digital Information Link J and Quick Reference Guide*; FM 11-55, *Mobile Subscriber Equipment (MSE) Operations*; FM 24-7, *Tactical Local Area Network Management*; FM 34-25-1, *Joint Surveillance Target Attack Radar System (JSTARS)*.

Notably absent from this review is any new title covering abstractions relating to dominating maneuver in the RMA. Content analysis of the specific coverage in FM 100-5 sheds some light on this, but for the moment the assessment of the Army's cognitive map indicates grudging acknowledgement that some fundamental changes attributable to an emergent RMA may be in motion in the Army's occupational jurisdiction. That acknowledgement is more a recognition of the impact of new technologies and systems than an exploration of new operational concepts.

The Army is conducting such an exploration of the potential for an RMA in its Army Transformation process. As a result of a decade of work, beginning with the Louisiana Maneuvers Task Force, progressing through a series of Army War-fighting Experiments, and continuing through the promulgation of Army Digitization Doctrine,[24] the Army has addressed some of the issues associated with the RMA, especially those concerning battle command. The Army is also working on its Interim Brigade Combat Teams doctrine that will eventually cover the Objective Force,[25] a design that will focus on architectures centering on the Future Combat System. The conceptual framework of this future doctrine is published in TRADOC Pamphlet 525-5, which has not been revised since its publication in 1994.

The most dramatically observable change in the Army's cognitive map is the emergence of joint doctrine. An entirely new meta-subject-area has been created in the renumbering of certain Army publications from field manuals to joint publications. This move is as radical a departure as the creation in 1923 of the Field Service Regulations that unified the Army's infantry, cavalry, and field artillery schools of warfare into a single consistent body of abstractions.

It is apparent from the 1998 General Officer Survey that the Army has begun to subordinate its doctrine, at least some significant components of it, to joint doctrine. The titles of several of the Army's 90- and 100-series manuals now include the term *Joint*. Most had no purely Army equivalent in previous editions. The most significant indicator of this trend is the redesignation of FM 100-5 as FM 3-0 to conform to the notation of JCS Pub 3-0, *Operations*. The listing of 25 joint doctrinal publications as equal in authority to Army field manuals is a radical departure from previous listings in which joint publications provided authority over a few very narrow technical subjects or a few areas so broad as to have no real impact on the conduct of Army operations.

The Army's cognitive map also reveals insight into how the Army has responded to jurisdictional challenge in non-war-fighting areas. The data indicate that the Army's adaptation to the 1990s requirements to conduct stability and support operations has been to re-invent what had been its traditional approach to such operations before the birth of AirLand Battle doctrine.

While they were not on the street in time for operations in the early 1990s in Haiti, Somalia, and the Balkans, the Army has quickly produced new doctrinal publications such as FM 100-23, *Peace Operations*, and FM 100-23-HA, *Multiservice Procedures for Humanitarian Assistance Operations*, both of which were promulgated in late 1994. It also published FM 7-98, *Operations in Low-Intensity Conflict*, in 1992; FM 90-29, *Noncombatant Evacuation Operations*, in 1994; and FM 100-19, *Domestic Support Operations*, in 1993. More significantly, the Army relied on a number of its older doctrinal publications, trying to make some of them more relevant to these challenges of the 1990s. Falling into this category are FM 100-20, *Military Operations in Low Intensity Conflict*, from 1990; FM 19-15, *Civil Disturbances*, born of the Army's role in domestic operations of the 1960s; and FM 90-8, *Counterguerrilla Operations*, FM 31-23, *Stability Operations*, and FM 31-20, *Doctrine for Special Forces*, each from the Vietnam-era doctrinal files.

The Army's cognitive map of its abstraction of occupational jurisdiction is revealing. It implies an evolutionary approach to incorporating the advances of the RMA into the heartland of its core competence for full and final jurisdiction over land warfare. It reveals an attempt to share in jurisdiction over joint operations. And it seems to be establishing a preference to serve in an advisory role in the nonwar jurisdiction in its approach to stability and support operations.

The Heartland of Army Core Competence: FM 3-0

The Army's foundational mechanism for claiming jurisdiction is Field Manual 3-0, *Operations* (2001). Where the Index of Publications identifies the cognitive *structure* of the Army's occupational domain, it is in this manual that the Army defines for society its cognitive *content* in such a way as to legitimize its claim to exclusivity over land warfare. The document has served that function at least since the 1982 version of FM 100-5 that was published in response to widespread external criticism of the 1976 version of the manual. Editions since 1982 have served as the Army's codification of its adaptation to jurisdictional competition as new tasks emerge. It has provided for the elaboration of Army knowledge at several layers of abstraction by means of amalgamation (absorbing new jurisdictions and groups) and division (creating new jurisdictions and groups to occupy them). Other manuals provide details of diagnosis and treatment of Army problems—this is what the Army means by its frequent colloquial reference to "how to fight." FM 100-5 provides the ordering of abstractions for inference.[26] It claims to supply enduring principles that can be applied to almost any problem that confronts military professionals.

The absence of significant change articulated across the 1982, 1986, 1993, and 2001 versions of *Operations*[27] in the Army's basic approach to offensive and defensive operations is stunning. Every one of these manuals states at the beginning of the section dealing with the offense, "The offense is the decisive form of war," although important modifiers and explanatory statements vary somewhat across the versions. The characteristics of the offense itself are almost unvarying (see Tables 6-3 and 6-4).

1982	1986	1993	2001 (FM 3-0)
Concentration	Surprise	Concentration	Surprise
Surprise	Concentration	Surprise	Concentration
Speed	Speed	Tempo	Tempo
Flexibility	Flexibility		
Audacity	Audacity	Audacity	Audacity

Table 6-3. *Comparative Characteristics of the Offense among the FM 100-5 Series.*

The forms and types of maneuver for the offense are identical in each edition.

Forms of Maneuver	Types of Offenses
Envelopment	Movement to Contact
Turning Movement	Attack (Hasty or Deliberate)
Infiltration	Exploitation
Penetration	Pursuit
Frontal Attack	

Table 6-4. *Forms and Types of Offensive Maneuver.*

There is no mention of jointness in these discussions of offense, nor is there any discussion of non-war-fighting operations. The 2001 version of Student Text 3-0 (the authoritative text used at the U.S. Army Command and general Staff College while FM 100-5 was under revision) does include a brief discussion of RMA-related issues in an appended one-page section on technology suggesting that intelligence, surveillance, and reconnaissance technological advances may allow commanders to lead from the front, avoid the movement-to-contact, increase the tempo of the offense, and create more options during the conduct of offensive operations.

FM 3-0 does reveal some fundamental change in the abstraction of the defense. Each version of the manual begins with the statement that the purpose of the defense is to defeat an enemy attack until the force can go over to the offense. There has been more change in description of the basic characteristics of the defense, in contrast to the consistency in the discussion of the offense. In particular, in the 1983 and 1986 versions, the emphasis in the defense is on detailed planning to allow the concentration of forces in adaptation to enemy actions as the defensive battle progresses. The 1993 version replaces the focus on detailed planning with greater adaptation in battle command. It refers to massing effects rather than forces, and agility in execution rather than detailed branches and sequels.

The forms of the defense also change by the time Student Text 3-0 is published in 2001. The 1982 version of FM 100-5 articulates the basic forms of the defense, while the 1986 edition creates a conceptual framework within which those forms take place (Deep Battle Area, Security Area, Main Battle Area, Rear, Reserve). The 2001 version of ST 3-0 proposes a radical departure in that framework to a higher level of abstraction (Decisive Operations, Shaping Operations—including Information Operations—

and Sustaining Operations in Depth). It asserts that defensive operations will be nonlinear and noncontiguous.

In the domain of defense operations there is greater evidence than in the offense that the Army is attempting to incorporate RMA concepts into its basic doctrine, but as with its coverage of the offense, this treatment of the RMA is implicit, not confronted head-on. Similarly, there is no content on jointness in the discussion of the defense, nor is there any treatment of non-war-fighting operations.

Clearly, if the Army believes that it faces jurisdictional competition in the emergent concepts of the RMA, jointness and non-war-fighting operations, it does not consider the challenges to its traditional knowledge domain to be serious enough to cause it to adapt or revise its core offense and defense concepts. To the extent that the FM 100-5 series recognizes an emerging RMA, the Army appears to be asserting that these phenomena will not result in any fundamental change in the Army's full and final jurisdictions.

The doctrinal evidence on Army professional abstraction changes dramatically once the analytic focus shifts away from the core competencies to the chapters of the FM 100-5 series that appear before and after this heartland material.

One set of changes reflects a similar departure that is observed in the cognitive map, that is, the growing role of logistics concepts in Army operations. This change has more to do with the changing nature of the Army's internal approach to conducting operations than with forming a response to jurisdictional competitions. Nevertheless, the impact of changing logistics requirements on Army doctrine is significant. The 2001 manual devotes 21 pages to logistics concepts, the previous manuals half or less than that. These changes are related to changing policy and strategy trends as the Army shifts from a forward-based posture to a force-projection approach.

More relevant to this analysis, the manuals reveal a growing concern with non-war-fighting operations. The 1982 FM 100-5 barely mentions that Army forces may again be called upon to conduct unconventional warfare operations in a veiled reference to the Vietnam War era. The 1986 version introduces the concepts of low-intensity conflict in addressing such operations as foreign internal defense, counterinsurgency, peacetime contingency operations, peacekeeping operations, and anti-terrorism. The 1993 manual devotes an entire chapter, for the first time in the FM 100-5 series, to such activities as "Operations Other Than War." That chapter provides greater specificity in the definitions of the types of non-war-fighting operations the Army must be prepared to conduct, but it makes a revealing argument when it maintains that such operations are subsumed within standard Army operational doctrine. (See Chapter 8 by Thomas McNaugher in the present anthology for a fuller treatment of this argument.)

Student Text 3-0 (2001) re-names these operations as "Stability and Support Operations," or SASO (actually a throwback to 1960s terminology), and devotes two chapters to the abstractions Army professionals need to apply. The types of operations included in SASO are expanded to include security assistance, support to insurgencies, support to counterdrug operations, arms control, show of force, and civil and domestic support operations. ST 3-0 maintains the argument, although in more sophisticated form,[28] that such operations are not the Army's primary business of war-fighting, but that Army forces are very good at them as lesser included capabilities.

The various editions of FM 100-5 are quite explicit in their treatment of the issues related to the increasing jointness of the Army's professional jurisdiction. The 1982, 1986, and 1993 versions of FM 100-5 maintain that there are two chains of command in joint operations, one for operations and a separate one for administration. The operational chain of command is usually joint, the administrative chain is nearly always service-specific. This bifurcation disappears in ST 3-0, which states that there is a single chain of command from the National Command Authority through the Joint Force Commander to forces provided by the services. This represents a fundamental shift in the Army approach to conceptualizing jointness.

It is significant that ST 3-0 makes this statement not in the context of core Army defensive and offensive operations, nor in its discussion of battle command. Rather, ST 3-0 makes this point about such a fundamental change in Army concepts in the context of a discussion on a higher level of abstraction about the levels of war—tactical, operational, and strategic. This point is further evidenced in a recent briefing posted to TRADOC's Doctrine Developer's Course web site. On slide number three of the briefing titled "Army Doctrine Hierarchy and Numbering Update," there is an interesting audio voice file that plays as the slide builds. In the panel of the slide that discusses how the existing doctrine numbering system needs to be revised to be compatible with the joint system, the narrator asserts that joint doctrine will become an extension of Army doctrine.

One way to interpret these statements about the relationship between Army and joint doctrine is as an attempt by the Army to share jurisdiction with emerging and competing joint organizations over land combat. By casting the jurisdictional conflict in terms of abstractions that the Army traditionally has mastered—the levels of war—the Army seems to be attempting to exert greater control or influence over the terms of the jurisdictional division that is emerging. By asserting that joint doctrine is really nothing more than a logical outgrowth of Army doctrine, the Army may be trying to establish the basis for later arguments. In doing so, it may be that the Army recognizes that it will inevitably lose some control over land combat and hopes to retain greater residual jurisdictional control by participating in the defining of terms relevant to sharing jurisdiction. In other words, rather than risk permanent loss through mutually exclusive claims of subordination or splitting jurisdiction, claims that the Army perhaps fears it would lose, it may be attempting to lay out a broader claim for shared jurisdiction with joint institutions. If it can win this claim for shared jurisdiction, the Army will at least have the opportunity to make a future argument for jurisdiction.

In sum, the institutional evidence seems to indicate that the Army is attempting to adopt an amalgamation strategy in its current jurisdictional competitions over the Revolution in Military Affairs, increasing jointness, and the re-emergence of non-war-fighting operations. The doctrinal evidence, especially as to the Army's core competence embodied in FM 3-0, suggests that the Army believes it can exert full and final jurisdiction over new warfare areas in the emerging RMA insofar as they continue to involve land combat. The Army's cognitive map, as well as FM 100-5, clearly reveals the Army's search for accommodation in the jurisdictional competition with joint organizations. And the available institutional evidence implies that the Army does not believe that emerging requirements for non-war-fighting operations represent a competition that it wants necessarily to win. Rather, the Army seems to be pursuing a strategy of making itself available in an advisory capacity for such operations.

The Individual Level Of Analysis

At the individual level of analysis, the role of Army doctrine in the ecology of professions is more subjective than it is at the institutional level. While there are some interesting data available at the individual level, it is problematic to extrapolate from the responses of a population sample to behaviors and concepts of the Army as a whole. Nevertheless, it is useful to examine the role of Army doctrine at this level for additional insight into the trends observed at the institutional level.

Two data sets provide these insights for this analysis. The first set of data comes from the investigation by the Center for Strategic and International Studies (CSIS) titled *American Military Culture in the Twenty-First Century.*[29] Second, a review of articles in the Army's professional journals, focused on the three jurisdictional competitions, provides some insight into the nature of the debate over professional jurisdiction.

Of the ninety-nine questions in the Ulmer-Campbell Military Climate/Culture Survey (MCCS) conducted for the CSIS project, and the eighty-eight questions in the companion Staff Survey, some related to the three competitions of interest in this analysis. Two MCCS questions related to the RMA:

> **No. 45.** Our organization can adjust to new technologies and changing doctrine (No. 30 in the MCCS Staff Survey).
>
> **No. 56.** Our leaders consider the future, exploring new doctrine, tactics, equipment, and procedures.

Both sets of respondents provided a highly positive response that their organization adjusts to new technologies and changing doctrine. Staff respondents were more positive (mean score of 4.92 on the survey 6-point scale, compared to 4.21 on the broader survey), but in both sets the response was in the top twenty most favorable responses on the survey. This indicates a willingness among soldiers to adapt to the RMA. But there is a significant difference in perception between staffs and the broader population about the willingness of Army leaders to adapt to such change. While the leaders themselves rated their organizations' receptivity to RMA-like change higher than the broader population did, the broader population itself, when asked to narrow their response to the willingness of the leaders to adjust to the RMA, rated their leaders at 3.84, a substantially lower score than the leaders gave themselves. In other words, Army leaders think they are adapting to the RMA, but soldiers do not think their leaders are changing fast enough. This finding lends support to the institutional data suggesting that the Army is not adapting its doctrine with regard to the RMA fast enough to compete in this area of professional jurisdiction.

Several MCCS questions also dealt with the issue of jointness:

> **No. 23.** I have confidence in the other American military services that we might work with in joint operations (No.29 on the Staff Survey).
>
> **No. 53.** This unit would work smoothly with units from other military services (No. 36 on the Staff Survey).
>
> **No. 92.** Emphasis on joint education, doctrine, and training has contributed to the effectiveness of my Service (No. 66 on the Staff Survey).
>
> **No. 81.** (Staff Survey Only) My future value to my service would be enhanced by my completing a tour on a joint or combined staff.

On both surveys, the confidence in other military services ranked very positively. For the total active Army this question received the 11th highest positive ranking, with a mean score of 4.35 out of 6, and on the Staff Survey it was the third highest positive score (5.45). Both surveys rated the question of working smoothly with units from other services positively (3.95 on the total active Army survey, 4.83 on the Staff Survey) as well as the question of the value of joint education, training, and doctrine (3.81 on the total survey, 4.22 on the staff survey). While these data do not shed any new light on the institutional question of the Army's attempt to negotiate a shared jurisdiction over land combat, it does indicate that service members would be supportive of such an outcome.

Three questions covered issues related to the Army's jurisdictional adaptation to non-war-fighting operations:

> **No. 28.** (Staff Survey No. 18) Members of this unit believe it is appropriate for us to be involved in a variety of operations—from "humanitarian" to combat.
>
> **No. 43.** (Staff Survey No. 28) The essential mission of America's armed forces is to be prepared to win in combat.
>
> **No. 90.** (Staff Survey No. 78) My service has the flexibility and resources to handle "peacekeeping" and other noncombat missions without significantly degrading its wartime readiness.

These three questions must be taken together. The responses to question No. 43 (Army mean = 5.08, Staff mean = 4.50) were more generally positive than those for questions No. 28 (Army mean = 3.63, Staff mean = 4.50) and No. 90 (Army mean = 3.77, Staff mean = 3.50). But there is a substantial difference in the salience of these perceptions between the total active Army respondents and the Staff respondents. Staff respondents seem to hold the view that the Army can simultaneously be prepared to win in combat and still be involved in a variety of operations from "humanitarian" to combat. The overall lower positive scores in this area tend to support the Army's institutional approach to this jurisdictional competition in seeking an advisory role.

The Army's professional literature contains additional insight into the individual level of analysis of Army approaches to its jurisdictional competitions. On the subject of the RMA, the Army's professional dialogue was lively for five years from about 1993 until about 1998.[30] The subject has now virtually disappeared from the pages of *Parameters* and *Military Review*. Interestingly, just before his retirement in 2001, Army War College Commandant Robert Scales published a provocative, though largely unnoticed article on the RMA in which he argued that the U.S. Army, far from dominating its opponents through mastery of the RMA, is in fact increasingly more vulnerable to counter-RMA approaches presently under study in the military forces of several foreign countries.[31] If the Army's institutional strategy in this jurisdictional competition is one of presumed dominance, then the absence of a continuing dialogue in the professional literature is indicative that Army professionals have bought in to the presumption.

Very few articles relate to issues of the jointness in the ecology of professions. Gen. Robert Riscassi argued persuasively in a 1993 article that the underlying abstractions of then-current Army doctrine formed the conceptual basis for emerging joint doctrine and should form the basis for the development of doctrine for combined operations as well.[32] David Keithly and Stephen Ferris make a similarly veiled case for employing Army abstractions about command and control as the underlying principles for sharing jurisdiction with joint organizations in joint operations, especially in a multinational context.[33] And the general officer doctrine survey undertaken by TRADOC in 1998 seemed to settle the issue for the Army's senior officers. Those senior Army leaders perhaps believed that in seeking to work out a system of shared jurisdiction, the Army would gain a competitive advantage in the struggle for jurisdiction at the abstract level.[34]

In the third area of jursidictional competition there is no lack of professional dialogue. It seems that one of the hottest topics among Army writers has been the concepts and issues associated with the non-war-fighting operations characteristic of the 1990s.[35] All seem to support the idea that such operations are a legitimate mission area for the U.S. Army. None disagree with the notion observed in the institutional assessment and the other individual data sets that the successful conduct of non-war-fighting operations, while requiring increasingly complex skills, can be accomplished exemplarily by Army units and soldiers well-schooled in combat operations.

Settlements in the Professional Competitions

The evidence supports the conclusions that in the three basic competitions the Army is presently engaged in, it is seeking a settlement of full and final jurisdiction over the RMA, it seeks to share jurisdiction over joint operations, and it is content to serve in an advisory role in non-war-fighting operations (see Table 6-5). Given these apparent settlement strategies, then, how will the Army fare in the ensuing competition for jurisdiction? I believe the prospects are troubling for the Army as a profession because pursuit of these strategies may lead to an erosion of the Army's ability to maintain legitimacy successfully with regard to its asserted and traditionally secured jurisdiction over land combat.

In the RMA domain, the Army is largely losing the intellectual battle over the definition of this emerging future of combat. It is losing this battle at the abstract level. While Army war-fighting concepts are steeped in the tradition of AirLand

Competitions	Possible Outcomes of Jurisdictional Competition					
	Full & Final	Subordination	Split	Shared	Advisory	Divided
RMA	X					
JOINTNESS				X		
NONWAR					X	

Note: The Xs indicate which outcomes the Army is seeking in the three competitions for jurisdiction over land warfare.

Table 6-5. *Outcomes Sought by the Army in the Competition for Jurisdiction.*

Battle, Air Force concepts are exploring innovative new areas such as Global Precision Strike and Effects-Based Operations.[36] Although Army concepts for the Interim Brigade Concept Team and the Future Combat System-oriented Objective Force may well rival such Air Force concepts at an abstract level, the Army has chosen to develop its RMA concepts entirely in-house, largely inaccessible to non-Army analysts. The Army will never achieve full and final jurisdiction over land combat during the RMA so long as it continues to choose not to engage in the intellectual competition over the meaning of the RMA. Based on my own observations of the joint concept development process, the Army no longer exerts its former leadership over the Dominant Maneuver component of Joint Vision 2020 and instead has acquiesced in an understanding of the RMA that is dominated by the proponents of Precision Strike and Information Warfare. This is not an approach that the Army institutionally is equipped to win in the abstract since the basis of the Army is the control of territory, people, or things.

The Army is engaged in a risky approach to the competition over jointness. The Riscassi argument has worked but it is almost too clever. The particular sharing arrangement that the Army prefers may not turn out to be the one the Army achieves. Rather than defining the abstraction upon which the arrangement will be based, the Army may find itself reacting to alternative concepts proposed by other competitors for professional jurisdiction.

Competing operational concepts are already emerging from joint sources, such as the Rapid Decisive Operations (RDO) concept developed by the Joint Forces Command Joint Futures Lab. The Army has taken a somewhat disdainful attitude towards this concept, which does not provide for an Army role that comes anywhere close to the war-fighting concept articulated in ST 3-0. But the Army cannot win that intellectual fight by avoiding getting into the ring with the Joint Staff as it has done in the RDO exercises to date. The Joint Staff has conducted three studies this year in support of the joint war-fighting capabilities analysis (JWCA), including studies on precision engagement, dominant maneuver, and command and control. In response the Army maintains an outdated briefing package on its approach to the Quadrennial Defense Review that does not come close to addressing the issues raised in the joint staff studies. This competition may also include not only the other services and joint organizations, but, perhaps even more likely, conceptualizations conceived and imposed by influential members of the Office of the Secretary of

Defense. It would not be the first time a Secretary of Defense has mastered the services as a result of superior intellectual powers of abstraction.

Even in the seemingly less risky domain of Stability and Support Operations (SASO), the Army's approach to jurisdictional competition carries not-so-hidden dangers for its claims to exclusive jurisdiction over the use of lethal force in land combat. Andrew Abbott argues that advisory jurisdiction is "the bellwether of interprofessional conflict. Where there is advice today, there was conflict yesterday or will be conflict tomorrow."[37] In my view, the Army's strategy of settling for an advisory role in non-war-fighting-operations is indicative of conflict both before and yet to come. In the professional conflict leading up to the mid-1990s, the Army attempted to avoid taking on SASO as much as it could. It largely viewed such operations as incompatible with the hugely successful combat organization it had revolutionized after Vietnam and that had secured the dramatic victory in the Persian Gulf War. As it became clear that the nation intended to call on its Army increasingly for such operations in the early 1990s, the Army accepted this enlargement of its preferred jurisdiction by adopting its Cold War approach to them—they would be a lesser included set of capabilities that would be offered in support of other organizations who would have the lead. In the case of domestic operations the Army would support some other designated lead federal agency. In the case of overseas operations, the Army held that other nations with more direct interests would take the lead.

While the Army has largely succeeded in this approach, and may yet be spared by the Bush administration from engaging in the number and tempo of such operations seen in the 1990s, at the abstract level the Army has made itself vulnerable to a multitude of rival claims for such advisory support in the future. Typical of such arguments is Mary Caldor's call for the U.S. Army to lead the way in a new global era of employing armed forces strictly for humanitarian interventions.[38] The Army cannot afford to brush off such arguments as so much drivel from the liberal left. These are intellectually powerful arguments, reinforced by the Army's own jurisdictional strategy. The Army needs recognizable intellectual giants of its own to respond, someone cut from the same mold as Canadian Brigadier Lewis MacKenzie, who has argued cogently in this area.[39] And those intellectual giants need abstractions that will empower them to articulate the Army's claims to legitimate monopoly over the use of lethal force in global conflict.

Toward a General Theory of War

So how might the Army go about establishing a firmer intellectual foundation for its abstraction about its professional jurisdiction? As argued earlier, the Army does not need more doctrine. It has more doctrinal publications now than it ever has had before and even more are on the way. Nor does the Army need a new approach to the development of doctrine. The interaction of TRADOC with the schoolhouses, the Combat Training Centers, and units in the field is working better than it ever has. These components of the Army doctrinal machine are not broken, so they need not be fixed.

What the Army needs is a higher level of abstraction to provide it with a stronger form of influence over its jurisdiction by means of greater control over its knowledge domain. The Army must seize the intellectual high ground in the current doctrinal competition. It needs to develop a general theory of war.

Although the Army as an institution generally disdains theory-building and theorists,[40] it is time for the institution to re-establish its intellectual curriculum vitae. This is an ideal time for such a development since there has not been much new thought at this higher level of abstraction for the post-Cold War era. Even Cold War military theory was dominated by early nuclear era theorists such as Bernard Brodie, Albert Wohlstetter, and Andrew Marshall.[41] Yet the Army theory of war is still steeped in the nineteenth-century concepts of Jomini and Clausewitz. It is time to shed the principles of war and develop new knowledge for the new era of warfare. This is not to say that the classic principles of war are irrelevant, any more than it is to say that in the era of quantum mechanics the laws of physics articulated by Isaac Newton no longer apply. It's not that the old laws have zero explanatory power, it is that those Newtonian laws are largely trivial or irrelevant to the problems of modern physics. Likewise, simply stating, for example, that the Somalia operation failed because it violated the principles of mission and unity of command is not helpful in approaching future such operations.

If the Army is to win the intellectual battle of abstraction upon which its future occupational jurisdiction depends, it must now begin to elevate the level of its debate. Presently such discussions have official recognition only in a very limited circle that includes the Army War College's Strategic Studies Institute, Leavenworth's School of Advanced Military Studies, and TRADOC's office of the Deputy Chief of Staff for Doctrine (DCSDOC). The Army needs to broaden its approach by pursuing a General Theory of War for the twenty-first century.

Such a pursuit would need at least three basic components. First, the Army should create a core institution for theory-building. The Army War College and the Command and General Staff College already make important contributions to the development of the "How to Fight" process at the strategic and operational levels. The TRADOC DCSDOC integrates these processes across the Army. Many of the people who accomplish these tasks for the Army are eminently capable of developing theory and implicitly do so in the course of their work. But these organizations have no mandate to focus on theory and certainly do not have the time to do so. The Army needs a new, separate organization dedicated to this function.

Two existing institutions could house such an organization. The United States Military Academy could be effective in this role. USMA now has the kind of interdisciplinary faculty possessed by institutions of higher learning, research, and theory-building in other successfully competing professions (e.g., medicine, engineering, law). Alternatively, the Army has a considerable investment in the RAND Corporation Arroyo Center, a federally-funded research and development center, that would provide similarly broad access to scholars with a well-established—and well resourced—bureaucratic organizational infrastructure already in place.

Wherever the theory-building function is housed, the organization would need to be empowered to reach within the existing Army research enterprises as well as out to other institutions. The theory center could be a funding source for Army graduate students pursuing advanced degrees and dissertations, command and staff college theses, and advanced military studies focusing on the theory of war. It could also commission outside scholarship and perhaps hold a biannual conference to debate and discuss research. The organization should also be about the business of collecting intellectual intelligence about the jurisdictional competitions confronting the contemporary Army. It should address such research questions as these: What are the dimensions and boundaries of the current competitions at the abstract level? Who is competing for what jurisdiction? For what purpose? The center should examine theoretical developments of foreign countries. Most importantly the center would serve to stimulate innovative thinking across the Army by identifying good theorists and encouraging them to think out loud through both personal communications and professional forums.

The second necessary component is the promotion of theory-building across the profession. This will require a cultural shift among Army professionals to recognize the intrinsic value to the profession of those who chose to pursue intellectual abstraction as a career goal over "muddy boots." Many medical and legal professionals who teach or conduct pure research do not concentrate on being practitioners because that is not where their interests primarily lie. They are creating the necessary inferential framework for future adaptations in diagnosis and treatment that will be required in the competition among professions. There is no reason why some of the Army's senior theory-building professionals could not likewise be set apart from its practitioners.

The third ingredient is that the Army must open its dialogue in the abstract to outside contribution and review. It should welcome rival claims by proponents of ideas from other services, the joint community, and even the Office of the Secretary of Defense. It cannot view every new idea as a potential threat to Army plans, programs, and budgets, but should welcome the opportunity to demonstrate the superior persuasiveness of Army doctrinal concepts at the intellectual level. As the Army allows outsiders to contribute, it must carefully control the rules of the game and focus initially on those issues that reinforce the Army's claims to legitimacy. This suggests that the debate should focus not on the peripheral issues of future battle in the RMA, jointness, or stability and support operations. Rather the Army should control the agenda of abstractions by challenging traditional notions of offense and defense in war with innovative contributions of its own in these Army heartland core competencies.

There have been some attempts at theory-building that could serve as a launching point for a new Army approach. These approaches have largely been derived from general systems theory[42] and have proven to fall short of providing the kind of generalized theory needed for jurisdictional competition. More recent attempts at thinking about a new theory of war have pursued certain biological metaphors such as complex adaptive behavior and complexity sciences.[43]

Success in creating a process for building a new theory of war would go a long way toward enabling the Army to escape the negative outcome I believe it faces in the present professional competition for jurisdiction.

Notes

1. Field Manual 100-5 is one source of the Army's definition of *doctrine*. The 1982 and 1986 versions of this manual discuss, almost exclusively, the war-fighting aspects of doctrine in its definition of the term. The 1993 edition expands its scope to include operations other than war and its domain to include how to think about war-fighting as well as how to fight. In the October 2000 version of Command and General Staff College Student Text 3-0, a surrogate for the new (June 2001) edition of Field Manual 3-0, *Operations*, which replaced the 1993 version of FM 100-5, the notion of doctrine is adjusted to connote greater joint-ness. Training and Doctrine Command Regulation 25-36 *Coordinating Draft*, *http://www-tradoc.monroe.army.mil/dcsdoc/*; Internet, accessed on 24 May 2001, also offers a view of the nature of doctrine, "which consists of principles and [Tactics, Techniques, and Procedures], and defines, in terms of existing capabilities, how the Army intends to conduct operations across the full range of military operations. It is the fundamental principles by which military forces...guide their actions in support of national objectives. These principles reflect the Army's collective wisdom regarding past, present, and future operations. It is the body of thought on how the military fights in the present to near-term with current force structure and material. They focus on *how to think* about operations, not *what to think.*"
2. Current manuals in force and under development are listed in a database at *http://doctrine.army.mil*. Authoritative complete text of Army doctrinal publications is available at the Reimer Digital Library, "Field Manuals," and "Joint Chiefs of Staff Publications," *http://155.217.58.58/cgi-bin/atdl.dll?type=ANY&school=ANY*; Internet; accessed 20 October 2000.
3. Headquarters, U.S. Army Training and Doctrine Command, "Joint/Army Doctrine Directorate," *http://www-tradoc.monroe.army.mil/dcsdoc/jadd_roster.html/*; Internet; accessed on 30 January 2001.
4. Brig. Gen. Stanley E. Green, Deputy Chief of Staff, Doctrine, Headquarters, U.S. Army Training and Doctrine Command, "Doctrine Study 00/01 Study Advisory Group," 27 September 1999, 11.
5. See, for example, U.S. Army Combined Arms Center, Center for Army Lessons Learned, "National Training Center Trends Analysis 4QFY94-2QFY96 NO.97," at *http://call.army.mil/call/ctc_bull/97-3anly/ta1.htm*; and *http://call.army.mil/call/ctc_bull/llcmtc96/c2.htm*; Internet; accessed 2 February 2001.
6. Samuel P. Huntington, *The Soldier and the State* (New York: Vintage Books, 1957); Andrew Abbott, *The System of Professions: An Essay on the Division of Expert Labor* (Chicago, IL: Chicago University Press, 1988) 4.
7. Abbot, 2.
8. U.S. Army Command and General Staff College Student Text 3-0, *Operations* (Fort Leavenworth, KS: U.S. Army Command General Staff College: 1 October 2000), 1-14.
9. My own exegesis of Abbott, 2,4.
10. The Army makes this case in its capstone manual, FM 1-0. It is excerpted in ST 3-0, "Army Forces are the decisive component of land warfare."
11. Abbot, 98-108.
12. Ibid., 69-79.
13. Jeffrey McKitrick et al., "The Revolution in Military Affairs," in *Battlefield of the Future*, eds. Barry R. Schneider and Lawrence E. Grinter, Air War College Studies in National Security No. 3 (Maxwell Air Force Base, AL: Air University Press, 1995).
14. Lt. Col. Mike Goodwin, "Update: Army Doctrine XXI," slide 7 of 12 , *http://www-tradoc.army.mil/jadd/adxxi/*; Internet; accessed 20 October 2000. Nota bene: this site is no longer accessible to the public. I will provide hard copy to any researcher who desires to examine this

document. Aggregate results of the survey are contained in the briefing "Doctrine Study 00/01 Study Advisory Group" cited in n.4.

15. The Army's view of the division of doctrinal turf is summarized in a table accompanying a briefing on the relationship between Army and joint doctrine. The table lists the numbers of joint doctrinal publications by proponent: Army 26, Navy 5, Air Force 14, USMC 5, USCG 3, CinCTRANS 8, CinCSPACE 1, CinCSOC 7, CinCSTRAT 1, CinCJFC 7, J1 2, J2 7, J3 7, J4 7, J5 1, J6 2, J7 10, PA 1, SP 1.
16. During this study, FM 100-5 was in the process of revision and was released after the completion of the research. Its number was changed to FM 3-0 as part of the complete revision to the Army Manual numbering system. As a surrogate for FM 3-0, I used USACGSC Student Text 3-0.
17. Abbott, 71.
18. U.S. War Department, FM 21-6, *Basic Field Manual List of Training Publications*, 16 March 1940.
19. William O. Odom, *After the Trenches: The Transformation of U.S. Army Doctrine, 1918-1939* (College Station, TX: Texas A&M University Press, 1999).
20. Eric Heginbotham, *The British and American Armies in World War II: Explaining Variations in Organizational Learning Patterns,* Defense and Arms Control Studies Working Paper, Defense and Arms Control Studies Program (Cambridge, MA: MIT Center for International Studies, February 1996). Timothy Lupfer has argued that a similar robustness gave the German Army a relative tactical advantage over the Allies during World War I: Timothy T. Lupfer, *The Dynamics of Doctrine: The Changes in German Tactical Doctrine During the First World War*, Leavenworth Papers No. 4 (Fort Leavenworth, KS: U.S. Army Command and General Staff College, July 1981).
21. Robert A. Doughty, *The Evolution of U.S. Army Tactical Doctrine, 1946-76*, Leavenworth Papers No.1 (Fort Leavenworth, KS: U.S. Army Command and General Staff College, Combat Studies Institute, August 1979); *Sixty Years of Reorganizing for Combat: A Historical Trend Analysis,* Combat Studies Institute Report No. 14 (Fort Leavenworth KS: U.S. Army Command and General Staff College, January 2000). John L. Romjue, *Prepare the Army for War: A Historical Overview of the Army Training and Doctrine Command 1973-1993* (Fort Monroe, VA: U. S. Army Training and Doctrine Command, 1993); John L. Romjue, *American Army Doctrine for the Post-Cold War* (Fort Monroe, VA: U.S. Army Training and Doctrine Command, 1996).
22. I have compiled a cross-referenced longitudinal database index of all Army FMs sampled from the following publications: War Department Field Manual 21-6, *List of Field Publications for Training*, March 1945; Department of the Army Field Manual 21-6, *List and Index of Department of the Army Publications*, 10 April 1948; Department of the Army Pamphlet 310-3, *Military Publications Index of Training Publications,* 14 March 1960; DA PAM 310-3, *Military Publications Index of Training Publications,* 1 October 1954; DA PAM 310-3, *Military Publications Index of Doctrinal,Training and Organizational Publications,* 31 August 1970; DA PAM 310-3, *Military Publications Index of Doctrinal, Training and Organizational Publications,* 1 January 1982; DA PAM 25-30, *Military Publications Index of Doctrinal Training and Organizational Publications,* 30 September 1990.
23. The Coordinating Draft of TRADOC Regulation 25-6 provides a table cross-referencing Functional Categories, Numbers Series, and Title for Doctrinal Publications between the old and new numbering systems. Among the new categories appearing for the first time in this table are: Management Information Systems, Organizational Effectiveness, Environmental Operations, and Research Development and Acquisition.
24. As of this writing, the Army has not granted public access to its ongoing development of doctrine for the digitized force. Before it became password-protected the following was available: Lt. Col. Ron Gregory, "Army XXI Issues Associated with Development of Doctrine and TTP

for the Digitized Force"; Internet; *http://www-tradoc.army.mil/jadd/adxxi2/*; accessed 20 October 2000.

25. Col. Michael Mehaffey, "Vanguard of the Objective Force," *Military Review* 80 (September-October 2000); Col. Kent E. Ervin and Lt. Col. David A. Decker, "Adaptive Leaders and the Interim Brigade Combat Team," *Military Review* 80 (September-October 2000). Although the Army does not allow public access to its doctrine development web sites for the Interim and Objective Forces, it is possible to obtain recent versions of briefings on those subjects provided to Doctrine Developers Course at the TRADOC and Combined Arms Center doctrine web sites: "The Foundations of Army Transformation," "The Interim Force: Organizations and Capabilities," 30 January 2001, and "L07 Doctrine Development," Doctrine Developers Course 13 February 2001.
26. Abbott, 48-52.
27. Again I am using ST 3-0 as a surrogate for the 2001 edition of *Operations* which has been re-numbered as FM 3-0.
28. ST 3-0, 9-1.
29. Joseph J. Collins, Project Director, *American Military Culture in the Twenty-First Century: A Report of the CSIS International Security Program* (Washington, DC: Center for Strategic and International Studies, February 2000).
30. Maj. Steven J. Mains, "Adapting Doctrine to Knowledge-Based Warfare," *Military Review* 77 (March-April 1997); David Jablonsky, "U.S. Military Doctrine and the Revolution in Military Affairs," *Parameters* 24 (Autumn 1994); Maj. Jon J. Peterson, "Changing How We Change," *Military Review* 78 (May-June 1998); Ryan Henry and C. Edward Peartree, "Military Theory and Information Warfare," *Parameters* 28 (Autumn 1998); Antulio J. Echevarria II, "Tomorrow's Army: The Challenge of Nonlinear Change," *Parameters* 28 (Autumn 1998); Steven Metz, "The Next Twist of the RMA," *Parameters* 30 (Autumn 2000).
31. Maj. Gen. Robert H. Scales, Jr., "Adaptive Enemies: Achieving Victory by Avoiding Defeat," *Joint Forces Quarterly* (Fall 1999).
32. Gen. Robert W. Riscassi, "Doctrine for Joint Operations in a Combined Environment: A Necessity," *Military Review* 73 (June 1993). This article also appeared in the first edition of *Joint Force Quarterly* (Summer 1993). General Riscassi retired in July 1993.
33. David M. Keithly and Stephen P. Ferris, "*Auftragstaktik*, or Directive Control, in Joint and Combined Operations," *Parameters* 29 (Autumn 1999).
34. See note 4.
35. Maj. John Robert Evans, "*Task Force 1-22 Infantry From Homestead to Port-Au-Prince*," (Master's thesis, U.S. Army Command and General Staff College, Fort Leavenworth, KS, 2000); Mark Edmond Clark, "US Army Doctrinal Influence on the War in Bosnia," *Military Review* 79 (November-December 1999); Maj. Mark A. Tolmachoff, "Is Army Aviation Doctrine Adequate for Military Operations Other Than War?" (Master's thesis, U.S. Army Command and General Staff College, Fort Leavenworth, KS, 2000); Lt. Col. Daniel Ward, "Assessing Force Protection Risk," *Military Review* 77 (November-December 1997); Lt. Col. Walter E. Kretchik, "Force Protection Disparities," *Military Review* 77 (July-August 1997); Thomas Knight Adams, "Military Doctrine and the Organization Culture of the United States Army," (Ph.D. Diss., Syracuse University, 1990); Robert Bunker, "Failed-State Operational Environment Concepts," *Military Review* 77 (September-October 1997); Brig. Gen. Stanley F. Cherrie, "Task Force Eagle," *Military Review* 77 (July-August 1997); Col. Benjamin C. Freakley et al., "Training for Peace Support Operations," *Military Review* 78 (July-August 1998); Lt. Col. Wray R. Johnson, "Warriors Without a War," *Military Review* 79 (December-February 1999); David Fastabend, "The Categorization of Conflict," *Parameters* 27 (Summer 1997); Col. Andrei Demurenko and Alexander Nikitin, "Concepts in International Peacekeeping," *Military Review* 77 (May-June 1997); Lawrence A. Yates, "Military Stability and Support Operations: Analogies, Patterns and Recurring Themes," *Military Review* 77

(July-August 1997); Col. Charles H. Swannack, Jr., and Lt. Col. David R. Gray, "Peace Enforcement Operations," *Military Review* 78 (November-December 1997); Capt. Gregory R. Sarafin, "UN Observer Mission in Georgia," *Military Review* 77 (November-December 1997); Lt. Col. John Otte, "UN Concept for Peacekeeping Training," *Military Review* 78 (July-August 1998); Maj. Charles J. McLaughlin, "US-Russian Cooperation in IFOR: Partners for Peace," *Military Review* 78 (July-August 1997); Lt. Col. Douglas Scalard, "People of Whom We Know Nothing: When Doctrine Isn't Enough," *Military Review* 77 (July-August 1997); Maj. Gen. Robert H. Scales, Jr., "From Korea to Kosovo: How America's Army Has Learned to Fight Limited Wars in the Precision Age," *Armed Forces Journal International* (December 1999).

36. On Effects-Based Operations (EBO), see the briefing by Dr. Maris McCrabb, "Effects-Based Operations: Examples & Operational Requirements," 14 June 2000, available on-line from the Air Force Research Laboratory, Wright-Patterson Air Force Base. The USAF is now conducting a three-year-long Advanced Concept Technology Demonstration to experiment with technology concepts associated with EBO.
37. Abbott, 76.
38. Dr. Mary Caldor, "New Wars in the Global Era," (keynote address at Tel Aviv University/IDF/AUSA Symposium on Martial Ecologies, March 2000); Internet; www.martialecologies.com; accessed 30 May 2001.
39. Lewis MacKenzie, "A Crucial Job, But Not One for a Superpower," *Washington Post*, 14 January 2001, B3.
40. In contrast to today, there were at least two periods in the post-World War II Army in which theory-building was valued and had a direct impact on Army professionalism. One was the development of AirLand Battle under Gen. Donn Starry, the other was the development of the 1993 FM 100-5 under Gen. Frederick M. Franks, Jr.
41. Edward Mead Earle, ed., *Makers of Modern Strategy: Military Thought from Machiavelli to Hitler* (Princeton, NJ: Princeton University Press: 1943, 1971).
42. Majors E.A. Bryla, M.S. Lancaster, and W.C. Rennagel, *Contending Concepts, Tactics & Operational Art*, Volumes I and II (Newport, RI: U.S. Naval War College, Center for Advanced Research, June 1979); Robert H. Scales, Jr., *Firepower in Limited War* (Washington, DC: National Defense University Press, 1990); Shimon Naveh, *In Pursuit of Military Excellence: The Evolution of Operational Theory* (London: Frank Cass, 1997); Zvi Lanir, *SRT and Military Innovation* (Tel Aviv, Israel: Praxis Ltd , March 1999); Maj. John W. Taylor, "A Method for Developing Doctrine," *Military Review* 59 (March 1979); not to mention the voluminous Soviet literature on the systems approach to a general theory of war, although it is steeped in ideological considerations that even at the time most Soviet theorists did not really believe.
43. *SWARM Marine Infantry Combat Model (C-SWARM)—Beta Version User's Manual,* Office of the Secretary of Defense (Net Assessment) in support of the Marine Corps Combat Development Command, 20 October 1998; Dr. Michael L. Brown, *Thinking Biologically: The Impact of Complexity Sciences on the Future of Warfare,* Report of a Workshop Conducted for the Office of Net Assessment, 18 February 1997; Michael Brown and Andrew May, *Defeat Mechanisms: Military Organizations as Complex Adaptive Nonlinear Systems*, Report Prepared for the Office of Net Assessment, 10 March 2000.

7 The Digital Battlefield: What Army Transformation Efforts Say about Its Future Professional Jurisdiction

Elizabeth A. Stanley-Mitchell

> The revolutions in information technologies (IT) and knowledge-based systems hold almost unimagined promise for the army that grasps them. IT will be a breakthrough in warfighting. . . . Discriminating sensors providing information on enemy and friendly forces will link to computers that display relevant information in real time in digestible bites. Using IT more than explosive weapons, forces will maneuver against and defeat their enemies more quickly and with less risk. Targeting the enemy's fighting forces and, more decisively, his command and control facilities will provide an unprecedented ability to defeat him.
>
> STEVE MAINS[1]

The contemporary Revolution in Military Affairs (RMA) promises quantum leaps in the war-fighting capabilities of the United States armed forces. Given such high-potential rewards, each service has made a concerted attempt to harness these new technologies. The Army is no exception—its efforts are collectively known as the Army Transformation program.[2] By 2030, the Army will create the Objective Force, which will be responsive and dominant across the spectrum of operations. Encompassing the entire Army, the Objective Force will combine heavy force lethality, survivability, and sustainability with light force deployability. These increased capabilities will come about principally through dominant battlefield awareness.[3]

Although frequently overlooked in today's discussions because the "Transformation Vanguard"—the Interim Brigade Combat Teams (IBCT)[4]—currently holds the spotlight, battlefield digitization is a key component of Army transformation efforts. Battlefield digitization is the process of automating all Army command and control systems. It promises to use modern information technology to create a common view of the battlefield for all forces at all command levels.[5] By minimizing the fog of war,[6] the technology should lead to increased lethality, survivability, and operating tempo for a digital force.[7]

This chapter will outline Army battlefield digitization efforts in light of its evolving professional jurisdiction. How the Army has chosen to adopt new information technologies says much about the Army's future professional jurisdiction. As this chapter will show, information technologies from the contemporary RMA are both an objective and subjective force for jurisdictional change—but not all of these changes are necessarily good news for the Army. Although the Army has been capitalizing on technological opportunities to enhance its conventional war-fighting capability, it seems unwilling or unable to address the wider threats that such technological change could bring to its traditional jurisdiction. This chapter will address some of those threats. The resulting analysis will have significant implications for how the Army should renegotiate its jurisdictional boundaries—and thus redefine its profession.

The chapter is divided into five sections and a conclusion. The first section reviews the theoretical considerations at the foundations of my analysis—the "system of professions" and contemporary RMA technologies that could affect this system. The second section outlines the Army's battlefield digitization efforts in more depth. The next three sections examine the implications of Army digitization in light of three concentric visions of the professional system. More specifically, the third section looks within the Army profession itself. The fourth section looks at the wider professional system of traditional warfare, while the fifth section looks at the widest professional system of "national security."

RMA Technologies and Professional Jurisdictions

In *The System of Professions: An Essay on the Division of Expert Labor* (1988), Andrew Abbott argues that an occupation's identification as a profession and its standing within society are outcomes of social competition within a system of professions for control over expert knowledge as applied to particular jurisdictions. Accordingly, he defines professions as "exclusive occupational groups applying somewhat abstract knowledge to particular cases."[8] Professional life has three characteristics.[9] First, professions should be seen for what they do, not just how they are organized to do it. In other words, the essence of a profession is its work—its legitimated claim to apply expert knowledge to a particular set of tasks, which Abbott calls jurisdiction. Second, professions operate in an interdependent system—the "system of professions." Professions compete for the control of work, and the jurisdictional boundaries among them are constantly disputed. Professions occupy a jurisdiction by filling a vacancy or fighting for legitimate control of it through a variety of channels—the legal system, public arena, and workplace. As a result, a move by one profession inevitably affects others. Third, many variables affect the content and control of work, including technology, organizations, culture, etc. Perhaps the most important of these is technology.

From a jurisdictional perspective, technology is a double-edged sword. On the one hand, technological change can create new jurisdictional opportunities by (1) causing existing professional competitors to disappear; (2) creating new tasks; or (3) providing a new way to perform existing tasks. On the other hand, technological change can destroy jurisdictional opportunities by (1) introducing new professional competitors; (2) allowing existing competitors to take over existing tasks; or (3) driving existing tasks into obsolescence. Moreover, although technology can create new jurisdictional opportunities, rapid jurisdictional expansion is very difficult because there is a qualitative challenge to institutionalizing new work. As a result, even jurisdictional opportunities can lead to an invasion by outsiders seeking to claim legitimacy over those new tasks.

The Army's current professional jurisdiction is fighting and winning the nation's wars. Certainly this jurisdiction is being changed by the technological innovations embodied in the contemporary RMA.[10] Before looking more carefully at the Army's changing jurisdiction, however, it is important to understand what the contemporary RMA is all about.

The contemporary RMA is generally postulated as the result of linking precision weaponry to knowledge to create a radical enhancement of capabilities for future warfare. The emerging picture of the future battlefield centers on an integrated system of battlefield assets—a reconnaissance-strike complex—that promises continuous, real-time, sensor-to-shooter links between all targets and all available weapons in the battle space. Technological innovations of the contemporary RMA fall into three categories: (1) intelligence, surveillance, and reconnaissance; (2) advanced command, control, communications, computers, and intelligence; and (3) precision strike weapons. Applications of these three technologies together form a "system of systems."[11] Through these new technologies, using conventional weaponry, the RMA's advocates promise rapid, decisive victory, very low casualties and collateral damage, and strategic results.

Scholars generally concur that at the operational level of warfare, RMA technologies provide three major improvements in capability: precision strikes, increased velocity, and information dominance. First, precision strikes not only allow the military to conduct operations at a significant distance from the enemy—what Michael Mazarr calls "disengagement"[12]—but they also will reduce the number of casualties and collateral damage associated with combat operations.[13] Second, increased velocity will create preemptive warfare "between cohesive, fast-moving friendly forces and unready, disrupted enemy forces."[14] RMA information technologies will increase velocity by allowing battlefield leaders to use their enhanced knowledge to eliminate irrelevant and counterproductive movement.

Third, and perhaps most importantly, information dominance promises to erode or destroy the enemy's means of collecting, processing, storing, and disseminating information.[15] The Army defines information dominance as "the degree of information superiority that allows the possessor to use information systems and capabilities to achieve an operational advantage in a conflict or to control the situation in operations short of war, while denying those capabilities to the enemy."[16] Military theorists believe that information dominance is comprised of three effects. First, it will enhance situational awareness by providing accurate, complete, real-time information about friendly and enemy forces and the surrounding environment. In other words, it will answer the three questions that have plagued soldiers in battle from time immemorial: "Where am I? Where are my buddies? Where is the enemy?"[17] Second, as a result of situational awareness, it will dissipate the fog of war so that all soldiers, at all levels, will share a common view of the battlespace at all times.[18] Retired Adm. William Owens calls this common view "dominant battlespace knowledge." He explains that "this kind of knowledge constitutes an insight into the future, for it enables us to understand how the enemy commander sees his own battlefield options, and therefore increases the accuracy of predicting what he will try to do."[19] Third, armed with this information, U.S. forces can operate within an enemy's decision and action cycles,[20] thus enmeshing the "adversary in a world of uncertainty, doubt, mistrust, confusion, disorder, fear, panic, chaos. . . . And/or fold [him] back inside himself so that he cannot cope with events/efforts as they unfold."[21]

Of these three potential improvements at the operational level of warfare—precision strike, increased velocity, and information dominance—this chapter will concentrate on the third. Fundamentally, information dominance requires that a military organization have both the technology and the doctrine for handling and processing information and empowering commanders with fused, real-time knowledge of the battlefield. This is what the Army's battlefield digitization effort is attempting to achieve.

The Army's Battlefield Digitization Efforts

The Army has been trying since the 1970s to incorporate at least some of the contemporary RMA technologies into its war-fighting capability. Beginning with Sigma Star in 1978 and continuing through various programs to today's Army Battle Command System,[22] the Army has tried to minimize the fog of war and leverage new information technologies to improve the way it fights. This section will outline the Army's recent efforts to digitize the battlefield.

The father of Army digitization efforts is former Chief of Staff of the Army Gen. Gordon Sullivan.[23] Sullivan first institutionalized a concept-based, long-term orientation in the Army with the creation of the so-called "Louisiana Maneuvers"—the first in a series of Army experiments that eventually became the Army Experimentation Campaign Plan (AECP).[24] In March 1994, he published a campaign plan for the Force XXI process—as the AECP was known at that time—which argued that the Army needed to embrace the power of information technology and implement the current RMA. "The high ground is information. Today, we organize the division around killing systems, feeding the guns. Force XXI must be organized around information—the creation and sharing of knowledge followed by unified action based on that knowledge which will allow commanders to apply power effectively."[25]

Until October 1999—when Chief of Staff of the Army Gen. Eric Shinseki announced his new Army Vision—the Army experimentation program was oriented along three axes: the Mechanized Contingency Force, the Strike Force, and the Light Contingency Force. The mechanized axis focused on heavy forces. Most of the Army's digitization experiments and equipment fielding have involved units of this type, as will be discussed further below. The light axis was to modernize units that can deploy rapidly, while the strike force axis was to create a medium-weight force to form a bridge between early-entry light forces and slower-to-arrive heavy forces.[26] The light and strike force axes appear to have been folded together as part of the Interim Brigade Combat Teams (IBCTs) at Ft. Lewis,[27] although the 2nd Armored Cavalry regiment remains an experimental strike force.[28] Light and medium weight forces have generally not been as involved with Army digitization efforts, although some units from the 10th Mountain Division did participate in a September 2000 joint exercise, in which some digital equipment was tested in an urban environment.[29]

Overall, Army digitization efforts have concentrated on heavy, or mechanized, forces. Since 1994, the Army has launched a series of exercises—called Advanced Warfighting Experiments (AWE)—to evolve the digitization concept. The Army's digitization strategy uses a bottom-up approach that experiments echelon by echelon with several experimental systems simultaneously. In early 1996, experimental equipment was fielded to the 1st Brigade, 4th Infantry Division (Mechanized) at Ft. Hood; that unit became the core of a brigade combat team designated the Experimental Force (EXFOR).[30] EXFOR experimented with 72 different initiatives—operational concepts and equipment prototypes—during its milestone, two-week AWE in March 1997 at the National Training Center, Ft. Irwin.[31] The EXFOR AWE was followed in November 1997 by an AWE at Ft. Hood based on a division-level simulation.[32]

Digitizing and equipping the EXFOR was a monumental effort. The complexity and complications of the process led an Army official to compare it to "giving birth to a bale of barbed wire."[33] The EXFOR received more than 7,000 individual pieces of equipment, with more than 900 vehicles modified into more than 180 different configurations.[34] Digital appliqués, consisting of a computer and a Global Positioning System (GPS) locator, were mounted on all vehicles and wired into a communications system of Army radios, reengineered to handle more digital traffic.[35] This communications infrastructure, called the Tactical Internet, is a key element of the digitized force. In the EXFOR brigade alone, there were more than 800 different internet protocol (IP) addresses.[36]

Most of the equipment initiatives tested in the various AWEs are components of the Army Battle Command System (ABCS). ABCS integrates the command and control systems found at each echelon—from dismounted soldier or individual weapons platform to the ground force commander at the theater or joint task force level. ABCS embraces the "single point of entry" concept, so that once data has been captured in digital form, it can be shared "across echelons and geographic boundaries without being retyped or otherwise reentered."[37] At the upper echelons, ABCS is interoperable with joint and multinational command and control systems across all battlefield functional areas.

ABCS has three major levels. First, at the highest level, the Army Global Command and Control System (GCCS-A) operates at the division, corps, and theater levels, overlapping with the DoD's GCCS. GCCS-A provides the communications underpinning for force tracking, host nation and civil affairs support, theater air defense, psychological operations, C2, logistics, and medical support.[38] Designed from existing Army-wide communications systems, GCCS-A was first fielded to Army units in Hawaii and South Korea in September 2000.[39]

Second, the upper level of the Tactical Internet (formerly known as the Army Tactical Command and Control System, or ATCCS) operates in the middle command echelons, from corps to brigade. This level is organized into five battlefield functional areas (BFAs): maneuver, field artillery, intelligence and electronic warfare, combat service support, and air defense.[40] Various automated systems associated with some of these five BFAs have existed in the Army since 1978, but it was only during the EXFOR AWE that all five were finally able to communicate laterally with each other.[41]

Third, the lower level of the Tactical Internet uses the Force XXI Battle Command Brigade and Below (FBCB2) to provide situational awareness and command and control to the lowest tactical echelons—from the brigade down to dismounted soldiers and individual weapons platforms. The FBCB2 system is comprised of (1) a computer that can display a variety of information, including a common picture of the battlefield overlaid with icons of friendly and enemy forces,[42] (2) software that automatically integrates GPS data, military intelligence data, combat identification data, and platform data (such as fuel and ammunition status), and (3) interfaces to communications systems.[43] Battlefield data are communicated to and from users of FBCB2 through the Tactical Internet, a radio network comprising a positional navigation and reporting capability[44] and a voice- and digital-converting radio.[45] For dismounted soldiers, these components are mounted into a man-portable system called Land Warrior.[46]

In August 1997, the Army announced that the 4th Infantry Division would become its first digitized division. This conversion was finished in 2000, and the 4th ID conducted its capstone exercise for two weeks at the National Training Center (NTC) in April 2001.[47] The second digitized division will be the 1st Cavalry Division (also at Ft. Hood), scheduled for digitization by the end of FY03. The first digitized corps will be III Corps—which consists of these two divisions, the 3rd Armored Cavalry Regiment at Ft. Carson, and other corps assets—scheduled for digitization by the end of FY04. The Army intends to digitize its remaining divisions by 2010-2012,[48] starting with the 3rd Infantry Division at Ft. Stewart.[49]

In sum, the Army has been trying since 1994—in a very determined and structured way—to digitize its heavy forces for major conventional combat. The Army has made less progress with lighter forces but asserts that many of the new digital assets can be used in a lighter environment as well. While the digitization process has not necessarily been a smooth one, these new technologies promise "increased lethality, survivability, and operating tempo"—irresistible benefits to a service whose current professional jurisdiction is "fighting and winning the nation's wars." Yet, as discussed in the first section, new technologies—like those embodied in the Army's battlefield digitization effort—have the potential to change professional jurisdictions. The rest of this chapter will address some of the implications that adopting digital technology could have.

Implications of Digitization within the Army Profession Itself

The U.S. Army profession draws its mandate from the National Security Strategy and the National Military Strategy. These documents define the national military objectives as "to promote peace and stability and, when necessary, to defeat adversaries." Given these objectives, the purpose of the armed forces is (1) "fighting and winning our nation's wars," and (2) "protecting U.S. national interests."[50] Given this guidance, the Army currently envisions its professional jurisdiction as "finding peaceful solutions to the frictions between nation-states, addressing the problems of human suffering, and, when required, fighting and winning our nation's wars—our non-negotiable contract with the American people."[51] However, as battlefield digitization continues, adopting new technologies will cause the Army profession to change, both internally and in relation to other professions in the wider national security system. This section will address potential changes within the Army profession itself.

There are four ways that the Army profession may be changed by battlefield digitization. First, the real-time sensor-to-shooter architecture associated with digitization may eliminate the need for some command echelons. Second, the basis for learning and professional development appears to be changing, as simulations become a more important—perhaps even the principal—way the Army trains its soldiers. Third, there is a real potential for degraded decision-making ability among tactical leaders. Fourth, top-down command centralization becomes a higher risk with the new digitization technologies.

Eliminating Echelons

Digitization and its resulting dominant battlefield awareness create a number of potential synergies that could have implications for the Army's organizational structure. As in the business community, the real-time information link among various echelons and between sensors and shooters may make it possible to flatten the Army's hierarchy by eliminating some command echelons.[52] Moreover, because information technology enhances the ability to reallocate combined arms assets quickly and more flexibly, it may no longer be necessary to have all capabilities at every echelon. Finally, the link between sensors and shooters may blur the traditional distinction between operations and intelligence.

Because of these synergies, most proponents of streamlining Army organizational structure—including Douglas Macgregor and John Brinkerhoff—argue that the division is too large and cumbersome to fit the needs of the digital battlefield. Instead, in their view the Army should adopt a brigade-sized combat group as its basic combined arms organization, and scale back or eliminate the divisional echelon.[53] A full discussion of eliminating command echelons is beyond the scope of this chapter. However, from the perspective of professionalism, it is important to note two risks associated with flattening the hierarchy. First, eliminating some command echelons implies that the span of control for senior leaders necessarily increases, thus reducing their ability to supervise their subordinates' activities or identify problem areas. Possessing the informational wherewithal to dispense with echelons does not in itself give the higher commanders the additional time and opportunity needed to deal on a personal basis with additional subordinate commands. Second, flattening the organization could have negative effects for "growing" future leaders, as the number of opportunities to command with ever-increasing responsibility is reduced.

Simulations

It used to be that learning and professional development occurred as a result of real-world experiences. Much of that learning and development now occurs with simulations. The Army conceives simulations training as a toolbox with three different tools: (1) live simulation, in which soldiers use assigned equipment with some form of simulator for weapons systems; (2) virtual simulation, in which soldiers and crews in simulators replicating combat vehicles "fight" as if they were in the field; and (3) constructive simulation, in which large-scale computer simulations replicate units at or above the battalion level.[54]

Simulations create a number of training efficiencies. First, and most obviously, they can be cheaper and safer than training in the field. Without burning fuel, firing live ammunition, or imposing wear and tear on equipment, crews and units can learn skills necessary to fit their actions into a broader combat context. Second, they create training synergies by allowing units to train together over a network or via remote conferencing. For example, during the AWE in November 1997, digital simulation tools allowed the 4th Infantry Division at Ft. Hood to compete against the Army's World Class Opposing Force at Ft. Leavenworth.[55] Third, simulations can be used for soldiers to complete online components of a professional development course before attending it and thus reduce the time they are away from home station. Finally, simulations can help soldiers prepare for fielding new equipment in a more efficient manner.

But simulations can also have some unintended adverse consequences from a professional perspective. First, before simulations were so common, professional learning used to be based on direct real-world experiences. Real-world experiences favored a professional development system that valued seniority; the more senior the soldier, the more experiences he or she had. Mentors were prized because they passed on tricks of the trade and provided institutional memory.[56] Learning through simulation can forfeit the advantages of seniority and experience, in a sense leveling the learning base among junior and senior officers.

Second, training with simulations ignores both friction and the human dimension to decision-making. Proficiency is developed by stopping and starting the simulation process at will, erasing mistakes. Simulation exercises can assume away vehicles breaking down or getting stuck in the mud, support showing up late, people getting disoriented and lost. Moreover, during simulation, soldiers get more rest and suffer less stress than they do training in the field. Simulations also disregard subordinate input or low morale. As Robert Bateman argues, even with digitization there are two constants on the battlefield that will not be changed: fear and leadership.[57] Yet simulations do not create opportunities for real training for either of these.

Finally, most Army simulations are still built with attrition-warfare models. Attrition models are part of the "old" way of waging war, as will be discussed further below. Although simulations create training efficiencies, these efficiencies are meaningless if the training portrays a battlefield environment that no longer exists. For example, as Robert Leonhard argues, these simulations "cannot, in their current state, simulate perpetual unreadiness or vulnerability to dislocation,"[58] both key issues for future warfare.

Degraded Decision-Making Ability

Digitizing the battlefield creates a real potential for degraded decision-making ability among tactical leaders. If digitization works as it is supposed to, soldiers will come to depend on the digital icons presented on their computer screens. Those icons will represent the "true" battlefield, and they will become the lifeline by which leaders make decisions and soldiers fight. Yet the digital picture will never fully lift the fog of war. One informed observer argues that the "best truth" we can expect from the digital picture is "80/80/50"—80 percent accuracy for friendly forces, 80 percent accuracy for the environment, and 50 percent accuracy for the enemy.[59] What happens if soldiers assume the picture is perfect when it is not or they freeze in its absence?

There are four ways that digitization can degrade tactical decision-making. First, and most obviously, over-reliance on the computer screen could degrade traditional warfare skills. Basic skills like navigating and calling for fire will become rusty, as the digital equipment automates navigation and target acquisition processes.[60] In a more general statement of this argument, Donn Parker tells us that today's information technologies create "noledge," which is "information that we do not know and that we may never know by study or experience."[61] For example, "noledge" would include the fire control formulas that are used to compute artillery fire sequences. The end user enters the relevant positions, but never sees—and possibly never even learns or remembers—the formulas used to plot trajectories or loads. This is fine until the "noledge" disappears or becomes unavailable as a result of human programming error, enemy hacking, or computer malfunction.

Second, the digital technology could create indecisive leaders who become overwhelmed by data they have not been trained to assimilate. On the one hand, soldiers could become so dependent upon their screens that they lose the ability to infer information from environmental clues. Soldiers will need to compare conflicting inputs from the electronic sensors that created the digital picture and inputs from their eyes and ears. When conflicting data arrive, soldiers may be unable to work out the cognitive dissonance between the screen and "reality" around them. The greater the dissonance, the slower soldiers will be to sort it out, and the less sure their resulting actions and reactions will be.

On the other hand, having a digital picture may obscure awareness of other key inputs to decision-making, especially inputs that cannot be measured with electronic sensors in the digital system. In certain situations, these other inputs—for example, political and environmental conditions—could be more important than the data provided by the screen. Most importantly, leaders will come to rely on the picture their screens present, but the screens may be displaying data, not processed information. Because of bypassing the staff that was supposed to filter and analyze the data for him, the commander now has to perform that task himself. This actually increases his or her cognitive workload, because the commander must perform both analytical and decision-making tasks.[62]

Third, a digital picture of the wider battlefield could actually be harmful—from a morale perspective—for small-unit leaders. This gets at the positive side of the fog of war: what soldiers don't know may not hurt them. In the past, without a wider view of the battlefield, soldiers had to rely on personal information and depend upon their buddies in the local fight. But as Robert Bateman asks, "Are we opening a new Pandora's box by allowing information that has not been available to the common soldier for more than one hundred years out of the realm of command and control and back onto the battlefield?"[63] In other words, what happens when all of the friendly icons in the area get wiped out and the screen is covered with enemy? For example, only a handful of intelligence analysts are authorized to remove an enemy icon from the screen if it is killed, but those analysts frequently are too busy to keep the screens up to date. This delay can make the enemy situation appear worse than it is.[64] Having the wider picture of the battlefield can be reassuring—until the wider battlefield looks worse than the "reality" of the local fight. Faced with this ugly picture, soldiers may go into shock and lose the ability to continue fighting. Or they may unilaterally decide to retreat. Regardless, it adds an additional challenge to the tactical decision-making process.

Finally, inherent in dependence on a digital picture of the battlefield is the risk that this digital picture will be taken away. Jamming and hacking could have disastrous consequences for tactical leaders who have come to rely on digital technology to fight. Yet the more units rely on such technology, the more likely it will become a target for enemy disruption.[65] For example, what happens if the enemy gets into the network and manipulates the digital picture that friendly forces are seeing? What happens when the screen is "lying"? Obviously, two other factors discussed above—cognitive dissonance and rusty non-digital skills (like navigation and calling for fire)—would compound the adverse effect.

Without overstating the threat, computer viruses, equipment failures, faulty software, enemy intrusion, casual hackers, data theft, and overloaded communications pipes all raise the possibility of having degraded digital capabilities. While the AWEs have simulated some disruption of friendly networks, most jamming and hacking has been innocuous so as not to interfere with new equipment testing.[66] As the Army's digitization effort continues, this enemy information warfare simulation should be stepped up, so that leaders can practice making decisions in a degraded information environment.[67]

Top-Down Command Centralization

The Army's nine tenets of battle command include initiative, agility, depth, integration, versatility, flexibility, judgment, intuition and empathy[68]—values embodied in what German military thinkers called *Auftragstaktik*, or tactics based on mission orders. In this method of command, only the outline and minimum goals of an effort are established in advance; the rest is left up to subordinate leaders. Executing mission orders require a mindset and value system that support independent thinking, decisive action, and risk-taking. Many theorists claim that mission orders are key to successful maneuver-based warfare in the fog of war. In other words, when the fog of war exists, the lower commander is better able to make decisions.

The digital battlefield has the potential to change all that. Digitization increases the risk of top-down command centralization, or what Martin van Creveld calls "command by direction."[69] As Thomas Czerwinski points out, the Army's digitization effort does indeed embody the first of van Creveld's "iron rules" for improving command performance—increase information processing capabilities and thus increase the "central directing organ." Czerwinski's analysis suggests, however, that the interventionist capabilities of command by direction, as represented in the Army's digitization effort, risk being self-defeating.[70] Czerwinski is right. In fact, the question is one of efficiency (from command by direction) versus effectiveness (from decentralized command).

The digital technologies being adopted by the Army can encourage top-down centralization in two ways. First, although in theory everyone has the same digital picture of the battlefield at the same time, lower echelons cannot see the whole battlefield as well as higher echelons. This is the rather prosaic result of the size of the computer screens in tanks, Bradleys, and Land Warrior headgear sets. Computer screens in individual weapons platforms are much smaller than in command vehicles and thus can show only a small portion of the battlefield at a time. Moreover, even if they wanted to scroll around and discern the bigger picture, soldiers in these platforms (and dismounted soldiers) have less time for interfacing with their digital screens than commanders and staff officers.[71] As a result, higher echelon commanders have a better picture of the digital battlefield than their subordinates and may choose to intervene to take advantage of the initiative that such digital information provides. In short, on the digital battlefield, the best place to see the battle may be from within the command post vehicle, next to all the "bubba-vision" screens.[72] The net result is that the commander who wants to have access to the best information will be tied to his or her C2 vehicle.

The second effect flows from the first: with their better picture of the battlefield, commanders risk micromanaging their subordinates. This, in turn, discourages lower-level leaders from thinking independently and taking initiatives, and instead trains them to be good at following orders. For example, Bateman in painting a scenario of maneuver at the company level notes that the higher commander is in better position to run the battle:

> Their commander knew as much about each of them as they themselves knew, and knew it at the same time or even before they did. While moving they had little time to look down, manipulate their computer interfaces, and access the same information sequentially that their commander could see simultaneously. Most importantly, they rarely got the chance to make independent decisions regarding the employment of their own units.[73]

The digital battlefield thus risks creating an overly-centralized organization, where the commander sits in his C2 vehicle and merely moves his subordinate puppets around the battlefield. The question then becomes: how does such an organization grow officers to direct the battle via bubbavision, if these same officers spend their whole junior careers obeying orders from the incumbent bubbavision warrior chief?

There is a tension between the abstract knowledge of the higher echelon commander—embodied in the bubbavision—and the particular knowledge of the local commander. Professions have been confronting this tension forever, and for good reason: professions need both kinds of knowledge—abstract and particular—for proper diagnoses and treatment. The top-down command centralization enabled by the digital battlefield risks missing the proper professional balance. What is most *efficient* from the perspective of winning the battle may not be what is most *effective* from the perspective of building a profession.

In sum, there are four possible ways the Army profession may be changed by battlefield digitization. First, the real-time sensor-to-shooter architecture associated with digitization may eliminate the need for some command echelons. Second, simulations are changing the way the Army trains and develops its officer corps—by devaluing real-world experiences, deemphasizing friction and the human dimension in decision-making, and depicting a battlefield built on outmoded assumptions about the nature of warfare today. Third, digital technology increases the potential for degraded decision-making ability—by atrophying traditional warfare skills, creating indecisive leaders unable to assimilate information, and making leaders too reliant on the digital picture in front of them. Finally, top-down command centralization increases the risk that the profession will value efficiency in battle more than effectiveness in training junior members.

Implications of Digitization within the Wider Professional System of "Traditional Warfare"

This section of the chapter will address how the Army's battlefield digitization efforts could change the Army's professional jurisdiction in relation to other professions that participate in conventional warfare. There are three ways that a digital Army could affect the wider professional system of traditional warfare. First, battlefield digitization will add a new player to the professional system—civilian contractors. Second, because the Army will not fight independently in the future, joint connectivity means that digitization cannot be an Army-exclusive jurisdiction. Finally, the Army's preferred approach to exploiting the digital advantage—precision strikes—may not be in its best long-term jurisdictional interest.

Reliance on Civilian Contractors

There is no doubt that information technology is complex to operate and maintain under the best conditions; the difficulty increases exponentially when one employs such equipment tactically, which entails taking it into the field. The equipment comprising the digital battlefield includes computers, radios, satellite terminals, switches, and software—all of it potentially faulty, weak, or insecure. This added complexity and fragility—entailing a special process of diagnosis and treatment—requires a new kind of battlefield professional: the civilian contractor.

Information technology has created a new jurisdiction that the Army is unable to cover by itself, and thus civilian contractors are rapidly becoming indispensable on the digital battlefield.[74] The digital Army relies on these contractors to train and equip its forces and keep its digital systems operational. For example, building the Tactical Internet required integrating the efforts of forty-eight different contractor vendors. During the Task Force XXI AWE in March 1997, 1,200 contractors from these forty-eight vendors were in the field at NTC with the EXFOR, providing advice, maintenance, and technical support.[75] Such support requirements are projected to increase, as the number of digital systems fielded to Army units rises.

Indeed, the whole process of fielding equipment has been changed by the digitization effort. To equip the 4th Infantry Division, the Army has brought together contractors, TRADOC developers, 4th ID soldiers, AMC acquisition officials, and ABCS program officers to create the Central Technical Support Facility at Ft. Hood. By placing all of the Force XXI systems in the same building, marrying up the end users with the contractors, and running 24-hour operations, "in 2 years the Army was able to accomplish what it usually does in 6."[76] In this new approach—called "spiral development"—the Army develops and fields a 20 percent solution and then gets immediate feedback before developing any further.[77] While spiral development certainly makes equipment development and acquisition more efficient, it virtually guarantees civilian contractors a professional jurisdiction in traditional warfare.

There are two major issues associated with this trend of ever-greater civilian involvement. First, civilianizing military functions raises the issue of whether civilians can be compelled to stay on the job in times of crisis.[78] What happens if the civilians decide to leave when the shooting starts? Second, increasing civilian involvement can cloud the combatant/noncombatant distinction that is key to the law of armed conflict. Operating high-tech systems moves civilian contractors from traditional support functions to what are arguably hostile activities, increasing the risk they will become characterized as "unlawful combatants" under international law.[79] To prevent such characterization, some authors suggest establishing a new type of part-time military, without "much of the military regimen" in dress and physical fitness standards.[80] Although a part-time military structure might create the necessary legal framework to compel civilians to remain on the job during crisis, it would not fully replicate the professional ethics, unit cohesion, and training of the Army profession.

Digitization Is Not an Army-Exclusive Jurisdiction

In a future conventional conflict, it is almost inconceivable that the Army would ever fight independently. At a minimum, other U.S. services would be involved; indeed, all U.S. armed forces are necessarily moving in the direction of joint connectivity.[81] Even more likely, the United States would fight as part of a coalition, with combined forces thus being present. Indeed, one of the major issues confronting NATO and other U.S. allies today is the question of interoperability with U.S. forces.[82]

From the perspective of professional jurisdiction, joint connectivity and allied interoperability raise two important points. First, and most importantly, if forces must be able to communicate, they must develop common protocols, operating procedures, and technical standards. Interoperability among the armed services will mean not just adhering to the same technical standards; it will require common doctrine and procedures in network management. Yet, common protocols and standards, by definition, will mean that digitization is not a U.S. Army-exclusive jurisdiction. While not necessarily a bad thing, connectivity with others works against the Army's digitization effort becoming its own niche for future warfare. If everyone can use digital equipment, what prevents other U.S. armed services from usurping this professional jurisdiction from the Army?

Second, because communications must be standardized across the allies, the digital forces must be able to accommodate least-common-denominator communication systems—what the Army calls "analog" or legacy (nondigital) systems. Even disregarding other armed forces, the "analog" question is an important one for the Army, for two reasons. On the one hand, when the Army gets its initial digitized lineup in 2010, the entire force will not be equipped. According to a 1998 plan, only one-third of each division—one fighting brigade slice—will be digitized. Another third will make do with appliqué systems—off-the-shelf commercial gear strapped onto legacy equipment. The third divisional combat brigade and its supporting arms will keep non-digitized legacy weaponry.[83] On the other hand, the "total army" vision raises the issue of digitizing Reserve and Guard components. Impending integrations of National Guard light infantry companies in active duty brigades, the formation of two integrated divisions, and the continued existence of separate enhanced brigades underscore the relevance of reserve component combat arms units.[84] Yet plans for digitizing the reserve components are still uncertain.

The question of digital and analog interoperability is much broader than merely communication among units. Primarily because of the vast differences in command and control capabilities, digital and analog units cannot seamlessly integrate and respond to orders in an equivalent manner. A digital commander can see the battlefield, and thus decides where and how to attack the enemy from a position of advantage before his own unit is seen. In contrast, an analog unit must still go forth and find the enemy physically. This has profound importance for how the forces are employed in battle.

Unless the Guard and Reserve are digitized concurrently with the active component, Army commanders risk having to employ subordinate units in a dissimilar fashion. Even if the Guard and Reserve are digitized in the near future, these units face training challenges that active units do not—they train at lower echelons, less frequently, and for shorter periods of time. These training disadvantages become a critical issue with digital equipment, because digital skills are very perishable.[85] As a result, from an effectiveness perspective, the Guard and Reserve may no longer be useful for digitized warfare—even if they are very useful from a political or budgetary perspective.

Recognizing these differences, David Fautua has argued against assigning *any* digitization-dependent role to the Guard and Reserve. Because of the inherent speed that digitization creates, a digitized force will be designed to conduct "burst operations" as opposed to sustained campaigns.[86] Yet burst operations will not require or even allow for mobilization, which will render reserve forces inconsequential. As a result, Fautua argues that the Reserve and Guard should develop competencies in "shaping" and peace operations missions, because these missions play to their strengths: small-unit cohesion, a comparative inclination for expeditionary-type missions, and less manpower-intensive training requirements.[87]

Exploiting the Digital Advantage: Precision Strikes

Given successful digitization, what will the Army do with its information advantage in conventional warfare? If the AWEs serve as a guide, the Army will choose to convert that information into targeting data to feed to its artillery and other precision strike weapons. During the AWEs, digitized spot reports reached the battalion in five minutes, as opposed to nine minutes under conventional communication means. Moreover, digitized messages needed repeating only four percent of the time, whereas one-third of conventional messages needed repeating. The conclusion was that digitized spot reports save time and can rapidly synchronize direct and indirect fires.[88]

> As a result, both experiments revealed an enormous increase in the logistical demand for more ammunition. Suddenly able to perceive, track, and identify literally thousands of targets, Experimental Force commanders reclined in their natural tendency

> toward caution and long-range fires. Information Age warfare degenerated into a turkey shoot. . . . Army officials pushed beyond rational limitations on available ammunition and allowed Experimental Forces the freedom to blast the enemy into nonexistence at extreme ranges to their hearts' content. Realistic limitations on transportation, ammunition, and the ability to fire into inhabited areas were tacitly ignored. The simulated enemy enthusiastically and obediently cooperated with the cybercarnage, stupidly charging into terrain that was easy to target, unit after unit, never learning and never adapting to fires like a real enemy would.[89]

Warfare is about killing the enemy with battlefield fires faster than he kills you. As James Blackwell observes in the present book (see Chapter 6), the Army's RMA doctrine seems to be insisting that the best way to do this is through long-range destructive fires—fires throughout the depth of the battlefield. If the new information technologies point toward "an increase in the depth, breadth, and height of the battlefield," then they also allow for "armies with high technology [to] place an increasing emphasis on simultaneous strikes throughout the battlespace."[90] In short, one of the key results of Army digitization has been to increase the power of long-range fires. As Leonhard notes, "The senior leaders of the Army have chosen to cast away all the advantages of digitization in favor of reinforcing the demonstrably inapplicable power of long-range fires."[91]

From the perspective of professional jurisdiction, there are two major points here. First, the effectiveness of precision strikes is contingent on information dominance. We cannot expect firepower to destroy enemy capabilities with pinpoint accuracy if we do not know the exact location of enemy and friendly forces. Is information dominance feasible, as digitization proponents insist? This question will be discussed further in the next section.

Second, even if information dominance *is* possible, precision strikes are not necessarily the Army's exclusive jurisdictional task. On the one hand, the Army is using its digital capability to expand its jurisdiction deeper into the battlefield, to take deep fire missions away from the Air Force and the Navy. On the other hand, it may still be more efficient for such long-range fires to be delivered by those services. Given the United States' proclivity to use military force in the most restrained and surgical manner possible, the continued attractiveness of air power alone is bad news for the Army. But the Army is arguably making it worse for itself by choosing to use its digital advantage to feed target data to long-range artillery. This choice may be self-defeating because land power through fire-supported maneuver, not precision strike, is arguably the Army's core jurisdictional task. At the very least, the Army should be aiming for a balance between fire and maneuver. As John Antal warns,

> The ability to win bloodless victories through firepower alone is a siren's call—an idea with the best intentions that has historically produced bad results. War is a complex event, and combat solutions are rarely purely technological. The ascendancy of fires and the dominance of precision strikes in U.S. Army doctrine are a dangerous case in point.[92]

Contrary to the wishes of many, the close fight may never disappear from future battlefields.

In sum, there are three possible ways that battlefield digitization could affect the Army's professional jurisdiction within the wider professional system of traditional warfare. First, information technology has created a new profession on the digital battlefield—civilian contractors. They are rapidly becoming indispensable, but they will usurp some jurisdictional tasks from the Army. Second, interoperability among U.S. armed services and allies will mean that digitization cannot become a U.S. Army-exclusive jurisdiction. Finally, if the AWEs are any indication, the Army is preparing to use its new information advantage to increase the power of long-range fires. Of the three points, this latter one is most critical because it is the one the Army can most control. The Army should be aiming for balance between fire and maneuver, because the close fight will never disappear from all future battlefields. From a jurisdictional standpoint, the Army should appreciate this—land power through maneuver may be the one jurisdictional task the Army can continue to hold exclusively in traditional warfare of the future.

Implications of Digitization within the Widest Professional System of National Security

Up to this point, I have looked at the Army's digitization efforts through the Army's preferred lens. How the Army has chosen to digitize is an expression of its view of its expert knowledge and professional jurisdiction. Army digitization efforts have focused on the high end of the conflict spectrum—conventional warfare. Yet the professional system of national security encompasses much more than conventional warfare. This final section of the chapter looks at the Army's digitization efforts within this wider professional system, arguing that the Army is woefully unprepared in this arena.

The Wider Conflict Spectrum

The Army's digitization efforts are not a new way for the Army to do business; rather, they are a way of doing existing business better. The digital battlefield, as the Army currently conceives it, incrementally improves mechanized and armor forces that still rely on maneuver as the basis of their operational art. With the Objective Force, the Army is looking at newer technologies and possibly radical reform of its organization and doctrine, but it continues to operate within the traditional parameter of warfare—dominating an enemy on the physical battlefield. "Components of [the battlespace] are determined by the maximum capabilities of friendly and enemy forces to acquire and dominate each other by fires and maneuver and in the electromagnetic spectrum."[93] This newest doctrinal definition expands the battlespace in depth, breadth, and height, but it does not fundamentally challenge the kind of conflict that one would find there.

Unfortunately, over the last decade, nontraditional adversaries and forms of conflict have evolved that challenge these assumptions underpinning the Army's vision of the future battlefield. For example, the National Intelligence Council forecasts that through 2015, the most common threats to stability around the world will be internal conflicts, transnational terrorism, and weapons of mass destruction.[94] The 11 September 2001 terrorist attacks on the Pentagon and World Trade Center lend weight to this prediction. Potential future adversaries acknowledge U.S. military superiority but will find ways to exploit vulnerabilities in the U.S. firepower-centered way of war.

> This perception among present and potential adversaries will continue to generate the pursuit of asymmetric capabilities against U.S. forces and interests abroad as well as the territory of the United States. U.S. opponents—state and such non-state actors as drug lords, terrorists, and foreign insurgents—will not want to engage the U.S. military on its terms. They will choose instead political and military strategies designed to dissuade the United States from using force, or, if the United States does use force, to exhaust American will, circumvent or minimize U.S. strengths, and exploit perceived U.S. weaknesses. Asymmetric challenges can arise across the spectrum of conflict that will confront U.S. forces in a theater of operations or on U.S. soil.[95]

Robert Bunker argues that one such asymmetric strategy may be for adversaries to move out of the conventional battlefield, which U.S. forces can see with their digital systems, and into the nonhuman sensing dimension of cyberspace.

> Given this perceptual lens, a terrorist in civilian garb who is standing five meters from a U.S. soldier and whom the U.S. soldier views as a noncombatant is at a much greater battlefield range than a hostile tank that is visible 1000 meters away—and yet is potentially far more dangerous to the soldier than is the tank.[96]

Another asymmetric strategy would be to force the United States into urban combat, with its inherent risk of higher casualties and collateral damage, its ability to degrade U.S. military advantages, and its intensive ground force manpower requirement.[97] Given that half of the world's population today lives in urban areas—and by 2025 that figure is expected to reach 85 percent—this is hardly an unlikely scenario.[98]

Yet the Army's digitization efforts are not enthusiastically addressing these new threats.[99] While the Army did test its Land Warrior dismounted soldier system in the September 2000 Joint Contingency Force AWE, Land Warrior remains significantly less developed than digital systems for heavy weapons platforms. Land Warrior weighs more than 90 pounds, in addition to the food, ammunition, and other gear that a dismounted infantryman carries.[100] As Daniel Bolger says, "Imagine carrying another guy on your back forever and you get the idea. You cannot fight like that no matter how much physical training you do."[101] Moreover, because each Land Warrior system costs about $167,000, they will not be issued to all dismounted soldiers.[102] Bolger argues that by not digitizing all friendly dismounted troops, the Army in effect is telling the enemy where to strike—presumably at the nondigitized soldiers. A more likely scenario, however, is that digitizing dismounted soldiers creates an easier targeting system for the enemy. For example, sniper weapons are being developed to focus on the frequencies transmitted by the Land Warrior system. These weapons would make the sniper's job easier—the sniper would not have to see a body but merely shoot at the source of an electromagnetic transmission at a particular frequency.[103]

Yet these second- and third-order effects of digitization on the lower-end of the combat spectrum are not getting the same attention that conventional warfare issues are. The Army should not assume that forces using current RMA information technology and precision weapons can meet all demands across the conflict spectrum.

Because of their focus on precision engagement and high-speed maneuver, digital forces may be very good at deterring and compelling. But they may not be at all effective when employed in many peacetime engagements, stability operations, or combat operations at the lower end of the conflict spectrum.[104]

By having chosen in its digitization effort to focus on the high end of the conflict spectrum, the Army is expressing its preferred view of its professional jurisdiction—conventional warfare in open terrain. In general, the Army has not focused on peace operations or urban combat because it does not *want* to focus them there. Yet if the Army chooses not to prepare for digitized military operations in these other parts of the conflict spectrum—and political leaders want these kinds of operations conducted—the Army risks having its professional jurisdiction poached. The U.S. Marine Corps' extensive experimentation with new technologies in the urban environment and the private security firms discussed in Deborah Avant's chapter titled "Privatizing Military Training: A Challenge to U.S. Army Professionalism" in the present book (Chapter 9), are merely two cases in point.

In contrast to the Army's heavy Land Warrior system, the Marines have developed a palm-top computer for the urban environment, based on Apple Computer's Newton system.[105] The Marines also completed a three-year series of experimental exercises, collectively called "Urban Warrior," culminating in a March 1999 amphibious urban assault into Oakland, California.[106] These experiments were designed to address the technological challenges of an urban environment—especially the interference caused by concrete walls, phone lines, electronic devices, and urban structures. Additional experimentation through 2002 continued in a program called Project Metropolis.[107] The Marines have also expanded their doctrine for the urban environment, so as to perform more effectively in what former USMC Commandant Gen. Charles Krulak called the "three block war." The idea is that Marines may be expected to provide humanitarian aid, administer the peace, and fight, all within three urban blocks.[108]

Whether such urban doctrinal and technological experimentation is more comprehensive than similar Army efforts is open to debate. Nonetheless, the Marines are certainly doing a much better job of selling their willingness and ability to perform such urban missions to the public. This is very important, because most professional jurisdictions in the realm of national security come from the public's acknowledgment of a particular group's expertise. In a telling explanation of the Marines' increased focus on urban conflict, Lt. Gen. Paul K. Van Riper, the former commander of the USMC Combat Development Command, observed,

> The nation doesn't need a Marine Corps. If the United States Army had disaster on the battlefield, we as a nation would have to re-create the Army. Same obviously for the Air Force or Navy. But if the Marine Corps ever was to fail in a mission . . . there is always someone seeming to say, do we need a Marine Corps? So you'd better offer something unique.[109]

The Marine Corps seems to be making it clear: if the Army won't go into the urban jungle, then the Marines are happy to oblige. In short, if there is a need for military operations at lower levels of the conflict spectrum, and the Army is not ready to perform them, other national security professions will certainly fill the void.

The Media Threat to Information Dominance

Even at the Army's preferred end of the conflict spectrum—conventional warfare—the Army may be missing an important development in the wider professional system of national security. Many of the promised capability enhancements that will accrue to the digital force presume that such a force will possess information dominance—"the degree of information superiority that allows the possessor . . . to achieve an operational advantage in a conflict or to control the situation in operations short of war, while denying those capabilities to the enemy."[110] However, information dominance is probably not possible on the future battlefield, and the Army has not yet realized this.

Even as Joint Vision 2020 insists that "the joint force must be able to take advantage of superior information converted to superior knowledge to achieve 'decision superiority,'"[111] the information explosion engendered by new technologies may not let any combatant achieve superiority, much less dominance. A major reason is the transformation of the news media as they exploit these new technologies as well. In the last five years, a qualitative advance in civilian communications technologies has occurred, and the ability for the military to maintain a secure information environment during conflict has eroded significantly.

While the U.S. Army was learning lessons from the Persian Gulf War, the news media were learning lessons, too. During the Gulf War, journalists had to take their stories to Allied Forward Transmission Units (FTUs), which had satellite links with London and Washington. In the opinion of many journalists, military dispatchers delayed physical transportation of stories to the FTUs and reviewed all stories before they were released for transmission.[112] After the war, journalists vowed never again to be beholden to such military "censorship." Thus, while the Army has used the last decade to create a digital battlefield—prompted in large part by the success of precision strikes and information dominance in the Gulf—the news media have used the same decade to become as independent from the military in case of a future war as possible.

A decade ago, mobile uplinks required a flatbed truck and came with a crew of five journalists. Today, a two-person journalist team would be able to go to war with a digital camera, a wideband cellular phone to uplink to a satellite, and a laptop computer to coordinate the transmission. The equipment fits into two cases and weighs about 100 pounds. "Live from the battlefield" will no longer be primitive or cumbersome—it will be routine.[113] For example, a Thrane & Thrane satellite phone, which can be set up anywhere in 30 seconds and retails for about $3,000, allows "voice and data transmission from any place on the planet outside the Polar zones."[114] Advent Communications offers an International Mobile Satellite (INMARSAT) system that is small enough to be handled by one person.[115] And Aerobureau of McLean, Virginia, already can deploy a self-sustaining flying newsroom. The aircraft is equipped not only with multiple video, audio, and data communications links, but also gyro-stabilized cameras, side- and forward-looking radars, and its own pair of camera-equipped remotely piloted vehicles.[116]

While the U.S. military has begun deploying its system of electronic sensors, so have the news media conglomerates. While most news media cannot own a high-resolution satellite themselves, they can purchase such products on the open market.

Imagery from these satellites is not prohibitively expensive. For example, SpaceImaging, Inc. offers "news pix" for about $500 each, and it will re-task satellite coverage for about $3,000.[117] Most new commercial companies have focused their efforts on supplying relatively high-resolution visible and infrared data (five meters or less).[118] Since five-meter resolution is enough to identify buildings accurately,[119] these satellites create a profound capability for news organizations and other paying customers—including potential adversaries. Moreover, advanced software, along with a cadre of expert ex-military consultants, will enable them to fuse the raw inputs into useful, real-time or near real-time reportage. In other words, the news media will become the "poor man's intelligence service."

In short, the Army's goal of seeking information dominance on the future battlefield is profoundly unrealistic. Charles Dunlap rightly argues that savvy militaries should "focus on developing doctrine and strategies for operating in an environment of information transparency or information parity."[120] At a minimum, the Army needs to recognize that there will be non-traditional professions competing for information on the future battlefield, and it must be ready to operate in an environment lacking the information dominance it always assumes it will have.

Strategic Information Warfare

Finally, the term "information warfare" (IW) is increasingly being used to encompass a broader set of information-age warfare concepts. These emerging concepts are directly tied to the prospect that the ongoing evolution of cyberspace—the global information infrastructure—could bring both new opportunities and new vulnerabilities. In this sense, information warfare is much broader than electronic warfare or anti-C2 network warfare. Instead, in future conflicts, battlefield C4I vulnerabilities may become much less significant than vulnerabilities to the national infrastructure, or what strategic IW theorists call the "zone of interior."[121] The essential U.S infrastructures—those "whose incapacity or destruction would have a debilitating impact on our defense or economic security"[122]—include systems like the public telephone network, securities and commodities exchanges, water-supply systems, utility networks, air transportation, highways, and the Internet.[123]

Today, strategic IW remains mostly a theory, although the 11 September 2001 attacks on the World Trade Center and the Pentagon, which temporarily crippled air travel and the financial markets, could be characterized as anti-infrastructure warfare. Until anti-infrastructure warfare is proven ineffective, states and nonstate actors with the capacity to attempt it probably will, because it appears potentially effective, less risky, and has lower entry costs than other forms of armed conflict. Obviously, the Army is not the right profession to handle these threats. When it becomes possible to wage war with a handful of computers, a whole group of new organizations will try to move into this professional jurisdiction—military services, other government agencies, security divisions of transnational corporations, and private security firms. What is important to note here is that the Army must stay abreast of these emerging forms of warfare—and of the competitors that such warfare will bring into the professional system of national security.

In sum, how the Army has chosen to digitize is an expression of its view of its expert knowledge and professional jurisdiction. Army digitization efforts have focused on the high end of the conflict spectrum—conventional warfare. Yet the professional system of national security encompasses much more than conventional warfare, as this section has shown. Emerging threats across the conflict spectrum suggest that conventional warfare in nonurban terrain—the way the Army would *choose* to fight—may not be what future warfare is all about. Moreover, even in its preferred part of the conflict spectrum, Army digitization assumptions about information dominance are profoundly unrealistic. Looking at digitization from the perspective of the wider professional system of national security suggests that the Army is unprepared for the future. By consciously or unconsciously ignoring nontraditional forms of warfare all along the conflict spectrum—and the potential professional competitors that such warfare will bring about—the Army risks being caught unaware and disregarded. The net result could be a bankrupt Army profession that is incapable of countering the threats that actually endanger our nation's security.

Conclusion

How the Army has chosen to adopt new information technologies says much about its future professional jurisdiction. As this analysis has shown, information technologies from the contemporary RMA are both a subjective and an objective force for jurisdictional change. This chapter has examined the implications of Army digitization in light of three concentric visions of the professional system—within the Army profession itself, within the wider professional system of traditional warfare, and within the widest professional system of national security. The analysis has suggested that although the Army has been capitalizing on technological opportunities to enhance its conventional war-fighting capability, it seems unwilling or unable to address the wider threats that such technological change could bring to its traditional jurisdiction.

By choosing to focus on the high end of the conflict spectrum in its digitization effort, the Army is expressing its preferred view of its professional jurisdiction—conventional warfare in nonurban terrain. In general, the Army has not focused on peace operations or urban combat because it does not *want* to focus on them. Yet, if the Army chooses not to prepare for digitized military operations in these other parts of the conflict spectrum—and political leaders want these kinds of operations conducted—the Army risks losing part of its professional jurisdiction.

In other words, the Army profession is focusing on a jurisdiction that the client may not want and may not need. And as Abbott warns, no profession delivering unwanted or ineffective services can stand indefinitely against competent outsiders, however powerful it may be.[124] Especially with military professions, jurisdictional contests are often decided by client choice. If there is no demand for a profession's services, the profession will inevitably lose its jurisdictional claims. What makes this conclusion so unsettling is that battlefield digitization was originally envisioned as an Army effort to regain its strategic relevancy in the post-Cold War world. Ironically, the way the Army has chosen to digitize may be making it more irrelevant for future conflict than ever.

Notes

1. Steve J. Mains, "Adapting Doctrine to Knowledge-Based Warfare," *Military Review* 77 (March-April 1997), available from *http://www.cgsc.army.mil/milrev/english.marapr97/mains.html*; Internet; accessed 6 March 2001.
2. "The Army Transformation," available at *http://www.army.mil/armyvision/transform.html*; Internet; accessed 12 October 2000. See also Scott R. Gourley and Mark Hewish, "U.S. Army Refines Plans for Medium-weight Combat Force," *International Defense Review* 33, no. 4 (1 April 2000).
3. Briefing by Chris Lamb, Acting Deputy Assistant Secretary of Defense for Requirements, Plans and Counterproliferation Policy, 14 December 2000 at Georgetown University.
4. The Interim Brigade Combat Teams are designed to bridge the gap in capabilities between today and the Objective Force. The first IBCTs are based at Fort Lewis, Washington. Preconfigured in ready-to-fight combined arms packages, each IBCT is intended to deploy within 96 hours. For more information about the "Transformation Vanguard" and the IBCT, see the September-October 2000 issue of *Military Review*. See also Henry S. Kenyon, "Army Sharpens Tip of the Spear," *Signal Magazine* (April 2001), available at *http://www.us.net/signal/currentissue/april01/army-april.html*; Internet; accessed 9 April 2001.
5. For an audio tour introduction to the digitization effort, see "Digitization 101," available at *http://www.armyexperiment.net/aepublic/digit_101/digi.html*; Internet; accessed 20 October 2000.
6. Karl von Clausewitz, *On War*, ed. and trans. Michael Howard and Peter Paret (Princeton, NJ: Princeton University Press, 1976), 101. In Clausewitz's words, "War is the realm of uncertainty; three quarters of the factors on which action in war are based are wrapped in a fog of greater or lesser uncertainty."
7. The hypothesis the Army has been testing during its digitization experiments is as follows: "If information age battle command capabilities and connectivity exist across all battle operating systems and functions, then enhancements in lethality, survivability, and tempo will be achieved." See Ron Gregory, "Army XXI: Issues Associated with Development of Doctrine and TTP [Tactics, Techniques, and Procedures] for the Digitized Force," available at *http://www-tradoc.army.mil/jadd/adxxi2/sld001.html*; Internet; accessed 20 October 2000.
8. Andrew Abbott, *The System of Professions: An Essay on the Division of Expert Labor* (Chicago, IL: University of Chicago Press, 1988), 8.
9. Abbott, chapters 2-4.
10. For a good introduction to the contemporary RMA debate, see Andrew Krepinevich, "From Cavalry to Computer," *The National Interest*, no. 37 (Fall 1994): 30-43; and Eliot A. Cohen, "The Revolution in Military Affairs," *Foreign Affairs* 75 (March-April 1996): 20-36.
11. William A. Owens, *Lifting the Fog of War* (New York: Farrar, Strauss, Cirouk, 2000), 118. See also Owens, "The Emerging US System-of-Systems," Institute for National Strategic Studies, *Strategic Forum*, No. 63, February 1996; and Joseph Nye, Jr., and Owens, "America's Information Edge," *Foreign Affairs* 75 (March/April 1996).
12. Michael Mazarr, *The Revolution in Military Affairs: A Framework for Defense Planning* (Carlisle Barracks, PA: U.S. Army War College Strategic Studies Institute, 10 June 1994), 16-21.
13. Steven Metz and James Kievit, *Strategy and the Revolution in Military Affairs: From Theory to Policy* (Carlisle Barracks, PA: U.S. Army War College, Strategic Studies Institute, 10 June 1994), 5.
14. Robert R. Leonhard, "A Culture of Velocity," in *Digital War: A View from the Front Lines*, ed. Robert L. Bateman (San Francisco, CA: Presidio Press, 1999), 146.
15. Alvin and Heidi Toffler, *War and Anti-War: Survival at the Dawn of the 21st Century* (Boston, MA: Little, Brown and Company, 1993), 65-79.

16. "The 2001 Trained and Ready Division," draft document, 13, available at *http://www-cgsc.army.mil/dao/fa30/New%20Information/DIGITAL%20TRAINING.html*; Internet; accessed 16 March 2001.
17. "Digitization 101," available at *http://www.armyexperiment.net/aepublic/digit_101/digi.html*; Internet; accessed 16 March 2001. TRADOC Pamphlet 525-5 (Fort Monroe, VA: Headquarters U.S. Army Training and Doctrine Command, 1 August 1994)defines situational awareness as: "The ability to have accurate and real-time information of friendly, enemy, neutral, and noncombatant locations; a common, relevant picture of the battlefield scaled to specific level of interest and special needs."
18. Chapter 3, "Force XXI Operations," TRADOC Pamphlet 525-5.
19. Owens, *Lifting the Fog of War*, 117-18.
20. This is based on the widely-quoted concept of Air Force Col. John Boyd's observation-orientation-decision-action (OODA) loop.
21. Edward Mann, "Desert Storm: The First Information War?" *Airpower Journal* (Winter 1994): 6-7.
22. For a more detailed history of Army attempts to automate C2 systems, see Elizabeth A. Stanley, *Evolutionary Technology in the Current Revolution in Military Affairs: The Army Tactical Command and Control System* (Carlisle Barracks, PA: U.S. Army War College, Strategic Studies Institute, 25 March 1998).
23. As one early digitization official noted, "Without General Sullivan, it would have never happened. He was a zealot." Interview with retired Army Col. Dominick Basil, former SINCGARS program manager and former Deputy Commander of CECOM, May 1997.
24. More information about early Army experiments is available at *http://www.armyexperiment.net/aepublic/previous_ae/default.html*; Internet; accessed 24 January 2001.
25. Gen. Gordon R. Sullivan, "Memorandum dated 7 March 1994, re: Building the Force for the 21st Century—Force XXI," in Gordon R. Sullivan and Anthony M. Coroalles, *Seeing the Elephant: Leading America's Army into the Twenty-First Century* (Cambridge, MA: Institute for Foreign Policy Analysis, 1995), 66.
26. "Army Force Posture FY00," available at *http://www.army.mil/aps/00/aps00.htmlpagepage51*; Internet; accessed 16 March 2001.
27. The two IBCTs are 3rd Brigade, 2nd Infantry Division and 1st Brigade, 25th Infantry Division, both based at Fort Lewis. General Shinseki has announced that there will eventually be between five and eight IBCTs. See "IBCTs are first step in creating objective force," available at *http://www.dtic.mil/armlink/news/May2000/a20000515gordonibct.html*; Internet; accessed 12 October 2000.
28. Briefing by Col. Wilt Ham, Chief Operations and Fielding Division, Army Digitization Office, 15 January 1999; Barbara Starr, "USA Identifies Force for Future Warfighting Tests," *Jane's Defence Weekly* 29, no. 13 (1 April 1998): 6; Greg Seigle, "US Regiment Will Become Digitized Strike Force," *Jane's Defence Weekly* 30, no. 16 (21October 1998): 17.
29. The Joint Contingency Force Advanced Warfighting Experiment (JCF AWE) was held 8-20 September 2000 at Ft. Polk. Most of the Army and Marine training involved military operations in an urban environment. The Land Warrior system was the premier digital equipment tested at the JCF-AWE. See "JCF-AWE mission begins," available at *http://www.dtic.mil/armylink/news/Sep2000/a20000908awebegins.html*; "Final JCF-AWE Attack Shows Equipment Works," available at *http://www.dtic.mil/armylink/news/Sep2000/a20000920finaljcf.html*; William A. Graves, "Land Warrior Night Live-fire Shed Light on Future," available at *http://www.dtic.mil/armylink/news/Sep2000/a20000918mout.html*. All available on Internet, accessed 16 March 2001. See also Jim Caldwell, "JCF AWE To Help Army Find Answers to Digitizing Light Forces," available at *http://www.tradoc.army.mil/pao/JCFRound.html*; Internet; accessed 12 October 2000.

30. Mark Hanna, "Task Force XXI: The Army's Digital Experiment," *Strategic Studies*, no. 119 (July 1997).
31. Mark Thompson, "Wired for War," *Time*, 31 March 1997, 72-73.
32. Elke Hutto, "Reaping the Battlefield Digitization Harvest," *International Defense Review Special Report*, Quarterly Report No. 2 (1 June 1998): 3.
33. Rupert Pengelley, "International Digitizers Wrestle with Reality," *International Defense Review* 30, no. 9 (1 September 1997): 38.
34. Hanna, "Task Force XXI."
35. Scott R. Gourley, "US Glimpses a Digitized Future," *International Defense Review* 30, no. 9 (1 September 1997): 49; Gourley, "US Army Conducts Task Force XXI Experiment," *International Defense Review* 30, no. 6 (1 June 1997): 6; "Army Digitization Experiment Shows Information Highway Is Bumpy," *C4I News* 5, no. 7 (10 April 1997).
36. Interview with then Maj. Marcus Sachs, 4th Infantry Division DAMO, May 1997.
37. Chapter 1, "Doctrinal Concepts," *The Army Battle Command System*, Field Manual 24-7, available at *http://www.adtdl.army.mil/cgi-bin/atdl.dll/fm/24-7/toc.html*; Internet; accessed 17 March 2001.
38. Gourley, "US Glimpses a Digitized Future."
39. Frank Tiboni, "Global Command System Speeds Planning," *Defense News,* 25 September 2000, 4.
40. Each battlefield functional area has a different automated system in ATCCS. They include the Maneuver Control System (MCS), the Advanced Field Artillery Tactical Data System (AFATDS), the All-Source Analysis System (ASAS), the Combat Service Support Control System (CSSCS), and the Air and Missile Defense Work Station (AMDWS, formerly known as FADC2I). For more information about ATCCS, see Stanley, *Evolutionary Technology*, or "The ABCS Primer," available at *http://www.armyexperiment.net/aepublic/abcs/default.html*; Internet; accessed 24 January 2001.
41. Interview with Sachs.
42. On weapons that already have an embedded computer, like the M1A1 Abrams tank and the Bradley infantry fighting vehicle, FBCB2 software is added to the existing computer system.
43. "Battlefield Automation: Acquisition Issues Facing the Army Battle Command, Brigade and Below Program," Report No. NSIAD-98-140 (Washington, DC: U.S. Government Accounting Office, 30 June 1998).
44. Today, this capability is embodied in the Enhanced Position Location and Reporting System (EPLRS), which constantly transmits an update of the soldier's or weapon platform's current location.
45. Most systems still have a Single Channel Ground and Airborne Radio System (SINCGARS) radio, although there has also been experimentation with newer radios capable of moving the large amounts of data that digital forces require. These other systems include the Near-term Digital Radio (NDTR), the Joint Combat Information Terminal (JCIT), and the Joint Tactical Radio System (JTRS). See Steve T. Wall, "Multifunctional Communication on the Battlefield," *Army Logistician* (July-August 2000), available at *http://www.amlc.army.mil/alog/JulAug00/MS507.html*; Internet; accessed 24 January 2001; David C. Isby, "US Army Considers Alternatives to JTRS," *Jane's Defence Upgrades* 3, no. 3 (9 January 1999): 3; "Improved Radio Eases Digitization Traffic," *International Defense Digest* 31, no. 10 (1 October 1998): 5; and Bruce D. Nordwall, "Software Radios Give Army Helo C2 Systems," *Aviation Week and Space Technology* 151, no. 11 (13 September 1999): 85.
46. Barbara Jezior, "The Land Warrior," in *AY97 Compendium Army After Next Project,* ed. Douglas V. Johnson, II (Carlisle Barracks, PA: U.S. Army War College, Strategic Studies Institute, 6 April 1998); "Army's Restructured Land Warrior Program Needs More Oversight," Report No. NSIAD-00-28 (Washington, DC: U.S. Government Accounting Office, 15 December 1999).

47. David Wood, "Digitized Unit Exploits Surveillance Technology," *Army Times*, 7 May 2001, 16; Greg Jaffe, "In the New Military, Technology May Alter Chain of Command: The Army Bets Its Battlefield Success On Soldiers Armed with Better Data," *Wall Street Journal*, 30 March 2001, 1; Victor Hallis, "4th ID shows information superiority at DCX," available at *http://www.dtic.mil/armylink/news/apr2001/a20010414dcx.html*; Internet; accessed 20 April 2001.
48. "Battlefield Automation: Performance Uncertainties Are Likely When Army Fields First Digitized Division," Letter Report No. NSIAD-99-150 (Washington, DC: U.S. Government Accounting Office, 27 July 1999). By 2005, one corps and five and one-half divisions should be digitized, according to Ron Gregory, "Army XXI: Issues Associated with Development of Doctrine and TTP for the Digitized Force," available at *http://www-tradoc.army.mil/jadd/adxxi2/sld001.html*; Internet; accessed 20 October 2000.
49. Scott Gourley, "4th Infantry Division Sets Trend in Motion," *Jane's Defence Weekly* 28, no. 13 (1 October 1997): 24.
50. Gen. John M. Shalikashvili, *National Military Strategy*, available at *http://www.dtic.mil/jcs/core/nms.html*; Internet; accessed 20 March 2001.
51. "The Army Vision," available at *http://www.army.mil/armyvision/armyvis.html*; Internet; accessed 20 March 2001.
52. Michael Evans, "Fabrizio's Choice: Organizational Change and the Revolution in Military Affairs Debate," *National Security Studies Quarterly* 7, no. 1 (Winter 2001): 15; Francis Fukuyama and Abram N. Shulsky, "Military Organization in the Information Age: Lessons from the World of Business," in *Strategic Appraisal: The Changing Rules of Informational Warfare*, eds. Zalmay M. Khalilzad and John P. White (Santa Monica, CA: RAND, 1999), 327-60.
53. Douglas A. Macgregor, *Breaking the Phalanx: A New Design for Landpower in the 21st Century* (Westport, CT: Praeger, 1997); John R. Brinkerhoff, "The Brigade-Based New Army," *Parameters* 27 (Autumn 1997): 60-72. See also David Fastabend, "An Appraisal of 'The Brigade-Based New Army,'" *Parameters* 27 (Autumn 1997): 73-81. For a history of the Army divisional structure, see Richard W. Kedzior, *Evolution and Endurance: The US Army Division in the Twentieth Century* (Santa Monica, CA: RAND, 2000).
54. National Simulation Center, *Training with Simulations* (Fort Leavenworth, KS: Combined Arms Center, November 1996), 34-35.
55. Leonhard, 135.
56. Paul T. Harig, "The Digital General: Reflections on Leadership in the Post-Information Age," *Parameters* 26 (Autumn 1996): 133-40.
57. Bateman, "Introduction," in *Digital War*, 2.
58. Leonhard, 150.
59. Daniel P. Bolger, "The Electric Pawn: Prospects for Light Forces on the Digital Battlefield," in *Digital War*, 118.
60. Ralf Zimmerman makes a similar point in his column, "In a technology-filled battlefield, let's not forget the basics of combat," *Army Times*, 7 May 2001, 62.
61. Donn B. Parker, *Fighting Computer Crime: A New Framework for Protecting Information* (New York: J. Wiley, 1998).
62. Lawrence G. Shattuck, "A Proposal for Designing Cognitive Aids for Commanders for the 21st Century," in *Future Leadership, Old Issues, New Methods*, ed. Douglas V. Johnson, II (Carlisle Barracks, PA: U.S. Army War College, Strategic Studies Institute, June 2000), 104.
63. Bateman, "Pandora's Box," *Digital War*, 22.
64. Jaffe, 16.
65. Robert H. Scales, Jr., "Adaptive Enemies: Achieving Victory by Avoiding Defeat," *Joint Forces Quarterly* No. 23 (Autumn/Winter 99-00): 7-14.
66. For example, when the Land Information Warfare Activity (LIWA) provided hackers for the DAWE, the hackers would leave "calling cards" reading "You've been hacked" so that oper-

ators would know when they had been invaded. These were used to avoid running the risk of interfering with the ongoing AWE. Hackers did not try to manipulate the friendly systems or use the information they obtained through the hacking to help the enemy forces. See Hutto, 3.

67. GAO made a similar point in its recent study, "Battlefield Automation: Opportunities to Improve the Army's Information Protection Effort," Letter Report No. NSIAD-99-166 (Washington, DC: U.S. Government Accounting Office, August 11, 1999).
68. Department of the Army, *Battle Command, Draft 2.1* (Fort Leavenworth, KS: Battle Command Battle Lab, April 1994).
69. Martin van Creveld, *Command in War* (Cambridge, MA: Harvard University Press, 1985).
70. Thomas J. Czerwinski, "Command and Control at the Crossroads," *Parameters* 26 (Autumn 1996): 121-32.
71. Hanna, 3.
72. "Bubbavision" is the term that many soldiers use to describe the large television monitors hooked up to the ATCCS systems in command posts. See also Katherine McIntire Peters, "Split Decision," *Government Executive*, October 1999, 43.
73. Bateman, "Pandora's Box," 20.
74. Katherine McIntire Peters, "Civilians at War," *Government Executive*, July 1996, 23; David Silverberg, "Crossing Computing's Cultural Chasm," *Armed Forces Journal International*, February 1997, 38-39; Bryan Bender, "Defense Contractors Quickly Becoming Surrogate Warriors," *Defense Daily*, 28 March 1997, 490.
75. Hanna, 4.
76. Information from Col. Steven Boutrelle about the TFXXI AWE at the Rayburn House Office Building on 4 June 1997.
77. For more information on spiral development, see Stanley, 48.
78. Charles J. Dunlap, Jr, "Organizational Change and the New Technologies of War," available at *http://www.usafa.af.mil/jscope/JSCOPE98/Dunlap98.html*; Internet; accessed 30 March 2001.
79. An unlawful combatant is an individual who is not authorized to take a direct part in hostilities but does. The term is frequently used also to refer to otherwise privileged combatants or noncombatants in the armed forces who use their protected status as a shield to engage in hostilities. Unlawful combatants are a proper object of attack while engaging as combatants, and if captured they may be tried and punished. See Dunlap, "Organizational Change."
80. Stephen Bryer, "New Era of Warfare Demands Technology Reserve Force," *Defense News*, 17-23 March 1997, 27; Brig. Gen. Bruce M. Lawlor, ARNG, "Information Corps," *Armed Forces Journal International*, January 1998, 26-28.
81. See, for example, the effort to develop a joint digital radio, n. 45.
82. For more information on allied efforts to digitize their forces, see Rupert Pengelly, "International Digitizers," and Pengelly, "Hooking up to the digital battlefield: BMS upgrades for armoured vehicles," *Jane's Defence Systems Modernisation* 9, no. 9 (1 December 1996): 20. On the wider issue of joint connectivity and its effects on military effectiveness, see Nora Bensahel, "The Coalition Paradox: The Politics of Military Cooperation" (Ph.D. diss., Stanford University, 1999).
83. Jim Tice, "How Will You Fare in XXI?" *Army Times*, 22 June 1998, 12, 15.
84. Gen. Dennis J. Reimer, *One Team, One Fight, One Future* (Washington, DC: Office of the Chief of Staff of the Army, 30 June 1998).
85. For more information about the challenges of digital training for National Guard units, see Mike Pryor, "Digitization, Simulations, and the Future of the Army National Guard," in *Digital War*, 81-112.
86. Steven Metz, "Which Army After Next? The Strategic Implications of Alternative Futures," *Parameters* 27 (Autumn 1997): 15-26.

87. David T. Fautua, "Transforming the Reserve Components," *Military Review* 80 (September/October 2000): 57-67.
88. John A. Antal, "The End of Maneuver," in *Digital War*, 161.
89. Leonhard, 137.
90. Headquarters Department of the Army, US Army Training and Doctrine Command, "Force XXI Operations," *TRADOC Pamphlet 525-5* (August 1, 1994), Section 2-3, available at *http://www.tradoc.army.mil/tpubs/pams/p525-5toc.html*; Internet; accessed 24 January 2001.
91. Leonhard, 147.
92. Antal, 164.
93. "Force XXI Operations," *TRADOC Pamphlet 525-5*.
94. National Intelligence Council, *Global Trends 2015*, available at *http://www.cia.gov/cia/publications/globaltrends2015/index.html*; Internet; accessed 26 March 2001.
95. "Reacting to US Military Superiority," *Global Trends 2015*.
96. Robert J. Bunker, "Advanced Battlespace and Cybermaneuver Concepts: Implications for Force XXI," *Parameters* 26 (Autumn 1996): 117.
97. James Kitfield, "War in the Urban Jungles," *Air Force Magazine* 81, no. 12 (December 1998); Jennifer Morrison Taw and Bruce Hoffman, *The Urbanization of Insurgency: The Potential Challenge to US Army Operations* (Santa Monica, CA: RAND, 1994); Daryl G. Press, *Urban Warfare: Options, Problems, and the Future*, Conference Summary, January 1999, available at *http://web.mit.edu/ssp/Publication/urbanwarfare/urbanwarfare.html*; Internet; accessed 30 March 2001.
98. DoD officials believe that urban operations may require nine times the ground forces of operations on open terrain. For more information about the Army and Marine efforts to prepare for urban conflict, see U.S. GAO, "Military Capabilities: Focused Attention Needed to prepare U.S. Forces for Combat in Urban Areas," Report No. NSIAD-00-63NI (Washington, DC: U.S. Government Accounting Office, February 2000).
99. Robert F. Hahn II and Bonnie Jezior, "Urban Warfare and the Urban Warfighter in 2025," *Parameters* 29 (Summer 1999): 74-86.
100. See GAO, "Battlefield Automation: Army's Restructured Land Warrior Program Needs More Oversight." See also Maj. Gen. (Ret.) John R. Greenway, "The Soldier Is the System," *Military Information Technology Online*, March 2001, available at *http://www.mit-kmi.com/features/5_3_Art3.html*; Internet; accessed 24 April 2001.
101. Bolger, 123.
102. In December 1999, the Army planned to buy 34,000 Land Warrior systems. See GAO, "Battlefield Automation: Army's Restructured Land Warrior Program Needs More Oversight."
103. Interview with Tim Rosenberg, CEO, White Wolf Consulting, an information security consulting firm, 18 March 2001.
104. For some force structure implications of this line of argument, see William T. Johnsen, *Force Planning Considerations for Army XXI* (Carlisle Barracks, PA: U.S. Army War College, Strategic Studies Institute, 18 February 1998).
105. John Stein Monroe, "Advanced Warfighting: Hunter Warrior Puts Technology to Test," *Federal Computer Week*, 17 March 1997, available at *http://208.201.97.5/pubs/fcw/1997/0317/hunter.html*; Daniel Verton, "Marines Take IT to Town," *Federal Computer Week*, 8 December 1998, available at *http://208.201.97.5/pubs/fcw/1998/1207/fcw-agmarines-12-7-98.html*. Both available on the Internet, accessed 30 March 2001.
106. John J. Stanton, "Training Marines for War in the City: Drills in Simulated Urban Zones Underscore the Need for New Equipment," available at *http://www.geocities.com/Pentagon/6453/jstanton.html*; Eric Brazil and Larry D. Hatfield, "Marines Open War," *The San Francisco Examiner*, 15 March 1999, available at *http://www.sfgate.com/cgi-bin/archive/1999/03/15/news8730.html*. See also the USMC Urban Warrior official website at

http://131.84.1.60/specials/urbanwarrior.html. All available on the Internet, accessed 30 March 2001.

107. "Project Metropolis," available at *http://www.mcwl.quantico.usmc.mil/promet.html*; Internet; accessed 30 March 2001.

108. Dominick Donald, "Rebooting the Marines," *Red Herring Magazine*, 1 April 2000, available at *http://www.redherring.com*; Internet; accessed 30 March 2001; Joel Garreau, "Point Men for a Revolution: Can the Marines Survive a Shift from Platoons to Networks?" *Washington Post*, 6 March 1999, A1.

109. Garreau, A18.

110. "The 2001 Trained and Ready Division," 13.

111. Joint Vision 2020, 11. Text-only version available at *http://www.dtic.mil/jv2020/jv2020.pdf*; Internet; accessed 30 March 2001.

112. Philip Taylor, *War and the Media: Propaganda and Persuasion in the Gulf War,* 2nd Edition (New York: Manchester University Press, 1998), 56.

113. For more information about the new media capabilities, see Steven Livingston, "Remote Sensing Technology and the News Media," in *Commercial Observation Satellites: At the Leading Edge of Global Transparency*, eds. John Baker, Kevin O'Connell, and Ray Williamson (Santa Monica, CA: Rand Corporation and the American Society for Photogrammetry and Remote Sensing, 2000); Livingston, "Transparency and the News Media," in *Power and Conflict in the Age of Transparency*, eds. Bernard Finel and Kristin Lord (New York: St. Martin's Press, 2000); Barrie Dunsmore, "Live from the Battlefield," in *Politics and the Press: The News Media and Their Influences*, ed. Pippa Norris (Boulder, CO: Lynne Rienner, 1997), 237-73; and Ed Offley, "The Military-Media Relationship in the Digital Age," in *Digital War*, 257-91.

114. Nicholas Kristof, "Have Adapter, Will Travel—A Foreign Correspondent Reflects on the Technotricks of Life on, and off, the Road," *New York Times*, 24 September 1998.

115. Livingston, "Transparency and the News Media," 275.

116. Charles J. Dunlap, "21st Century Land Warfare: Four Dangerous Myths," *Parameters* 27 (Autumn 1997): 27-37.

117. *http://www.spaceimaging.com/aboutus/corpFAQ.htmlpricing*; Internet; accessed 14 May 2001.

118. Derek D. Smith, "A Double-Edged Sword: Controlling the Proliferation of Dual-Use Satellite Systems," *National Security Studies Quarterly* 7, no. 2 (Spring 2001): 31-68; Ann M. Florini and Yahya Dehqanzada, "Commercial Satellite Imagery Comes of Age," *Issues in Space and Technology* 16, no. 1 (Fall 1999): 45-52; John C. Baker, and Ray A. Williamson, "The Implications of Emerging Satellite Technologies for Global Transparency and International Security," in *Power and Conflict in the Age of Transparency*, 221-55; Vipin Gupta, "New Satellite Images for Sale," *International Security* 20, no. 2 (Summer 1995): 94-125; and George J. Tahu, John C. Baker, and Kevin M. O'Connell, "Expanding Global Access to Civilian and Commercial Remote Sensing Data: Implications and Policy Issues," *Space Policy* 4 (August 1998): 179-88.

119. Bruce D. Berkowitz and Allan E. Goodman, *Best Truth: Intelligence in the Information Age* (New Haven, CT: Yale University Press, 2000), 53.

120. Dunlap, 31.

121. Roger C. Molander, Andrew S. Riddile, and Peter A. Wilson, "Strategic Information Warfare: A New Face of War," *Parameters* 26 (Autumn 1996): 81-92.

122. Roger C. Molander, Peter A. Wilson, and Robert H. Anderson, "US Strategic Vulnerabilities: Threats Against Society," in *Strategic Appraisal*, 257.

123. For more information about the essential U.S. infrastructures, see the recent study by the President's Commission on Critical Infrastructure Protection (PCCIP), *Critical Foundations: Protecting America's Infrastructures* (Washington, DC: GPO, 1997).

124. Abbott, 135.

8 The Army and Operations Other Than War: Expanding Professional Jurisdiction

Thomas L. McNaugher

Introduction

No one familiar with the Army's Vietnam experience doubts the organization's deep attachment to its core professional tenet: "We fight and win the nation's wars." In the early 1970s, an Army deeply scarred by its experience in Vietnam focused its attention on "winning the land battle," especially in Europe against Soviet forces.[1] It effectively turned its back on the Vietnam experience; military and political leaders alike were understandably reluctant to engage again in Vietnam-like political-military missions lying largely outside the boundaries of popular support. To be sure, the Army was always willing to do what the nation asked of it, but nonwar missions came to be seen as "lesser included cases" for a war-fighting Army—that is, missions that could be handled, easily enough, by an Army organized and trained for major war. With the Soviet Union supplying focus and the Reagan administration supplying ample funds, the U.S. Army prepared to fight and win the nation's wars with a passion—a passion that paid off, one might say, in Operation Desert Storm.

Historians would point out that the Army of the 1970s was merely returning to its roots in affirming a war-fighting creed. They would cite as evidence the Army's 1923 *Field Service Regulations*, which focused on a mission the Army had undertaken but once since the Civil War, namely, decisive war against a sophisticated, well-armed and trained enemy force.[2] To the contrary, for nearly half a century the Army had done mainly nonwars—or "military operations other than war" (MOOTW), to use the contemporary expression: occupying the postbellum South, securing the nation's expanding frontier, and stabilizing imperial outposts in places like the Philippines. Yet the Army made no effort to codify the lessons of that experience, leaving that task to the U.S. Marine Corps, which published its famous "small wars" manual in 1940. In 1923 as well as in 1990, these missions were peripheral to the core job of preparing for major war.

It is against this backdrop that the Army's post-Cold War experience takes on importance, especially to the student of the Army profession. For the Army, the twentieth century ended as it began, with the organization focused principally on war-fighting or, more specifically, remaining prepared to fight two major theater wars (MTWs) nearly simultaneously, while its forces actually conducted a series of nonwars in places like Somalia, Haiti, Bosnia, Kosovo, and the Sinai. In contrast to its behavior at the beginning of the last century, however, in the 1990s the Army's senior leaders came to embrace MOOTW as an important part of the organization's professional jurisdiction. As Gen. Montgomery Meigs, commanding Army forces in Europe, said of the young officers deployed to Bosnia and Kosovo, "What you're getting out of this is a corps of leaders in the Army, young leaders coming up, who are very, very, very tough in experience. That is worth its weight in gold."[3] A chapter on MOOTW has crept into FM 100-5 (now FM 3-0), the Army's capstone operations field manual, albeit not without controversy.[4] And in launching a major transformation of the Army in October 1999, the Chief of Staff, Gen. Eric Shinseki, made clear that he was seeking a force that will be "persuasive in peace" as well as "compelling in war." He seeks a "full-spectrum" force, "dominant across a range of missions" stretching from "humanitarian assistance and disaster relief to peacekeeping and peace-making to major theater wars."[5]

Finally, the organization has consistently opposed further internal role specialization as well as the establishment of a national "constabulary force," directed (perhaps) by the State Department, to handle peacekeeping and other nonwar missions.[6] While fighting and winning wars remains the Army's "non-negotiable contract with the American people,"[7] the Army evidently wants and is competing for the "lesser" missions, and is preparing its forces to handle them.

Much of the change so far is rhetorical—not a trivial change in itself given the Army's past attachment to war-fighting. Yet precisely because the Army's focus on war-fighting has been so deeply ingrained, even in the face of nonwar experiences in the past, those concerned about the Army profession want to know if the change is taking hold of the organization in reality as well as rhetorically. Is the Army's conception of its expert knowledge and its professional jurisdiction, so long confined to "fighting and winning the nation's wars," really expanding to include the full spectrum of conflict? To what extent is the Army really preparing to be a full-spectrum force?

This chapter seeks answers to these questions based on the Army's experience with MOOTW over the past decade. It looks at the kind of hybrid units the Army put together to handle these missions, at how these units were assembled and trained, and at the effects these processes had on the wider Army. The chapter highlights the Army's own views of the demands these nonwar operations place on Army leaders. And it outlines the steps the organization took during the 1990s to prepare itself better for future MOOTW.

There is no dramatic performance failure in this empirical record. Although success in these murky political-military undertakings is hard to measure, the Army has handled all of them more or less successfully. Yet success has not come easily. Each of the MOOTW has imparted a degree of stress to established organizational patterns and routines. Somewhere in its organization the Army has virtually all the capabilities required to be a full-spectrum force. Yet capabilities organized for war-fighting are not easily assembled into a MOOTW force. Nor is a force generally ready for war pliantly disposed to offer up units ready for specific small deployments, especially those that continue indefinitely. Finally, people—and above all leaders—trained for war-fighting are not entirely ready to handle tasks associated with delivering humanitarian relief supplies, keeping the peace, or promoting political stability. There is clearly an overlap between what an army needs for war and what it needs for "other than war," yet these are not the same thing and an army geared for one cannot effortlessly shift between the two. Rather, it confronts a series of compromises in the way it organizes and trains that must be made on the basis of expectations regarding the nature of future Army missions.

The U.S. Army did not, in fact, compromise much during the 1990s. That is, it did not fundamentally change its training routines, its approach to readiness, or its organizational structure over the course of the decade. Instead, it handled MOOTW by preparing the units about to deploy to such an operation as they neared their departure date, and then returning these units to war-fighting readiness quickly after they returned. In this sense the Army continued to treat MOOTW as "lesser included cases" while it remained geared primarily to fighting and winning the nation's wars. Clearly this approach "worked," albeit not without stress. If the future continues to offer up a significant menu of MOOTW, however, and if the Army wishes to make good on its desire to become a full-spectrum force—that is, if it truly wishes to "own" this mission area—it very likely will have to adapt its organizational structure, its training and leader development, and its approach to readiness in the years ahead.

The Challenge of "Other Than War"

What happens to a war-fighting Army when it is sent repeatedly on "other than war" missions? While unsatisfying, of course, the answer is "it depends," insofar as each of the MOOTW is unique along dimensions like mission, operational environment, enemy (if any), and distance from supporting bases. Yet the Army acquired enough experience with MOOTW in the 1990s to highlight some common challenges to the organization's well-honed war-fighting structure, training routines, and personnel assignment policies.

Flattening the War-fighting Hierarchy

MOOTW tend to be relatively small operations that often unfold under a politically imposed cap on the number of Army personnel allowed in the area of operations. Yet they demand substantial command and control assets. Consider, for example, the hybrid unit the Army sent to Somalia late in 1992 on a "humanitarian intervention" mission. The Tenth Mountain Division (Light) was given this job, which involved establishing sufficient local security to deliver humanitarian aid to starving Somalis.[8] The division commander and his staff deployed with one of the division's two brigades, for a total of roughly 4,000 division personnel. As Figure 8-1 shows, about twice that number of people were drawn from higher echelons of the Army and attached to the division to prepare it fully for its mission. The situation challenged the division's war-fighting command and control in at least three ways:[9]

- **Increased span of control.** Numerous attachments confronted the division commander and his staff with a much greater span of control than they face in their war-fighting structure. True, rigged for war, the division would be almost as large as the hybrid division that deployed to Somalia. But divisional elements

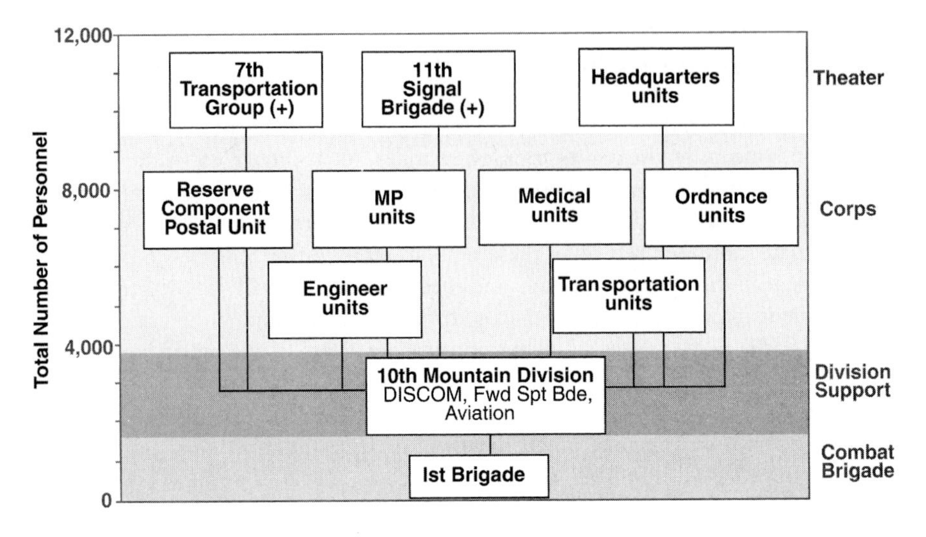

Figure 8-1. *Army Forces in Somalia.*

would normally be contained within subordinate brigades, whereas in this case a diverse array of nondivisional assets connected directly to the division headquarters. In fact, the array was larger than that shown in Figure 8-1, since several small units belonging to friends and allies of the United States were attached to the 10th Division for support of one kind or another. Although the international non-governmental organizations (NGOs) working to feed the Somalis did not belong to the command, they needed to be consulted, if not supported, and this too added to the span of control challenge.[10]

- **Increased range of operation.** Normally expected to cover roughly a 30-kilometer front in wartime, the 10th Division was assigned a square of Somalia some 200 kilometers on a side. This placed outlying units well beyond the range of the division's essentially line-of-sight communications gear while demanding considerable independent judgment from subordinate commanders. From the 11th Signal Brigade the division thus drew some 28 TACSAT (tactical satellite communications) detachments capable of linking units over these distances, although not while the units were moving.
- **New duties and skills.** As the highest-echelon Army unit in Somalia, the 10th Division became the Army Forces (ARFOR) component of a Joint Task Force (JTF) commanded by a U.S. Marine Corps general. The division commander and his staff thus faced a range of unfamiliar tasks that in war would normally be handled by higher-echelon staffs.

To be sure, the Somalia operation was an extreme case—a light division moved with little warning to a relatively undeveloped and distant theater and was given a mission covering thousands of square miles of countryside. Figure 8-2 depicts a less problematic case—the deployment of a brigade of the Army's 1st Cavalry Division

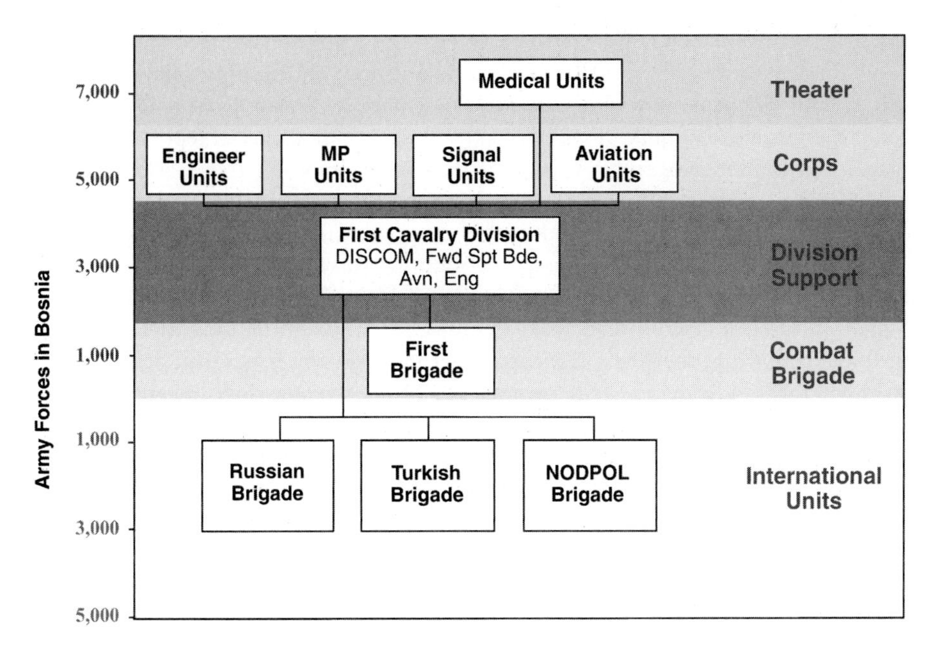

Figure 8-2. *1st Cavalry Division in Bosnia, 1998.*

to Bosnia in 1998. As a heavy unit, the 1st Cavalry contains more "beef"—more organic support—than the 10th Division. It deployed to a more developed locale with considerably more infrastructure than was found in Somalia. The mission unfolded conveniently just off NATO's southern flank, allowing much of the operation's logistics to be headquartered in Tazar, Hungary, just north of Bosnia. Finally, the initial mission—keeping the peace among Bosnia's warring factions—more closely resembled the division's combat mission than the 10th Division's humanitarian intervention mission.

As might be expected under these circumstances, the 1st Cavalry Division needed fewer higher-level attachments than the 10th Mountain—although it still carried attachments from both corps and theater army totaling some 2,000 soldiers. Beyond these distinctions, however, there are many parallels to the earlier example. Like the 10th Mountain Division, the 1st Cavalry deployed its forces over a much wider area than the division would confront in wartime; happily, local infrastructure allowed for communications among units without substantial signal augmentation. Like the 10th Mountain, the 1st Cavalry also served as the ARFOR, with a direct link to JTF headquarters. The 1st Cavalry Division's commander also acquired responsibility for units of friends and allies—in this case, three full brigades, from Turkey, Russia, and the Nordic countries.

Each of these operations forced the Army to push command and control technologies from higher to lower levels of its war-fighting organization. The requirement in each case was driven partly by the sheer size of the organization and the nondivisional assets linked directly to the division headquarters. In this sense, the hybrid units the Army sends to MOOTW tend to be flatter, in organizational terms, than the usual combat organization. The need for more command and control at lower levels was also driven partly by new responsibilities; ARFOR duties, for example, normally require a set of specialized communications gear and software to meet special reporting requirements. Finally, the geographic sprawl of these units imposed a special demand for longer-range communications gear than that normally found at division level.

For the Army to assemble and deploy these elements on very short notice testifies to the fact that, somewhere in its war-fighting structure, the Army has virtually all the capabilities needed for a diverse array of MOOTW. Yet these capabilities are widely diffused and are not coaxed easily out of the war-fighting hierarchy. The problem is not that the Army is incapable of reorganizing itself; task organizing or so-called force tailoring is a core competency of the organization. But these phrases normally refer to tactical-level adjustments—adding a tank company to an infantry battalion, for example, to create a combined arms team. Depending on the mission, the degree of adjustment for MOOTW can stress the entire organization, including reserve components. MOOTW in this sense call for force tailoring on a larger scale than the Army normally undertakes. And that in turn has forced the organization to redistribute command and control, disrupting the normal hierarchy of its war-fighting organization.

Training, Leader Development, and the Force Structure Balance

With the transfer of command and control technology comes the need to train for its use. This is especially the case when a division picks up ARFOR responsibilities; division staffs have had to undertake training in the specialized skills needed to communicate with the JTF headquarters. The same principle applies when the division picks up other higher-level capabilities and responsibilities requiring special skills normally assigned to higher commands.

Obviously the training and doctrine challenges MOOTW pose to a war-fighting Army are greater than these examples suggest. But they are not straightforward. Clearly, MOOTW ask war-fighters to take on missions and tasks that lie outside their usual training. Infantry are trained to shoot and kill, not contain, isolate, apprehend, or mollify (much less secure election facilities or run well-baby clinics). The same can be said for armor or artillery branches. Yet there remains considerable overlap in skills. In General Meigs' view, "about 70 percent of the collective tasks [performed by combat soldiers in MOOTW] are the same things you do in combat."[11]

Meanwhile, as the organization charts in Figures 8-1 and 8-2 suggest, the Army owns many capabilities that are well suited to MOOTW. One might expect the Army's communicators to thrive in a MOOTW environment, for example, given the heightened demands these operations often place on command and control. And military police are trained to do exactly the sort of things many MOOTW demand.[12] Meanwhile, for several Army specialties—logistics, administration, and so forth—the demands of war and MOOTW are roughly the same.[13]

To this extent, the training issue meshes subtly with the issue of force structure or, more specifically, the balance of skills available within the Army's force structure. An army that expected to confront frequent MOOTW with little prospect of war would presumably want to have more military police, transportation specialists, political-military experts (civil affairs, psyops), and perhaps logisticians than a war-fighting army, which would of course want relatively more infantry, armor, artillery, and combat aviators. Conversely, an army that preserved its war-fighting structure would face a commensurate need for more training, to turn war-fighters into peacekeepers. Exposed to continuous MOOTW, a war-fighting army would probably also experience stress on units that are crucial to MOOTW but underrepresented in an Army geared for war-fighting. (In Defense Department parlance, the latter problem forms part of the so-called high-demand, low-density (HDLD) problem, referring to units in high demand for MOOTW missions that are present in relatively low density in the active force structure.[14]) An Army that confronted both war and MOOTW would have to find a balance between the two extremes.

Determining where that balance lies is further complicated by the prevailing judgment that even a peacekeeping force needs war-fighting skills. On the basis of the Army's experience with MOOTW in the 1990s, U.S. Army leaders argue that the ability of a peacekeeping force to rise to "the next step"—that is, to take on a serious fight if it is forced on them—is a prerequisite for keeping the peace:

> Time and time again, the generals in Bosnia pointed to occasions when the former warring factions would test, prod, push, and hope that the U.S. forces would misstep. In each case the former warring factions watched America's senior military leadership counter their every move with swift deployment of forces and decisive action. These responses averted potential conflict.[15]

To be sure, these arguments might not apply to highly structured peacekeeping situations like the one that has separated Greek Cypriot and Turkish forces on Cyprus for many years. But these are not missions normally given to U.S. forces—in part perhaps because they do not require the capability to move to "the next step." Whatever other nations might field for peacekeeping, the U.S. Army sees good reason to field a force that can go to war, if necessary, to keep the peace. And that heightens the need for combat forces, even for peacekeeping operations.

Whatever the balance of skills within its force structure, the Army faces in this situation a separate and substantial challenge to what it calls "leader development" (in contrast to training, which applies to troops and units). Rising leaders need to know *both* war-fighting and peacekeeping. In the terminology of this anthology, the Army's "expert knowledge" has to expand to include skills and knowledge applicable to both activities. They are not the same skill sets.

Indeed, there can be no doubt that war-fighters have much to learn as they go off to various MOOTW. On the basis of interviews with virtually all senior Army leaders involved in the Army's peacekeeping deployment to Bosnia, for example, Col. Howard Olsen and John Davis concluded that "U.S. Army doctrines were largely inadequate in an environment that forced American commanders to wrestle with the political, diplomatic, and military demands of stability operations."[16] The current Army Chief of Staff, Gen. Eric Shinseki, who commanded the Stabilization Force in Bosnia in 1997-98, was quoted as saying:

> There wasn't a clear doctrine for stability operations. We are developing it, using the Bosnia experience, to define a doctrine for large stability operations. But it is this absence of a doctrine for a doctrine-based institution that you walk into in this environment. There you are in a kind of roll-your-own situation.[17]

"Having trained for thirty years to read a battlefield," said Maj. Gen. William Nash, commander of the First Armored Division when it entered Bosnia at the end of 1995,[18] "general officers were now asked to read a 'peace field.'"[19]

Some of the special training required for MOOTW has to do with the specific mission in question. Units and particularly leaders heading for Bosnia need to understand the identity and history of competing ethnic and religious groups, the geography of the country, the roots of the conflict, and the legal strictures of the Dayton Accords. But much of the training required goes beyond specifics to broader skills in negotiation and conflict resolution, and the ability to deal with nonmilitary organizations, civilian officials, and allied commanders and forces. Finally, observers note the need for particular personality characteristics—confidence, patience, and awareness of and sensitivity to other cultures—that are helpful to Army leaders in any situation, but are especially valuable in MOOTW.[20]

Clearly a full-spectrum force would feature training in the special needs of MOOTW, especially for rising leaders from early in their career. Also, officers would be advanced to higher levels partly on the basis of their prior experience with joint and allied commands, peacekeeping assignments, and demonstrated personal character traits. Howard Olsen and John Davis see the senior service college level (lieutenant colonel to colonel) of military education as "the place where leadership training for peace operations must be conducted and the place that needs the most curriculum development."[21] But they also highlight the need for early training of future leaders, starting with entry-level basic course training for new officers.

Nor is this leader development challenge confined merely to senior commanders. Junior officers and NCOs, occasionally even junior enlisted personnel, can take on tasks in MOOTW that are more common to mayors or police chiefs than to soldiers. Army officials often speak of the "strategic sergeant," the NCO who confronts a mob or a thorny local distribution issue in full glare of CNN's cameras, knowing that his or her choices could easily have much broader ramifications for the nation's overall interests. This is not to say that Army commanders or soldiers in wartime confront only war-fighting choices; in the run-up to Operation Desert Storm, for example, commanders at all levels dealt with local authorities and migrating bedouin almost daily. But clearly in almost any conceivable MOOTW the balance between political and military calculations shifts dramatically toward the former.

The Tempo Problem

Over the course of the 1990s, frequent MOOTW deployments added considerably to the stress on Army personnel and the organization's personnel assignment system. Army officials normally have referred to this as a "tempo" problem, invoking visions of soldiers traveling back and forth to overseas deployments, mainly but not exclusively MOOTW, with adverse consequences on morale and, ultimately, retention in the Army. In fact, the problem is both less and more than this. While it has its roots in the reduced size of the Army, it springs as well from force structure and personnel assignment policies set in place during the Cold War.[22]

In some broad sense, tempo is a simple function of the size of the Army and the size and frequency of its deployments. In the 1990s, long-standing commitments to Europe and Northeast Asia shrank from Cold War levels, but did not disappear. In 2000, the Army rotated individuals through some 65,000 slots in Europe and 33,000 slots in the Republic of Korea. These set the backdrop for deployment responsibilities acquired during the 1990s. Some of the latter have not been MOOTW but are more properly classified as "deterrent" or war-fighting deployments. In the aftermath of the Gulf war, for example, the Army began to rotate combat battalions and the occasional brigade through Kuwait, where it prepositions a brigade's worth of combat equipment and supplies. It also has consistently supplied Patriot air and missile defense units to Saudi Arabia. In contrast to the *individual* rotations used to fill slots in Europe and Korea, these involve *unit* rotations lasting up to six months.

MOOTW have also been serviced with unit rotations lasting six months. Early post-Cold War deployments to Somalia and Haiti have shrunk or disappeared entirely. But in the second half of the decade the Army consistently rotated units to both Bosnia and, after 1999, Kosovo, in the Balkans. The Bosnia operation began with the deployment of the First Armored Division in 1996, but has shrunk since then to roughly 4,000 soldiers in 2000. Roughly 6,000 Army soldiers now serve in Kosovo at any given time. Each area is serviced by units rotated in from the United States.

The numbers in these cases are fairly small, prompting many observers and policy-makers to wonder why an Army of nearly a million soldiers (including reserve component forces) experiences problems with tempo. Part of the answer lies in the fact that for every brigade deployed overseas there are generally two other brigades involved, one preparing to replace the deployed unit, one recovering from its deployment. Thus at the very least one must take into account a "three-for-one" rule in assessing the effects of deployments on the Army.

Rough comparisons of size do not capture the full dimensions of what is going on, however, even when the three-for-one rule is applied. An additional factor is linked to the way the Army normally fills its combat units. During the Cold War the Army used an individual replacement system designed to spread soldiers and officers more or less evenly across the combat force. This in effect distributed readiness across the entire active force, as would be expected of an Army that was preparing to fight a large war on a moment's notice.[23]

Significantly, Army units, even combat units, were rarely fully manned during the Cold War. With the exception of ready-reaction units like the 82d Airborne Division, the personnel fill of most Army divisions fell 5, 10, or even 20 percent[24] below levels called for in their Table of Organization and Equipment (TO&E), the staffing document for each kind of unit. Overall, the active Army's *authorized* end-strength in 2000—480,000 soldiers—fell roughly 10 percent below its *required* end-strength—the roughly 520,000 soldiers required to fill its active force structure to 100 percent. This practice too could be linked to the enormity of the Soviet threat, which encouraged the service to "overstructure" itself on the assumption that it could quickly push people from various parts of its overhead entities into the unfilled combat structure to make a slightly larger fighting Army when war came. In fact, however, the practice has much deeper roots; in the aftermath of major wars of the twentieth century the Army's leadership has routinely sought to retain more force structure than could be supported by prevailing end-strength.[25]

The Army continued these practices into the 1990s, arguably for the same reason; it expected to need all of its active force structure (reduced roughly 40 percent in size from Cold War levels) to meet the demands of two major theater wars. Yet the rising frequency and duration of MOOTW, especially the need to service repetitive deployments to the Balkans, brought the system under considerable stress. A continuous stream of relatively small but non-trivial deployments, whether for MOOTW or for deterrence, demands *focused* readiness—readiness that shifts from unit to unit across the force as particular units chosen for deployment are brought up to full readiness before they deploy.

The demand for focused or concentrated readiness on a personnel system accustomed to spreading readiness evenly caused, among other things, a substantial increase in personnel turbulence surrounding units about to deploy. These had to be brought up almost to full strength quickly, while older or damaged equipment was replaced with equipment fully ready for operations. People and equipment were normally stripped from nearby units, leaving the latter even less fully equipped and staffed.

The personnel turbulence provoked by this process was exacerbated substantially by the use of peacetime rules to govern individual deployments. If war approaches, the Defense Department normally puts a "stop-loss" on all departures from the services and a freeze on personnel transfers. Thus nearly everyone in each service when the crisis begins is available for deployment and remains available, often in the same position (barring injury and the like) for the duration of the crisis. In the case of the Army's Balkans deployments, however, soldiers just returning from unaccompanied tours in, for example, Korea have been excused from another deployment for a specified period of time (up to a year) after their return. Soldiers with orders to leave their unit, or the Army, soon after their unit would otherwise deploy to Bosnia or Kosovo have also been deemed nondeployable. In practice, roughly 35 to 40 percent of the soldiers in deploying units have been nondeployable by these standards.[26] These have had to be replaced by "deployable" soldiers, normally from nearby units but in some cases from distant posts. The turbulence surrounding deployment preparation thus is substantial; up to 6 or 7 soldiers have been jostled in some fashion for each one who deployed. Given that divisions were understaffed in the first place, the process of readying one of the division's brigades for deployment has also occasionally left the rest of the division with too few soldiers in critical skill specialties for it to go to war.

Clearly, then, while the deployment tempo for soldiers has increased,[27] focusing on tempo in the narrow sense of soldiers endlessly deploying and redeploying overseas misses the turbulence, readiness, and workload problems that MOOTW deployments impose on the units and soldiers who stay behind. Units at the same post, or in the same division, as a unit that deploys to the Balkans pick up home-station duties left to them by the departing unit. They also pick up the additional and often substantial workload associated with preparing their sister unit for deployment—helping to train the departing unit, bring its equipment up to standard, and the like.[28] With fewer than usual soldiers doing more than the usual tasks, those who remain behind face a tempo problem that is in some ways as burdensome and disruptive as the tempo of the deploying unit.

Meanwhile, home-station training almost inevitably suffers.[29] As we saw in the case of the Somalia and Bosnia deployments, the Army assigns its MOOTW to a division headquarters, often augmented with communications equipment from higher echelons. As a rule, most of that headquarters deploys to the MOOTW. This leaves behind a truncated division staff much less able to command and control stay-behind units. Units too are truncated, as the departing unit takes with it parts of units—platoons but not companies, partial battalions and the like—in an effort to remain within mandated caps on end-strength in the area of operations. Partial units commanded by truncated staffs generate a highly stressful home-station situation far from ideal for training.

A final complication associated with deployments in the 1990s was rooted in the 1973 decision to increase Army divisional structure without commensurately increasing active strength. This was done by moving much of the Army's combat support and combat service support (CS/CSS) into the reserve components.[30] This meant that no major commitment to war could be made without mobilizing the reserves. In the 1970s this may have seemed like a good idea to those with the Vietnam War fresh in mind, since it ensured that national decision-makers could not send the active force off to major war without mobilizing the reserves and ensuring a major popular commitment to the conflict. In the 1990s, however, the CS/CSS policy aggravated the HD/LD problem in the active force, since many of the capabilities most useful in MOOTW are precisely the combat support skills that were moved into the reserve component forces many years ago. The policy made it almost inevitable that reservists would have to be called up involuntarily to support MOOTW deployments.

Thus, while the number of U.S. Army soldiers actually deployed is usually fairly small, the ripple effects of even small, repeated deployments are larger and more perverse than the deployment numbers alone might suggest. The Army has serviced all deployments, but the costs to training and quality of life have been significant, even if they are often hidden or poorly understood. A general officer panel, which the Chief of Staff specifically tasked to examine problems in Army systems for training and leader development, found that the internal costs to the institution of increased MOOTW deployments have been immense—particularly to junior officer morale, trust within the officer corps, and Army culture.[31]

Alternatives and Choices

The fact that MOOTW demand significantly different training and skills than the war-fighting Army has prompted some observers to favor the creation of a "constabulary force" to handle MOOTW missions, a force wholly separate from the Army itself.[32] This would leave the Army free to train solely to fight and win the nation's wars. This is the nation's choice to make, of course, not the Army's. But Army leaders and civilian officials are being consulted as the idea is considered. Despite its primary allegiance to war-fighting, the Army as an institution has opposed the constabulary idea.

There are good reasons for this position.[33] For example, a constabulary force would lack the training and equipment necessary to go to "the next step." It thus might be useful in a highly structured and constrained peacekeeping environment, but not in tougher, potentially violent situations that are likely to require U.S. leadership and involvement. Meanwhile, as is clear from the examples of hybrid units deployed to Somalia and Bosnia, forces deploying to conduct MOOTW normally take with them a sizable contingent of communications, engineering, medical, and administrative units, all of which are lodged somewhere in the Army's force structure. If the constabulary force "borrows" such support from the active Army, it is not truly a separate force. Yet to replicate these units in a separate force would involve wasteful duplication. A relatively small constabulary force would experience enormous tempo problems servicing repetitive deployments to, say, Bosnia. Or it might simply deploy and remain overseas, in the manner of the French Foreign Legion—an option many Americans are likely to find unpalatable. Finally, the resourcing of a constabulary force would almost inevitably result in the further drawdown of the Army. For all these reasons the Army is right to resist the creation of a separate constabulary force.

Instead, the Army prefers to continue performing the bulk of the nation's MOOTW missions, adapting its organizational routines to bridge the demands of war and other than war. The adaptation has come slowly, paralleling the nation's slowly increasing commitment to MOOTW. Only with the Bosnia deployment, for example, did the Army finally acquire a sizable, long-term MOOTW commitment, and this despite initial assurances that U.S. forces would be withdrawn from the Balkans within a year after they deployed. In this sense, the organization's adaptation to MOOTW continues; the story of the 1990s is only the initial chapter. What it shows, however, is that few of the organization's war-fighting routines and structures have been altered more than temporarily to accommodate the new missions.

Redistributing Command and Control

To meet the need for greater command and control at lower levels of the organization, the Army has continued the practice, begun with the Somalia deployment, of temporarily attaching higher-level C2 assets to the deploying headquarters. Experience with MOOTW has led to steps that facilitate this process. The 11th Signal Brigade, for example, created packages of needed communications gear and personnel that could be assigned ("chopped") rapidly to any division facing a MOOTW mission. Joint Forces Command came to understand better what a deploying division would need to function as an ARFOR headquarters. The Army has turned to commercial satellite communications suppliers when necessary to increase the range of communications available to its deploying units. Finally, in the

Balkans in particular more permanent communications systems and some logistics facilities have been installed, often staffed by the civilian engineering firm Brown & Root rather than by Army personnel.

In 1997, then Chief of Staff Gen. Dennis Reimer proposed the creation of a "strike force," which, despite its title, was mainly a robust headquarters, lacking any permanently assigned combat units but specifically designed to pick up various functional components depending on the mission it was given.[34] This would have had the additional benefit of freeing the Army's war-fighting division headquarters from MOOTW deployment, depending of course on the number of MOOTW the Army was asked to take on. The strike force idea was short-lived, however, leaving the distribution of command and control technologies across the Army's long-standing echelons fixed throughout the decade. Significantly, the idea of a standing headquarters has been revived at the joint level by the Bush administration, which has set as a goal the establishment of "a prototype for a Standing Joint Task Force (SJTF) Headquarters."[35]

In consideration of its present transformation effort, the Army would appear to be deferring change in this area to the future. Early designs of the "interim brigade combat team" (IBCT) suggest that the Army plans to push at least some command and control technologies down to lower levels of the organization. Organic to the IBCT headquarters element, for example, will be three TACSAT teams for longer-range communications, plus three SMART-T units to link to space-based assets. Given that the 10th Mountain Division needed 28 TACSAT teams to fully perform its mission in Somalia, the IBCT may still not be adequately equipped for MOOTW. But the elements will be there, and with them come the skills to integrate more teams into the IBCT more easily. There are also more communications devices in the IBCT than are available in today's combat brigade headquarters.

This would appear to be a case where the demands of MOOTW and the demands of future war-fighting overlap. Along with the other services, the Army sees in the future a more distributed form of warfare in which lower-level Army units enjoy much enhanced situational awareness and access to much deeper-reaching fires relative to what is available in the present combat hierarchy.[36] The result, at least from initial designs, seems to be more robust headquarters all the way down the Army's combat organization. This is expected to produce a more effective war-fighting organization. But it is likely to benefit the organization's performance in MOOTW as well.

Presumably, pushing more command and control to lower levels of the organization will have the added benefit of producing a greater number of deployable headquarters elements. The Army's headquarters of choice for MOOTW deployments has been the division headquarters, as we saw in the Somalia and Bosnia cases. With only ten active divisions and perhaps a few other headquarters (armored cavalry regiments, for example) that might serve this purpose, it takes very little time to reduce wartime readiness, not by deploying personnel to MOOTW—the numbers are fairly small—but rather by deploying separate headquarters. In this sense, the divisional headquarters may itself be a "high-demand, low-density" item for the Army. Producing a larger number of such elements thus may be more important than fielding more combat units. Fortunately, this seems to be the way in which the Army is headed.

Training, Leader Development, and Structural Adjustment

As we saw in the earlier section on training and leader development, there is a third issue meshed with these two, namely, that of structural balance between combat units and support units whose skills fit in naturally with MOOTW—Military Police, Transportation, Civil Affairs, and similar units. In none of these three areas does there appear to have been much change compared to the 1990s.

Training. In the 1990s, the Army solved the training demands of MOOTW largely by delivering "just enough [training] just in time."[37] Units slated for deployment to MOOTW generally received training on the mission in the months and weeks before deployment. As it became clear that the Bosnia mission would last far longer than the year originally predicted for it, the Army transferred this mission from units in Europe to stateside units, which then began to use the Joint Readiness Training Center (JRTC) at Fort Polk, Louisiana, for training. Ultimately, JRTC established a Bosnia-like town and used opposing forces developed to confront training units with Bosnia-like challenges. The goal was to pull units together for a deployment, train them for it, and then return them to war-fighting status upon their return from MOOTW duty.

Indeed, the main training concern over the 1990s, within the broader defense policy community as well as the Army, was not how to train better for MOOTW, but rather how to minimize the deleterious effects of MOOTW on the Army's overall readiness for war. The two nearly-simultaneous wars for which all U.S. military forces had to be prepared throughout the decade were generally agreed to require virtually all of the Army's active combat units. Taking roughly a division off line to deploy to Bosnia or Kosovo thus arguably reduced the Army's ability to support the National Military Strategy. Occasionally, an Army division commander would admit publicly that his unit, though "ready," was not prepared for war because it was already deployed to an operation other than war. Overall, however, the Army worked hard to keep as much of its force as possible ready for war.

Leader Development. The Army has used a similar "just in time" approach for preparing its leaders, providing them with some weeks of training in the history of the target region plus workshops to develop political, peacekeeping, and negotiating skills.[38] But it is not clear how far training for MOOTW has made its way into the developing officer's staff college and senior service college curriculums. The Army's Command and General Staff college (for rising captains and junior majors) devotes 50 of the 293 total hours of its two core operational courses (C300, C500) to conflicts below major theater war on the conflict spectrum. Two classes of the Army War College course (for colonels and rising lieutenant colonels) on the history of warfare deal with the Army's experience in Haiti and Bosnia.[39] Most of the courses on MOOTW at both USAC&GS and the USAWC remain electives. The principal focus remains war-fighting.

In this area, as in the realm of training, the principal worry has to do with the loss of war-fighting skills that may accompany MOOTW deployments. To be sure, senior Army leaders understand that MOOTW teach officers how to make decisions under stress, how to build morale and cohesion in their units, and certain basic soldier skills. All of these are valuable to future Army leaders in any role. But they cannot substitute for war-fighting training. At least under current constraints on career and assignment length, combat skill suffers seriously when officers are deployed to MOOTW.

Frequently, senior officers cite the example of a young captain who takes command of an infantry company and soon deploys to Bosnia, where his experience yields excellent training in handling the political-military demands typical of MOOTW. But by the time that captain returns home there is rarely enough time left in his command tour to ensure that he and his unit rotate through the National Training Center, where they would hone their war-fighting skills. Senior officers thus wonder how this officer will handle war-fighting demands he may face some years later, when he assumes command of a battalion or brigade. Arguably the laws, regulations, and de facto criteria governing officer career development create a virtual zero-sum game for rising officers; they have time to learn one set of skills, but not both (see Chapter 24, by Andrew Abbott, of the present volume for discussion of the professional implications of the Army's efforts to cling to two or more separate skill sets for officers).

The future in this case remains ill defined. Unquestionably, the Army's leadership is concerned about training future Army leaders. But a large part of the challenge they see ahead lies in the complexity inherent in their vision of future warfare. Low-level units of the future will contain several of the combat arms, for example; company commanders will thus have to know far more than the professional lore of their own branch. Lower-level units will also have control of deeper fires than today's units. With the expert knowledge required to wage future war expanding substantially, it will take care and attention to ensure that sufficient allowance for MOOTW training is made available in the future leader's development program.[40]

Force Structure Balance. If the Army has resisted imparting more robust and formalized training in MOOTW skills to its units, it has also resisted rebalancing the active force in the direction of capabilities more compatible with MOOTW, despite suggestions that it do so.[41] The Army had one active duty civil affairs battalion during the Cold War, and continues to have one today in its active force structure. The ratio of Military Police units to the overall size of the active force changed very little during the 1990s. Indeed, the overall distribution of skills across the active enlisted force remained roughly constant throughout the decade, save in the case of Patriot crew members, who are needed for deterrence and missile defense on the Arabian Peninsula.[42]

Thus the Army has had to turn to its reserve components for help with MOOTW. While it has long used volunteers from the reserve components as well as reserve soldiers serving on their annual two-week training commitment, the demands of MOOTW have resulted in involuntary call-ups of reserve soldiers, hence a tempo problem in the reserves that can raise problems with civilian employers and families.[43] Although this approach has worked so far, the burden MOOTW deployments have placed on these non-volunteer "civilian" soldiers continues to stir debate about the nature of the reserve component's "contract" with the active force and the nation.[44]

Ameliorating Tempo Problems

The full scope of the Army's tempo problem came as a surprise even to many in the service itself, not to mention outside observers; few foresaw the many ripple effects that small, repetitive deployments would have on the greater organization. Part of the reason, of course, is that no one foresaw, or was allowed to plan for, repetitive deployments. The nation's commitment to Bosnia was billed as a one-year event when it was launched, and even after the first anniversary passed without a U.S. withdrawal no policy-maker was willing to predict an enduring commitment. Indeed, in the 2000 presidential election campaign, candidate George W. Bush spoke frequently of the need to withdraw U.S. forces from such commitments so they could focus on "fighting and winning the nation's wars." Yet, at least up to this point, the Bush administration has committed itself to keeping those forces in the Balkans until European forces withdraw—perhaps a very long time off.

Partly because creating a nominal command in Bosnia and rotating individual soldiers through it would smack of permanence, the Army instead has continued to use unit rotations of relatively short duration to service both of its Balkan commitments. This has forced it to find other ways to reduce tempo and turbulence. As the reality of a long-term commitment became clear, the personnel deployment system began to funnel deployable soldiers to units due to deploy. Local personnel distribution also sought to minimize the assignment of nondeployable soldiers to units on the deployment roster. Both moves were aimed at reducing local turbulence associated with readying a brigade to deploy.

In another effort to reduce the local turbulence caused by deployments, Chief of Staff Eric Shinseki directed late in 1999 that all combat (TO&E) units be filled to 100 percent strength, starting with armored cavalry regiments and divisions and slowly moving up toward corps.[45] This policy will increase the fill and hence readiness of stay-behind units. It should thus partially reduce local turbulence when a unit is readied to deploy. But it cannot fully eliminate turbulence so long as peacetime deployment rules remain in place and the service continues to distribute personnel more or less evenly across its force. Meanwhile, given no overall increase in active Army end-strength, filling combat units to full strength has extracted a price elsewhere, notably in the Army's Training and Doctrine Command, which has complained for some time about falling readiness due to "money and people shortages."[46]

Just as it has had to reach into its reserve component forces for combat support units, so the Army has also turned to its reserve components to help reduce tempo. The Army has mobilized and deployed whole sections of the Army National Guard to Bosnia largely to take some pressure off the active force. In Fiscal Year 2000, the command element of the Texas 49th ARNG Division took over command responsibilities in Bosnia, with the 10th Mountain Division supplying combat forces.[47] Plans call for the ARNG to fill still more slots in Bosnia and Kosovo in future rotations.

The next step in reducing turbulence would probably involve a more fundamental shift toward a rotational force management strategy. In this scheme, units would move sequentially through periods of low, medium, and high readiness, during the last of which they would be available for deployment if the nation needed them. Personnel assignments would be geared to the scheme, meaning those returning from a deployment would be assigned to units just recovering. Tempo would not necessarily decrease, since in the end it is a function of the size of the Army compared to the size and number of its deployments. But deployments would be made more predictable, and local turbulence would be reduced substantially. On the other hand, this approach amounts to "tiered readiness"—some units would be very ready to deploy, others would be quite unready to do so—and that would reduce wartime readiness of the whole force.[48] Although the Army has studied rotational deployment strategies, so far it retains its traditional personnel assignment system.

In lieu of other rotational deployment strategies, the Army has argued for more end-strength. In fact, over the 1990s, as the Army came to understand better the complex effects of deployments on its various routines, it grew better able to articulate the link between MOOTW and end-strength. In 1993, for example, the so-called "Bottom-up Review" (BUR) paid only passing notice to the need for Army forces besides those required for war. In contrast, in the run-up to the 2001 Quadrennial Defense Review (QDR) Army officials were more effective in arguing for force structure on the basis of its deployment requirements, including but not limited to deployments to MOOTW. While it did not gain any force structure in that review, it did not fall victim to the cuts it was rumored to be facing.

Still a Lesser-Included Case?

It is difficult to identify fundamental organizational change as the Army has adapted to the relatively full menu of MOOTW presented to it over the 1990s. It has preserved the war-fighting hierarchy of its combat forces despite "flattening" pressures from its MOOTW experience. It has preserved the balance of skills in its active force, and between its active and reserve components, that it inherited from the Cold War, forcing it to turn to the reserves for key supporting units for MOOTW. It has delivered training for MOOTW to units and leaders on a "just in time" basis—that is, in the months before they deploy—without significantly changing the focus of the curriculum of its officer development schools. In effect, the "just in time" approach applies to more than just training: in the 1990s the Army organized and equipped units for MOOTW as they prepared to deploy, and then sought to return them quickly to war-fighting shape after they returned. It is hard not to conclude that MOOTW remained a lesser-included-case for an Army that continued to focus the bulk of its attention on "fighting and winning the nation's wars."

The Costs of Current Practice

"So what?" one is tempted to ask. Although the Army confronted more MOOTW in the 1990s than it did during the Cold War, the "real" wars it planned for in the 1990s—two nearly simultaneous "major theater wars"—were seen as highest in priority and all-consuming. These were national priorities, enshrined in formal planning guidance. Perhaps the service would have entertained more fundamental change had it experienced dramatic failure in one or another of its MOOTW deployments. But in fact, it performed its MOOTW missions reasonably well, bringing calm, for example, to what began as a very edgy situation in Bosnia. Indeed, policy- makers have been able to reduce the size of the overall multinational commitment to Bosnia from a full division in 1996 to a mere 4,000 soldiers today. If what the Army has done amounts to treating MOOTW as a peripheral concern, then perhaps that is the way they *should* be treated.

In the short run this argument is difficult to refute. If there are problems in the way the Army has chosen to handle MOOTW, clearly they are subtle and unfold over a longer period than can be measured by looking at unit performance in the missions the Army has confronted thus far. One potential problem has to do with the accommodation the active Army has reached with its reserve component forces. Many in the reserves appear to welcome their more active engagement in national security and with the active Army. But others worry that more frequent activation of reservists will slowly squeeze valued members between their reserve commitments and the demands of their civilian careers, not to mention their families. Presumably such a squeeze would produce reserve component recruiting and retention problems. That these have not appeared thus far does not rule out their appearance in the future.

Another potential problem has to do with the tolerance of the present approach to MOOTW for still more nonwar missions in the future. In the early 1990s Army MOOTW came and went. Starting with the "one-year only" deployment to Bosnia in 1996, however, they have come and stayed, producing a steady drain on Army resources. It is not clear that the Army could sustain another major, long-term deployment to, say, Macedonia, the Golan Heights, or the West Bank, all of which have been broached in recent years as possible sites for Army peacekeepers. Although the war on terrorism dating from 11 September 2001 remains new and unpredictable in scope and content, it too may portend a series of long-term, relatively small Army force deployments, whether for combat or for MOOTW. At some point the sheer number of MOOTW commitments is likely to strain the Army's current approach to servicing them.

A more immediate problem stems from the Army's decision to fill its combat divisions to 100 percent in order to reduce turbulence related to MOOTW deployments. In the absence of an increase in end-strength, the additional soldiers needed for this purpose have come from other parts of the Army, notably its schools. In effect, this policy favors the near-term demands of current missions over the long-term need for education and training. This is an odd choice for an Army seized by the need to transform itself and beset by the new training demands entailed by that transformation.

In fact, tension between the Army's current approach to MOOTW and its view of its own future goes beyond staff shortages in the institution's schoolhouses. The current approach is not likely to yield the Chief of Staff's goal: "forces which will, with minimal adjustment and minimum time, generate formations which can dominate at any point on the spectrum of operations."[49] In truth, it is difficult to imagine truly rapid transitions under any circumstances. A lean, rapidly deployable force is likely to be stripped of all but absolutely essential supporting elements, when of course supporting elements are precisely what most MOOTW call for. Even in a mature theater with plenty of Army forces present, some time will be consumed exchanging units to meet the demands of shifting missions. Putting aside these structural problems, however, the "just in time" approach to preparing today's units for deployment is of questionably value to an Army of rapid full-spectrum transitions. The Army's training and education system thus will require more fundamental expansion to produce troops and leaders—especially leaders—able to move effortlessly from mission to mission.

Between the potential and the present problems associated with the Army's current approach to MOOTW lies the broader issue to which this chapter contributes: the Army as a profession. While the Army has embraced the need to be a full-spectrum force, it continues to treat MOOTW as a lesser-included case. This inevitably blurs the issue of professional jurisdiction and leaves open the question of just what "expert knowledge" the profession expects its members to have. More tangibly, it creates professional dissonance for those whose chief professional experience in the Army has been one or perhaps several operations other than war. Army rhetoric is "Right on!" so far as embracing MOOTW is concerned. But young officers returning from MOOTW deployments will confront an organization dedicated narrowly to war-fighting—indeed, to a concept of war-fighting that is dauntingly complex and demanding.

Conclusions

Anyone who talks to senior Army leaders today will recognize that they have embraced MOOTW as a legitimate part of the U.S. Army's business. Yet it is hard to see this appreciation reflected in the underlying routines and structure of the organization. Clearly the Army can *do* MOOTW. But it has not let the doing of it affect its war-fighting orientation more than temporarily.

In a sense, the Army has bet on war not only as its core mission, but as its likely future mission. In terms of national politics as well as the Army's own deeply ingrained ethos, this looks like a reasonable bet. The core of the National Military Strategy throughout the 1990s, after all, was the need for U.S. military forces to win two nearly simultaneous major theater wars. U.S. policy-makers have not been anxious to deploy forces for MOOTW. Even when they have done so, as in Bosnia, they have promised relatively quick retreat—a year in Bosnia, for example. And the current president, George W. Bush, campaigned on, among other things, the promise to remove U.S. forces from nagging deployments so they could get back to the really important work of training to "fight and win the nation's wars." The political signals the Army has gotten over the decade have favored war over MOOTW.

In terms of actual experience, however, the focus on war has been less than a good bet. MOOTW deployments have become an important part of the Army's operational diet, especially in the second half of the 1990s. Nor is this likely to end. Although the nation has resisted entreaties to commit troops to Macedonia, further unrest in that country could yet give the Army a third deployment in the Balkans. Meanwhile, notwithstanding campaign rhetoric to the contrary, the Bush administration's first Quadrennial Defense Review, altered in the wake of the terrorist attacks of 11 September, favors continued engagement overseas, working to reassure friends and allies. Ironically, the "war" the nation confronts after 11 September 2001 seems likely to pose to the Army a series of small deployments, including but not limited to MOOTW; the years ahead, in other words, may look like the years immediately past. Possible large wars remain on the horizon, to be sure. But the Army's operational tasks, and thus its professional jurisdiction, seem destined to include MOOTW for the foreseeable future.

If this proves to be true, and especially if the Army picks up more MOOTW without losing some of the ones it now has, stress on the organization is likely to mount, along all the dimensions outlined in this chapter. Under these circumstances, the Army's senior leaders are likely to face three broad choices. First, they can continue to embrace MOOTW in the manner and to the extent that they embraced it in the 1990s—as effectively a "lesser included case" for war-fighting forces. This will exacerbate the stresses noted above. Second, they could choose to withdraw from competition for this particular professional jurisdiction. This would of course require negotiation with political leaders who might veto the choice. Yet there are certainly those among the nation's political leaders who have said all along that MOOTW should not be part of the Army's jurisdiction. Finally, they could seek more fundamental adjustment to the organization and its training, the better to confront the full spectrum of missions. This too would require negotiation with political leaders, who have stressed major war while repeatedly assigning Army units to "other than war."

What would fundamental adjustment involve? Most of the Army's senior leaders feel that a significant increase in the strength of the active force is an essential element of this adjustment. This would almost inevitably reduce tempo problems, while allowing the service to fill shortfalls in its training apparatus. And it would give the Army the resources to handle a more demanding homeland security role, should the nation decide to use Army forces, rather than or in addition to civilians, for this mission.

But an increase in end strength alone leaves many of the problems associated with handling MOOTW unresolved. It would not in and of itself eliminate the local turbulence associated with readying Army units for a MOOTW deployment. This is the result of the long-standing policy of distributing people and readiness fairly evenly across the Army's combat units. While that policy made sense during the Cold War, a unit rotation system with a significant degree of tiered readiness may be a better bet for an Army that mostly faces multiple, relatively small global deployments—especially if the active Army gets no larger in the years ahead.

In any case, an increase in end-strength used simply to fill out required force structure would not do much to alleviate the "high-demand/low-density" problem the Army confronted in the 1990s. The Army may well have to revisit the issue of force mix within the active component and between its active and reserve components—the more so, one suspects, if the demands of homeland security shift reservists from overseas to stateside deployments. Given an increase in strength, on the other hand, it is not clear that simply fleshing the organization out to its *required* force structure is the best policy. Rather, the Army might want to use 40,000 more personnel spaces to buy a disproportionate number of HD/LD skills. The numbers involved may not be very large. But they represent a departure from the established force structure.

Nor would getting more people deal with the other structural issues highlighted in this chapter. The Army may need more headquarters elements. It is already beefing up the headquarters it sends to MOOTW, and might want to make that augmentation more permanent. It is being forced to tailor its deploying units at a level and to a degree that lie beyond routine practice. What would it take, structurally, to create an Army of "building blocks" that could be fit together under a reasonably robust high-level headquarters, to handle a diverse range of missions? Some, perhaps all, of these structural changes might be part of the broader "Army Transformation," now underway. But they may be needed well before the transformation has run its course.

Nor, finally, would more end-strength resolve the training and leader development issues that lie at the core of the challenge confronting the Army as a profession. In both areas, but especially in the latter, the Army needs to seek a better balance between war-fighter training and training in skills needed for nonwar. This will not be easy, given the sizable demands on leader development inherent in the Army's vision of future warfare. But if the experience of the 1990s endorsed the argument that training as a war-fighter is a necessary ingredient for leadership in many MOOTW, it also made clear that war-fighter training is not sufficient for these operations. If the Army wishes to be a full-spectrum force, it will have to give its future leaders the broad political-military understanding as well as an appreciation of the doctrine and nonwar skills these operations will demand.

Notes

1. See the discussion in David E. Johnson, *Modern U.S. Civil-Military Relations: Wielding the Terrible Swift Sword,* McNair Paper No. 57 (Washington, DC: National Defense University, Institute for National Strategic Studies, July 1997), 64-73.
2. Ibid., 2-3.
3. "U.S. Army Europe Commander Says 'Constabulary' Force Is a Bad Idea," *Inside the Army*, 29 May 2000, 11.
4. The 1993 edition of FM-100-5 included a chapter on MOOTW that was, by all reports, the subject of considerable discussion within the Army. The nomenclature has since been changed to FM 3-0 to bring Army field manuals into line with the numbering of Joint Staff publications. The Army's new FM 3-0, *Operations*, published in June 2001, contains separate chapters on "stability operations" and "support operations." For some sense of the debate, see Col. David Fastabend, "The Categorization of Conflict," *Parameters* 27 (Summer 1997), 75-87. In some writings on the subject, MOOTW is shortened to OOTW ("operations other than war").
5. Gen. Eric K. Shinseki, "The Army Vision: Soldiers on Point for the Nation: Persuasive in Peace, Invincible in War," October 1999. Available at www.army.mil/vision/The%20Army%20Vision.PDF.
6. See, for example, "U.S. Army Europe Commander Says 'Constabulary' Force Is a Bad Idea," 1, 10.
7. Shinseki, "The Army Vision."

8. There were several operations in Somalia in the early 1990s, including an initial effort, UNOSOM I, that failed to provide humanitarian aid in the presence of opposition from warlords. Data in this example are taken from so-called UNITAF, or Operation Restore Hope, which replaced UNOSOM I with a military intervention that successfully deterred (or in some cases fought off) warlord opposition to deliver humanitarian aid.
9. This section draws on Thomas McNaugher, David Johnson, and Jerry Sollinger, *Agility by a Different Measure: Creating a More Flexible U.S. Army*, IP-195 (Santa Monica, CA: RAND, 2000). See also Jennifer Morrison Taw and John E. Peters, *Operations Other Than War: Implications for the U.S. Army*, MR-566-A (Santa Monica, CA: RAND, 1995), 30; *Operation Restore Hope, Revised Final Draft,* 16 August 1993 (Fort Leavenworth, KS: Center for Army Lessons Learned, 1993); and Kenneth Allard, *Somalia Operations: Lessons Learned* (Washington, DC: National Defense University Press, 1995).
10. For a discussion of the latter problem, see Jennifer Morrison Taw, *Interagency Coordination in Military Operations Other Than War: Implications for the U.S. Army*, MR-825-A (Santa Monica, CA: RAND, 1997), especially 17-20.
11. "U.S. Army Europe Commander Says 'Constabulary' Force Is a Bad Idea," 10.
12. Thomas E. Ricks, "U.S. Military Police Embrace Kosovo Role," *Washington Post*, 25 March 2001, A21.
13. See especially Jennifer Morrison Taw, David Persselin, and Maren Leed, *Meeting Peace Operations' Requirements While Maintaining MTW Readiness,* MR-921-A (Santa Monica, CA: RAND, 1998), 31-33.
14. In fairness, the HDLD problem, to the extent that it has existed, owes as much to special war-fighting demands as it does to the demands of MOOTW. Patriot missile operators, for example, are often seen as the quintessential HDLD skill. This reflects the need to supply Patriots for missile defense to allies on the Arabian and Korean Peninsulas rather than the demand for Patriots in various MOOTW.
15. See the statement by Gen. George Joulwan, NATO commander when U.S. forces first entered Bosnia, in Howard Olsen and John Davis, *Training U.S. Army Officers for Peace Operations: Lessons from Bosnia*, U.S. Institute of Peace Special Report (Washington, DC: 29 October 1999), 5.
16. Ibid.
17. Ibid.
18. The 1st Armored Division's first elements entered Bosnia on the very last day of 1995. See "Transcript of Press Briefing held on 01 January 1996," at *http://www.nato.int/ifor/trans/t960101a.htm.*
19. As quoted by Olsen and Davis, "Training U.S. Army Officers for Peace Operations," 5.
20. Ibid., 5-9.
21. Ibid., 11.
22. This section draws heavily on extensive RAND work on turbulence, summarized in J. Michael Polich, Bruce R. Orvis, and W. Michael Hix, *Small Deployments, Big Problems*, IP-197 (Santa Monica, CA: RAND, 2000).
23. To be sure, some units were more ready than others. The 82d Airborne Division, for example, was always more fully staffed and funded than, say, the 4th Infantry Division, given the 82d's role as the Army's rapid reaction unit. But these gradations in readiness were fairly small, meaning the active force was more or less uniformly ready for war.
24. Enlisted strength of the 10th Mountain Division was held at roughly 80 percent authorized fill during the 1990s, for example.
25. See Michael James Meese, "Defense Decision Making Under Budget Stringency: Explaining Downsizing in the United States Army" (Ph.D. diss., Princeton University, May 2000), especially 273-75. Strategically, a larger structure has been seen as easing the expansion of the

Army in future crises, while perhaps also better deterring adversaries. Politically, maintaining more force structure than people allowed the Army to argue for more people, while sacrificing structure to match end-strength risked simply losing more end-strength. "The difficult choice that army leaders think they face," Meese argues, "is not between a robust force and a hollow [that is, understaffed] force. It is a choice between a hollow force and an even smaller force that . . . may itself become hollow" (274).

26. Polich et al., 1-2.
27. Army data show that "for soldiers in operational units, the average time deployed rose nearly 30 percent between 1997 and 2000." Still, tempo was distributed very unevenly; some units deployed a great deal, but others rarely deployed at all. Overall, "the fraction of the force affected remains relatively small. Out of 1,400 units reporting . . . only 222 units had more than 120 days away from home station in the year ending October 2000; only 89 units had more than 180 days away." Ronald E. Sortor and J. Michael Policy, *Deployments and Army Personnel Tempo*, MR-2372-1-A (Santa Monica, CA: RAND , 2001), xii.
28. See Taw et al., *Meeting Peace Operations' Requirements While Maintaining MTW Readiness*, 11-30.
29. See Ronald E. Sortor, *Army Forces for Operations Other Than War*, MR-852-A (Santa Monica, CA: RAND , 1997), 94.
30. See the discussion in Martin Binkin, *Who Will Fight the Next War? The Changing Face of the American Military* (Washington, DC: The Brookings Institution, 1993), 107-111. Also Johnson, *Modern U.S. Civil-Military Relations*, 69-70.
31. The general officer panel was constituted as the Army Training and Leader Development Panel (ATLDP). For a full discussion of the ATLDP, see Case Study No. 3 by Joe LeBoeuf in Chapter 22 of the present anthology.
32. A notable recent example came from The United States Commission on National Security/21st Century," better known as the Hart-Rudman Commission. In the second of its three-volume report, the commission called for the creation of "*humanitarian relief* and *constabulary* capabilities" (italics in original). Noting that these operations "require forces different from those designed for major theater war," the commission encouraged the administration "to adapt portions of its force structure to meet these demands." The United States Commission on National Security/21st Century, *Seeking A National Strategy: A Concert for Preserving Security and Promoting Freedom*, 15 April 2000, 14-15.
33. See also Taw et al, *Meeting Peace Operations' Requirements While Maintaining MTW Readiness*, 59-60.
34. See *United States Army Posture Statement FY00*, 38. This speaks of a strike force headquarters "that can assimilate light, airborne, air motorized, joint, and combined forces to create a tailored force package," but foresees the development of an actual Strike Force, including force elements as well as a headquarters, at a later date. See also U.S. Army Training and Doctrine Command, "Information Paper: U.S. Army Strike Force" (Fort Monroe, VA: U.S. Army Training and Doctrine Command, n.d.).
35. Department of Defense, *Quadrennial Defense Review Report*, 30 September 2001, 33-34.
36. See, for example, Department of the Army, *Army Vision 2010*, 9-17.
37. Department of the Army, FM-100-23, *Peace Operations* (30 December 1994), 86.
38. Olsen and Davis, "Training U.S. Army Officers for Peace Operations: Lessons from Bosnia," 3-4.
39. Drawn from David E. Johnson, "Preparing Potential Senior Army Leaders for the Future: An Assessment of Leader Development Efforts in the Post-Cold War Era" (unpublished ms, RAND, Washington, DC, 2000), 15-16.
40. See Sean D. Naylor, "Rock Bottom: Training Centers Report They Can't Meet Mission," *Army Times*, 11 September 2000, 8, which notes the creation of a "common" officer basic course aimed at teaching entering officers more than just their basic branch. This represents a start on the broad leader development challenge posited here.

41. See, for example, Congressional Budget Office, "Making Peace While Staying Ready For War: The Challenges of U.S. Military Participation in Peace Operations," CBO Paper (Washington, DC, December 1999), 57-59, which suggests converting one active duty heavy division entirely into support units. Alternatively, the CBO study proposed adding to the Army's active end-strength roughly the equivalent of a division's worth of support units.
42. MP soldiers comprised 3.9 percent of the Army's active duty enlisted force in 1990, 4.1 percent in 1995, and 3.8 percent in September 2000. Overall the distribution of skills across the active duty enlisted force changed very little for *any* specialty, save Patriot crewmembers.
43. Sortor, *Army Forces for Operations Other Than War,* 30-32.
44. For a thoughtful assessment of the disruptions to civilian careers induced by frequent and prolonged deployments, see James T. Currie, "Remember, They're Not Replacements," *Washington Post*, 25 March 2001, B3.
45. See Jim Tice, "Fully Manned," *Army Times*, 7 August 2000, 8, 10. Announced late in 1999, the goal was to fill divisional and cavalry units to 100%, matching rank and specialty, by January 2001. The policy was part of a broader push to fill the entire active force to 100 percent. Given traditional understaffing of the active force, this would require an addition to Army end-strength. In fact, General Shinseki pushed for additional strength during the last year of the Clinton administration. Early in the Bush administration, however, he grew more interested in fending off cuts. See Jim Tice, "Size of Military Top Issue in QDR Talks," *Army Times*, 20 August 2001, 10.
46. Naylor, "Rock Bottom: Training Centers Report They Can't Meet Mission," 8.
47. See the discussion in *United States Army Posture Statement FY00*, 77-78.
48. In the late 1990s the Army was driven to "tiering" readiness, mainly for lack of sufficient training money. Ibid. 79. It has also begun to consider the possibility of rotating units, rather than individuals, through Korea and Europe. Jim Tice, "A Move toward Stability," *Army Times*, 18 December 2000, 14.
49. The Army Vision.

9 Privatizing Military Training: A Challenge to U.S. Army Professionalism?[1]

Deborah Avant

The wave of privatization that has washed over the United States since the Reagan years has also hit the U.S. military. Some argue that private contractors are cheaper, others that using contractors gives the government more flexibility, and still others that privatization allows the military to focus on its core missions. Though there may or may not be truth to these justifications, the privatization current still surges ahead. The Navy hires ship painters to relieve sailors of tedious duty in San Diego, the Air Force puts 1,459 jobs at Offutt Air Force base up for bid, and the Army deepens its reliance on contractors for Reserve Officer Training Corps (ROTC) instruction.[2]

I will argue that privatization efforts both offer insights into what the Army sees as the core of its profession and promise to affect the future direction of the profession. Looking at the privatization of training, perhaps the activity most central to the future of the profession, suggests that the Army is at once quite representative of the American society and increasingly divorcing itself from key tasks surrounding the use of violence. The Army reflects American society's prominent mindset that looks positively at the marketing of previously public goods. The Army's ready delegation of training missions to the private sector so that it can focus on those tasks at the pointy end of the spear, however, makes it increasingly focused on the least likely use of violence.[3]

Professionalism and the U.S. Army

There are several levels at which professionalism manifests itself. We can think of army professionalism, for instance, at the national level (e.g., the U.S. Army) or at the international level (all armies). We can also see that military professionals are common to each of the armed services, and that there might be sub-professions within an individual service (e.g., the Infantry branch within the Army). Andrew Abbott argued in 1988 that the key variable in the development of a profession is the relationship between the profession and its work, or its jurisdiction.[4] Professionalization is the pursuit of jurisdiction in competition with rival professions. Depending on the level of professionalization examined, the Army as a profession could buttress itself against competition from other services in war or against other professions specializing in the control of violence—e.g., police. Analysis could focus on the jurisdictional battles between services within an individual country or could be undertaken internationally to see how all armies protect (or extend) their jurisdictions vis-à-vis competitors.

A variety of contextual factors—the state of technology, the state of organizational development, the surrounding politics, the power basis of society—affects the success of the jurisdictional claims a profession like an army makes. In the midst of changes to any of these contextual factors, professions may find themselves fending off challenges to their jurisdiction. As a profession in a formally subservient position to government, armies pursue their jurisdiction by interpreting objective, external information through their professional lenses based on established knowledge systems.

Some specific qualities of an army as a profession must be considered when thinking about that army's jurisdiction and professional competition. In the last 150 years, most armies—and most military forces—have generally been public forces created by political entities. Thus, the U.S. Army constitutes its tasks by balancing the norms and standards of the profession with the political goals of its masters. If it leans too far in the direction of Army norms and standards to the detriment of civilian goals, it risks retaining control of a jurisdiction decreasingly in demand. The profession might then die because there would be no one to pay those who practice it. If it leans too far in the direction of the political goals of its masters, it may lose its ability to pursue its jurisdiction at all, and thus its professional status.

Changes in the Army's operational context in the post-Cold War era promised to affect both the Army's definition of its jurisdiction and challengers to the Army's jurisdiction. The Army's decisions about privatizing training reflect its definition of its core tasks. Competition with contractors over military training is further affecting the professional jurisdiction of the Army.

Privatization in the Post-Cold War

In the wake of the Cold War, changes in both the type of competitors the United States expected to face and the competition strategy the United States embarked upon were bound to affect the Army and its jurisdictional claims. Both have also influenced the privatization or outsourcing of tasks by the Army. Privatization refers to the delegation of public duties to private organizations, [5] for example, selling state assets, deregulating, contracting out, and dispensing vouchers redeemable for private services. In some cases privatization refers to using private means to achieve public goals. In others it means shrinking the public sphere.[6]

While major threats have diminished in the post-Cold War period, and U.S. forces have been downsized, such complications as ethnic conflict, humanitarian emergencies, and the engagement strategy have caused the number of operations involving the U.S. military to rise. In scrambling to meet more and different requirements with fewer personnel and a competitive labor market, policy-makers (and military leaders) have increasingly turned to private contractors to carry out their foreign military training programs. The general push toward privatization and outsourcing of government functions has only encouraged this response.

The idea that "private" is more efficient than public has surfaced as a prominent movement in many Western states,[7] and the Defense Department has not been immune to the effect of this movement. Throughout the 1970s and 1980s, U.S. military services also launched several initiatives to transfer education and training of recruit and enlisted forces to private, or at least nonmilitary, entities.[8] The Commission on Roles and Missions of the Armed Forces chartered in 1994 looked specifically into privatization as a way to improve defense logistics.[9] In 1996, the Department of Defense (DOD) Report, "Improving the Combat Edge through Outsourcing," focused on several additional areas for which privatization was to be examined, including education and training. Later that year, the Congressional Research Service (CRS) set out a framework for discussing further privatization of DOD functions.[10]

In 1987, the widespread interest in privatization and outsourcing led a group of retired American military officers to capitalize on what they saw as an increasing demand in the United States for private companies to supply noncore military services by establishing Military Professional Resources, Inc. (MPRI). Other established contractors have deepened their capacities in the post-Cold War era for undertaking work formerly performed by the military services. Though some of these contractors hire from all walks of life, an increasing number rely on retired military personnel for military contracts. The military downsizing that accompanied the end of the Cold War (about 30 percent across the board) has insured that there are skilled personnel available to be recruited by private companies.

Part of the Army's justification for privatization has been to insure that, in the midst of downsizing, core roles stay within the military by privatizing noncore missions. In some cases, the tasks privatized make sense with this justification—why have scarce military personnel producing paychecks if you can contract out to a specialized firm? In other cases, such as training, the logic is less obvious—training officers, it would seem, is a key task for the profession. Still, the Army has proceeded with privatizing a variety of training missions for U.S. troops. At the same time, the Army has also delegated responsibility for training foreign troops to private contractors. On their own, companies that provide training services have also begun to sell them directly to foreign governments.[11]

The privatization of these tasks promises to affect the Army's professional jurisdiction in several different ways. Outsourcing the educating and training of U.S. forces can be seen as a loosening of the Army's control on the profession. Because educating and training its own members are critical tasks for maintaining the profession's internal control system, outsourcing this mechanism may lead to less control of the profession on the part of professionals. In other words, the Army may be limiting its ability to define its jurisdiction by outsourcing the training of its own forces. Few in the Army have looked at the privatization of training in this way. Indeed, training is not considered a core task (the core tasks being those closest to war-fighting). As it feels itself being asked to do more and more with fewer bodies, the Army is not competing with contractors, but happily ceding these training tasks. Andrew Abbott's model of professional development and demise would have the Army reconsider this perspective.

The privatization of foreign military training poses a different set of issues. First, the political importance of the Army is tied to its role (or jurisdiction) in U.S. defense. Training foreign armies came to play a significant role in the defense strategy of the United States in the 1990s—training was a prime component of the past administration's engagement strategy, according to the Office of the President's 1999 statement "A National Security Strategy for a New Century." Foreign military training was said to further U.S. contact with other countries, to aid in the spread of democracy and good civil-military relations, and to express specific U.S. strategic concerns with such issues as counternarcotics and counterterrorism. U.S. Special Forces have accordingly trained with over 100 countries annually in the 1990s. Many argue that this focus should continue. If military training becomes a more important part of national defense and private companies provide more of the training, it follows that the Army's jurisdiction in national defense may shrink. To see why this is so, consider the goals of foreign military training and the likely impact of privatization. If one goal is to further U.S. contact, we must realize that this contact will be performed by private entities, not military officers. The model of civil-military relations we are presenting for emulation by other countries will be a model with a significant private component. And those pursuing missions in counternarcotics and counterterrorism will be private actors.

There is little doubt that all of these tasks are less critical to the Army as a profession than the professional training of its own members. There is a relationship, however, between training foreign militaries and the development of abstract professional knowledge that must also be considered. Having companies that sell military training opens the way for a new process by which military knowledge is shared and spreads. Companies like MPRI, Cubic, or DynCorp offer military knowledge outside the structure of the Army. Their position outside the U.S. government may give them a vantage point from which to offer different variants of knowledge that competes with the Army's line. Or they simply may sell knowledge derived from the U.S. Army in a way that the Army has little influence on (unclassified field manuals, for example, are in the public domain). Either way, the process by which professional knowledge and norms spread is altered. If the Army gives up its unique claim to the development of abstract knowledge, thereby enabling others outside the profession a legitimate position from which to offer knowledge, it may find itself competing for its unique claim in the future.

Thinking about all of these issues requires an examination of whether (or under what circumstances) retired Army officers working as private military contractors are Army professionals. What kind of argument would exclude retired officers or those now working in the private sector from a professional status? If retired personnel are considered professional, how does their professional status relate to that of active duty personnel? Does it make sense to distinguish between organizational interest and professional jurisdiction? Finally, if the Army is entering a competition with (potentially professional) contractors for training missions, it should think about the implications for recruitment, retention, and morale within the service.

Outsourcing Internal Training

The Army outsources a number of education and training programs. MPRI alone provides support to the U.S. Army Force Management School, U.S. Army ROTC program, U.S. Army Combined Arms and Services School, and TRADOC Pilot Mentor Program at the Command and General Staff College.[12] The Army also uses MPRI's support in the development of doctrine—including, ironically, the doctrine for using contractors on the battlefield.[13] Outsourcing of education and training is quite consistent with DOD guidelines, which state that education and training are commercial activities that could be (but not necessarily should be) outsourced.[14] Is the Army giving away its capacity to control the profession internally by having private contractors provide these tasks?

Perhaps most controversial among these examples is privatization within the ROTC program. In this program future officers are shaped through their educational careers at civilian universities. In a typical ROTC program as recently as 1997, there were five authorized officers (though typically the positions were staffed at only four), three training noncommissioned officers (NCOs), and one general schedule (GS) civilian. Among the officers, one lieutenant colonel served as the professor of military science and commander of the battalion, while one major (often an active Guard or Reserve officer) and two captains served as assistant professors of military science. Two of the NCOs were responsible for organizing and conducting training while one handled supply and logistics.[15] A plan designed by the RAND Arroyo Center proposed that two of the assistant professors of military science and two NCOs—one trainer and the one responsible for logistics and supply—could be contracted out with uncertain cost impacts (in fact, some assumptions resulted in increased cost).[16] In the 1997-98 school year, the Army began experimenting with outsourcing portions of ROTC staff. According to MPRI's website, 220 colleges across the country are outsourcing a portion of their ROTC training in the year 2001.[17] The majority of positions that have been contracted out are assistant professors of military science. Trainers and logistics positions have also been outsourced, but not to such a degree, and administrative positions have been outsourced the least.

There are clear standards that retired officers who serve as assistant professors of military science or enlisted instructors must meet. They must have been retired no more than two years, wear a uniform during performance of their duties, meet height and weight standards, pass the Army Physical Fitness Test, and participate in ROTC advanced and basic camps during the summer.[18] The personnel MPRI has hired have met these standards continually.

Given the importance of the few military personnel that staff these programs as role models for future officers, however, some question the wisdom of having for-profit companies staff the programs with retired officers. Questions do not arise over whether MPRI fields good retired officers, but over whether the normative framework the program is supposed to instill in these future officers may be compromised by the use of a commercial company to perform this function for a profession whose defining tenet is self-sacrifice. Several have noted the "deprofessionalization" of the military in recent years—charting the trend of military service being seen by those serving as just a job or a means to achieve side payments (like education, travel, acquiring technical skills, etc.).[19] Outsourcing ROTC training turns over the job of educating and modeling professional behavior to profit-seeking firms. This practice may further promote the de-professionalization of the Army. Interestingly, in designing the plan for the Army, the Arroyo Center considered and rejected the possibility of having ROTC programs hire retired military personnel directly because of administrative hurdles.[20]

Much more research is required to determine the impact of outsourcing ROTC training and then to evaluate its relative benefit. This research should obviously focus on whether the program saves money and meets standards. It should also, however, evaluate deeper issues regarding the way in which the Army is communicating its standards of professionalism by using contractors to train officers. Can a contractor who is a profit-seeker impart the same standards of service as an active duty officer or NCO? If not, what are the losses and how is this likely to affect the attitudes of military personnel?

Private Training of Foreign Militaries

It makes sense to divide privately provided foreign training into two categories: those services paid for by the United States and those paid for by other countries.[21] These two categories are regulated by different laws and overseen by a different process.[22] The first category represents a classic form of outsourcing. The contracts operate according to Federal Acquisition Regulation (FAR) and additional rules unique to the Defense Department as outlined in the DOD FAR Supplement (DFARS).

Contracts with other countries are not really outsourcing at all, from a legal perspective. They are a category of military export. The State Department must approve such contracts under the International Traffic in Arms Regulations (ITAR). Like arms exports, the issues involved are wrapped up in how to regulate the distribution of a product. Also as with arms exports, the same companies are selling both to the U.S. government and to foreign governments. Though related, these different activities pose different dilemmas to Army professionalism.

Outsourcing Foreign Military Training

Outsourcing foreign military training affects the role the Army plays in U.S. defense. Foreign military training is a component of U.S. defense policy and clearly falls within the realm of potential Army jurisdiction, considering that the Army's role in training is important for thinking about what the U.S. Army should look like in the future. Some have argued that, given the post-Cold War downsizing, the Army should shed responsibility for noncore tasks such as training. To the degree it sheds these tasks, it is better able to focus on its core mission to fight and win the nation's wars. Others argue that if the Army sheds responsibility for missions that are close to the core of U.S. grand strategy, it marginalizes its importance. Questions over the proper long-term shape of the Army's jurisdiction underlie the debate over what a core mission is and therefore what best lends itself to privatization.

Privatizing foreign military training will weaken the U.S. Army's capacity for participating in engagement activities. When the U.S. government pours money for training into companies rather than into its national forces, it encourages private rather than public expertise. Also, given that private expertise is drawn directly from U.S. forces, buying from private providers undermines the long-term supply of the expertise in general. Companies like MPRI have an impressive list of personnel with varied language and experience portfolios because these individuals were trained in the U.S. Army. Every contract that MPRI wins for training takes away from developing comparable internal training experience in the Army.

Hiring private companies to train foreign militaries also signals to the members of the Army that training is not a "core" task. There has been ambivalence within the Army (and the other services) as to whether its job is to perform some of the new post-Cold War missions. To the degree that missions are privatized, they are considered further from the core and therefore less important for people's careers and for the organization. Privatizing training makes it more likely that the Army (as an organization) will take training less seriously.

In 1976 Congress created the International Military Education and Training (IMET) program to strengthen mutual understanding between the United States and its friends and allies and to facilitate these countries' ability to insure their own security. The IMET legislation separated military training from the Military Assistance Program (MAP) and set up funding to help countries that could not afford to purchase U.S. military training. The general program to sell U.S. training to personnel from other countries operates under the Foreign Military Sales (FMS) Act. Basically, the IMET program provides scholarships for students to come to Professional Military Education (PME) programs in the United States or subsidizes countries that cannot afford the entire program. Those that can afford it send students to the PME courses under FMS at their own expense.

IMET was amended in 1978 to increase the participating countries' awareness of human rights. In 1991, an amendment titled Expanded IMET (E-IMET) focused on defense resource management, respect for and grasp of democracy and civilian rule, military justice in a democracy, and human rights. Additional legislation made possible civilian and nongovernmental organization participation in the program.[23] General foreign military training under the umbrella of IMET and FMS takes place either on U.S. soil at one of the PME establishments or in short-term on-the-job-training, orientation tours, or instruction by a Mobile Training Team (MTT) outside U.S. territory.

The burgeoning of democracies and democratizing states in the post-Cold War era has led to an increased demand for professional military education in the United States. The United States has managed (in many eyes) to balance operational competence with firm civilian control under democratic principles. Both political and military officials are struggling with these issues in many democratizing countries and want to expose their promising personnel to the education and practices of the U.S. system. Such countries are interested in U.S. operational insight into combat and humanitarian operations. Indeed, many governments around the world share the current expanded U.S. security agenda covering democracy, drugs, arms control, peacekeeping, and the like. Finally, there is a sense among many that participating in such education makes them part of a larger global professional security community. As it was put by one Australian officer, "I don't think we seek 'expertise'—its more education, exposure, and breadth of understanding. Nonetheless, the management skills discussed . . . and the intellectual exposure . . . *must* make students *much* better [officers]."[24]

The U.S. foreign education and training missions have not gone without criticism. The Western Hemisphere Institute for Security Cooperation (formerly School of the Americas), once located in Panama but now at Fort Benning, Georgia, has long been criticized as the "School of Assassins and Dictators."[25] Furthermore, it has not always been clear that military education equates to U.S. influence. Proponents of foreign military training point out obvious benefits of IMET such as access to foreign personnel and subjective ties—those that participate in U.S. PME develop lasting ties with their colleagues. Though this debate is unsettled, IMET, E-IMET, and foreign military education for both civilian and military personnel have become a cornerstone of U.S. foreign policy since 1991. It is received wisdom in the Pentagon and elsewhere (though wisdom that demands validation through empirical research) that "low-visibility defense instruction has exceptional value in promoting both democracy and military cooperation."[26]

A similar focus on training and establishing relationships with foreign military personnel has been pursued within Joint Combined Exchange Training (JCET). These missions are primarily designed to train U.S. Special Forces personnel, preparing them to operate around the world against the variety of new threats the post-Cold War offers. Commanders of Special Operations Forces are authorized to deploy and pay for the training of U.S. and foreign troops if the primary purpose of the training is to train the special operations forces of the combatant command. As Special Operations Forces have come to be seen as increasingly valuable and appropriate for combating many post-Cold War threats—witness their participation in Afghanistan—the training needs for these forces has escalated. Indeed, the Special Operations Forces have also escaped much of the military downsizing of the earlier part of this decade and have a larger force now than they have had in the past.

The side benefit is that these operations also provide military-to-military contact with foreign troops. As former Secretary of Defense William Cohen put it, "In those areas where our forces conduct JCETs, they encourage democratic values and regional stability. In the future, we can expect our forces to confront threats posed by an increasingly diverse set of actors, placing a premium on the skills our forces developed in JCETs."[27] Increasingly, Special Operations Forces under the rubric of JCETs have operated as a new kind of diplomatic tool. JCETs have been used as a prelude to other kinds of military-to-military contact: "There is definitely a political card played with these JCETs. They are a direct instrument of U.S. foreign policy. They may be the most involved, tangible, physical part of U.S. foreign policy in certain countries."[28]

Despite the prevailing wisdom about the value of military engagement, however, there has been some question within the Army about how central this mission should be. Some who believe in the centrality of this mission to American security are clearly behind it. Others are behind these missions because they come with funds. Many others, however, suggest that in an era of more tasks and less people something has to go, meaning that training other armies is a good thing to cede to contractors. This is an example of an internally contested jurisdiction where the Army has not yet decided whether to embrace these missions or not.

The African Crisis Response Initiative (ACRI) and the African Center for Strategic Studies (ACSS) grew out of the same ideas that are prevalent in IMET, FMW, and JCET. The idea is to reward democratizing countries, enhance the capacity of their militaries, and encourage military professionalization. In both programs, however, private companies have come to play a substantial role. The ACRI began as a program run with U.S. troops which later privatized some of its training.

Former Secretary of State Warren Christopher launched the ACRI 1996. The purpose was to work with international partners and African nations to enhance African peacekeeping and humanitarian relief capacity. Particularly, the United States would offer training and equipment to African nations dedicated to democracy and civilian rule that would like to improve their peacekeeping abilities. The United States intended to use the ACRI to coordinate its African peacekeeping approach with the Organization of African Unity (OAU) and other interested parties (Britain, France, etc.). And this has worked. In December 1997, representatives from these countries, interested donors, and troop contributors met at the United Nations (UN) Department of Peacekeeping Operations in New York for the first peacekeeping support group meeting ever. Initial plans called for a five-year program run out of the State Department whereby U.S. Special Operations Forces from Central Command and European Command would provide equipment and training to 10,000-12,000 African soldiers before 2002.[29]

The United States has completed initial training for battalions from Senegal, Uganda, Malawi, Mali, Ghana, Benin, and Cote d'Ivoire. As the program developed, however, Special Forces personnel were stretched thin. New battalions needed to be trained and those already trained needed to be kept current. Thus, the State Department began to look for help. Both MPRI and Logicon won contracts to support the ACRI. The program continues to be run through the State Department, though many of the personnel in the ACRI office are also contractors—employees of USATREX.[30]

The ACSS was first announced during President Bill Clinton's trip to Africa in April 1998.[31] It is modeled after the Marshall Center in Garmisch, Germany. The Marshall Center acts as a forum and training center to address security and defense issues in the North Atlantic region.[32] The ACSS, which does not yet have a permanent location, is to provide a forum for senior military and civilian officials to explore together complex defense policy issues and provide training to strengthen civil-military relations in burgeoning democracies. It began as two-week seminars given at different locations around Africa. It is hoped that eventually someone will offer a permanent location for the center. Unlike the ACRI, the ACSS was designed with contractors in mind. It is being run out of the Pentagon. Four contractors competed for the project. MPRI and DFI won the contract for the first session, which began in October 1999 and continued for the second session in May 2000.

Privatization of these programs seemed to happen for purely pragmatic reasons. There was no dissatisfaction with the job that U.S. Army personnel were doing, and there was no clear choice on the part of Army to discard these jobs. The Army simply did not have the numbers of personnel to staff the programs. Partly due to beliefs about the peace dividend that was not fully realized and partly due to a preference among the military for core missions, the U.S. military has been downsized despite an increase in operational tempo. Pragmatic policy-makers have scrambled to come up with solutions that satisfy the U.S. desire to maintain an international presence and influence in many unstable parts of the world without unduly stretching a smaller military. Private companies have seen this coming and organized themselves to take advantage of a new kind of government demand. Using private companies to staff these programs provided a flexible solution that did not require increasing the size of U.S. military forces. They may also provide some cost savings, though this is less clear.

Most people overseeing these programs have expressed strong support for MPRI's, DFI's, and Logicon's services in ACRI and ACSS. Many of the personnel involved are former U.S. armed forces personnel. They have been indoctrinated into a system of public service and know the standards of appropriateness expected from military professionals. On the other hand, these firms are clearly different from the Army. They operate for profit, and the profit motive underlies their search for contracts and is bound to influence their decisions about appropriate behavior in borderline situations. MPRI, Logicon, and others are not identical to the Army. They are providing a separate voice from the Army in offering military advice to foreign countries.

Small steps taken for pragmatic reasons can end up initiating large changes. In this instance, the decisions to privatize may end up further reducing the capacity of the Army to perform these kinds of tasks. More and more of the U.S. presence in unstable parts of the world may be left to private hands. This may or may not be a wise jurisdictional choice, but it is wise to raise questions about the implications of this shift both for the Army as a profession and for civil-military relations:

- Is it cost-efficient to pay to train personnel in the U.S. armed forces who take their retirement only to sell their service back for profit?
- Is it wise for foreign policy goals if more and more of the U.S. presence abroad is private?
- Do we want to give up our Army's capacity for conducting foreign military training?
- Might there be a connection between foreign military training and expert military knowledge in general? Once the Army cedes its unique role in generating expert knowledge (by privatizing foreign military training), will it be able to compete effectively with contractors?

U.S. Firms Contracting Directly with Foreign Militaries

Contracts for military training between private U.S. firms and foreign governments have also been on the rise. The instance receiving the most press has been MPRI's contracts with the Croatian government and the Bosnian government. But there are many others. Hungary has contracted with Cubic for help in reorganizing to meet the standards for entering NATO. MPRI also has been licensed to conduct training in Angola (though the contract was frozen by an informal policy hold by the State Department), has a signed contract with Equatorial Guinea, and is being paid for some of its work in Nigeria directly from the Nigerian government (there are similar prospects in Colombia).

These contracts are regulated within the International Transfer of Arms Regulations (ITAR), which are monitored by the Department of State's Office of Defense Trade Controls. Before a license is granted, the appropriate regional office, political-military bureau, desk officer for the country, and others (such as the Bureau of Democracy, Human rights, and Labor) all have input. In the event that a country with no State Department restrictions wants to buy nonlethal training (such as leadership training or instruction in appropriate civil-military relations), the license is not controversial. At the other extreme, for a country on the State Department Embargo Chart, a license would not be issued.

For in-between cases, where a country is making the headlines for one reason or another or has some informal policy holds against it, a variety of offices make their case and the Assistant Secretary makes a final decision. We've already noted the State Department's policy hold on MPRI's Angola contract. In another instance, the African Regional office rejected MPRI's initial application for a license with Equatorial Guinea the first time around. Upon appeal, consideration of the nature of the activity (evaluating the need for a Coast Guard and restructuring the defense department so as to professionalize the military) and the amount of U.S. investments in the country ($3 billion by U.S. oil companies) led the regional office to change its mind. It concluded that improvement of security for U.S. citizens and their investments was worth the risk.[33] At that point, however, the Bureau of Democracy, Human Rights, and Labor held the contract up over human rights concerns. MPRI appealed to the Assistant Secretary for Human Rights and lobbied in Congress. In the end, MPRI received the license and has carried out its initial assessment in Equatorial Guinea.[34]

Once a license is issued, there is no formal oversight process required by the ITAR. In its day-to-day work, of course, the State Department monitors the behavior of other countries. If a country licensed to receive military advice and training commits egregious acts, the State Department can freeze the contract (just as it can freeze weapons transfers). In some cases, as with the Train and Equip program in Bosnia, significant oversight structures are put in place. In others, there are no reporting or oversight mechanisms aside from routine monitoring by embassy staff. Finally, only contracts of more than $50 million in defense services require congressional notification before they begin.

MPRI's work with Croatia provides a useful example. As mentioned above, MPRI was founded in 1987 by top-ranked retired U.S. generals.[35] It boasts a database of over 10,000 potential employees, all having significant experience in the U.S. armed forces, and has managed to bring in over $12 million in annual revenue.[36] The company began by contracting with the U.S. government to train (and provide other support for) the U.S. armed forces and has gradually moved overseas. Part of what makes MPRI advice so sought-after is that this is the same advice that is sold to the U.S. armed forces.[37] MPRI's endeavors in the Balkans have received much attention. The Train and Equip program in the Bosnian Federation was a cornerstone of the Dayton Accords. Because of its relationship to the Dayton Accords, however, the contract with the Bosnian government is very specific to that conflict and therefore is probably not the most useful to generalize from. The company's work with Croatia is more typical; thus that contract will be the illustrative focus here.

The genesis of MPRI's contact with the Croatian government is disputed. Some have argued that the United States sent the Croatians to MPRI. MPRI claims that the Croatians had seen a presentation MPRI made to states eager to get into NATO. For whatever reason, the Croatian Defense Minister Gojko Susak sent a letter in March 1994 requesting permission from the Pentagon to negotiate with MPRI to obtain training in civil-military relations, programs, and budget.[38] Croatia and MPRI signed a contract in September 1994 and the State Department licensed the project in December. MPRI then teamed its experts with English-speaking Croatians to set up courses in budgeting, leadership development, military ethics, and relations with civilian society. According to a Pentagon official, MPRI personnel were briefed for days before they went to Croatia and closely monitored once they were there (MPRI officials claim to have no contact with the Pentagon).[39]

However, many find the coincidence between MPRI's presence and the turn-around in the Croatian Army's performance and tactics to be noteworthy. Shortly after MRPI started working with the Croatian Army, the Croats began to experience success in their quest to regain their territory (Serb forces were occupying about 30 percent of Croatian territory when MPRI arrived). In May 1995, the Croats retook the areas to the southwest of Zagreb, then recaptured parts of western Slovenia, and finally, in August 1995, the Croatian Forces launched Operation Storm to retake the Krajina region.[40] The offensive in Krajina looked (to observers in the region) like a textbook NATO operation, leading to endless speculation about whether MPRI was instrumental in Croatia's operational success.[41] According to Roger Cohen, "The Lightning five-pronged offensive, integrating air power, artillery, and rapid infantry movements, and relying on intense maneuvers to unhinge Serbian command and control networks bore many hallmarks of U.S. Army doctrine."[42] There has also been speculation about whether the United States shared satellite information with the Croats.[43]

According to MPRI spokesman Lt. Gen. (ret.) Harry Soyster in 1995, the instruction "has no correlation to anything happening on the battlefield today."[44] He explained that MPRI had far too much to lose to violate the terms of its license by advising the Croats on battlefield strategy, weapons, or tactics.[45] Of course, this assumes that MPRI would have lost its license. More likely is the speculation that if MPRI did violate its license, it was with the knowledge of officials in the Pentagon and elsewhere.[46] As Juan Zarate argues, MPRI's training (even as it was touted as a long-term venture) *was meant* to have a positive effect on the Croatian military—otherwise, why would they have landed the job?[47]

The strongest evidence supporting MPRI's claims that it was not responsible for the Croatian success is simply the fact that there were only fourteen MPRI personnel in the field during this time. It seems far-fetched to attribute the Croatian Army's success to those few people alone. Whether MPRI did violate the terms of the arms embargo is something that is hard to learn for sure, as is the impact of having advice from a retired U.S. general on Croatian morale.

As it was, the Croatian successes changed events on the ground such that strategic bombing by NATO was able to push the Serbs over the edge and to the negotiating table—the results of which were the Dayton Accords (and another contract for MRPI). By licensing MPRI, the United States retained its neutral status while influencing events on the ground. An assessment of the costs and benefits of the policy is complicated by the Croatian human rights abuses in the Krajina region in the wake of Operation Storm. It was, up until Kosovo, the largest incident of ethnic cleansing since the end of the Cold War. It is estimated that Croatian forces uprooted between 150,000 and 170,000 Serbs from their homes.[48]

MPRI continues to operate in Croatia, with its work there headed by retired Maj. Gen. John Sewall. The company is working closely with the Croatian government to reorganize its Ministry of Defense and military forces to better reflect a modern professional military. While the State Department must approve any new contract, MPRI has no reporting requirements, and there is no particular oversight of the company's activities.

In this case, the controversy has arisen over the degree to which MPRI may have allowed the United States to skirt public debate and encourage unofficial actions that would be officially untenable. It would have been very difficult to engender support for U.S. troops to train the Croats in 1995. Moreover, had U.S. troops been engaged in such training while the Croats were engaged in ethnic cleansing, it is likely that the United States would have been held responsible at some level. Firms like MPRI, and the services they offer, may allow governments to avoid traditional processes and accountability in the creation of foreign policy. It is simply more cumbersome and opens up more public debate when countries send national troops abroad—or even hire contractors to work abroad. Licensing a commercial company provides a way to stay below the radar.

While this controversy is mostly about foreign policy efficiency and democratic accountability, it has implications for Army professionalism as well. Having private companies responsible for transmitting army professionalism causes the Army (and the U.S. government) to relinquish its monopoly on the spread of knowledge about an army's proper behavior. Though it may be the case that MPRI works closely with the U.S. Army, to the degree that those who work for MPRI develop a slightly different perspective, this perspective potentially competes with the Army's. Furthermore, though MPRI hires retired Army officers precisely because they are imbued with the normative and professional standards of the Army, as Andrew Abbott points out these standards are most clearly and practically put forward during work. While MPRI is doing work, it is also potentially influencing these standards—and doing so from a different perspective. Just as in the case of private foreign military training with U.S. dollars, MPRI's work is undermining the Army's claim to unique expert knowledge.

This is not to suggest that private foreign military training is undermining military or Army professionalism generally. Indeed, some would argue that private companies operating in Africa and other parts of the world are doing more to spread professional norms—about the proper role for army personnel, about the proper role for police, about proper civil-military relations, etc.—than professional militaries are. To the degree that private companies are spreading these norms, however, it is increasingly general military professionalism rather than U.S. Army professionalism. The increasing capacity of the private sector in offering military advice to foreign countries may thus be putting at risk the U.S. Army's leadership position in establishing jurisdiction for the system of army professions.

Private Professionals?

One of the hallmarks of private military and security services in the 1990s is their claim to professionalism. This claim is based on the extensive training and education their personnel carry as well as the fact that most companies try to operate within national and international rules and norms. They claim to be neither rogue companies nor mercenaries, but rather professional providers of military and security services.

Any argument that might exclude retired officers from working in the private sector in their professional status would most likely be based on the public nature of military service in the last 150 years. That is a difficult argument to sustain simply because the public nature of military service is frequently tied to the notion of a citizen army rather than a professional army.[49] Rather than ipso facto assuming that private companies are not professional, it may be more useful to think about the ramifications of different types of professionals competing for jurisdiction.

Thinking about it this way requires us to distinguish between the organizational interest of the U.S. Army and the professional jurisdiction of armed service. The U.S. Army may be competing with different entities such as private contractors for the rights to establishing jurisdiction. The fact of this competition alone may change the process of establishing jurisdiction. Perhaps professional officers working in a variety of capacities—in private firms, in public organizations, in international organizations—are all struggling to define the profession of armed service. How this bodes for the U.S. Army may or may not matter for how it bodes for the profession of armed service. Indeed, Abbott speaks of the contrast between professions that are highly structured and thus need to mobilize themselves against an attack, on the one hand, and those that are loosely organized and therefore more flexible and competitive, on the other. If the Army remains committed to its traditional core jurisdiction, it may be less adaptive for use in the world. Perhaps the proliferation of private military companies (as well as innovations by other national militaries) will maintain the relevance of the military in general vis-à-vis other competitors in the use of violence for political means.[50]

Contested Jurisdictions and Individual Choices

How will individual members of the military (and potential members) respond to contractors competing for professional status in the world of armed service? Anecdotal evidence suggests a mix of reactions. During the initial deployment of U.S. soldiers to Bosnia, the efforts of private firms such as Brown and Root received much attention. Newspapers routinely quoted soldiers as saying they could not wait to retire and then return to work for Brown and Root where the real money was. One could argue that contracting opportunities could be recruitment plugs because military experience can more easily transfer to the private market.

Most accounts are less optimistic. Many members of the U.S. military are not pleased to be competing with contractors. Contractors frequently receive greater financial rewards and exercise greater freedom while they are at it. Contractors get regular days off, get paid overtime, and can always quit. Military personnel do not have those luxuries. Members of the U.S. military are subject to strict codes of conduct, a separate judicial system, and a set amount of pay. Anecdotes from various bases where much "gruntwork" has been outsourced suggest that military personnel are frustrated to be doing the same jobs as contractors for less pay and with fewer benefits.

While military training is hardly gruntwork, there is evidence that competition with contractors over training also creates problems. Specifically, some reports suggest that if a contractor can be hired to do a job, active duty personnel immediately view the job as less important. The very fact that the Army privatizes noncore tasks means that if a task is privatized it is noncore. Members of the Army take pride in conducting missions that only active duty personnel can do. Missions that are contracted out become inherently less respected and valued among the members of the service.

Contracting out some jobs may be a form of signaling about what is core and not in the Army. It may also have an impact on morale, recruitment, and retention. If more and more jobs are contracted out, issues of equity between military and civilians are bound to arise. These are additional issues that the Army should consider as it evaluates privatization options.

Conclusion

Relying on private contractors for training has undoubtedly provided flexibility for the Army to meet the many demands of the 1990s. It appears that the Army has made these decisions, however, as a cash-strapped and oversubscribed bureaucracy rather than as a profession. The implications of privatization for Army professionalism should also be considered. The Army needs to think about how much it values training its members, the interaction between internal and foreign training missions, the multifaceted impacts of ceding expertise to the private sector, and the impact of competing with contractors on the individual professionals within the service.

In the end, decisions about what to privatize should depend on how the Army views its future, and how civilians view the Army's future. The Army would do best to attune its vision with that of the civilian leadership and the demands of the world so as to avoid clinging to its traditional professional jurisdiction at the cost of its relevance. If the Army sheds its capacity for training missions at the same time that training missions become a greater part of U.S. grand strategy, it may hold on to a tradition without a constituency. Of course, the Army is part of the society and economic system in which it operates. Resisting the tides of privatization could turn out to be useless and counterproductive. Thinking clearly about its future and the implications of different strategies for privatization, however, may help the Army manage these changes more effectively. Though the ultimate decisions are (and should be) civilian decisions, the Army will have considerable influence on them and would be wise to deliberate fully on professional costs and benefits in both the wider context and the longer term.

Notes

1. The author wishes to thank the John D. and Catherine T. MacArthur Foundation for its generous support of the research for this chapter.
2. Tony Perry, "Chipping Away at a Navy Tradition," *Boston Globe*, 10 February 01; Joe Dejka, "Private Contracting to Affect 1,459 Offutt Jobs," *Omaha World-Herald*, 18 February 2001; http://www.mpri.com/subchannels/nat_ROTC.html.
3. Training is only one manifestation of this focus. In the high-technology area, as well, the private sector looms large. During the March 1997 Task Force XXI Army War-fighting Exercise, 1,200 contractors from 48 vendors participated in the field with the EXFOR. See Mark Hanna, "Task force XXI: The Army's Digital Experiment," *Strategic Studies* 119 (July 1997).
4. Andrew Abbott, *The System of Professions: An Essay on the Division of Expert Labor* (Chicago, IL: University of Chicago Press, 1988).
5. John Donahue, *The Privatization Decision* (New York: Basic Books, 1989), 3.
6. Harvey Feigenbaum, Jeffrey Henig, and Chris Hamnett, *Shrinking the State: the Political Underpinnings of Privatization* (New York: Cambridge University Press, 1998).
7. Feigenbaum et al., *Shrinking the State.*
8. Most of these programs were in concert with established institutions, community colleges, technical institutes, or proprietary schools. A 1991 RAND report commissioned by DOD (in concert with the GAO) found little firm evidence on which to base a conclusion about the cost savings associated with these programs or their relative benefits. See Lawrence Hanser, Joyce Davidson, and Cathleen Stasz, *Who Should Train? Substituting Civilian Provided Training for Military Training* (Santa Monica, CA: RAND, 1991).
9. Loren Thompson, "Privatization of Defense Support Functions: A Public Sector Case Study" (presentation to Kennedy School of Government, Cambridge, MA, 28 April 1995).
10. Gary Pagliano, "Privatizing DOD Functions Through Outsourcing: A Framework for Discussions," *CRS Report*, 6 August 1996. As of this time, DOD had already outsourced 25 percent of base commercial activities, 28 percent of depot maintenance, 10 percent of finance and accounting, 70 percent of army aviation training, 45 percent of surplus property disposal, 33 percent of parts distribution, and portions of other functions.
11. This is a disputed claim. Many argue that U.S. contractors that have sold military training in places like Croatia are actually acting at the behest of the American government, albeit informally. See Ken Silverstein, *Private Warriors* (New York: Verso, 2000), 145.
12. *http://www.mpri.com/subchannels/nat_business.html.*
13. Ibid.
14. *http://www.acq.osd.mil/inst/icim/icim/newpriv/newpriv.htm.* Cites Appendix 2, Office of Manpower and Budget Circular No. A-76, Revised Supplemental handbook, "Performance of Commercial Activities," March 1996.
15. Charles A. Goldman et al., *Staffing Army ROTC at Colleges and Universities: Alternatives for Reducing the Use of Active-Duty Soldiers* (Santa Monica: RAND, 1999), 7-8.
16. Ibid., p. 12.
17. *http://www.mpri.com/subchannels/nat_ROTC.html.*
18. Ibid.
19. See Charles Moskos and Frank Woods, eds., *The Military: More than Just a Job?* (McLean, VA: Pergamon-Brassey's, 1988).
20. Hiring former military members as civilians would subject the Army to limited civilian personel ceilings, the requirements of which would negate any cost savings from the outsourcing effort. See Goldman et al., *Staffing Army ROTC*, 3.

21. The latter category also includes training associated with the sale of weapons and equipment, but I will not focus on these contracts here.
22. The U. S. military contracts out many services for its own forces—from base security in some areas to ROTC training at many universities. For the purposes of this chapter, I am focusing on private provision of advice and training to foreign militaries.
23. John A. Cope "International Military Education and Training: An Assessment," *McNair Paper* 44 (Washington, DC: Institute for National Strategic Studies, National Defense University, October 1995), 5-6.
24. Ibid., 15. There is some tension among foreign participants in the U.S. education and training programs. Some, including Central Europe, the former Soviet states, and Africa, are quite interested in the restructuring efforts promoted by E-IMET focusing on civil-military relations, enhancement of democratization, human rights, etc. Other long-standing IMET customers would prefer greater attention to traditional military and operational missions and expanded opportunities for officers as opposed to civilians. See Cope, "International Military Education," 17.
25. In July 1999 the House voted to eliminate funds for training foreign officers at the facility. See John Lancaster, "House Kills Training Funds for School of the Americas," *Washington Post*, 31 July 1999.
26. Cope, "International Military Education," 41; "U.S. Training of Foreign Military and Police Forces: the Human Rights Implications and Issues for Consideration by AI," Interim Report prepared by Lora Lumpe, March 2000.
27. Dana Priest, "Special Alliances: the Pentagon's New Global Engagements," *Washington Post*, 12 July 1998, A1.
28. Controversy over these programs emerged when it was reported that JCET missions are not governed by the same State Department regulations that govern other military training. In a series of articles, the *Washington Post* reported that U.S. Special Operations Forces frequently conduct JCETs with foreign troops that would be precluded from receiving military training otherwise. For instance, JCETs proceeded in Colombia during 1996 and 1997 even as the Clinton administration "decertified" Colombia for military assistance because of its failure to comply with U.S. anti-narcotics policy. In Indonesia a U.S. congressional ban on training Indonesian officers and a checkered human rights record has not prevented 41 JCET exercises since 1991. Similar stories in Papua New Guinea, Equatorial Guinea, Rwanda, Suriname, and elsewhere have led some to suggest that these programs undermine, or work at cross-purposes with, U.S. diplomatic efforts. See Priest, "Special Alliances." Others have questioned the Pentagon's commitment to human rights when the United States continues military-to-military relationships with leaders such as Kagame in Rwanda. Referring to military-to-military contact with Rwanda in 1995, Lynne Duke asserts, "Clearly the focus of Rwandan-US military discussion had shifted from how to build human rights to how to combat an insurgency." "Africans Use Training in Unexpected Ways," *Washington Post*, 14 July 1998, A1.
29. The sum of $15 million was allocated the first year, and $20 million each for the next two years.
30. The program is funded through 2001. It is unclear whether it will continue to be funded through the Bush administration. Scott E. Brower and Anna Simons offer support for the program but argue that it must develop a command and control system to ultimately succeed. Part of MPRI's contract is to develop just such a system. See Scott E. Brower and Anna Simons, "The ACRI Command and Control Challenge," *Parameters* 30 (Winter 2000-2001).
31. See discussion by Susan Rice, Assistant Secretary of State for African Affairs (speech to the City Club of Cleveland, Cleveland, Ohio, 17 April 1998). When first announced, it was the African Center for Security Studies.

32. See speech by Warren Christopher before the North Atlantic Cooperation Council, Athens, Greece, 11 June 1993.
33. Interview with State Department official in African Regional Affairs Office.
34. Interview with Lt. Gen. (Ret.) Harry E. Soyster, 1 December 1998.
35. Gen. Frederick Kroesen (Chairman of the Board), Maj. Gen. Vernon Lewis, Jr. (President, CEO), Gen. Robert Kingston, Gen. Robert Sennewald, and Lt. Gen. Richard West. Equally impressive are those who have since joined: Lt. Gen. Harry E. Soyster (Vice President for International Operations) and Gen. Carl E. Vuono (Vice President and General Manager of the International Group).
36. Jakkie Cillers and Ian Douglas, "The Military as Business: Military Professional Resources, Incorporated," in *Peace, Profit or Plunder? The Privatisation of Security in War-torn African Societies*, eds. Jakkie Cillers and Peggy Mason (South Africa: Institute for Security Studies, 1999); Mark Thompson, "Generals For Hire," *Time*, 15 January 1996, 34-36; Juan Carlos Zarate, "The Emergence of a New Dog of War: Private International Security Companies, International Law, and the New World Disorder," *Stanford Journal of International Law* 34 (1998): 75-162.
37. Interview with Lt. Gen. (Ret.) Harry E. Soyster, 1 December 1998.
38. Bradley Graham, "US Firm Exports Military Expertise: Role In Training Croatian Army Brings Publicity and Suspicions," *Washington Post*, 11 August 1995, A1.
39. Personal interview with Pentagon official, October 1997.
40. Zarate, "The Emergence of a New Dog of War," 106-107.
41. Robert Fox, "Fresh War Clouds Threaten Ceasefire: Secret US Military Advice Helps Cocky Croats Push Toward Eastern Slovenia," *Sunday Telegraph*, 15 October 1995, 26.
42. Roger Cohen, "U.S. Cooling Ties to Croatia After Winking at Its Buildup," *New York Times*, 28 October 1995, 1.
43. Charlotte Eager, "Invisible U.S. Army Defeats Serbs," *Observer*, 5 November 1995, 25.
44. Ibid.
45. Ibid.
46. Cohen, "U.S. Cooling Ties," Cohen reports that senior State Department officials admitted that Croatia became our de facto ally—that arms flowed in despite the embargo and top retired American generals were allowed to advise the Croatian Army.
47. Zarate, "The Emergence of a New Dog of War," 108.
48. Cohen, "US Cooling Ties"; Mark Danner, "Endgame in Kosovo," *New York Review of Books*, 6 May 1999, 8.
49. Eliot Cohen, *Citizens and Soldiers: The Dilemmas of Military Service* (Ithaca, NY: Cornell University Press, 1985); Martin Van Creveld, *The Transformation of War* (New York: Free Press, 1991).
50. Though it would be unwise to stretch the analogy too far, the transition from the age of feudalism to the age of military contractors was also accompanied by restrictions on the use of feudal knights. In England, for instance, feudal obligations to do war service was forty days per year and did not extend overseas. This led kings as early as Edward I (1274-1307) to contract for service with willing warriors. See Fritz Redlich, *The German Military Enterpriser and his Workforce*, vol. I (Wiesbaden: Franz Steiner Verlag GMBH, 1964), 22.

III

The Political-Social Arena[1]

The Army profession serves its collective client, the American people, through interactions with the citizenry's elected and appointed leaders and the nation's other government agencies. Such interactions with the clients' surrogates include competing for vital resources and receiving directives regarding where and when to apply the Army's expertise. Traditionally, Army leaders have viewed these political and social interactions narrowly, fearing the politicization and tainting of the institution. While those remain concerns, today's global and national circumstances—for example, America's preeminent global role, the increased role of joint military commands and agencies, as well as that of the staffs and agencies of the Executive and the Congress—demand expanded political and social literacy of Army professionals of all ranks as well as increased interactions between America's military and the larger defense community. However, these vitally important activities, particularly those of the profession's strategic leaders, require areas of expert knowledge that remain under-appreciated within the Army profession. This must change if the Army is to compete successfully within its jurisdictions in the 21st century.

Thus the four chapters in this section examine the effects of politics on the Army profession and the role of professional norms in the practice of political and social affairs, both past and present. These chapters focus on the subtle but critical divide between appropriate and inappropriate behavior by the profession's strategic leaders as they work in these milieus. For example, a topic of concern is the difference between rendering professional advice and advocating for a particular policy, or, said another way, rendering military advice in a manner seen to be above the parochial interest of advancing the Army's cause. While difficult, the Army profession must clarify appropriate norms for its members' actions and develop a larger cohort of strategic leaders with real political expertise if it is to be perceived as a profession rather than a political interest group. Since many of the profession's jurisdictional competitions are often fought with just such expertise, Army practice and culture must be adapted to value that expert knowledge so that the profession may be fully represented in deliberations vital to its ability to serve the American people.

Note

1. The authors express their appreciation to Dr. Meena Bose of the USMA Department of Social Sciences, who served as Panel Rapporteur for this section's topic during the USMA Senior Conference, June 2001.

10 Rules of the Game? The Weinberger Doctrine and the American Use of Force[1]

Suzanne C. Nielsen

The early 1990s saw a resurgence of concern about the condition of civil-military relations in the United States. Perhaps first strongly articulated in Richard Kohn's article, "Out of Control: The Crisis in Civil-Military Relations," a significant issue was whether senior military leaders were playing an inappropriate role in the policy-making process.[2] Kohn especially focused on Chairman of the Joint Chiefs of Staff Colin Powell's public positions against the integration of homosexuals into the armed services and against U.S. military intervention in Bosnia.[3] Although Kohn himself later acknowledged that the term "out of control" overstated his case with regard to the military as a whole, he was concerned about senior military leaders' influence and military politicization.[4]

After Gen. John Shalikashvili took over as Chairman in 1993, some observers felt that he initiated significant progress toward restoring "a functional balance" to civil-military relations.[5] He served less as a policy advocate, and seemed to focus more on giving purely military advice. However, a look at the complete record suggests that General Shalikashvili also at times publicly took positions about the proper ends of policy. When he was selected to be Chairman, General Shalikashvili had a reputation as critic of the Clinton administration's inaction in the Balkans.[6] General Shalikashvili also made some fairly broad policy statements as Chairman. For example, in remarks at George Washington University in 1995, he said the following:

> When we respond to a humanitarian tragedy and our hearts are most affected, we must be wary of the impulse to do more than provide relief. . . . No matter what we think we agree on, it should not be to rebuild or restructure other nations—that takes generations of sustained effort.[7]

Is this statement, which is intrinsically about policy goals, an appropriate comment from a professional military officer? In other words, is it based on professional military expertise? If not, what are the implications of military leaders acting in an extra-professional capacity?

In this chapter, I will suggest a framework for thinking through these questions focused on one particular issue—what is the appropriate role for a professional military officer in the context of use-of-force decision-making? In addressing this question, the term "professional" is significant, and has a particular meaning. The view of professions taken here is fundamentally informed by Andrew Abbott's *The System of Professions* (1988). Abbott defines professions as "exclusive occupational groups applying somewhat abstract knowledge to particular cases," with the tasks they perform being "human problems amenable to expert service."[8] In this light, then, professional military officers provide expert service in meeting society's needs in the realm of military security. This view of officers as the expert providers of a service is a starting point for establishing norms for their role in state decision-making about the use of force.

Based upon Abbott's definition of a profession, the argument below will proceed in four steps. I will begin by suggesting the relevance of Abbott's discussion of professional practice in thinking through the role of the professional military officer in decision-making on the use of force.[9] Second, I will draw on works dealing with strategy and civil-military relations to fill in the content of the military officer's expertise. Once this groundwork has been laid, I will use the insights from these two sections to analyze an attempt to establish guidelines for decision-making on the use of force in the United States. I will focus on the Weinberger Doctrine, using it as a vehicle to discuss the professional officer's special expertise and role in the decision-making process. Finally, I will conclude by suggesting implications for the future of the military profession that flow from the manner in which senior military leaders conduct themselves in rendering advice and making decisions on the use of force. Although many have examined the role of military officers in this process, their analyses have usually attempted to draw implications relevant to the health of civil-military relations or civilian control.[10] Here, instead, the focus will be on implications for the future of the military profession itself.

Before beginning this analysis, it is important to note that my conclusions about the Weinberger Doctrine do not necessarily reflect a consensus view among today's military officers. Below I will argue that the Weinberger Doctrine constitutes a problematic guide for a professional military officer to adopt on use-of-force questions. This assessment is in tension with some currently prevalent military attitudes as depicted by the recent survey findings of the Triangle Institute for Security Studies Project on the Gap Between the Military and Civilian Society. For example, that study found that a majority of elite officers believe it is appropriate for the military to "insist" on use-of-force measures such as the development of an exit strategy.[11] The study then goes on to suggest that the choice of "insist" over the other alternatives of "be neutral," "advise," "advocate," or "no opinion," may reflect lingering distrust from the Vietnam period. While some may see a focus on exit strategies as a lesson from the Vietnam experience, the perspective developed here will challenge the appropriateness of its adoption as a requirement by military leaders. It will do so by examining classic works on strategy and American civil-military relations to derive propositions about the proper role of the military professional in use-of-force decision-making in the United States.[12]

Professional Practice

As stated in the introduction, the first step in the argument is to suggest that as senior military officers participate in use-of-force decision-making, they are engaging in professional practice. Abbott's discussion of the three tasks of professional practice—diagnosis, inference, and treatment—provides a useful set of concepts to use when analyzing the manner in which military leaders provide advice and carry out decisions once they are made.[13]

The first task, that of diagnosis, involves taking information into the professional knowledge system. The professional sorts available facts according to professional rules of relevance, and evaluates them in accordance with rules of evidence. In so doing, the professional is attempting to structure the situation. In describing this process, which Abbott calls "colligation," he gives the example of generals who "must ask leading questions of civilian politicians who 'don't know what they want.'"[14] Rules of relevance may be strict:

> If the client is an individual, such extraneous qualities often include his or her emotional or financial relation to the "problem." If the client is a group, they include irrelevant internal politics, financial difficulties, and so on. (A diagnosed problem may still be ambiguous, but the ambiguity will be profession-relevant ambiguity—ambiguity within the professional knowledge system.)[15]

An additional step in the diagnostic process is to classify a problem, which involves referring it "to the dictionary of professionally legitimate problems."[16] Classification is often related to expected successful treatments.

The second professional task is inference, which consists of reasoning about the connection between diagnosis and treatment. According to Abbott, "Inference is undertaken when the connection between diagnosis and treatment is obscure."[17] This would seem to be the case with many problems of matching military means to political ends. The costs and reversibility of treatment also affect the type of inference needed: "Where costs are prohibitive or there is no second chance, a professional must set a strategy from the outset."[18] Abbott argues that this is the case in military tactics: "Classic military tactics . . . work by construction. The tactician hypothesizes enemy responses to gambits and considers their impact on his further plans."[19]

The third and final task is treatment. Abbott describes treatment in this way: "Like diagnosis, treatment is organized around a classification system and a brokering process. In this case brokering gives results to the client, rather than takes information from the client."[20] In the military context, treatment consists of executing the use-of-force decision that has been made. Performed in combination, the three tasks discussed above—diagnosis, inference, and treatment—constitute professional practice and provide a useful way to think about the role of military leaders in use-of-force decision-making.

However, Abbott's discussion of professional practice is useful for an additional reason as well. It suggests mechanisms for future change:

> The central phenomenon of professional life is thus the link between a profession and its work, a link I shall call jurisdiction. To analyze professional development is to analyze how this link is created in work, how it is anchored by formal and informal social structure, and how the interplay of jurisdictional links between professions determines the history of the individual professions themselves.[21]

In the last sentence above, Abbott's emphasis is on interprofessional competition as a source for change, but a profession can affect the strength or shape of its jurisdiction in the course of its own work. Associated with each of Abbott's professional tasks are characteristics that may affect the strength of a given profession's jurisdiction. Recalling that the focus of this chapter is on how the role that military leaders play in use-of-force decision-making may affect the future of their own profession, it is useful to review these jurisdiction-related characteristics briefly here.

Beginning with diagnosis, Abbott argues that there are three aspects of this task that can affect jurisdictional vulnerability. First, there is the restriction of information: "To the extent that a profession restricts the relevant information or specifies the admissible types of evidence it risks competition with groups whose standards are less restrictive."[22] This could provide professionals with an incentive to meet the client on the client's own terms and not according to professional expertise.[23] The second two properties involve both the clarity and encompassing nature of classification. Professions are more vulnerable to losing functions that they do not identify as central to their work. "Since they are not claimed explicitly . . . on the main list of professional tasks, but rather implicitly by the dimensions of professional jurisdiction, they are held only weakly. Redefinition under a new system of abstractions can easily remove them."[24] One can easily imagine the relevance of such redefinition to the U.S. Army with reference to tasks such as foreign military training, antinarcotics operations, and perhaps even small-scale contingency operations more generally.

The nature of inference can also affect jurisdictional strength. Perhaps most relevant here, Abbott argues that the strength of a profession's hold on its jurisdiction is enhanced by logical inferential chains that are neither too short (in which case the matching of diagnosis to treatment becomes routine, thus obviating professionals), nor too long and too abstract. Perhaps an example of this latter case is the tendency described by Peter Feaver for civilians to exercise assertive control over nuclear weapons policy and operations.[25] Not only are there extraordinarily high levels of cost and risk associated with the use of nuclear weapons, but also reasoning about nuclear deterrence and escalation can be highly abstract and has (fortunately) never been tested. An extended and tenuous inferential chain could be one reason that military attempts to achieve autonomy in nuclear operational matters based on claims of expertise have not been more successful.

The third and final professional task, treatment, also has properties that affect jurisdictional strength. Abbott lists the following: measurability of results, specificity of treatment (generally the more specific, the stronger the hold of the profession), acceptability of the treatment to clients, and, perhaps most important, efficacy. Abbott argues that the prestige of professionals is enhanced when they conduct treatment on their own terms, but that in doing so they risk competition from those more willing to meet clients on their own ground:

> So it is that many professions meet clients on their own grounds—phrasing their treatments in common language, offering advice on professionally irrelevant issues, indeed promising results well beyond those predicted by the treatment structure itself. If they didn't do it, clients would take their problems to someone who would. A similar process forces professionals to keep a weather eye to treatment costs.[26]

If this dynamic holds in the civil-military relationship, it implies that both parties must accept and prefer that the military officer play the role of professional advice-giver. If political leaders do not, the professional officer's incentive to abandon strictly professional grounds to maintain influence or relevance in the policy process becomes clear.

In sum, Abbott's discussion of professional practice is useful both because it provides a way to think through the role of military officers in use-of-force decision-making, and also because it suggests ways in which the conduct of professional work could affect the profession itself. According to Abbott:

> Every profession aims for a heartland of work over which it has complete, legally established control. . . . Every profession aims not only to possess such a heartland, but to defend and expand it. The few who are content with limited jurisdictions—actuaries, veterinarians—are quite atypical."[27]

Perhaps U.S. military services are "atypical" professions that are content with limited jurisdictions; perhaps not. The important point is that if Abbott is right that professions shape their jurisdictions in the course of their work, it would be valuable for strategic military leaders to be conscious of this process and consider the implications of the choices they make in this light.

The Military's Professional Expertise

To this point, I have argued that Abbott's discussion of professional practice provides a useful way to think through the work of professional military officers as they participate in the state's decision-making about the use of force. Abbott's discussion of this concept also provides useful insights concerning the manner in which that role is played for the military itself. However, what has not yet been specified is the nature of the special expertise that the professional officer brings to bear. In other words, what is the content of expert knowledge in a military context? In order to answer that, it is useful to supplement Abbott by drawing upon Carl von Clausewitz's magisterial *On War* (1832). In addition, because Abbott tells us that the work societies believe properly belongs to experts can vary from society to society and from time to time, it is useful to check those insights against classics of American civil-military relations. The two texts that I will use for this purpose are Samuel Huntington's *The Soldier and the State* (1957), and Morris Janowitz's *The Professional Soldier* (1960).[28] I will draw on these works in an attempt to attach substance to the idea of military expertise, and to draw out implications for the strategic roles of U.S. military leaders in use-of-force decision-making.

Clausewitz: Knowledge and Experience

One of the reasons Clausewitz's work has been so enduring is the timelessness of his insights into the nature of war as both a military and political phenomenon. His fundamental argument is that war is always political in nature, "*merely the continuation of policy by other means.*"[29] Since war is fundamentally rooted in politics, politics provides the source of war, governs (at least to some extent) the means to be used, and also establishes the ends to be sought. War is a subordinate phenomenon, whose logic is deducible from the always-ascendant politics. However, within its subordinate realm, war is also unique; or, as Clausewitz says, war has its own grammar.[30] War has its own tools and, because its means are violence, takes place in a unique environment of danger, fear, physical exertion, and uncertainty. The uniqueness of war also stems from the fact that it is a form of human interaction—it takes place between living forces that react to each other's actions. These properties create the complexity of warfare, ensuring that "absolute, so-called mathematical, factors never find a firm basis in military calculations."[31] Mere abstract calculations cannot be relied upon to predict outcomes.

Because of these characteristics, Clausewitz argues that the most proficient experts in the grammar of war must have a combination of knowledge, experience, and even certain character traits. With regard to knowledge, Clausewitz focuses on what he calls the only means of war, combat: "However many forms combat takes . . . it is inherent in the very concept of war that everything that occurs *must originally derive from combat*."[32] The following excerpt elaborates on this point:

> Warfare comprises everything related to the fighting forces—everything to do with their creation, maintenance, and use. Creation and maintenance are obviously only means; their use constitutes the end. . . . If the idea of fighting underlies every use of the fighting forces, then their employment means simply the planning and organizing of a series of engagements. The whole of military activity must therefore relate directly or indirectly to the engagement.[33]

For Clausewitz, the military leader must be especially expert in the conduct of war (Clausewitz seems to classify the creating, maintaining, and training of military forces to be subordinate activities).[34] This involves expertise in both tactics, "*the use of armed forces in the engagement*," and strategy, "*the use of engagements for the purpose of the war*."[35] To get an even more complete picture of the specific knowledge Clausewitz would require of the competent commander, one need only examine the level of detail that Clausewitz devotes to the topics of Books Four though Seven of his work, which are the engagement, military forces, the attack, and defense. A critical study of military history can aid in the development of this expertise.[36]

When discussing the highest level of command, Clausewitz adds new requirements without dropping any of those that have previously been levied. The commander-in-chief must have a grasp of national policy, while remaining aware of what he can achieve with the military means at his disposal:

> He must be familiar with the higher affairs of state and its innate policies; he must know current issues, questions under consideration, the leading personalities, and be able to form sound judgments . . . he must know the character, the habits of thought and action, and the special virtues and defects of the men whom he is to command . . . he must be able to gauge how long a column will take to march a given distance under various conditions.[37]

In order to use military means to achieve the political purposes being sought, the commander-in-chief must have both the skills of a military leader and a sound grasp of national policy.

When Clausewitz discusses the expertise of the military leader in the grammar of war, it is consistent with his thought to conclude that it is a transformational grammar he is discussing.[38] In the context of language, transformational grammar allows for the identification of underlying structure or a deeper relation between sets of concepts so that even if words are ordered differently or appear differently in different sentences, in their "deep structure" one can find similar meanings.[39] Clausewitz's conception of war as "*an act of force to compel the enemy to do our will*" is broad enough to allow many different combinations of military means and political ends to share the same underlying meaning.[40] Clausewitz acknowledges this when he argues that in war many roads lead to success, shaped by both the ends being sought and the means chosen. Since even the personality of statesmen and commanders matters, "these questions of personality and personal relations raise the number of possible ways of achieving the goal of policy to infinity."[41] As long as there is decision by combat, whether or not fighting actually takes place, the military commander's expertise in the grammar of war is relevant.[42] A great advantage of this is that Clausewitz's concept is then open to being updated. For example, there is nothing inconsistent with Clausewitz's thought in the idea that knowledge of joint military operations (a topic that Clausewitz, who only discusses ground forces, does not touch on) is part of the expertise of today's grammar of war.

Returning to Abbott for a moment, it is useful to recall that he defines professionals to be those who apply abstract knowledge to particular cases. In this sense, is the knowledge Clausewitz describes professional? Clausewitz does see benefit, though limited, in the military leaders' use of theory as long as it is closely related to (derived from and tested against) the historical record. At a minimum, it is valuable from an educational perspective:

> Theory exists so that one need not start afresh each time sorting out the material and plowing through it, but will find it ready to hand and in good order. It is meant to educate the mind of the future commander, or, more accurately, to guide him in his self-education, not to accompany him to the battlefield.[43]

Clausewitz clearly believes that those in senior levels of command must have strong intellectual abilities which theory can help develop but should not constrain when it comes to practice. This seems very much in line with Abbott, who recognizes the instructional value of what he calls "academic knowledge," but also recognizes that mastery of "prestigious academic knowledge" does not necessarily indicate an ability to excel in professional work.[44]

This point leads to what Clausewitz sees as an additional component of the special expertise of the military leader—experience. It is consistent for Clausewitz to emphasize experience given what he describes to be the special environment of warfare. The following passage is revealing:

> If one has never experienced war, one cannot understand in what the difficulties mentioned really consist, nor why a commander should need any brilliance and exceptional ability. Everything looks simple; the knowledge required does not look remarkable, the strategic options are so obvious that by comparison the simplest problem of higher mathematics has an impressive scientific dignity.[45]

Clausewitz argues that what uninitiated analysts tend to miss are all the aspects of war, to include physical danger, fear, physical exertion, uncertainty, and even the vagaries of weather, that create "friction." Friction is what distinguishes "real war from war on paper," and makes true the claim that "everything in war is very simple, but the simplest thing is difficult."[46] For Clausewitz, the only lubricant for friction is experience—ideally in combat, but at a minimum in demanding peacetime exercises. Clausewitz argues:

> Action in war is like movement in a resistant element. . . . The good general must know friction in order to overcome it whenever possible, and in order not to expect a standard of achievement in his operations which this very friction makes impossible. . . . Practice and experience dictate the answer: "This is possible, that is not."[47]

A significant component of the professional expertise military leaders bring to bear is the experience they have gained through years of professional practice during peace and war.

A final requirement that Clausewitz requires of those at the top of their profession, or those with "military genius," has to do with character. For Clausewitz, genius in any complex activity requires "*a harmonious combination of elements*" which includes "appropriate gifts of intellect and temperament."[48] Because war is a dangerous realm, courage and self-confidence are essential traits in a military leader. For Clausewitz, the highest kind of courage is a combination of courage in the face of personal danger and courage to accept responsibility. This special temperament also includes an urge to act rationally, regardless of circumstances.

> Strength of character does not consist solely in having powerful feelings, but in maintaining one's balance in spite of them. Even with the violence of emotion, judgment and principle must still function like a ship's compass, which records the slightest variations however rough the sea.[49]

Although it is hard to deny the significance of this insight, it is also difficult to generalize from it. Perhaps the most that can be said is that the past or present demonstration of this attribute could certainly contribute to a professional officer's credibility in future practice.

To sum up the argument to this point, Clausewitz argues that the special expertise of the military officer is derived from a combination of experience and specialized knowledge in the grammar of war. To clarify this further, it is useful to look at what Clausewitz excludes as well as includes when discussing military expertise. He excludes at least three types of matters that are relevant to this chapter: domestic politics, the ends of policy, and interpretation of the international political environment. To see this in reference to domestic politics, it is useful to examine the manner in which Clausewitz bounds his conception of strategy. He argues:

> Strategy . . . does not inquire how a country should be organized and a people trained and ruled in order to produce the best military results. It takes these matters as it finds them in the European community of nations, and calls attention only to unusual circumstances that exert a marked influence on war.[50]

In other words, domestic political organization is beyond a theory of war. A second way Clausewitz makes this point is through his insight that one key to Napoleon's great success was that he transformed war itself by engaging the population in a fundamentally new way. This was a political change, and it was a failure on the part of governments not to recognize it. As Clausewitz observed, "The tremendous effects of the French Revolution abroad were caused not so much by new military methods and concepts as by radical changes in policies and administration, by the new character of the French people, and the like."[51] These changes were primarily political, and therefore the failure to grasp and adapt to them was political. Even if a military leader had perceived these changes, he could not have acted on this perception since a response required the mobilization of society itself.

A second issue that is beyond military expertise, according to Clausewitz, is the ends of policy. Clausewitz makes this clear when discussing his conception of war as a trinity made up of the people, the commander and his army, and the government. The people are identified with the passions of war, and the commander and his army represent the creative element because they operate in "the realm of probability and chance." However, his strongest statement is about the government. He opines that "the political aims are the business of the government alone."[52] In the following comment, Clausewitz reveals both his views on the irrelevance of domestic political matters to the military officer, and the proper location of authority over policy goals:

> It can be taken as agreed that the aim of policy is to unify and reconcile all aspects of internal administration as well as of spiritual values, and whatever else the moral philosopher may care to add. Policy, of course, is nothing in itself; it is simply the trustee for all these interests against other states. That it can err, subserve the ambitions, private interests, and vanity of those in power, is neither here nor there. In no sense can the art of war ever be regarded as the preceptor of policy, and here we can only treat policy as representative of all interests of the community.[53]

In other words, even if policy is based on the selfish interests of the ruler alone, or even if it is mistaken, from the military commander's perspective it must be assumed to be in the interests of the entire community. The need to adjust political purposes based on available military means may on occasion slightly modify these purposes, but will not change them fundamentally.[54]

Finally, Clausewitz seems to place the primary responsibility for interpreting the international political environment on political rather than military leaders. Governments can contribute to their states' success in war by accurately interpreting the character of international relations. For example, Clausewitz notes that changes in international alignments can drastically affect the success of an offensive operation: "All depends on the existing political affiliations, interests, traditions, lines of policy, and the personalities of princes, ministers, favorites, mistresses, and so forth."[55] The ability to analyze these factors is the special expertise of political leaders, not military commanders.

Given this discussion of the special nature of military expertise, what does it imply for the role of military leaders in use-of-force decision-making? First, Clausewitz does expect military commanders to bring to bear a special expertise, since he recognizes that political leaders may lack detailed knowledge of military matters. Clausewitz does not believe that this is necessarily a problem because above all the policy-maker needs "distinguished intellect and strength of character," and "can always get the necessary military information somehow or another."[56] A second clear expectation is that commanders will remain aware, from the initiation of hostilities to their completion, of the subordination of military means to the political ends being sought. In Clausewitz's words, "If we keep in mind that war springs from some political purpose, it is natural that the prime cause of its existence will remain the supreme consideration in conducting it."[57] Finally, Clausewitz seems to expect close and continuous cooperation between political leaders and their senior military commander. In fact, he argues that if events allow it the supreme military commander should sit in on cabinet meetings so that the cabinet can be involved in the commander's activities. This is necessary because "no major policy can be worked out independently of political factors."[58] In sum, the military commander's role is to be the expert advisor and executor of policy whose function is always subordinate to the political purposes of the government.[59]

Do Clausewitz's Insights Hold in a Liberal, Democratic Context?

Despite the universality many attach to Clausewitz's work, it is nevertheless necessary to ask whether his insights remain valid in the context of a liberal democratic society such as the United States. To address this question, it is helpful to turn to the signal works of Samuel Huntington and Morris Janowitz—two authors who have written authoritatively about the military profession in the United States in the twentieth century. Relying primarily on Huntington's *The Soldier and the State* and Janowitz's *The Professional Soldier*, I will address the question raised by this section in three parts. First, is Clausewitz's affirmation of the special expertise of the military professional valid in the American context? Second, do these authors agree with Clausewitz that the matters of politics, policy, and international relations should be excluded from the military officer's realm of special expertise? Third, do they agree with Clausewitz on the nature of the role that military officers should play in use-of-force decision-making?

So far as the military officer's special expertise is concerned, Huntington's and Janowitz's conceptions seem very close to those of Clausewitz. Huntington accepts the importance of both knowledge and experience; his summary of the central skill of the officer is captured in the idea that officers are experts in the "management of violence." This includes organizing, equipping, and training the force, planning for its use, and directing the force in and out of combat. He also agrees with Clausewitz that these tasks require a special expertise that can be acquired only through considerable training and experience. It is neither purely an art nor a craft, but "an extraordinarily complex intellectual skill requiring comprehensive study and training."[60] The development of military expertise requires both outside education and vocational training.

Janowitz seems basically to agree. This is interesting given that one of the hypotheses of *The Professional Soldier* is that there was a convergence underway between the skills of soldiers and civilians. However, he points out that "the military professional is unique because he is an expert in war-making and in the organized use of violence."[61] Later, he resumes this train of thought: "Despite the rational and technological aspects of the military establishment, the need for heroic fighters persists. The pervasive requirements of combat set the limits to civilianizing tendencies."[62] In the 1974 prologue to a new edition Janowitz notes the continued preeminence of combat in the military's view of its work, adding that another distinguishing aspect of military professional work is the normality of risk: "Even when the military are engaged in essentially defensive, protective, and constabulary types of activities, the element of risk and uncertainty is viewed as part of the standard operating procedure."[63] In contrast to other activities, where risk is due to the possibility of failure, it is viewed by the military as a normal component of its activities. In sum, with perhaps a few minor additions such as the idea of normality of risk, Clausewitz's conception of military expertise receives validation from authorities writing in an American context.

Regarding the second question, do these authors agree with Clausewitz that domestic political affairs, the ends of foreign policy, and interpretation of the international political environment are areas outside the military officer's special expertise? Huntington strongly agrees on the first two, arguing:

> Politics deals with the goals of state policy. Competence in this field consists in having a broad awareness of the elements and interests entering into a decision and in possessing the legitimate authority to make such a decision. Politics is beyond the scope of military competence, and the participation of military officers in politics undermines their professionalism, curtailing their professional competence, dividing the profession against itself, and substituting extraneous values for professional values.[64]

Janowitz seems to agree that both of these are beyond the military's special competence, but suggests that members of the military might be tempted to try their hand anyway. Janowitz argues that because some military leaders are suspicious of the efforts of politicians to formulate national goals, "military literature is replete with self-generated efforts to ascertain the consensus of American society, as a basis for long-range planning."[65] Janowitz also notes the tensions between what he sees as two conflicting ideologies in the military over the use of force. One group, which he calls the absolutists, holds that only total victory is an acceptable military outcome. Members of the other group, which he labels the pragmatists, display greater flexibility with regard to acceptable military outcomes. Janowitz's preference for the pragmatists places him directly in Clausewitz's camp.[66]

The two authors together also raise a point about policy ends that Clausewitz does not discuss. Huntington argues that the professional "is judged not by the policies he implements, but rather by the promptness and efficiency with which he carries them out."[67] If so, this would make it easier for military leaders to be neutral about the ends of policy and act in a strictly professional manner. However, Janowitz points out in a discussion of the use of force in both Korea and Vietnam that "opposition to the expansion of the President's war-making powers has 'spilled over' in political criticism of the armed forces."[68] As Janowitz as well as many Vietnam veterans could attest, public judgments of the military will at times be political rather than based solely on military "promptness and efficiency." The fact that military officers may be held responsible in the public eye for policy failures does not mean that the ends of policy are within their professional purview, but it does explain why refraining from addressing policy goals could be extremely difficult.

In sum, Huntington and Janowitz seem to agree with Clausewitz that domestic political affairs and the ends of policy fall outside the domain covered by professional military expertise. However, on the question of responsibility for the interpretation of the international political environment, things are not quite as clear. Huntington argues that one of the functions of the military officer is to "represent" security needs within the state's machinery based on "what he considers necessary for the minimum military security of the state in the light of the capabilities of the other powers."[69] However, this seems problematic given Huntington's logic elsewhere. First, though Huntington does specify "within the state's machinery," it is difficult to draw the line between professional and political behavior for a military officer while fulfilling this function. Second, this representative function would appear to involve advocating trade-offs between uses of resources that only political leaders are really positioned to make. Finally, it is dangerous to attempt to separate an analysis of enemy capabilities from an analysis of political factors such as one's own national interests, as well as the likely alignments of affected states in times of hostility. Once factors such as these are taken into account, it is doubtful that an assessment of military security needs could be founded on purely military expertise.

Janowitz also argues, although with a different emphasis, that part of the military officer's special expertise should be awareness of the political and social impact of the military's actions on international affairs. With an argument that seems more affected by the impact of the Cold War than the nature of the political system, Janowitz urges that "a 'political warfare' dimension has come to permeate almost every type of military operation in a 'no-war-no-peace' period."[70] In this environment, officers must be sensitive to and able to manage the political consequences of their actions. Janowitz believes that this has put a new strain on civil-military relations because "the more technical-minded officers are a hazard to the conduct of foreign policy; the more political-minded require more elaborate direction than is supplied by traditional forms of civilian supremacy."[71] Given that Janowitz's insight also seems to hold in the post-Cold War period, this addition to the content of military expertise originally laid out by Clausewitz seems useful when thinking about the special expertise of professional officers in the globally-engaged military of the United States.

The final question asked at the beginning of this section relates to the role military officers play in use-of-force decision-making. Here it would seem that the context of a liberal, democratic society is quite significant. It is important to note that Clausewitz's arguments about political control over the use of force are based purely on the logic of strategic success. In a democratic context, civilian supremacy is based not just on strategic requirements, but also on political ideology. Peter Feaver lays out this logic very clearly: in a society that embraces democratic values, "the prerogatives of the protectee are thought to trump the protectors at every turn."[72] The citizens of a democracy temporarily delegate authority to a political leader, and wish to maintain control over both this individual and the military, the military being the political leader's agent. Feaver argues that "although the expert may understand the issue better, the expert is not in a position to determine the value the people will attach to different issue outcomes . . . only the civilian can set the level of acceptable risk for society."[73] Political leaders are then accountable for these choices to the citizens they represent.

Beyond recognizing the fundamental requirement for civilian supremacy in a democratic society, do Huntington and Janowitz agree with Clausewitz's views on the role of military officers in use-of-force decision-making? The answer is basically yes, although both Huntington and Janowitz go further in discussing potential challenges that may arise. For example, Huntington discusses a list of potential grounds on which the military officer might disobey a political leader, including political ineptitude, illegality, immorality, and lack of professional competence. On the first of these grounds, political ineptitude, Huntington disclaims the legitimacy of military disobedience. On the second two, illegality and immorality, he argues that the political leader's interpretations deserve great weight. It is only on the last issue, lack of professional competence, that Huntington provides clear guidelines as to when disobedience can be justified. However, even here it is only in cases where a president orders an officer to "take a measure which is militarily absurd when judged by professional standards and which is strictly within the military realm without any political implications."[74] Given that significant actions will rarely be free from any political implications, the window opened here is either small or nonexistent. In any event, Huntington does go further than Clausewitz in addressing possible civil-military tensions.

Janowitz's concerns run in a different direction, stemming in part from his disagreement with Huntington's assertion that a professional military is by its very nature apolitical. Instead, Janowitz argues that "the military is a unique pressure group because of the immense resources under its control and the gravity of its functions."[75] He sees this as relatively inevitable:

> Their activities as a pressure group, if responsible, circumscribed, and responsive to civilian authority are a part of the decision-making process of a political democracy. Yet, at a point, knowingly or unknowingly, efforts to act as a leadership group can transcend the limits of civilian supremacy.[76]

In other words, some political activity by the military in a democracy is not shocking, and up to a point not abnormal or even a cause for concern. However, his idea that the military may become a political pressure group does more to explain the challenges facing senior military officers than it does to suggest a different conception of expert knowledge or a different appropriate role in use-of-force decision-making.

In sum, it would appear that Clausewitz's conceptions of the special expertise of military leaders and their role in use-of-force decision-making carry over into a democratic context relatively unscathed. The special expertise of a professional military officer is in the grammar of war, broadly defined. This is developed through both education and experience. Having a particular competence also implies that some arguably relevant matters are excluded. Some of the most important of those excluded are domestic political matters and the derivation of policy goals. Finally, the professional military officer brings this special expertise to bear in the state's use-of-force decision-making, serving both as an adviser and executor of policy. At the same time, the professional officer never forgets that the ultimate objectives are political, and the use of military means must always be in support of those ends.

The Weinberger Doctrine—Rules of the Game?

To this point, I have argued that it is useful to think of the participation by military officers in use-of-force decision-making as part of their professional practice.

Abbott's discussion of the tasks of diagnosis, inference, and treatment provides a way of thinking about strategic military leaders' work, and about how the performance of that work affects the future of the profession. Beyond that, I have drawn insights from works of strategy and civil-military relations to lend substance to the idea of professional expertise in a military context, and also to gather propositions about the role of military officers in use-of-force decision-making. This section will use such groundwork to examine the Weinberger Doctrine, which is one policy-maker's attempt to establish guidelines for use-of-force decision-making in the United States. The main question that this section seeks to address is whether, based on their professional military expertise and role in the process, military as well as civilian leaders will find the Weinberger Doctrine a suitable guideline for adoption. In the concluding section of this chapter, I will set forth the implications that military adoption of use-of-force guidelines that are not clearly based on professional criteria could have for the military profession itself.

The Weinberger Doctrine

On 28 November 1984, then Secretary of Defense Caspar Weinberger gave a speech to the National Press Club entitled "The Uses of Military Power." This speech was part of a debate among key players within the Reagan administration over the appropriate uses of military power. It was not the first public address by one of President Reagan's cabinet officials on the topic. On 3 April 1984, Secretary of State George Shultz gave a speech entitled "Power and Diplomacy in the 1980s," whose main message was that effective diplomacy must be backed by military power. Reacting to the Long Commission's report on the bombing of the Marine barracks in Beirut the previous year and its assertion that more options based on diplomacy should have been pursued, Shultz argued that this not only was an unfair depiction of U.S. efforts in that case, but also reflected an erroneous American tendency to believe that diplomacy and military options are "distinct alternatives." In fact, he said, "The lesson is that power and diplomacy are not alternatives. They must go together, or we will accomplish very little in this world."[77] He also argued that the position and responsibilities of the United States in the 1980s demanded a robust form of American engagement:

> Whether it is crisis management or power projection or a show of force or peace-keeping or a localized military action, there will always be instances that fall short of an all-out national commitment on the scale of World War II. . . . It is highly unlikely that we can respond to gray-area challenges without adapting power to political circumstances or on a psychologically satisfying, all-or-nothing basis.[78]

Turning to Central America, Shultz also acknowledged the complex moral, social, and economic issues associated with American engagement there, but argued against those who would turn these issues into "formulas for abdication, formulas that would allow the enemies of freedom to determine the outcome."[79]

Some six months later, Shultz gave another address in which he laid out his position on the necessity to respond to the problem of terrorism with an "active strategy."[80] This strategy relied on recognition that terrorism had become an instrument of warfare used by enemies of the United States, and, to combat it, "we must be willing to use military force."[81] Shultz argued that it was important to try to build consensus for U.S. action in this area, but in the event of specific incidents "the decisions cannot be tied to the opinion polls."[82] In his memoirs, he explained that part of his motivation for this speech was what he perceived as the unwillingness by Weinberger and the Joint Chiefs to seriously consider a military response to terrorist incidents even in the face of credible intelligence.[83]

In was in this context that Secretary Weinberger gave his speech in November 1984 that laid out "six major tests to be applied when we are weighing the use of U.S. combat forces abroad."[84] These six tests are as follows:

1. The United States should not commit forces to combat overseas unless the particular engagement or occasion is deemed vital to our national interest or that of our allies.
2. If we decide it is necessary to put combat troops into a given situation, we should do so wholeheartedly, and with the clear intention of winning. If we are unwilling to commit the forces or resources necessary to achieve our objectives, we should not commit them at all.
3. If we do decide to commit forces to combat overseas, we should have clearly defined political and military objectives. And we should know precisely how our forces can accomplish those clearly defined objectives.
4. The relationship between our objectives and the forces we have committed—their size, composition and disposition—must be continually reassessed and adjusted if necessary.
5. Before the United States commits combat forces abroad, there must be some reasonable assurance we will have the support of the American people and their elected representatives in Congress.
6. The commitment of U.S. forces to combat should be a last resort.[85]

In his memoirs, Weinberger explains that he found it necessary to address a State Department view that "an American troop presence would add a desirable bit of pressure and leverage to diplomatic efforts, and that we should be willing to do that freely and virtually without hesitation." He goes on to note that "the NSC staff were even more militant . . . spend[ing] most of their time thinking up ever more wild adventures for our troops."[86] In Weinberger's view, the option of using force should be turned to only under much more restrictive circumstances.

Weinberger's speech was not met by universal accolades, nor was it the last public round in the Shultz-Weinberger debate. One of the most scathing popular responses to the speech was a *New York Times* article by William Safire called "Only the 'Fun' Wars." In that piece, Safire argues that "the military mind has been brought to this surly, don't-call-us-until-you're-ready-to-abdicate-to-us philosophy by a serious of failures."[87] Safire critiques Weinberger's doctrine for its "popularity requirement," rejection of the need for limited military action, vital interest criterion, and "moral blindness." He approvingly cites Edward Luttwak's comment that Weinberger's Defense Department is "like a hospital that does not want to admit patients."

Shultz's own public response came in a speech entitled "The Ethics of Power" delivered the following month.[88] His logic again runs counter to Weinberger's in several ways. First, the requirements that a "vital interest" be at stake and that force should be used only as a "last resort" sit uneasily with Shultz's repeated view that "power and diplomacy must always go together." One difficulty with the concept of vital interests is that, "for the world's leading democracy, the task is not only immediate self-preservation but our responsibility as a protector of international peace."[89] On the question of last resort, Shultz does acknowledge the value of resorting to force only when other means of influence are not adequate. However, he also points out that "a great power cannot free itself so easily from the burden of choice. It must bear responsibility for the consequences of its inaction as well as of its action."[90] In his memoirs, Shultz also argues that "the idea that force should be used 'only as a last resort' means that by the time of use, force is the *only* resort and likely a much more costly one than if used earlier."[91]

The third of Weinberger's tests that Shultz criticizes is the requirement for public support. In Shultz's view,

> There is no such thing as guaranteed public support in advance. Grenada shows that a president who has the courage to lead will *win* public support if he acts wisely and effectively. And Vietnam shows that public support can be frittered away if we do not act wisely and effectively.[92]

Again, Shultz supplements this comment in his memoirs, arguing that the requirement for public support "was the Vietnam syndrome in spades, carried to an absurd level, and a complete abdication of the duties of leadership."[93] Shultz continues with a bit of philosophy about the American political system:

> My view is that democratically elected and accountable individuals have been placed in positions where they can and must make decisions to defend our national security. . . . The democratic process will deal with leaders who fail to measure up to the standards imposed by the American people and the established principles of a country guided by the rule of law.[94]

Shultz goes on to acknowledge that, while he had supporters, there were many others who disagreed with his views on the necessity for tying together power and diplomacy. It seems likely that a lack of consensus on issues such as these is not unusual in American politics.[95]

Towards A Professional Military Perspective on the Weinberger Doctrine

With the Weinberger-Shultz debate in mind, as well as the earlier discussion of professional practice, military expertise, and the professional military officer's role in use-of-force decision-making, is it feasible to arrive at the content of a purely professional military perspective on the Weinberger Doctrine? Here I will make the attempt, looking at each of Weinberger's six tests in turn. The central question is whether a uniformed officer, judging from a purely professional military perspective, could embrace each of these tests.[96]

1. *The United States should not commit forces to combat overseas unless the particular engagement or occasion is deemed vital to our national interest or that of our allies.*

Given the arguments above, an officer according to purely professional criteria cannot embrace this tenet for several reasons. First, there is the highly subjective task of prioritizing national interests. This would seem to be a task involving trade-offs among values that only political leaders are positioned to make. When it comes to the vital interests of allies, the task seems to become even more distant from the professional military officer's expertise. Second, this is inherently a question about which policy ends or goals are worth pursuing. Again, this task does not fall within the military leader's special expertise.

Another way to think through this is to consider the task of diagnosis. Is the level of national interest at stake relevant to the military officer's professional work? (Or, perhaps, what would professional rules of evidence look like?) Perhaps the best answer here is both yes and no. In terms of the initial decision to use force, it is relevant only in that a political leader may take the military professional's estimate of costs and decide to moderate policy goals. However, once use of force has been decided upon, the level of national interest at stake is clearly relevant. As Clausewitz argues, "The political object is the goal, war is the means of reaching it, and means can never be considered in isolation from their purpose."[97] Ends will invariably affect the means. Acknowledging this fact, however, does not entail accepting the idea that force can be used only when vital interests are at stake.

2. *If we decide it is necessary to put combat troops into a given situation, we should do so wholeheartedly, and with the clear intention of winning. If we are unwilling to commit the forces or resources necessary to achieve our objectives, we should not commit them at all.*

Although this seems to be mere common sense, there are some concepts here that are very subjective and need careful interpretation. First, there is the idea of committing forces "wholeheartedly." What does this mean? If it means relating military means to political ends in such a way as to achieve objectives at the lowest possible cost, it makes sense. Otherwise, its meaning and implications are unclear.

Second, and more problematic, is the idea of "winning." Is winning the same thing as achieving one's political objectives, even if the mission is peace enforcement or deterrence? The important point is that a professional who accepts the subordination of military means to political ends would also have to recognize that the political end is not always victory. Clausewitz is very clear on this point, noting that because war may have purposes that vary infinitely in scale, the only way to generalize about their purposes is to recognize that they all constitute a politically desired peace.[98] In other words, paraphrasing and at the same time disagreeing with Gen. Douglas MacArthur, there is a substitute for victory—the politically desired peace.

One reason this is worth highlighting is that in his memoir Weinberger seems to associate success with the departure of troops.[99] This hearkens to contemporary debates about exit strategies and timetables. Although these may be political requirements, there does not seem to be a basis within the special professional expertise of military officers to argue that they are a valid test to be passed when considering the use of force.

A second problematic association with the word "win" is that it could be taken to mean the military undertakes only operations like Desert Storm in which clear military victory is possible. General Shalikashvili notes that this could be a temptation for some, but rejects the idea of putting "a sign outside the Pentagon that says 'We only do the big ones.'"[100] Again, it would be difficult to find a basis within the military's professional expertise for an argument that the military should be given only missions in which there will be a decisive outcome.

3. If we do decide to commit forces to combat overseas, we should have clearly defined political and military objectives. And we should know precisely how our forces can accomplish those clearly defined objectives.

4. The relationship between our objectives and the forces we have committed—their size, composition and disposition—must be continually reassessed and adjusted if necessary.

These two tests are appropriately discussed together because they are closely related, and also because they are very consistent with the special expertise and professional practice of military officers. As strategists, professional military officers could reasonably be expected to assist in the refinement of political purposes and the development of military objectives to support those purposes. Asking for clarity, or helping to develop it, is not the same thing as embracing certain goals. Weinberger himself cites Clausewitz on this point: "No one starts a war—or rather, no one in his senses ought to do so—without first being clear in his mind what he intends to achieve by that war and who he intends to conduct it."[101] Military leaders should also provide estimates of military costs and risks so that political leaders can decide whether their goals are worth pursuing with military means. Even though the responsibility for developing the ends of policy belongs to political leaders, these political aims may need to be modified based on the ability of military means to meet them.

Military requests for a clear statement of goals—or military help in developing them—are not the same thing as military leaders embracing only certain goals. The first two are appropriate; the third is not. A clear political objective, for example, could be something like "show resolve," or "give the peace process a chance to work." The greatest challenge to military professionalism here may not be in the test itself, but rather in avoiding the temptation to label as "unclear" political objectives that are merely disliked.

The second point seems even more unexceptionable. During an ongoing use of force, the situation will continue to develop and even political objectives may change. The tasks of diagnosis, inference, and treatment may take place in a continuous cycle. In fact, this serves as a necessary supplement to the preceding provision about clarity of political and military objectives before committing forces.[102]

5. Before the United States commits combat forces abroad, there must be some reasonable assurance we will have the support of the American people and their elected representatives in Congress.

As Clausewitz recognized in *On War*, military leaders at the highest levels must be broadly familiar with the country's political affairs. To serve effectively, these senior leaders must have sensitivity to their civilian masters' political needs.[103] Such sensitivity may be extremely valuable in fostering the type of civil-military communication most likely to produce a coherent relationship between military means and political ends. However, that does not mean that it is appropriate for military professionals to set forth requirements for public support prior to operational deployments. As outlined above, the special expertise of the military professional does not include public opinion and political mobilization. While it is appropriate for military leaders to appreciate the potentially important role of popular support in gaining strategic success, it is not their particular area of expertise and therefore should not be the basis of their arguments.

If public support is in fact essential, it seems far more likely that political leaders have some familiarity with what is necessary to gain or sustain it. Also, still granting for the moment the idea that public support is essential, it is not entirely clear that public support is necessary prior to, rather than attained and sustained during actual hostilities. While this test has some appeal, it is more of a political matter than a purely military one.

It does seem possible that a military officer could argue that public support is essential to troop morale, and that troop morale is essential to victory in combat. However, this line of reasoning contains several significant claims that, as far as I am aware, have yet to be empirically established. This test seems to lack any solid grounding in military professional expertise.

6. The commitment of U.S. forces to combat should be a last resort.

Again, this tenet goes beyond the appropriate ground on which military professionals should be standing when participating in use-of-force decision-making. In any particular situation, senior military leaders may be the most familiar with the death and destruction that are part of the environment of warfare. They also have the responsibility to take care of their troops as well as fulfill national purposes. This combination of knowledge and responsibility gives military leaders a special obligation to state potential costs and risks in any military operation. Nevertheless, the requirement that force be used only as a last resort pushes this obligation too far. Though senior military leaders should have familiarity with all the instruments of national power, including the diplomatic, informational, and economic as well as military, it is up to the political leadership to decide what mixture of instruments best serves national policy goals in any given case. There are risks, costs, and values at stake in such choices that only political leaders have the authority to make.

In sum, it would seem unexceptionable for the professional military officer to embrace two but only two of the six tests, specifically the third one which relates to establishing clear objectives, and the fourth which argues for the maintenance of a coherent relationship between military means and political ends. In other words, if military leaders were to seek a professional code governing the participation of the military officer in use-of-force decision-making, the Weinberger

Doctrine would be a problematic set of criteria to adopt. Military leaders who adopt all of Weinberger's tests are taking positions that stretch proper conceptions of military professionalism—actions that could have implications for the military profession itself. The next and final section will suggest what some of those implications might be.

Why Clarity about the Role of Professional Military Officers Matters

There are two ways in which military participation in use-of-force decision-making could affect the future of the military profession. First, it could adversely impact on future interactions between political leaders and military officers. In other words, it could degrade the military professional's role as advisor. Second and more speculatively, it could bring into question the military's jurisdiction itself by shaping the scope of what is thought to be legitimate professional work. I will discuss each of these in turn.

The Professional Officer as Advisor

If a military officer expresses preferences among policy goals while acting in an official capacity, that officer may come to be seen as more a political figure than a military expert. To some extent, the danger here is the same as could arise from politicization. Huntington argues that politicization can happen either when military leaders espouse policies that have nonprofessional sources (the substance is political), or when military leaders are used to publicly advocate policy ends regardless of content (the manner of presentation is political). As an example, he asserts that the Joint Chiefs played political roles in the Truman administration. They became policy advocates at the behest of President Truman, who found that they could play a useful role in justifying his foreign policies before Congress and the public. Eventually, Huntington argues, the influence of the Chiefs dissipated. He cites Senator Taft, speaking of the Joint Chiefs in the spring of 1951: "I have come to the point where I do not accept them as experts, particularly when General Bradley makes a foreign policy speech. I suggest that the Joint Chiefs of Staff are absolutely under the control of the Administration."[104] This is particularly interesting because it shows that the benefits of behaving according to professional criteria go beyond enhancing civilian control. When military leaders base their positions on professional expertise they are also safeguarding their expert status, and perhaps preventing the discounting of military advice.[105]

A second possibility, with potentially even more negative consequences, is that the public stances of military officers could lead these officers to be disregarded in use-of-force decision-making. This could happen if officers were known to embrace certain policy ends, and political leaders came to believe they could not trust those officers' opinions or behavior. It could also happen if military leaders strongly embraced preconditions for the use of force such as tests for vital interests, exit strategies, public support, and last resort. In Abbott's language, this relates to diagnosis in the sense that these could be seen as constraints on what the military includes in its "dictionary of professionally legitimate tasks." If this "dictionary" has been formulated in an extremely restrictive way, political leaders may not seek military advice because they know in advance it would not be useful. Political leaders seeking to conduct coercive diplomacy might not consult the leaders of an Army that does only "the big ones."

In the U.S. political system, with its rule of law and civilian supremacy, such an outcome may not mean that forces are allowed to lie fallow, it may just mean that military leaders end up having less input as to how forces are used. Again with regard to Abbott's conception of professional practice, the military may be called upon only to provide a treatment conceived without benefit of military input. In this unfavorable scenario, military leaders are prevented from performing the tasks of diagnosis and inference because their conception of professional work was overly formulaic.

Positions based on either an overly narrow or an overly broad conception of the military's professional expertise could ultimately have negative consequences. The input of military officers could come to be seen either as irrelevant to the needs of the policy-maker, or as having dubious professional credibility. This implies that for the professional to be standing on ground that is as firm as possible, his positions and arguments should be solidly based on professional expertise. This also suggests that establishing a solid understanding within the profession of what that professional expertise includes would be of value.

A second implication of the discussion above is that both the substance and the manner of presentation are important. The officer whose argument is based on professional knowledge or experience is perhaps on firm ground, but that same officer's position would be all the more solid if the argument were expressed in a professional way. Since views on what constitutes professional behavior may vary, again a validated understanding of standards among the strategic leaders of the profession would be of value.

Impact on the Military Institution Itself

Although in the short run it seems likely that the biggest impact of the manner in which military leaders do their professional work will be on their role as military advisors, over the long run it may impact more directly on the armed forces themselves. In considering this possibility, it is useful to return to Abbott's concept of jurisdiction and the idea of jurisdictional claims. Abbott argues that a profession claims monopoly over a certain set of tasks—its jurisdiction. In the case of the U.S. military, it seems most reasonable to assume that the audience to whom it is making jurisdictional claims is the state itself. Settlements of these claims as well as jurisdictional boundaries may change over time. It is thus clear that both restrictive and expansive claims of jurisdiction could have an effect.

The first possibility, then, is that military leaders take a restrictive view of their jurisdiction and seek to limit the scope of what are seen to be professionally legitimate tasks. According to Abbott, one possible effect is that new actors will appear and claim those disclaimed tasks. Abbott suggests the following dynamic: "A powerful profession ignores a potential clientele, and paraprofessionals appear to provide the same services to this forgotten group."[106] While the idea of multiple clients does not carry over into the military context, the idea of new groups entering the scene to claim neglected tasks seems plausible. In the case of the U.S. Army on the issue of use of force, those new groups could be either other armed services or private military corporations. This is not to say that change along these lines is necessarily good or bad, but it does suggest that the conception of professional work implicit in the arguments made by military leaders in use-of-force decision-making could have this sort of effect, and that these leaders had best be aware of it.

A second possibility is that military leaders could develop an extremely expansive conception of military jurisdiction. However, Abbott sounds an interesting note of caution:

> No profession can stretch its jurisdiction infinitely. For the more diverse a set of jurisdictions, the more abstract must be the cognitive structure binding them together. But the more abstract the binding ideas, the more vulnerable they are to specialization within and to diffusion into the common culture without.[107]

Moreover, given the difficulty of being effective at a very large number of tasks at once, the danger of laying claim to an overly extended range of tasks is clear. For, as Abbott points out, the strength of a profession's jurisdictional claims rests on efficacy.

In sounding these cautionary notes, I am not making a case for any particular position between overly restrictive and overly expansive jurisdictions. My purpose rather is to suggest that the dynamics Abbott points to provide a helpful way of thinking through the implications of particular institutional choices. In the course of their work, service leaders collectively make explicit choices about force structure and doctrine. The argument here is that senior military leaders may also be making implicit or unwitting claims about the professional work of their institutions as they participate in decision-making about the use of force.

Conclusion

Max Weber once stated, "Politics is a strong and slow boring of hard boards. It takes both passion and perspective."[108] Perhaps the same can also be said about how challenging it would be to arrive at a consensus among the strategic leaders of the military profession about the issues raised here—the content of military professional expertise and the norms that should govern the professional military officer's participation in decision-making about the use of force. However, boring these hard boards would be worth the effort. Failure to be clear on these points could have seriously adverse consequences for the profession itself, however unintended.

Notes

1. The views expressed herein are those of the author and do not purport to reflect the position of the U.S. Military Academy, Department of the Army, or Department of Defense.
2. See Richard H. Kohn, "Out of Control: The Crisis in Civil-Military Relations," *The National Interest* 35 (Spring 1994): 3-17. See also Russell F. Weigley, "The American Military and the Principle of Civilian Control from McClellan to Powell," *The Journal of Military History* 57, no. 5 (October 1993): 27-58.
3. Kohn, "Out of Control," 13.
4. Richard Kohn in Colin Powell, et al., "An Exchange on Civil-Military Relations," *The National Interest* 36 (Summer 1994): 29. For another view suggesting the Joint Chiefs of Staff under President Clinton played an improper role in "circumscribing or preemptively vetoing policy options," see A.J. Bacevich, "Tradition Abandoned: America's Military in a New Era," *The National Interest* 48 (Summer 1997): 16-25.
5. See Lyle J. Goldstein, "General John Shalikashvili and the Civil-Military Relations of Peacekeeping," *Armed Forces and Society* 26, no. 3 (Spring 2000): 403.
6. Goldstein, "General John Shalikashvili," 393.
7. John M. Shalikashvili, "Employing Forces Short of War," *Defense* 3 (May/June 1995): 4.
8. Andrew Abbott, *The System of Professions: An Essay on the Division of Expert Labor* (Chicago, IL: University of Chicago Press, 1988), 8, 35.
9. Abbott, *The System of Professions*, 58.
10. Two recent examples are Deborah D. Avant, "Are the Reluctant Warriors Out of Control? Why the U.S. Military is Averse to Responding to Post-Cold War Low-Level Threats," *Security Studies* 6, no. 2 (Winter 1996/1997): 51-90; and Michael C. Desch, *Civilian Control of the Military: The Changing Security Environment* (Baltimore, MD: The Johns Hopkins University Press, 1999).
11. See *Project on the Gap Between the Military and Civilian Society: Digest of Findings and Studies*, First Revision (Durham, NC: Triangle Institute for Security Studies, 2000): 9-10. Accessed at *http://www.poli.duke.edu/civmil/summary_digest.pdf* on 12 August 2001. The study termed as "elite" those "officers at various ranks who have been identified for the education to prepare them for promotion and advancement."
12. An interesting novel that examines these tensions in the modern American military is Thomas E. Ricks, *A Soldier's Duty* (New York: Random House, 2001).
13. The following discussion of professional practice is drawn from Chapter 2, "Professional Work," in Abbott, *The System of Professions*, 35-58.
14. Ibid., 41.
15. Ibid.
16. Ibid.
17. Ibid., 49.
18. Ibid., 48.
19. Ibid., 49.
20. Ibid., 44.
21. Ibid., 20.
22. Ibid., 44.
23. A book suggesting that military leaders have done this in the past, and that this constituted a military failure, is H.R. McMaster's *Dereliction of Duty*. My interpretation of McMaster's argument is that the Chiefs did not abide by professional rules of relevance in their dealings with President Johnson. They altered their opinions based on factors that, though important

to Johnson, were irrelevant to the military situation in Vietnam. See H.R. McMaster, *Dereliction of Duty* (New York: HarperCollins Publishers, Inc., 1997), especially 323-34.

24. Abbott, *The System of Professions,* 44.
25. See Peter D. Feaver, *Guarding the Guardians* (Ithaca, NY: Cornell University Press, 1992), 9.
26. Abbott, *The System of Professions,* 48.
27. Ibid., 71.
28. These two works have the advantage of discussing the military profession in an American context, and in the twentieth century. However, a disadvantage of using them is that they, at least in their original editions, are over forty years old. Despite this limitation, I would argue that they are still useful for both their comprehensiveness and their provision of a useful baseline from which one can measure change.
29. Carl von Clausewitz, *On War,* trans. and eds. Michael Howard and Peter Paret (Princeton, NJ: Princeton University Press, 1976), 87. All citations from *On War* will be from this edition. Clausewitz used italics frequently; when quoting him, I will keep to his usage.
30. Ibid., 606.
31. Ibid., 86.
32. Ibid., 95.
33. Ibid.
34. Ibid., 128.
35. Ibid.
36. Ibid., 164. Clausewitz writes: "In the study of means, the critic must naturally frequently refer to military history, for in the art of war experience counts more than any amount of abstract truths."
37. Ibid., 146.
38. I am indebted to Professor James Burk for bringing this concept to my attention.
39. *Encyclopedia Britannica*, britannica.com 1999-2000, s.v. "transformational grammar."
40. Clausewitz, *On War*, 75.
41. Ibid., 94.
42. Clausewitz argues that the only means of war is combat, but acknowledges that combat takes many different forms. In all military actions, the outcome rests on the assumption that if it came to fighting, the outcome would be favorable: "The decision by arms is for all major and minor operations in war what cash payment is for commerce. Regardless how complex the relationship between the two parties, regardless how rarely settlements actually occur, they can never be entirely absent." Ibid., 97.
43. Ibid., 141.
44. Abbott, *System of Professions*, 54.
45. Clausewitz, *On War,* 105.
46. Ibid., 119.
47. Ibid., 120.
48. Ibid., 100.
49. Ibid., 107.
50. Ibid., 144.
51. Ibid., 609.
52. Ibid., 89.
53. Ibid., 607.
54. Ibid., 87.
55. Ibid., 569.

56. Ibid., 608.
57. Ibid., 87.
58. Ibid., 608.
59. These arguments are further developed in Suzanne C. Nielsen, *Political Control Over the Use of Force: A Clausewitzian Perspective*, The Letort Papers (Carlisle Barracks, PA: U.S. Army War College, Strategic Studies Institute, 2001).
60. Samuel P. Huntington, *The Soldier and the State* (Cambridge, MA: The Belknap Press of Harvard University Press, 1957), 13.
61. Morris Janowitz, *The Professional Soldier* (London: Collier-Macmillan Limited, 1960; reprint, New York: The Free Press, 1964), 15.
62. Ibid., 33.
63. Morris Janowitz, "Prologue: The Decline of the Mass Armed Force," *The Professional Soldier* (London: Collier-Macmillan Limited, 1960; reprint, New York: The Free Press, 1964, 1974), xv.
64. Huntington, *Soldier and the State*, 71.
65. Janowitz, *The Professional Soldier*, 272. This military skepticism about the values of political elites and their military expertise still exists today. See *Project on the Gap Between the Military and Civilian Society: Digest of Findings and Studies*, First Revision (Durham, NC: Triangle Institute for Security Studies, 2000), accessed at *http://www.poli.duke.edu/civmil/summary_digest.pdf* on 1 February 2001. Interestingly, even Caspar Weinberger expressed some skepticism about the role of political leaders. In justifying his "political support" test for the use of force, he argues: "Our government is founded on the proposition that the informed judgment of the people will be a wiser guide than the view of the president alone, or of the president and his advisers, or of any self-appointed elite." Caspar Weinberger, "U.S. Defense Strategy," *Foreign Affairs* 64 (Spring 1986): 689.
66. Janowitz, *The Professional Soldier*, 257-77.
67. Huntington, *Soldier and the State*, 73.
68. Janowitz, "Prologue," xlix.
69. Huntington, *Soldier and the State*, 72.
70. Janowitz, *The Professional Soldier*, 342.
71. Ibid.
72. Peter Feaver, "The Civil-Military Problematique: Huntington, Janowitz and the Question of Civilian Control," *Armed Forces & Society* 23, no. 2 (Winter 1996): 153.
73. Feaver, "Civil-Military Problematique," 154.
74. Huntington, *Soldier and the State*, 77.
75. Janowitz, "Prologue," xlviii.
76. Janowitz, *The Professional Soldier*, 343.
77. George Shultz, address before the Trilateral Commission on 3 April 1984, *Department of State Bulletin* 84, no. 2086 (May 1984): 13.
78. Shultz, Trilateral Commission address, 13.
79. Ibid., 14.
80. George Shultz, address before the Park Avenue Synagogue in New York on 15 October 1984, *Department of State Bulletin* 84, no. 2093 (December 1984): 16.
81. Ibid., 16.
82. Ibid.
83. George Shultz, *Turmoil and Triumph* (New York: Charles Scribner's Sons, 1993), 648.
84. Caspar Weinberger, remarks to the National Press Club on 28 November 1984, *Defense* 85 (January 1985): 9.

85. These are verbatim excerpts (though I changed his bullets to numbers for notational clarity) from Weinberger, National Press Club remarks, 10.
86. Caspar Weinberger, *Fighting for Peace: Seven Critical Years in the Pentagon* (New York: Warner Books, Inc., 1990), 159.
87. William Safire, "Only the 'Fun' Wars," *New York Times*, 3 December 1984, sec. A, 23. Among the failures, Safire mentions Vietnam, Desert One, and Lebanon. For another interesting press critique of the Weinberger Doctrine, see George J. Church, "Lessons from a Lost War: What Has Viet Nam Taught about When to Use Power—and When Not To?" *Time*, 15 April 1985, 40.
88. George Shultz, address at the convocation of Yeshiva University in New York on 9 December 1984, *Department of State Bulletin* 85, no. 2095 (February 1985): 1-3.
89. Ibid., 2.
90. Ibid., 3.
91. Shultz, *Turmoil and Triumph*, 650.
92. Shultz, Yeshiva University address, 3.
93. Shultz, *Turmoil and Triumph*, 650.
94. Ibid.
95. Weinberger again responded publicly, rejecting among other things the idea that as a justification for risking soldiers in combat, "the hope that a limited U.S. presence might provide diplomatic leverage is not sufficient." Caspar Weinberger, "U.S. Defense Strategy," 689.
96. Two interesting point-by-point discussions of the Weinberger Doctrine can be found in Christopher M. Gacek, *The Logic of Force* (New York: Columbia University Press, 1994), 265-69; and Michael I. Handel, *Masters of War*, 2nd ed. (London: Frank Cass, 1992; reprint, 1996): 185-203. The analysis here has been informed by their insights, but has a different focus.
97. Clausewitz, *On War*, 87.
98. Ibid., 143.
99. See Weinberger, *Fighting for Peace*, 137, 150.
100. Shalikashvili, "Employing Forces Short of War," 5.
101. Cited in Weinberger, National Press Club remarks, 10.
102. I would like to thank Professor Peter Feaver for helping me to clarify my thinking on this point. Part of the expertise officers bring to bear is an awareness of the history of conflict that will lead them to recognize that objectives often shift as operations progress. Therefore, while it is very helpful to start out with a clear set of political and military objectives because they foster an integration of political ends and military means, it is likely that these objectives will shift during the course of military operations.
103. On this point, see Lloyd J. Matthews, "The Politician as Operational Commander," *Army*, March 1996, 29.
104. Huntington, *Soldier and the State*, 386.
105. As Lloyd J. Matthews points out, the American political system requires senior military officers to manage a constant tension "between the internal demands of conforming one's speech to service on the Commander in Chief's national defense team, and the external obligation for honesty and candor before the nation, Congress, and the citizenry." Matthews gives valuable guidelines for managing this tension in "The Army Officer and the First Amendment," *Army*, January 1998, especially 31, 34.
106. Abbott, *System of Professions*, 91.
107. Ibid., 84.
108. Max Weber, "Politics as a Vocation," *From Max Weber: Essays in Sociology*, eds. and trans. H.H. Gerth and C. Wright Mills (New York: Oxford University Press, 1953), 128.

11 Army and Joint Professionalism after Goldwater-Nichols: Seeking a Balance

David W. Tarr and Peter J. Roman

"Jointness" has become the core of the American way of war. It dominates the way we operate across the full spectrum of conflict—the joint commander is in charge. It dominates the way we train—joint exercises are the capstone of the training hierarchy. It is becoming dominant in our doctrine, which is continually becoming more based on joint concepts and less based on simple combinations of service concepts. It is asserting an ever-stronger role in our acquisition decisions.

VICE ADM. DENNIS BLAIR[1]

Joint institutions have been an important feature of American defense organization for over a half-century. Defense organization in these years has been characterized by a fitful struggle between the four armed services, on the one hand, and joint institutions on the other. The Goldwater-Nichols Act of 1986 altered the terms of the debate but did not resolve it by any stretch. The Army and other services continue to search for the right balance between them and joint institutions. This will be a continuing struggle because it addresses the soul of these military institutions: the professional development of officers.

This chapter assesses the balance between joint and service professionalism, particularly the Army, in the wake of the Goldwater-Nichols Act. We focus on how that act altered the professional development of senior military leaders. We reject the tendency of some military commentators to argue in favor of *either* service *or* joint professionalism. In our view, the security challenges confronting the United States in the 21st century will demand excellence in many forms of military professionalism. Thus, beyond debates about revolutions in military affairs and force structure, the United States must confront the question of what types of military leaders it must cultivate. Our objective in this chapter is to contribute to a discussion of how the military can develop the finest service *and* joint military professionals.

Our investigation is divided into five parts. First, it presents our conception of three different forms of military professionalism. Second, we examine the relationship between service and joint professionals in the pre-Goldwater-Nichols era. Third, we outline the key sections of Goldwater-Nichols that are related to military professionalism. Fourth, we assess how implementation of Goldwater-Nichols has altered the relationship between joint institutions and the services, particularly the Army. Finally, we offer recommendations for improving the cultivation of joint and service professionals.

The Forms of Military Professionalism

It is widely accepted that the American military has been a profession since the mid- to late-nineteenth century. Although armies and navies have existed for thousands of years, militaries did not begin to organize themselves as professional institutions until the 1800s. Pioneering studies by Samuel Huntington and Morris Janowitz, among others, traced the emergence and consequences of the professional military officer.[2] These officers are "managers of violence" who share a common expertise, responsibility, and corporateness.[3] The professionalization of officers emerged within the context of the history, tradition, and cultures that had been built up for centuries. Throughout the twentieth century, the American military continued its professional evolution while maintaining elements of its pre-professional culture and traditions.

Military professionalism in the United States takes different forms and types. Military officers distinguish themselves in many ways that make the military a community of professional groups. One of the most significant distinctions is between combat and noncombat military professions. "War-fighters" share the bond of danger, risk, and self-sacrifice. Often, the worst thing that an officer can say about another is that the officer "never heard a shot fired in anger." The ribbons and decorations an officer wears, and doesn't wear, enable other officers to size-up his or her career. This is but one example of how military professionals distinguish themselves from one another. Our point is that the military is a community of professionals who have a shared profession of arms to varying degrees.

From our perspective, military professionalism in the United States takes three principal forms: service, joint, and national security. Each constitutes a distinctive body of knowledge and expertise for the professional military officer. Further, the distinctions between them are manifested and reinforced in the organization and functional division of labor within the military. It should not be construed that these forms of military professionalism are equal or effectively balanced. In fact, service professionalism has been dominant throughout American history for good reason. Until the twentieth century separate service functions, based on the distinction between land and sea warfare, were natural divisions of labor. However, technology and changes in the international role of the United States have altered the relationship between the three forms of military professionalism.

Service professionalism involves the ability to think and operate within a particular service. It is inherently based in the ability to fight within a particular medium—land, sea, or air. This core function of each of the armed services defines them, is rooted in historical experience and traditions, and provides an enduring raison d'être. Consequently, each armed service manifests a distinctive culture that molds and socializes its officers to varying degrees. Officers develop their core professional competencies within their service branch out of combinations of indoctrination, education, training, experience, and assignments. In this sense, the officer's professionalism is something he or she possesses, and both the quality and the character of that professionalism vary with the individual. Each armed service, as a corporate entity, can reasonably be regarded also as a professional institution akin to such hallmark professions as medicine and law, with culture, traditions, ethics, jurisdiction, credentialing, and other characteristics of professional systems.[4]

Further, each service has refined its own professional conception in response to technological change over time by creating new sub-service branches and eliminating others. There is competition to varying degrees between the branches of each service for billets, resources, and, ultimately, the service's direction.[5] Intraservice distinctions may be as important as interservice ones for the professional development of officers. This was made clear to us after an Army three-star general made a derogatory comment about another Army general. He noted the surprise on our faces and said, "What would you expect? He's artillery. You know what they say about the artillery? Better a sister in a whorehouse than a brother in the artillery!"[6] This colorful statement, however extreme, illustrates how officers within a service distinguish themselves from one another. The technological complexity of modern warfare serves only to increase the intra- and interservice distinctions between professional officers because it takes many years for them to master their particular forms of service professionalism.

Each armed service provides for its officers the institutional "home" that is their formative source of professional knowledge. The highest-ranking officers in a service are expected to be that service's foremost professionals, in full possession of the expertise associated with their experience, training, and understanding concerning the service's contribution to national security. Officers who serve as commanders or staff officers of service combat units are continuously engaged in the application of what we call "service professionalism." The apex of the service professional career is the position of chief of staff. Service professionalism is, always has been, and always will be the fundamental form of military professionalism for military officers.

The second form, joint military professionalism, involves the ability to think and operate across the armed services. It is, by definition, an integrative professional skill that emerges most clearly in operations and planning. First, joint professionalism is necessary to insure unity of command in modern military operations. Absent a unified military command that shares a common professional foundation, military operations risk a lack of focus and coordination that could result in excessive loss of life or even failure. While military officers have long recognized the principle of unity of command, they often resist its application across the services out of concern that an officer from another service might be in command. The necessity of joint unified command was recognized by Gen. George C. Marshall only weeks after the attack on Pearl Harbor: "I am convinced that there must be one man in command of the entire theater—air, ground, and ships. We cannot manage by cooperation. Human frailties are such that there would be emphatic unwillingness to place portions of troops under another service."[7] Unfortunately, achieving unified command across the services was not achieved for many more decades.

Second, the development of joint military professionalism is crucial for the effective coordination and integration of service programs and plans. Civilian officials, particularly the Secretary of Defense or his collective Office (OSD), can perform this function—unlike the unity of command function. However, reliance on civilian leaders and institutions as the sole integrators of military programs denies professional officers the opportunity to contribute their service expertise. We believe that joint military professionalism can and should make an important and distinctive contribution to this process.

Our assertion that joint professionalism is a distinctive form of military expertise is not universally accepted. The most prominent alternative view asserts that joint professionalism is simply the "sum of its parts."[8] Similarly popular is the attitude among some officers that joint professionalism is simply the ability "to pick the right tool from the toolbox." Views such as these de-emphasize to varying degrees the concept of a unique body of joint military professionalism. In many respects, they recall the pre-Goldwater-Nichols era when "joint" meant little more than compiling the different perspectives from the services and, in effect, stapling them together, a point we will pursue at greater length below.

These disagreements are manifestations of the state of joint professionalism. Unlike service professionalism, joint professionalism is the result of a "top-down" process in which joint institutions were created before the supporting professional foundations were developed within the officer corps. Even today, the *Department of Defense Dictionary* has twelve pages of "joint" entries, but "jointness" is not defined.[9] Further, joint professionalism has neither the history nor the institutions to cultivate and protect professionalism in the same manner that the services have done. Finally, because joint professionalism is inherently integrative, it will by definition lag behind service professionalism. As each service develops new capabilities, technologies, and doctrines for fighting within their particular environments, joint professionalism will struggle to achieve the highest degree of integration. In effect, the military's joint neural system will lag behind the service muscle.

The third form of military professionalism, national security professionalism, is the ability to think and operate across the range of governmental institutions in the national security field. This expertise, acquired in a variety of ways, manifests the most haphazard development pattern of the three types. The most common outgrowth of this kind of expertise occurs through graduate studies at institutions like the National War College, by participation in the interagency policy-making process, by cultivating regional knowledge, and through assignments to other agencies such as the National Security Council, State Department, Central Intelligence Agency, etc. This domain is frequently referred to as "pol-mil" in the military. Unlike service and joint professionalism, national security professionalism is not the exclusive domain of the military officer; rather, it is shared by professionals from the various national security bureaucracies, military and civilian.[10]

Military and civilian professionals in the national security bureaucracy have the opportunity, normally beginning in mid-career, to develop the skills and knowledge that are sorely needed in the national security policy-making arena.[11] Flag officers experienced in the ways of Washington's policy community are highly valued people who enhance the quality and policy relevance of military advice and staff support. This was illustrated in an interview with a Joint Staff flag officer: "I spent 30 days in a row last year with the Vice President working Bosnia and trying to explain to the House of Representatives why I thought we could go to Bosnia and make it work. I was over there working a political issue. If there was no Congressional support, we knew we would not be effective."[12]

Civilian and military leaders seek out such people because they need their military *and* national security professionalism. Since the end of the Second World War, a number of senior officers have manifested this kind of national security expertise, including Gen. George C. Marshall, Gen. Andrew Goodpaster, Adm. Stansfield Turner, and Lt. Gen. Brent Scowcroft, among others. Gen. Colin Powell is the archetypal contemporary product of this form of proficiency. His path to appointment as the Secretary of State was paved with a series of positions that cultivated a broad national security perspective: Office of Management and Budget staff, Military Assistant to the Secretary of Defense, National Security Advisor to President Reagan, and Chairman of the Joint Chiefs of Staff. The fact that fewer civilians currently have such skills and experience makes this an even more important consideration.

National security professionalism is even less-recognized as a form of military professionalism than is joint professionalism. Since national security professionalism is integrative across the government, no single institution or organization can claim it as its own. Consequently, no entity has the responsibility for establishing and maintaining professional norms, standards, education, and development. In contrast to service professionalism, national security professionalism remains informal and ad hoc. Additionally, national security professionalism requires that senior officers have an appreciation of the political-military dimensions of issues and work alongside political leaders and their civilian staffs. In this process, these officers are sometimes exposed to accusations that they are "political generals"—in effect, that they are self-promoting officers who have advanced their careers by sacrificing their military professionalism, straying outside the bounds of the traditionally defined military domain.

These three forms capture the different types of professional military officers, particularly at the more senior ranks. Each is at a different point of development and institutionalization. Service professionalism will always enjoy a position of primacy because it serves as the foundation for joint and national security professionalism. This is reflected in the popular aphorism that "without the services there is nothing to be joint about." Still, changes in technology as well as in America's global mission and technology have created great uncertainty for the continued evolution of each. This is an important factor in understanding the balance between the forms of military professionalism. To do this, we must turn our focus to defense organization.

Defense Organization and Military Professionalism

Defense organization is an important factor for assessing the state of the relationship between the three forms of military professionalism. Organization can be one indicator that a new profession has emerged or reached a point of maturation. It also serves to institutionalize the profession, define its jurisdiction, and protect it from external challenges. The experience of air forces and special operations forces in the twentieth century illustrate the role of organizational change in a profession's development.[13]

The role of organization within the military is one critical way that the military differs from civilian professions. Most civilian professions organize and conduct their professional affairs with minimal governmental interference. However, because the military is a profession wholly subsumed within the government, defense organization cannot be determined solely by the professionals. Military leaders have no authority on their own to create new organizations or abolish outmoded ones. Rather, defense organization results from an innately political process in which military professionals must gain the approval of political leaders on a range of organizational issues. It is the prerogative of the political leadership, especially Congress, to authorize the institutions, allocate responsibilities, provide budgets, and even determine billets. Contrary to Huntington's concept of "objective control," military professionalism cannot be maintained independent of politics since the organizations that manifest and maintain that professionalism result from the determinations of political leaders.[14]

Defense organization and military professionalism have been the subject of intense conflict between political and military leaders for almost a century. The stakes can be significant since organizations can legitimate professional specialties as well as demand resources. Further, the conflict cuts across political and military boundaries through loose alliances containing members of both groups. The current state of the balance between service and joint professionalism owes much to this political struggle over military professionalism and organization.

Joint without Jointness: 1942-1986

The first joint institutions, the Joint Chiefs of Staff and Joint Staff, were created out of necessity by President Franklin Roosevelt for the Second World War. With their value confirmed during the war, the National Security Act of 1947 gave legal status to the JCS and the unified command system. Among the act's intentions were "to provide for [the armed forces'] authoritative coordination and unified direction under civilian control but not to merge them; to provide for the effective strategic direction of the armed forces and for their operation under unified control and for their integration into an efficient team of land, naval, and air forces." Two years later, Congress, in an amendment to the act, added a significant qualifier: "but not to establish a single Chief of Staff over the armed forces nor an armed forces general staff."[15] While Congress desired unified control and integration, it would not unify the armed forces to achieve them. Thus, the legislatively established joint institutions were significantly limited in the degree of unification they could impose.

The services successfully subordinated joint institutions to service professionalism in the decades that followed. The services treated joint institutions as entities that should either be captured or kept weak. The services used their "executive agent" function to divide up the unified commands and to treat them as their wholly-owned subsidiaries; the Army owned the European Command, the Navy the Pacific and Atlantic Commands, and the Air Force owned Strategic Air Command. The component commanders frequently circumvented the regional CINC if he was from a different service and would often take orders, literally and figuratively, from their own service chief. This situation undermined the ability of the CINCs to prosecute their missions. In 1958, Gen. Lyman Lemnitzer, then Army Vice Chief of Staff and former CINC Far East (CINCFE), told the Army War College class that "it might be said that our structure leans heavily on cooperation, to some extent on coordination, but only slightly on command. In areas where there is a basic difference of opinion between the services this is a real weakness. . . . The authority of CINCFE was not directly proportional to his responsibilities, since he did not control all the means required to carry out his mission."[16]

Joint institutions that could not be captured presented a threat to the services and their professionalism and, therefore, were kept weak. This was particularly true of the JCS and the Joint Staff since, by definition, no single service could control them. The JCS was organized and operated in a manner that respected service norms and interests. Each service had the power to bring JCS processes to a halt if its interests weren't represented. The Joint Staff received lower quality officers who functioned as referees of the internal policy debates. Neither the JCS Chairman nor the Joint Staff was empowered to act independently or proactively in this era. Such incapacity simply manifested the dominant view that "joint" meant the sum of each service's contribution. There could be no body of joint knowledge or expertise that existed independently of the services. This JCS system protected service interests, but it also made the JCS less valuable to presidents and less effective in providing policy advice than it should or could have been.[17] More importantly, joint professionalism could not develop in support of joint institutions. It was an era of joint without jointness.

The subordination of joint institutions to service conceptions of military professionalism had negative consequences for both force planning and operations. Service programs developed independently of each other without any coordination; no joint institution, including the JCS, had any budgetary role. For service programs to be integrated and coordinated, a civilian political leader would have to impose it over the service-generated programs. Of course, this enhanced civilian control by elevating trade-offs between the service programs to the political leadership. However, when a president or a defense secretary made hard choices about competing service programs, it usually precipitated a political uproar led by the losing service. More often, political leaders avoided conflict by accepting service programs that may have been the result of compromise, but were not integrated or coordinated force plans. Consequently, for many decades the United States purchased forces that suffered from undesired redundancies and inefficiencies.[18] Importantly, interoperability and other facets of integration went unattended due to the absence of a joint military professional perspective.

The dominance of service professionalism over joint institutions translated into an array of problems when the United States employed military force. This is best illustrated by theater command and national strategy in the Vietnam War. Rather than creating a unified command in the theater operating under the Pacific Command, each service fought its own war with a shocking lack of coordination. A complex chain of command developed in the theater that defies logic and description.[19] The military never sought, nor would McNamara impose, unity of command on the Vietnam theater. Secretary of State Dean Rusk later recalled that "our military always talked about unity of command, but we never achieved it."[20] Consequently, the United States fought multiple wars in Vietnam, each driven by the services' institutional preferences, instead of a single, concentrated effort.[21] These theater command problems were compounded by the failure of the JCS in

Washington to provide integrated policy advice on the war in Vietnam. The dominance of service professionalism over the JCS and Joint Staff resulted in policy advice that promoted each service's preferences instead of a coherent national strategy. JCS advice throughout the Vietnam War resembled a laundry list more than a coordinated plan for fighting the war.[22] The flawed nature of JCS advice in Vietnam is best illustrated by the discussions between the JCS and Clark Clifford just after he became Secretary of Defense in 1968. "The military was utterly unable to provide an acceptable rationale for the troop increase" after the Tet Offensive, Clifford wrote. "Moreover, when I asked for a presentation of their plan for attaining victory, I was told that there was no plan for victory in the historic American sense."[23]

The JCS and unified commands, weakened by their subordination to service professionalism, could not provide integrated command or direction in the Vietnam War. Service control over officer career paths further insured that there would not be a leadership cohort of joint professionals with the skills to help achieve integration across the services. Further, the integrated force that was the goal of the 1958 Defense Reorganization Act was effectively stymied by the services during implementation.[24] Thus, the United States was woefully unprepared organizationally and professionally to fight jointly in Vietnam. With the military unwilling and unable to provide the necessary integration, the civilian leadership had to try to fill the vacuum even though it lacked the required professional expertise. In other words, the lapse in military cohesion at the top forced McNamara to act as his own CINC—something that no Defense Secretary, much less one with the leadership flaws of McNamara, should ever do. It is important to recognize that the military helped to bring about this sad state of affairs by opposing the development of joint professionalism for decades.

By the early 1980s, the United States had made little progress in developing joint military professionalism despite decades of joint institutions and a staggering defeat in Vietnam. "The system was seriously flawed," Gen. Colin C. Powell would later observe. "The deck was stacked against the very thing these dual-hatted leaders [the service chiefs] were supposed to achieve. . . . Jointness . . . was more often produced out of the necessity of the moment rather than built into the machinery."[25] Pressure for institutional reform had gained momentum.

The Goldwater-Nichols Act: Encouraging Jointness Through Legislation

In 1986, Congress passed the Goldwater-Nichols Department of Defense Reorganization Act to remedy a variety of perceived organizational deficiencies in the department. The act was the culmination of a five-year defense reform movement that included members of Congress and their staffs, civilian policy analysts, academics, and a few, mostly retired, military officers. Operational failures in Lebanon and Grenada, reports of ineffective JCS policy advice, and revelations of wasteful defense spending helped build momentum for the widest defense reform in four decades. The legislation passed both houses of Congress almost unanimously despite the opposition of the Defense Department's civilian and military leaders.[26] The Act's legislative intent included improving the quality of military advice, increasing operational effectiveness, and enhancing the Department's planning processes. The best way to achieve these goals, the Congress determined, was by strengthening jointness.

The Goldwater-Nichols Act contained two types of provisions relating to jointness.[27] The first type was designed to increase the authority of key joint positions, primarily the Chairman of the JCS (CJCS) and the unified commanders-in-chief (CINCs). The Act expanded the powers of the CJCS to include serving as principal military adviser to the National Command Authorities, as a statutory adviser to the National Security Council, and as the CINCs' representative in Washington. The law allowed the Chairman to provide civilian leaders with his own advice even if the other JCS members disagreed. Further, the Chairman was directed to provide his personal advice on a list of seventeen specific topics, from joint doctrine to alternative budget proposals. To help the Chairman perform these duties, the act made him manager of the Joint Staff and gave him a four-star deputy, the Vice Chairman of the Joint Chiefs of Staff (VCJCS). Together, these provisions were designed to improve the quality of military advice to the civilian leadership by placing the bulk of the burden squarely on the Chairman's shoulders. In the event that anyone might mistake the intent of Congress, the act made the Chairman the highest ranking officer in the United States military.[28]

The act increased the authority of the unified commanders by clarifying the chain of command so that they were directly subordinate to the Secretary of Defense. Under Goldwater-Nichols, the JCS no longer had legal authority over the unified commanders. Further, the act gave unified commanders greater powers over the service components within their commands, including the authority to select component commanders. It also included provisions that allowed the unified commanders to submit budget requests.[29] Collectively, these sections of Goldwater-Nichols significantly reduced the traditional service constraints above and below the unified commanders.

Congress's remedy for inadequate professional military advice and operational ineffectiveness was to increase the authority of the military's senior joint officers, that is, the JCS Chairman and the unified commanders. Generally, it was intended that the Chairman would provide better military advice in the policy process while stronger unified commanders would translate into greater operational effectiveness. Reducing the service constraints on each would free them to provide the best joint perspective in their respective areas. In doing so, they would be true joint professionals. In this respect, Goldwater-Nichols was designed to cultivate joint professionalism.

The second way that Goldwater-Nichols promoted joint military professionalism was by creating the Joint Officer Personnel Policy (Title IV) in the act. Military leaders considered this to be the most objectionable section of the act, both at the time and afterwards. Senior officers labeled it "bullshit," "a goddamn stupid idea," and "counter-productive."[30] Such opposition left Congressional reformers little choice but to draft the provisions without service input.[31] At the core of Title IV was the creation of the "Joint Specialty Officer" (JSO) to improve the quality of officers in joint assignments. The law mandated joint professional military education (JPME) requirements for JSOs, the designation of critical joint specialties that can be filled only by JSOs, mandated joint duty assignments prior to a first star, and even specified promotion rates for officers on joint and service staffs. The act also gave the JCS Chairman an oversight role in joint personnel policies and promotions for the first time.[32] Cumulatively, Title IV established the process for cultivating and promoting joint military professionalism among officers. Even more than the other parts of Goldwater-Nichols, it came about only because Congress imposed its views concerning military professionalism on the professionals themselves.

The Goldwater-Nichols Act's combination of strengthening the Chairman and unified commanders along with establishing the joint personnel policies set the conditions for the development of joint military professionalism. Stronger, more influential joint institutions would draw officers who wanted to be part of those important activities. However, to be competitive for these assignments, officers would have to be "purpled" through the joint personnel system with its JPME and other assignments. Thus, Goldwater-Nichols combined top-down directives for jointness with career incentives that, over time, might foster joint military professionalism through a "bottom-up" process. The act insured that, for the first time, "joint" would mean both more effective institutions and better professional development.

It was not clear that this attempt to promote jointness would succeed when the legislation passed in 1986. The act made only minor changes in defense organization, relying instead on reallocating the responsibilities of existing institutions.[33] Some reformers viewed the changes as cosmetic at best, with only a marginal chance of producing major changes within the military. Senator Gary Hart (D-CO), a leading proponent of defense reform, complained that "it would be a mistake to refer to Goldwater-Nichols as reform legislation. All it did was change a few boxes at the top. Real reform changes the way people think, and Goldwater-Nichols did nothing in that area."[34] Like any legislation of this type, the impact of Goldwater-Nichols would be determined not just by what the legislation said, but by how it was implemented.

A Shifting Balance: 1986-2001

Joint military professionalism has developed substantially in the fifteen years since the passage of the Goldwater-Nichols Act. This growth, however, has been uneven, with jointness being institutionalized in some areas and not making much of an impact on others. The Goldwater-Nichols Act did not create jointness, but established the conditions that allowed for its development over time. Consequently, the balance between service and joint military professionalism is continually shifting as the boundaries between them conflict over new issues or revisit old ones. Joint professionalism may be here to stay, but the services continue to view it uneasily or even suspiciously.

Joint professionalism has made its greatest impact on military advice to policymakers and operations, as intended by Goldwater-Nichols. The process of formulating and transmitting military advice to political leaders is now dominated by the JCS Chairman by virtue of his responsibility as principal military adviser. Military representation on interagency committees is through either the Chairman or one of his deputies, usually from the Joint Staff, sitting as his representative.[35] The service chiefs and their staffs are on the sidelines of the interagency process.[36] Military expertise on broad national security or "pol-mil" issues has gravitated to the joint community, especially the Joint Staff, in the post-Goldwater-Nichols era. This is a major change. Before Goldwater-Nichols, military advice in the interagency process was generated by service-dominated JCS processes.

Post-Goldwater-Nichols operational planning and command have also shifted from the services, or service-controlled mechanisms, to the joint community. The act's intent for a clear, short chain of command and stronger unified commanders has been realized. General Powell and other senior military leaders of the Vietnam generation recognized the necessity for *joint* unity of command; they implemented the Goldwater-Nichols reforms in this area quickly and fully.[37] Unified commanders now hold the authority over forces as well as freedom from service interference to conduct their missions. This, when combined with their augmented soldier-statesmen roles, explains the elevation of the geographic unified commanders after Goldwater-Nichols.[38] Unified commanders must still cope with civilian and military interference from Washington during operations, as evidenced in the Kosovo campaign. However, interference rarely emanates from the services as it did before Goldwater-Nichols.

Joint military professionalism has made the least impact, not surprisingly, on force planning and weapons development. Under Goldwater-Nichols, the primary responsibility of the services is to organize, train, and equip forces (called "Title 10" for that section of the law).[39] Having suffered a reduced role in providing policy advice and operations, the services have understandably jealously guarded their force development prerogatives. Service leaders are quick to object whenever the joint community is treading on their Title 10 responsibilities. They point to them, one Joint Staff flag officer told us, like they're a line painted on the Pentagon floor.[40] Joint leaders have created a number of instruments to provide an integrated perspective on service force development plans. These include the work of the Joint Requirements Oversight Council, the Chairman's Program Guidance, the Chairman's Program Recommendation, the CINCs' Integrated Priority Lists, and even documents like Joint Vision (JV) 2010 and JV 2020. A strong proponent of these mechanisms, Adm. William Owens, concedes that these efforts have had a limited impact.[41] He has recommended additional reforms in this area, although other senior joint officers have been reluctant to endorse his proposals. As one Joint Staff flag officer succinctly puts it, "Put a dollar on the table and purple fractures into green, blue, and khaki."[42]

The Goldwater-Nichols Act's overall effect has been to establish a rough division of labor between the services and the joint community. The services have the responsibility for developing and maintaining forces. Joint institutions provide the interagency policy advice and unity of command in operations. It is noteworthy that joint institutions are the reservoir for joint and national security professionalism in the military; the constraints of service professionalism over each have been reduced significantly. In this conception, jointness is not merely an additional layer of military bureaucracy atop the services. Rather, joint military institutions and professionalism make unique, distinctive contributions that cannot be made by the services. Thus, the reallocation of responsibilities in the Goldwater-Nichols Act led to a division of labor which has resulted in a more effective military and has enhanced the further development of joint and national security professionalism within the military.

While service and joint institutions make their own unique contributions, as do their subunits, each must depend on others in order to accomplish their missions successfully in peace and war.[43] It is no longer possible, if it ever was, for a single military service or institution to prepare for war, conduct operations, and manage political-military diplomacy *independently* of other military institutions. Instead, Goldwater-Nichols has precipitated an interdependence within the military. Military organizations exert influence, or even control, over each other. This interdependence is a source of conflict between military professionals. It is endemic to the system created by Goldwater-Nichols. In effect, the law has added the dimension of service-joint conflicts to the perennial interservice and intraservice conflicts within the military.

Service and joint professionals hold understandably differing perspectives on both the balance and the conflicts between them. They each hold their own lists of grievances toward the others. For service professionals, the rise of joint professionalism since Goldwater-Nichols has accelerated the decline of service independence and preeminence.[44] The services have lost control over essential functions such as operational command and political-military advice into the interagency process. In interviews, many civilian and joint leaders characterized the service chiefs and their staffs as marginal participants in these activities.[45] These changes have been reflected in officers' career patterns, particularly those of flag and general officers. As a result, senior officers are now more likely to seek joint assignments, even repeated ones.[46] It is through joint assignments that military officers have the opportunity to apply their strategic art at the highest level. The fact that unified commands, and not the services, fight wars is enough to alter career patterns of many senior officers. In short, the services have lost control over three things of value in the aftermath of Goldwater-Nichols: operational command, political-military advisory roles, and officers' career patterns.[47]

These changes have made the services and their advocates acutely sensitive to any joint activities that intrude on their Title 10 responsibilities to organize, train, and equip the military. Service concerns have focused on the work of the Joint Requirements Oversight Council (JROC) and the creation of Joint Forces Command (JFCOM), a unified command rather than a joint agency. The JROC is supposed to provide a joint assessment of service requirements and weapons programs in order to reduce unnecessary duplication and to produce a more interoperable force.[48] According to one Joint Staff flag officer, the JROC was intended as a vehicle to place the power over weapons decisions back in the hands of the military and away from the civilians in the Office of the Secretary of Defense.[49] After much prodding from Congress, JFCOM was created in October 1999 to develop joint training and experimentation for advanced war-fighting techniques.[50] This command is responsible for promoting transformation within the military that will achieve the goals established in JV 2010 and JV 2020.[51] The JROC and JFCOM are the two most prominent examples of how the joint community has tried in recent years to shape a joint force by determining how that force is developed.

Even though neither has yet to have the impact that proponents hoped or that detractors feared, the services continue to be concerned about both, and other similar efforts. Any joint institution that holds the authority to impose itself over service Title 10 responsibilities holds the potential, over the long term, to subordinate the remaining core of service professionalism to joint professionalism.[52] The ultimate result would be, in effect, a mirror-image of the pre-Goldwater-Nichols system. The services would have the *responsibility* for managing force programs but hold none of the *authority* for making decisions about those programs. Under such circumstances, service professionalism would wither. Consequently, it is not difficult to see why joint activities outside of operations or political-military advice are viewed as threats to service professionalism.

Joint professionals see the balance, or imbalance, between the services and joint community in very different terms. For them, joint military professionalism still has far to go despite the successes after Goldwater-Nichols. They believe that the services have constrained the further development of joint professionalism through their control of two essential resources: money and personnel. The services continue to control most of the budget authority and funds in the military with only a few exceptions.[53] Proposals during the 1980s reform movement to provide joint institutions, particularly the unified commanders, with significant budget authority proved to be far too controversial. Joint institutions usually have small budgets and depend upon one of the services to manage their budgets for them. From time to time, senior joint officers have made proposals to change this only to have their proposals dismissed as unreasonable.[54] Consequently, the budgetary support for joint institutions remains a remnant of the pre-Goldwater-Nichols system that is akin to the executive agent relationship. The activities of joint institutions are frequently forced to compete with service programs when the services allocate scarce resources. This is a prescription for conflict even under the best of circumstances; it is an inappropriate method for determining trade-offs between service and joint programs.

Joint professionals similarly point to the services' control over personnel as a factor inhibiting the continued development of joint professionalism. Certainly, Title 4 of Goldwater-Nichols mandated that all senior officers have some joint experience while the strengthening of the Joint Staff and unified commands meant that the services would send higher-quality officers to those assignments. Still, the services control the promotions process well into the senior flag ranks. As a result, professional development within the service remains the key for promotions at least until the flag and general officer levels. The services, particularly the Navy, consistently claim that it is detrimental to release their best mid-career officers in certain specialties for joint assignments. It is asserted that O-5s and O-6s need all of their assignments to cultivate their professional service expertise. This attitude has had pernicious effects on joint professional development. Service personnel policies have determined that joint professional development usually begins late in an officer's career and is not as robust as it should be. This effect is compounded by the continuation of the "quota system" for distributing joint positions. Joint institutions must maintain a rough balance among the Army, Navy (including Marine Corps), and Air Force, lest one service protest that it is being slighted. While joint positions are ostensibly open to the best candidates from any service, in reality these assignments frequently go to those service candidates that maintain the balance among the services. While this is not a problem in key joint institutions like the Joint Staff or the geographic unified commands, the services often fill assignments in other joint institutions with officers who either lack the necessary experience, have not distinguished themselves professionally, are on the verge of retirement,[55] or are simply being "dumped" by the service. At the mercy of the service personnel systems, joint institutions must make do with whomever they are sent. For example, some unified commanders have complained to us that too many officers on their staffs lacked a rudimentary understanding of joint capabilities and warfare.[56]

These issues of funding and personnel are indicative of the state of joint professionalism *writ large*. Unlike the services, the joint community has no person or entity whose sole mission is to cultivate and promote joint professionalism—and has the resources to do so. The creation of Joint Forces Command and the aggregation of transformation efforts under its aegis are a response to this institutional vacuum in joint professionalism.[57] The fact that these responsibilities were assigned to JFCOM indicates that joint professionalism has reached its limit under the Goldwater-Nichols system. Placing the responsibility for leading transformation, joint doctrine, or joint professionalism under a unified command blurs the distinctions between the services and unified commands as well as between war-fighting and supporting unified commands. In short, there is a need for a joint institution that serves the function of JFCOM, but a unified command is the wrong place for it.[58] More likely, a new type of joint institution—not a unified command or part of the Joint Staff—is needed to develop this aspect of joint professionalism. However, no such organizational innovation can be achieved without significant amendments to Goldwater-Nichols.

A mature post-Goldwater-Nichols system has clearly emerged in the fifteen years since the act was passed. Its main feature is a division of labor between the services and a stronger joint community. The act's organizational restructuring has precipitated changes in service and joint professionalism. Service professionalism has been forced to adjust to the loss of important operational and political-military functions to joint institutions. Joint professionals have had to build the components of their professionalism almost from scratch. Service and joint professionalism are now more interdependent than ever before, as evidenced by the ongoing conflicts between them. Certainly, both will continue to grow and evolve. However, the balance *between* them is unlikely to change without further Congressional reorganization legislation. Until that time, service and joint professionalism will remain in their state of mutual dependence and conflict.

Army Professionalism and Jointness

Joint military professionalism poses special challenges to Army professionalism—more than it does the other three services. The rise of joint professionalism has coincided with the shift in international politics from Cold War bipolarity to post-Cold War unipolarity. The combination of joint professionalism and the current international security environment has forced each service to adjust its mission, identity, and professionalism.[59] The Navy, Air Force, and Marine Corps have been able to maintain their idealized service identities into the post-Cold War era fairly successfully. The Navy and Air Force have focused on power projection strategies to maintain continuity in their service's core identity. For the Marines, their post-Cold War missions are their traditional missions; it is not unusual to hear Marine leaders emphasize this by saying "We are your Marine Corps—*and we do windows*!"[60]

The Army has struggled to translate its core mission of heavy, high-intensity conventional warfare to the post-Cold War security environment. Changes in international politics and the basis of economic power have decreased the probability of large-scale warfare in the near future, although the utility of military force has not been reduced.[61] The ground warfare in the style of the Second World War or the Persian Gulf War that is at the heart of the Army's identity and professionalism is also the least likely way that Army forces will be employed. In the next decade, the primary function of the Army's heavy mechanized forces may very well be as a deterrent and not much else.[62] This has been a traumatic adjustment for generations of Army officers steeped in the tradition of Grant, Pershing, and Patton. It reverberates in the internal conflict over Gen. Eric Shinseki's Army transformation initiatives and in the statements of countless officers that "it's not the Army that I joined." Of all the services, the Army has the greatest gap between its idealized professional identity and post-Cold War reality.

The Army's post-Cold War transition has been made more difficult by the growth of joint professionalism, particularly the incorporation of national security professionalism into joint institutions. The Army's function to win and subsequently occupy territory created a soldier-statesman tradition that far surpasses those of the other services. George Marshall, Douglas MacArthur, Dwight Eisenhower, Maxwell Taylor, Matthew Ridgway, and Andrew Goodpaster, among others, became Army icons for their political-military mastery, not for their mastery of ground warfare per se. The Army was able to institutionalize the soldier-statesman tradition in large part because career patterns were kept within the Army institution. Goldwater-Nichols moved the soldier-statesmen and their professionalism from the services to the joint world. The Army's soldier-statesmen of the post-Goldwater-Nichols era—Colin Powell, Norman Schwarzkopf, George Joulwan, and Wesley Clark—made their marks outside the confines of the Army and its institutions. Significantly, none of these officers ever returned to the Army in an official capacity after they became soldier-statesmen. In fact, since the passage of Goldwater-Nichols, no geographic unified commander has become a chief or vice chief of service—something that was fairly common before the act.[63] Thus, the shift of the soldier-statesman role to the joint community is a great loss because it departs from one of the Army's great strategic traditions. Further, having lost this tradition, the Army now seems to place less value on joint or political-military assignments in the promotions process. Many Army officers in joint assignments, especially at the O-5 and O-6 ranks, feel that they are at a disadvantage in the competition for promotions because "all that the Army values is troop time."[64]

As the new century begins, the Army's loss of its solder-statesmen and the unlikelihood of major conventional war have robbed Army professionalism of its strategic core. Functionally and philosophically, the Army is more tactically-oriented than it has been in a very long time. The result is an uncertain path for Army professionalism, potentially for a long period of time.

Conclusion

This chapter has identified and described three forms of military professionalism in the United States: service, joint, and national security. Traditionally, military professionalism was defined solely in terms of service professionalism. The failure of the United States to develop joint professionalism and interservice unity of command resulted in numerous operational inefficiencies and setbacks throughout the Cold War. The Goldwater-Nichols Act addressed these problems, and its remedies have largely been implemented. We view this evolution of the armed services from autonomous units toward greater operational and organizational unity as logical, progressive, and necessary.

The post-Goldwater-Nichols system that has emerged is based on a division of labor between the services and joint institutions that fosters interdependence between them. Conflict is thus an inherent element of the post-Goldwater-Nichols system. Since this is rooted in legislation, only new legislation can alter the relationship. Until then, the services and joint community will have to manage their relationship as best they can. Presumably, they will have to do so with limited resources, which will make this difficult to do.

The United States will require the finest service, joint, and national security professionals to be prepared for future conflicts. Inadequate development of one of the professional forms will be manifested in the other two, as the pre-Goldwater-Nichols experience shows. With this in mind, we offer the following recommendations for strengthening military professionalism:

- *Recognition and development of national security professionalism.* The services need to cultivate the development of national security professionalism among their officers. Officers' careers should not be adversely affected if they take such assignments.
- *Institutionalize a process within the services to benefit from joint and national security professionalism.* Each service needs institutional processes to incorporate the joint and national security expertise of their officers back into the service. The departure of soldier-statesmen from the Army since Goldwater-Nichols does not necessarily have to deprive the service of one of its strategic traditions.
- *Create more flexibility in officers' career patterns.* The problems of career patterns and strains have received much attention in recent years. It is difficult for officers to develop joint or national security professionalism while still maintaining their own service expertise. More creativity will be necessary in career patterns. This might include longer careers, fewer requirements before promotions, or even allowing some nonservice assignments not to count against an officer's time in rank.
- *Redesignate Joint Forces Command as a joint agency.* This new joint agency would have responsibility for stewardship of joint professionalism. It would correct the current organizational problems of placing these responsibilities within a unified command.
- *Create a National Security College.* This new interagency would be the primary educational institution for senior national security professionals from across the

government. The faculty and curriculum would represent all national security bureaucracies in the government; none would dominate it.

- *Strengthen the control of joint institutions over their own budgets.* The services should not use their executive agent functions to make trade-offs between service and joint functions.

Notes

1. Vice Adm. Dennis Blair, Senate Armed Services Committee, 1 April 1998, "Answer to Pre-confirmation Questions," available from http:www.senate.gov/~armed-services/statement/9809db.htm.
2. Samuel P. Huntington, *The Soldier and the State: The Theory and Politics of Civil-Military Relations* (Cambridge, MA: The Belknap Press of Harvard University Press, 1957); Morris Janowitz, *The Professional Soldier: A Social and Political Portrait* (New York, NY: The Free Press, 1960).
3. Huntington, *Soldier and the State*, 8-11.
4. See Don M. Snider and Gayle L. Watkins, "The Future of Army Professionalism: The Need for Renewal and Redefinition," *Parameters* 30 (Autumn 2000): 5-20.
5. Carl H. Builder, *The Masks of War* (Baltimore, MD: The Johns Hopkins University Press, 1989), 24-27; Edward Rhodes, "Do Bureaucratic Politics Matter? Some Disconfirming Findings From the Case of the U.S. Navy," *World Politics* 47 (1994), 1-41.
6. Confidential interview with an Army three-star general. Our interviews, largely concentrated on flag officers and high civilian officials, were conducted on a not-for-attribution basis from 1994 to the present. We have conducted over 140 interviews. We use the term "flag officer" throughout as a generic reference to both admirals and generals. Our repeated reference to "confidential interviews" is to remind the reader of the nature of our information and our promise not to disclose our sources.
7. He continued: "If we make a plan for unified command now, it will solve nine-tenths of our troubles. There are difficulties in arriving at a single command, but they are much less than the hazards that must be faced if we do not achieve this." Quoted in Laurence J. Legere, *Unification of the Armed Forces,* Harvard Dissertations in American History and Political Science (New York: Garland Publishing, 1988), 210. Shortly after the end of the Second World War in Europe, Gen. Dwight Eisenhower observed that "there is no such thing as *separate* land, sea, and air war." Robert J. Watson, *History of the Office of the Secretary of Defense: Volume 4, Into the Missile Age, 1956-1960* (Washington, DC: Historical Office, Office of the Secretary of Defense, 1997), 243.
8. Lawrence B. Wilkerson, "What Exactly is Jointness?" *Joint Force Quarterly*, no. 16 (Summer 1997): 66-68.
9. It defines joint as "connoting activities, operations, organizations, etc., in which elements of two or more Military Departments participate." Joint Pub 1-02, *Department of Defense Dictionary of Military and Associated Terms* (Washington, DC: Joint Chiefs of Staff, 10 June 1998), 237-48.
10. Some readers will no doubt associate our conception of national security professionalism with the "fusionist" model of civil-military relations. While we recognize the two concepts as very closely related, we see a fine but distinct difference between them. Fusionism, as attacked by Samuel Huntington, denies the existence of a sphere of military professionalism. Our view of national security professionalism affirms that sphere, but does not extend it to the shared sphere of policy-*making* where civilian and military professionals and political leaders must come together. Huntington, *Soldier and the State*, 350-54; Richard K. Betts, *Soldiers, Statesmen, and Cold War Crises*, (Morningside Edition) (New York: Columbia University Press, 1991), 32-42.
11. For more on this subject see Peter J. Roman and David W. Tarr, "Military Professionalism and Policymaking: Is There a Civil-Military Gap at the Top? If So, Does It Matter?" in *Soldiers and Civilians: The Civil-Military Gap And American National Security,* eds. Peter D. Feaver and Richard H. Kohn (Cambridge, MA: MIT Press, 2001), 403-28.

12. Confidential interview with a Joint Staff flag officer.
13. Herman S. Wolk, *Planning and Organizing the Postwar Air Force, 1943-1947* (Washington, DC: Office of Air Force History, 1984); Susan L. Marquis, *Unconventional Warfare: Rebuilding U.S. Special Operations Forces* (Washington, DC: The Brookings Institution, 1997).
14. Huntington, *The Soldier and the State*, 83-85.
15. Alice C. Cole et al., *The Department of Defense: Documents on Establishment and Organization, 1944-1978* (Washington, DC: Office of the Secretary of Defense Historical Office, 1978), 36, 86. Also see: Demetrious Caraley, *The Politics of Military Unification: A Study of Conflict and the Policy Process* (New York: Columbia University Press, 1966); Gordon Nathaniel Lederman, *Reorganizing the Joint Chiefs of Staff: The Goldwater-Nichols Act of 1986* (Westport, CT: Greenwood Press, 1999), 10-20; Amy B. Zegart, *Flawed by Design: The Evolution of the CIA, JCS, and NSC* (Stanford, CA: Stanford University Press, 1999), 109-30.
16. Gen. Lyman Lemnitzer, "Command, Coordination, and Cooperation at the Department of the Army Level," Lecture to the Army War College, 19 May 1958, Lyman Lemnitzer Papers, National Defense University Library, Washington, DC; Alice C. Cole et al., *The History of the Unified Command Plan, 1946-1993* (Washington, DC: Joint History Office, 1993), 11-29.
17. Peter J. Roman and David W. Tarr, "The Joint Chiefs of Staff: From Service Parochialism to Jointness," *Political Science Quarterly* 113 (Spring 1998): 93-97.
18. Admittedly, redundancies across the services brought great benefits for the United States during the Cold War, particularly with respect to nuclear deterrence. However, our point is that this system continued unwanted redundancies and wasteful defense spending.
19. Maj. Gen. George S. Eckhardt, *Vietnam Studies: Command and Control, 1950-1969* (Washington, DC: Department of the Army, 1991).
20. Dean Rusk, with Richard Rusk and Daniel S. Papp, *As I Saw It* (New York: W.W. Norton, 1990), 453.
21. This is reflected in the scholarship on the operational dimensions of the Vietnam War. Few studies have been produced that assess the totality of the war effort in the theater. The tendency of scholars is to concentrate on one aspect of the war—the ground war, the air war, counterinsurgency, etc.
22. H.R. McMaster, *Dereliction of Duty: Lyndon Johnson, Robert McNamara, the Joint Chiefs of Staff, and the Lies That Led to Vietnam* (New York: HarperCollins, 1997); Eliot Cohen, "Enough Blame to Go Around," *The National Interest*, no. 51 (Spring 1998), 103-108; James J. Wirtz, "A Review Essay of H.R. McMaster's *Dereliction of Duty*," *Political Science Quarterly* 114 (1999): 131-36.
23. He continues: "Although I kept my feelings private, I was appalled: nothing had prepared me for the weakness of the military's case." Clark M. Clifford with Richard Holbrooke, *Counsel to the President: Memoir* (New York: Random House, 1991), 493-94.
24. In his reorganization message to Congress in 1958, President Eisenhower wrote that "separate ground, sea, and air warfare is gone forever. If ever again we should be involved in war, we will fight it in all elements, with all services, as one single concentrated effort." Quoted in Watson, *Into the Missile Age*, 257.
25. Colin L. Powell, with Joseph E. Persico, *My American Journey* (New York: Random House, 1995), 410-11.
26. *The Goldwater-Nichols Act, U.S. Code,* Vol. 50; Senate Armed Services Committee Staff Report, *Defense Organization: The Need for Change* [the "Locher Report"], 99th Congress, 1st Session (Washington, DC: GPO, 1985); Barry M. Goldwater with Jack Casserly, *Goldwater* (New York: Doubleday, 1988), 334-59; Thomas L. McNaugher, with Roger L. Sperry, "Improving Military Coordination: The Goldwater-Nichols Reorganization of the Department of Defense," in *Who Makes Public Policy?* eds. Robert S. Gilmour and Alexis A. Halley (Chatham, NJ: Chatham House, 1994), 219-58; Lederman, *Reorganizing the Joint Chiefs of Staff*, 51-85.
27. The Goldwater-Nichols Act contained measures that related to other issues besides jointness. We do not address those because it is outside the scope of this chapter.
28. *The Goldwater-Nichols Act, U.S. Code*, Vol. 50, sections 151-55.

29. *The Goldwater-Nichols Act, U.S. Code*, Vol. 50, sections 161-65, 601.
30. Quoted in Gregory G. Gunderson, "In Search of Operational Effectiveness: Military Reform in the 1980s," (Ph.D diss., University of Wisconsin-Madison, 1997), 193.
31. As a result, these sections are considered to be some of the most poorly drafted in the Goldwater-Nichols Act. Title 4 was amended numerous times in subsequent years, although the intent remained unchanged.
32. *The Goldwater-Nichols Act, U.S. Code,* Vol. 50, sections 661-667; Howard D. Graves and Don M. Snider, "Emergence of the Joint Officer," *Joint Force Quarterly*, no. 13 (Autumn 1996), 53-57; Douglas C. Lovelace, Jr., *Unification of the United States Armed Forces: Implementing the 1986 Department of Defense Reorganization Act* (Carlisle Barracks, PA: U.S. Army War College, Strategic Studies Institute, 1996), 52-56.
33. Indeed, the pre-Goldwater-Nichols structure is almost identical to the post-Goldwater-Nichols one. The only significant difference is the creation of the Vice Chairman of the JCS. The major changes came in distributing power across the institutions.
34. He continued: "In fact, the Goldwater-Nichols legislation was the military's response to the pressure being put on by the reformers. The military felt it could head off real reform by having a reorganization, sponsored by individuals who were friendly to the military, and calling it 'reform.'" United States Congress, House of Representatives, Committee on Armed Services, *Goldwater-Nichols Department of Defense Reorganization Act of 1986: Conference Report to Accompany H.R. 3622* (Washington, DC: GPO, 1986), 96. Similarly instructive is Carl Builder's excellent study, *The Masks of War*. Published in 1989, Builder ignores Goldwater-Nichols almost entirely in his study of service cultures and approaches to war. This is indicative of the expectations of the act's impact in the years immediately following its passage. Carl H. Builder, *The Masks of War* (Baltimore, MD: Johns Hopkins University Press, 1989).
35. Roman and Tarr, "Military Professionalism and Policymaking."
36. Importantly, defense budgets and force planning are not considered within an administration's interagency policy-making process.
37. James Kitfield, *Prodigal Soldiers: How the Generation of Officers Born of Vietnam Revolutionized the American Style of War* (New York: Simon & Schuster, 1995); Powell, *My American Journey*, 447-48, 471-80.
38. This has been discussed in Dana Priest's series on the unified commanders in the *Washington Post*. Priest, "A Four Star Policy," *Washington Post*, 28 September 2000, A1; Priest, "An Engagement in Ten Time Zones," *Washington Post*, 29 September 2000, A1; Priest, "Standing Up to State and Congress," *Washington Post*, 30 September 2000, A1.
39. *The Goldwater-Nichols Act, U.S. Code*, Vol. 50, Title V.
40. Confidential interview with a Joint Staff flag officer.
41. William A. Owens, "Making the Joint Journey," *Joint Force Quarterly*, no. 21 (spring 1999), 92-95; William A. Owens, with Edward Offley, *Lifting the Fog of War* (New York: Farrar, Straus and Giroux, 2000), 231-35.
42. Confidential interview with a Joint Staff flag officer.
43. This is manifested in such documents as *Joint Vision 2020* (Washington, DC: GPO, 2000).
44. Technology and America's global mission have also been very important factors in reducing the independence of the services in the twentieth century.
45. This view has held throughout our confidential interviews.
46. There is a clear and dramatic difference between the career patterns of JCS chairmen and unified commanders before and after Goldwater-Nichols. The path to these top assignments used to run exclusively through the services; now it runs through the joint community. See Roman and Tarr, "Military Professionalism and Policymaking," 411-14.
47. The command function was formally reduced in the 1958 Reorganization Act, but the services retained command power through informal mechanisms.
48. Chairman of the Joint Chiefs of Staff Instruction, "Charter of the Joint Requirements Oversight Council, CJCSI 5123.01A, 8 March 2001; Michael B. Donley, *Joint Requirements*

and Resources Processes: Phase I Report to the J-8 (McLean, VA: Hicks & Associates, November 1999).
49. Confidential interview with Joint Staff flag officer.
50. Jim Garamone, "Joint Forces Command Faces the Military's Future," *Armed Forces Information Service*, 8 October 1999.
51. Gen. Henry H. Shelton, *Joint Vision 2020* (Washington, DC: GPO, June 2000).
52. For different perspectives on how the JROC should evolve, see Owens, *Lifting the Fog*, 231-36; John M. Shalikashvili with Bruce Rember, Phil Her, and Thomas Longstreth, "Keeping the Edge in Joint Operations," in *Keeping the Edge: Managing Defense for the Future,* eds. Ashton B. Carter and John P. White (Cambridge, MA: MIT Press, 2001), 37-48; Michael B. Donley, *Joint Requirements and Resource Processes: Key Issues In the JROC-JFCOM Relationship* (Phase 2 Report to the J-8) (McLean, VA: Hicks & Associates, November 2000).
53. Most importantly, Major Force Program (MFP) 11 was established to insure that special operations forces received sufficient resources. As a result, the Special Operations Command is similar to the services in some respects that also distinguish it from the unified commands. Marquis, *Unconventional Warfare*, 204-14. The role of the joint community in budgeting is described in Chairman of the Joint Chiefs of Staff Instruction, "Chairman of the Joint Chiefs of Staff, Commanders In Chief of the Combatant Commands, and Joint Staff Participation in the Planning, Programming, and Budgeting System," CJCSI 8501.01, 1 April 1999.
54. Confidential interviews with unified commanders.
55. Mid-career officers use the acronym "ROAD"—"Retired On Active Duty"—contemptuously to describe some of the senior officers in their service.
56. Confidential interviews with unified commanders.
57. Harold W. Gehman, Jr., "Progress Report on Joint Experimentation," *Joint Force Quarterly*, no. 25 (Summer 2000), 77-82.
58. The United States tried on two earlier occasions to use a unified command to perform these functions. Neither of the two commands, Strike Command and Readiness Command, succeeded in developing joint training or doctrine. Ronald H. Cole et al., *The History of the Unified Command Plan, 1946-1993* (Washington, DC: Joint History Office, 1995), 4-6, 32-33, 38-43.
59. For more on the relationship between service identities and professionalism, see Builder, *The Masks of War*; Carl H. Builder, *The Icarus Syndrome: The Role of Air Power Theory in the Evolution and Fate of the U.S. Air Force* (New Brunswick, NJ: Transaction Publishers, 1994). For more on the relationship between service identity and organizational behavior, see Andrew F. Krepinevich, Jr., *The Army and Vietnam* (Baltimore, MD: The Johns Hopkins University Press, 1986).
60. This is obviously a reference to the reluctance of some Army leaders to conduct military operations other than war.
61. John E. Mueller, *Retreat From War: The Obsolescence of Major War* (New York, NY: Basic Books, 1989); Carl Kaysen, "Is War Obsolete? A Review Essay," *International Security* 14 (Spring 1990): 42-64; John Orme, "The Utility of Force in a World of Scarcity," *International Security* 22 (Winter 1997/1998): 138-167.
62. Organizationally and professionally, the closest analogy to the Army's current situation may be the Air Force in the 1960s. From its inception to the early 1960s, the Air Force had been based on deterring and, if necessary, fighting a strategic nuclear war. With the advent of assured destruction, the Air Force found that it possessed, and would have to continue to possess, an arsenal of forces whose only function was to deter. The shift away from a focus on nuclear war-fighting as the Air Force's core professional mission was a painful one. The main difference, of course, between the Army's current situation and the Air Force of the 1960s is that large-scale conventional war lacks the universal condemnation that nuclear war has.
63. Betts, *Soldiers, Statesmen, and Cold War Crises*, 56-73.
64. Confidential interviews with Army officers.

12 Infusing Civil-Military Relations Norms in the Officer Corps[1]

Marybeth Peterson Ulrich

Introduction

The backdrop of American military professionalism is the American political system. The heart of the American political system is embodied in the democratic institutions established in the Constitution. The Constitution's division of powers and authority, along with its system of checks and balances, "has succeeded not only in defending the nation against all enemies foreign and domestic, but in upholding the liberty it was meant to preserve."[2] George Washington in his farewell address noted that the American political system was based on "the right of people to make and to alter their constitutions of government"[3] and that adherence to the constitution is a sacred obligation of all until it is changed "by an explicit and authentic act of the whole people. The very idea of the power and the right of the people to establish government presupposes the duty of every individual to obey the established government."[4]

The American founders chose to establish a republic as the best way to uphold liberty and ensure the security of its citizens. "A Republic, by which I mean a Government in which the scheme of representation takes place, opens a different prospect, and promises the cure for which we are seeking."[5] In a republic, the fate of the democratic citizenry is entrusted to "the medium of a chosen body of citizens, whose wisdom may best discern the true interest of their country, and whose patriotism and love of justice will be least likely to sacrifice it to temporary or partial considerations."[6] Representative democracy entrusts the management of governmental affairs to those elected by virtue of their demonstrated aptitude or desire to take on the burden of being responsible for the "people's business."

Most of democratic theory is focused on ensuring that these political agents remain accountable to the polity. The founders' preference for republican democracy was rooted not in the belief that a system based on a "scheme of representation" and shared powers would produce the most efficient outcomes, but in confidence that such democratic processes embodied the best chance for the preservation of liberty.[7]

Civil-military relations in a democracy are a special application of representative democracy with the unique concern that designated political agents control designated military agents.[8] Acceptance of civilian supremacy and control by an obedient military has been the core principle of the American tradition of civil-military relations.[9] U.S. military officers take an oath to uphold the democratic institutions that form the very fabric of the American way of life. Their client is American society, which has entrusted the officer corps with the mission of preserving the nation's values and national purpose. Ultimately, every act of the American military professional is connected to these realities—he or she is in service to the citizens of a democratic state who bestow their trust and treasure with the primary expectation that their state and its democratic nature will be preserved.

A military professional cannot operate independently from the state or the society he or she serves. Whether the service is to an authoritarian or democratic state or to something in between, the service is embedded in a societal context which forms the basis for a set of relationships. These civil-military relations reflect the absence or presence of democratic institutions and the nature of the particular national security system in place to formulate and implement national security policy.

This chapter seeks to spark a dialogue on the current state of civil-military norms within the U.S. military profession. A framework laying out key principles to consider when exercising professional judgment in this area is presented. Underlying the framework is the assumption that the character of a democratic state's civil-military relations has implications for the quality of its national security policy, the preservation of the democratic values on which the state was founded, and the relationship between the democracy's citizens and the military. A crucial factor in this process is the development of a brand of professionalism within the officer corps that is consistent with these underlying principles.

Democratic Military Professionalism

The military profession is unique because of the distinct function that society has entrusted to it, that is, to direct, operate, and control an organization whose primary function is the threat or use of deadly military might against enemy forces or targets which the political leadership designates. Military professionals in all political systems share a mandate to be as competent as possible in their functional areas of responsibility in order to defend the political ends of their respective states.[10] However, military professionals in service to democratic states face the added burden of maximizing functional competency without undermining the state's democratic character. These military professionals must practice a brand of professionalism which takes this into account.

Samuel Huntington in his classic work, *The Soldier and the State,* posited that there is an inherent tension between the state exercising its responsibility to provide for the security of its democratic polity and militaries established to fulfill this function. Indeed, the requirement to balance the functional imperative (providing for the national defense) with the societal imperative (preserving and protecting democratic values),[11] calls for the development of democratic military professionals.[12] Officers comfortable with their roles as democratic military professionals will be better equipped to navigate the complex terrain of civil-military relations.

The Nature of National Security Communities in Democratic States

Democratic military professionals do not pursue their responsibilities to the state in isolation. They are part of a broader national security community comprised of national security professionals[13] from both the civilian and military spheres, other societal actors such as journalists and academics who contribute intellectual capital and foster debate, legislative bodies with constitutional responsibilities to oversee and provide resources for national security policy, and, finally, the public at large to whom all of the above are ultimately responsible. National security policy is the product of the overlapping participation of all members of a state's national security community.

National security professionals, however, have a unique role because they are charged with the responsibility to formulate and execute national security policy within the prescribed bounds of a democratic policy-making process. These officials, who may come from both the civilian and military spheres, have diverse functions requiring mutual cooperation in order to make and carry out policy.[14] Each national security professional's "home sphere" emphasizes different areas of competence. The civilian national security professional's career will be characterized by greater experience at the strategic and political levels while the military national security professional may be more rooted in technical expertise and operational knowledge related to the use of force. In order to craft effective national security policy, civilian and military national security professionals must develop overlapping areas of competence.

Scholars and practitioners alike recognize that the lines separating the competencies of military professionals and political leaders have become increasingly blurred. There is no clear threshold between peace and war marking the point where political and military leaders hand off responsibility.[15] A strategic environment replete with military operations other than war, coalition partners, nongovernmental actors, and tenuous public support has resulted in an ambiguity of roles across the civilian and military spheres. Even if clear lines between peace and war could be drawn, the era when war was regarded mainly as the business of soldiers, and international politics the exclusive domain of diplomats, has long since passed.[16] The existence of a core group of national security professionals, comprised of capable and respected colleagues with overlapping competencies in political and military affairs, is instrumental in achieving balanced civil-military relations and effective national security policy outcomes.

Implications for Democratic Military Professionals

Democratic military professionals must understand the breadth and depth of their participation in the national security process. They must recognize that, while participation within a national security community is often a collaborative process requiring the expertise and inputs of various actors, there are distinct differences in responsibilities stemming from one's constitutional role in the process. These differences may dictate certain limits upon the various legitimate actors in the process. While national security in democracies is conducted within the context of civil-military relationships, these civil-military relations necessarily have a specific structure that channels participants' competencies and responsibilities in order to maximize security at the least cost to democratic principles.

The scope of this chapter is limited to the development of a set of comprehensive norms for military professionals in their civil-military relations. Implications for other members of the national security community will emerge throughout this chapter, but my charge is to improve the quality of participation of democratic military professionals in the national security policy-making process. My focus is on officers' political-social expertise, which includes behaviors related to their participation in the process as policy collaborators and in their participation in the political process in general.

A starting assumption is that professional development programs in the civilian and military spheres, as well as the conventional wisdom extant in society at large, inadequately address officers' political-social expertise. Studies reviewing the curricula at the pre-commissioning and senior service college levels of professional military education reveal that fundamental principles related to civil-military norms are poorly understood at the undergraduate level.[17] Furthermore, twenty plus intervening years of professional socialization to include attendance at a senior service college do not equip future senior military leaders with a thorough understanding of civil-military norms sufficient to navigate the ambiguities of "advice," "advocacy," "insistence," and "political participation."[18]

Even in the most advanced democracies such as the United States, participants in the national security process are continuously engaged in improving the competencies required to adequately exercise their national security responsibilities. When an area of competence is under-developed or the competency levels are not sufficiently balanced across the civilian and military spheres, the achievement of both the functional and societal imperatives is threatened.

First Principles for Military Professionals in Service to Democratic States

My professional experience as a cadet, officer, and civilian scholar with twenty-one years of socialization to the military profession informs me that there is no commonly accepted theoretical framework upon which to evaluate various civil-military behaviors. Cadets' and officers' understandings vary widely regarding what they regard as professional civil-military relations norms. Meanwhile, military professionals at various stages of development observe a confusing range of behaviors illustrating the profession's failure to espouse a particular set of civil-military relations norms.

The specific recommendations that follow as each specific issue area is explored in this chapter are rooted in a conceptual framework founded on two fundamental theoretical concepts that govern civil-military relations in democratic states. The first part of the framework examines the match-up between the competencies and responsibilities of actors from the civilian and military spheres who participate in the national security process. The second key piece of the framework stems from Samuel Huntington's contention that "military institutions of any society are shaped by two forces: a functional imperative stemming from the threats to the society's security and a societal imperative arising from the social forces, ideologies, and institutions dominant within the society."[19] This competition for preeminence between societal and functional imperatives is a primary source of tension in civil-military relations.

This conceptual framework governing democratic civil-military relations serves as the underlying theoretical basis to guide officers through the maze of civil-military issues that confront them throughout their careers. The competency-responsibility match-up and the balancing of the societal and functional imperatives will underpin the recommendations that follow for the development of civil-military norms for military professionals across the relevant issue areas.

Issue Area One: The Role Of Military Professionals In The Policy-Making Process

A fundamental concept guiding national security professionals as they carry out their respective national security duties is that each position carries unique competencies and responsibilities. Distinctions between competencies and responsibilities are related to the nature of the position and the constitutional authority upon which it is based. National security outcomes are optimized when civil-military relations are in balance, that is, when participants maximize their respective competencies and appropriately channel these competencies through their respective responsibilities. Conversely, sub-optimal policy outcomes are often the result of an imbalance between participants' competencies and their decision-making responsibilities. Such conditions often result in strained civil-military relationships.

It may periodically, or even frequently, be the case that military professionals perceive that their competence or expertise in a given issue area is superior to that of civilian authorities with the responsibility to make policy decisions. Military professionals may perceive that civilian decision-makers have set aside or discounted their expert advice in favor of counsel from national security professionals within the civilian sphere. Military professionals may perceive that the resultant policy outcome is poor. Indeed, the policy outcome *may be* poor as a result of such an imbalance between competencies and responsibilities within the national security community.

However, democratic civil-military relations are characterized by military professionals who tolerate such periodically poor policy-making outcomes in order to preserve the fundamental long-term interest of upholding the democratic character of the state. Military institutions in service to democratic societies should espouse as a fundamental norm of civil-military relations that the profession's first obligation is to do no harm to the state's democratic institutions. Usurping or undermining the decision-making authority of civilian decision-makers is a clear violation of the responsibilities inherent in each actor's constitutional role.

An officer's oath is to "support and defend the Constitution of the United States against all enemies, foreign and domestic." An officer's allegiance, therefore, is not just to the state, but to the democratic character of the state as embodied in the institutions established in the Constitution. Professional judgment in this area depends on having in place a system of officer professional development that encourages the incorporation of democratic values in an officer's overall set of internal values and the cultivation of a sense of duty, honor, and professional obligation that links the special requirements of service to a democratic state to an officer's overall professionalism.[20]

Table 12-1 below highlights the variations possible across the key factors of competence and responsibility, illustrating which variations deviate from first principles of civil-military democratic norms.

Competence ↑		
	4. High Competence + Low Degree of Appropriately Exercised Responsibility = Effective But Non-democratic Policy Outcomes	1. High Competence + High Degree of Appropriately ExercisedResponsibility = Effective and Democratic Policy Outcomes
	3. Low Competence + Low Degree of Appropriately Exercised Responsibility = Ineffective and Non-democratic Policy Outcomes	2. Low Competence + High Degree of Appropriately Exercised Responsibility = Ineffective and Democratic Policy Outcomes
	Responsibility →	

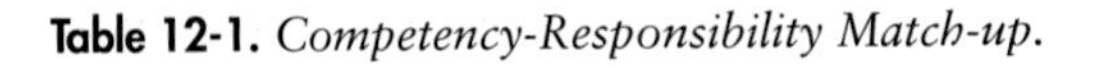

Table 12-1. *Competency-Responsibility Match-up.*

The ideal match-up maximizes competence *and* responsibility so as to get both the benefits of effective policy and compliance with norms of national security decision-making in a democracy. This option is found in Box 1 of Table 12-1. In this scenario, national security professionals pool their respective expertise in a collaborative process that culminates with the appropriate national security professional in the civilian sphere making the national security decision.

George C. Marshall's service at the highest levels of government from 1939 to 1950 epitomized the civil-military norms operating in Box 1. As Chief of Staff of the Army, Secretary of State, and Secretary of Defense, Marshall was the standard-bearer for a generation of officers that unwaveringly accepted the concept of civilian control. Indeed, professional judgment in this respect was so firm that American military leaders consciously and quietly accepted a not inconsiderable number of policy decisions at variance with their professional strategic judgment. These instances ranged from the decision to support a strategy to hold the Philippines in 1941 despite the perceived lack of military means to do so, to acquiescence in the decision to invade North Africa in 1942 despite their concerns that such a diversion of military resources would dangerously prolong the initiation of a decisive cross-channel invasion.[21]

Whether or not these military leaders exploited the limits of dissent within democratic policy-making processes will be addressed further within Issue Area Two—"civil-military norms and dissent." The relevant point here is that American military professionalism operative in the World War II era did not tolerate behaviors that threatened the democratic character of national security policy outcomes.

Perceived deficiencies in competency on the civilian side were not, however, completely ignored. At times when Marshall or his staff perceived strategic competence to be lacking within the political leadership, he encouraged his staff to see such instances as opportunities as opposed to liabilities. For instance, he viewed President Franklin Roosevelt's shortfalls as a strategic thinker as an opportunity to educate the President on how he could best support the war effort.[22]

Marshall's frankness coupled with his unquestioning acceptance of civilian authority gained him the full confidence of Roosevelt. Roosevelt repeatedly offered Marshall command roles in the war, which were never accepted due to each man's understanding that Marshall's contribution as expert adviser and honest broker in Washington was irreplaceable.[23] Marshall's rock-bound sense of duty, selflessness, and honesty were also highly prized assets in his service to President Harry Truman as Secretary of State. Truman remarked, "He was a man you could count on to be truthful in every way, and when you find somebody like that, you have to hang on to them."[24]

Marshall possessed an ability to refrain both from capitalizing on his popularity in order to prevail in the policy-making process and from spending his enormous political capital to further his ambition or to advance his personal judgments. Such restraint is evidence that he understood the distinct responsibility that civilian political leaders had in the decision-making process. Such behavior also ensured that he was never viewed as a competitor in the national security decision-making process, but as a trusted source of counsel able to accept the rejection of his professional advice on one occasion in order not to compromise his status as an objective adviser in the next.

In Box 1, civil-military tension is low due to the trust between expert military adviser and civilian policy-maker. Political objectives are communicated to democratic military professionals, who in turn apply their strategic expertise to help shape the policy through particular applications of the military instrument of power. Having participated in forging the military dimensions of the policy, the military professionals are far more likely to embrace the policy. The resultant policy is characterized by the combined expertise of national security professionals from both the civilian and military spheres, the democratic accountability inherent in the civilian policy-maker's position, and democratic institutions undiminished by officers intent on pressing their policy preferences beyond the bounds of the policy-making process.

The next best choice is found in Box 2 of Table 12-1. In this scenario, competency is lower, but decisions are at least made within the norms of democratic national security decision-making. The short-term policy decision may be ineffective, but no long-term damage has been inflicted on democratic institutions. This is the option that democratic military professionals may find the most difficult to carry out, particularly if the competence deficit is perceived to be present in the civilian sphere.

Examples of civil-military relations characterized by Box 2 can be found in states in the process of democratic transition from authoritarian rule. Democratic institutions are not yet fully developed, and competencies across the civilian and military national security spheres are uneven. Often the military possesses the lion's share of strategic expertise, but this expertise may be lacking in its appreciation of the broader political issues at play in strategy formulation.

The post-communist states of Central and Eastern Europe a decade into their transitions from authoritarian rule had largely mastered the task of subordinating their military institutions to civilian rule. However, in many respects national security policy outcomes remained ineffective due to the dearth of national security professionals with expertise in both political and military strategic competencies. Defense ministries and general staffs have been unable to develop national security planning processes that effectively set priorities, coordinate resources to focus on the achievement of the stated objectives, and ensure that the means allocated to defense are optimized. The political leadership, on the other hand, may often lack the interest, expertise, or both that are necessary to direct the overall formulation of national security policy. What has been particularly lacking is the ability of the political leadership to provide a strategic vision that can serve as the basis of a national security strategy and a national military strategy.

As a result, the actual capabilities of post-communist armed forces are low. Progress has been slow, to be sure, but democratic institutions have not suffered from the intervention or overreach of military leaders in search of short-term gains for their institutions. Over time, competency in both spheres has grown. The political leaderships have assumed a heightened role in the national security planning process, and national security institutions have begun to respond to the demands for more rational defense planning. Real collaboration in the national security policy-making process is beginning to take place with the potential to effect positive change in defense capabilities.

While the particular circumstances of democratizing states have been highlighted to illustrate the dynamics of Box 2, insufficient expertise channeled responsibly can at times plague even consolidated democracies. In Box 2, the trust between the military and political leadership will not be as high as what is possible within more collaborative conditions of policy-making. Indeed, respect between the two spheres may be low due to the perceived incompetence that each sphere attributes to the other. The lack of expertise within the policy-making process will inevitably lead to substandard decisions that will adversely affect the military's ability to carry out the functional imperative of providing the state with effective national security. However, the societal imperative maintains its hold within the officer corps' professionalism. Democratic decision-making processes and institutions remain intact so that when a balance of competencies across the civilian and military spheres is either achieved or restored, an outcome with the characteristics of Box 1 is possible.

The worst outcome is represented in Box 3 of Table 12-1. Decisions characterized as reflecting low competencies that are then inappropriately channeled to circumvent democratic processes will result in poor policy outcomes that simultaneously harm democratic institutions. This particular choice should be a rare phenomenon in advanced democracies with developed national security communities. This outcome may, however, be common in states with both weak democratic institutions and unprofessional military forces.

Military rule is characteristic of Box 3. Pakistan is an example of a state that has been plagued since its establishment in 1947 with questionable competencies on both the civilian and military sides as manifested in chronically poor governance, weak democratic institutions, and unprofessional military forces in terms of their lack of restraint toward undermining civilian governments. The Pakistani army has overthrown a legitimately constituted government four times in Pakistan's fifty-two-year history, which has severely strained the political stability and viability of democracy in the country.[25] The most recent takeover, in October 1999, posed particular challenges for the international community because it was the first time that nuclear weapons came under control of a military regime. During periods of military crisis, the caution, deliberation, and accountability that contribute to sustaining stability in democracies may be lacking in such contexts. Box 3 represents the worst outcome because the lack of competence inevitably leads to poor policy outputs coupled with the undermining or elimination of democratic processes.

Finally, Box 4 of Table 12-1 represents an outcome that threatens advanced democracies with unbalanced civil-military relations. Overall national security competence may be high, but it may reside more dominantly in one sphere or the other. Military leaders may perceive that their professional judgment necessarily points to a particular policy decision. The political leadership, not concurring with this judgment, may select a different course of action. Such a mismatch of competence as perceived by the military, who lack responsibility for decision-making, may tempt military professionals to exert various aspects of their informal power in order to prevail in short-term policy decisions.

The post-Cold War era in U.S. civil-military relations has featured a series of incidents that one civil-military relations scholar, Peter Feaver, has called military "shirking"—that is, various degrees of military noncompliance in the face of the political leadership's desired policy preferences.[26] These incidents include the military's reluctance to embrace the Somalia, Bosnia, and Haiti missions, the resistance to President Bill Clinton's initiative to allow homosexuals to serve openly in the military, and efforts to resist dramatic changes in the military services' roles, missions, and force structure in the strategic reviews of the Clinton and George W. Bush administrations.[27]

Noncompliance may take various forms. One technique in the post-Cold War era has been to present only a limited range of options for the use of the military instrument of power, with each known to be unacceptable to the political leadership. When advising the administration of the elder George Bush on courses of action available for a potential Balkan intervention to achieve the political objective of reining in the violence in 1991, Chairman of the Joint Chiefs of Staff, Colin Powell, put the price tag at 250,000 troops.[28] This calculation, however, was based on assumptions interjected by Powell himself into the policy-making process, namely, that victory must be decisive and that overwhelming force must be used (the so-called Powell Doctrine). This preference for decisive force was incompatible with the political leadership's search for a course of action that could employ limited force to achieve limited political objectives. The military's actions, in this case, delayed action on a top-priority U.S. foreign policy initiative until conditions met the military's terms.[29] Other methods of military "shirking" include manipulating the defense bureaucracy[30] to shape outcomes more in line with preferred policy ends and expending political capital in such a way that undermines the responsibility of the political leadership to implement their preferred policy end.

Exercising the option in Box 4 may also negatively affect the role of the military profession in affairs of state. The role of the military institution as a trusted policy collaborator that possesses expertise unique to the military profession may be compromised if Box 4 is repeatedly chosen. Civilian policy-makers may come to view the military actors in the national security community as competitors in the decision-making process, more concerned with promoting their own institutional interests than the national interest as discerned by civilian decision-makers. Furthermore, a collaborative policy-making environment is impossible to achieve when either side works to intimidate the other. Fear on the part of the political leadership and lack of respect on the part of the military as an institution characterized civil-military relations in the Clinton administration. It was reported that President Clinton was more intimidated by the military than any other political force. According to a former senior National Security Council official, "I don't think there was any doubt that he was out-and-out afraid of them."[31] But such fear-induced victories come at the cost of democratic institutions because the national security actor who has inappropriately made or influenced policy is neither directly nor indirectly accountable to society at large via the ballot box.

Recommended Civil-Military Norm: *The military profession's first obligation is to do no harm to the state's democratic institutions and the democratic policy-making processes that they establish. The civilian political leadership sets political objectives that the military supports in good faith. The military leadership should apply its expertise without "shirking" or taking actions that, in effect, have a determinative effect on policy outcomes. Military professionals must develop a clear sense of the distinction between national security competency and the responsibility to exercise competency through distinct roles in the national security policy-making process.*

The incorporation of such a principle does not mean that low levels of competency in either the civilian or military spheres are acceptable as a permanent condition of the civil-military relationship. Indeed, living with the result of ineffective short-term policy decisions resulting from low competency levels should inspire both sides to redouble efforts to close competency gaps. This could involve military and civilian actors in reassessing their national security competencies internally, with each side also identifying strategies to imbue their counterparts with either the military or political competencies that are lacking.

Issue Area Two: Civil-Military Norms Vis-à-Vis Dissent

An issue area closely related to civil-military norms in the policy-making process is the problem of dealing with dissent. Dissent is normally considered to issue only from the military sphere, since the ultimate power over decision-making in a democracy lies with the civilian leadership. However, disagreement is a regular visitor to any collaborative decision-making process. Norms should exist in both spheres to encourage healthy debate while recognizing that disparate responsibilities mandate ultimate deference to the civilian decision-maker.

Civilian policy-makers should encourage military professionals to offer their best advice and not punish military participants who work within the established bounds of dissent in democratic national security decision-making processes. History is rife with examples of military professionals whose professional expertise was not advanced with sufficient candor and vigor within the bounds of collaborative decision-making to influence civilian leaders' national security decisions. H.R. McMaster's *Dereliction of Duty,* a widely read book among the current generation of American military officers, makes the argument that although the Joint Chiefs recognized that the graduated pressure strategy in Vietnam was fundamentally flawed, they did not vigorously express their professional judgment in the policy-making process. As a result, McMasters contends, that strategy was set without the best advice of the President's principal military advisers.[32] More recently, Congressional leaders chastised the Joint Chiefs for continuing to support the fiscal 1999 defense budget instead of forthrightly advocating a budget that adequately ensured the readiness of the armed forces.[33]

Military professionals in service to democratic states should muster all of their national security expertise relevant to the achievement of a particular political objective that the civilian leadership sets forth. This is especially true when more strictly military competencies are at stake, such as judgments related to evaluating risks to soldiers and the tactical conduct of ongoing operations.[34] Civilian decision-makers rightfully expect that military professionals under their command be forthright and thorough, yet ultimately compliant to civilians with regard to the determination of national political objectives and the ways to achieve them.

Policy advocacy has its place within the bounded limits of a collaborative policy-making process, but advocacy actions counter to the civilian leadership's known preferences may begin to usurp the civilian leader's distinct responsibilities. For instance, efforts to influence the terms of debate in public forums while national security professionals are still at odds in intra-governmental discussions may shift the military professional's status from objective expert to suspect competitor. Such actions both poison a collaborative decision-making environment that is critical to the achievement of optimal national security outcomes and demonstrate a willingness to overlook distinctions in responsibility across the civilian and military spheres to achieve short-term political gain.

Resignation in protest is often discussed as an acceptable means of dissent that permits policy advocacy to spill beyond the normal channels of discourse between military and civilian actors in the national security policy-making process. Resignation, however, may not be the panacea that those in favor of a more assertive role for military leaders in the policy-making process sometimes suggest. Military leaders should consider both the positive and negative consequences of resignation before resorting to this action. First, resignation is the extreme method of dissent. The intent is to publicly express disagreement with the responsible political leaders, thus pitting the resigning actor's political capital against the policy-maker's. A resignation followed by news media outbursts aimed at advancing institutional interests over national interests may harm the objective status of the military profession in the democratic political process.

On the positive side, resignation represents a clear withdrawal from future participation in the policy-making process and consequently removes the threat that the resigning actor may continue to undermine the role of the political leadership within the official system. The possibility of playing the role of "competitor" vis-à-vis the elected political leadership within the policy-making process is diminished, although a competitor role could be assumed outside government in retirement.

Recommended Civil-Military Norm: *Military participants in the national security policy-making process should expect a decision-making climate that encourages a full exchange of expertise across the military and civilian spheres. Military professionals, furthermore, should have the expectation that their professional judgment will be heard in policy deliberations. However, military participants must develop the professional judgment to recognize when the bounds of the policy-making process have been breached. When acts of dissent take them outside these bounds, military leaders must acknowledge that they have gone beyond the limits of their roles in terms of offering advice and have begun to carry out behaviors that directly challenge the role of political leaders with the responsibility to make policy.*

Issue Area Three: Civil-Military Norms And Partisan Politics

Much attention has been paid in recent years to the increasing willingness of U.S. officers to label themselves as conservative Republicans. While Morris Janowitz's research in the 1950s documented military officers' self-identification as political conservatives,[35] partisan identification as conservative Republicans is a development of more recent decades. Despite Janowitz's finding that most officers of the pre-1960s were moderately conservative,[36] the ideological gradations within the two parties were such that these officers could have found a home in either the Democratic or Republican party at the time. Most officers of that era, however, chose not to profess a particular party affiliation.

Ole Holsti's research has contributed empirical data to the ongoing professional and scholarly dialogue concerning the growing gap between U.S. society and its military with regard to party identification. Using survey data resident in the Foreign Policy Leadership Project database, Holsti showed that between 1976 and 1996 the proportion of officers who identified themselves as Republicans doubled from one-third to two-thirds.[37] Tom Ricks's interviews with junior officers led him to conclude, "Today's junior officer seems to assume that to be an officer is to be a Republican."[38]

Ideological self-identification also points to a rise in professed conservatism in the officer corps. Holsti's data show that the ratio of conservatives to liberals in the officer corps went from four to one in 1976 to 23 to 1 in 1996.[39] The shift is due to the decline in the self-identification of officers admitting to be liberals of any variety—even moderate liberals. In 1996 only 3 percent of the officers surveyed reported an ideological self-identification as somewhat liberal or very liberal, while 73 percent reported that they were somewhat conservative or very conservative.[40]

While it may be reasonable to expect the ratio of conservatives to liberals in a culturally conservative, hierarchical institution such as the U.S. military to be something less than a mirror of American society, a 23 to 1 conservative to liberal ratio as evident in Holsti's data may begin to raise questions about whether or not the military reflects the society it serves. By way of contrast, the data gathered in the 1999 Triangle Institute for Security Studies (TISS) survey revealed that civilian elites self-identified more evenly between conservatives and liberals, with 32 percent reporting that they were either somewhat conservative or very conservative and 37.5 percent weighing in as either very liberal or somewhat liberal (in contrast, only 4.4 percent of the officers in the 1999 survey reported an ideological self-identification as somewhat liberal or very liberal).[41]

To fulfill its primary obligation to its client, society must grant the military the legitimacy to carry out its solemn function. Legitimacy is enhanced when the military institution is perceived to be "of society" in terms of being comprised of a representative cross-section of the population. The alternative outcome is to increasingly become a distinct group representing only limited demographic characteristics and attitudes of the society at large. As the perception grows that the military's political ideology is dramatically divergent from the society it serves, the parallel assumption will grow that the military is an institution apart from the society it serves. How an officer votes should remain an officer's prerogative in a democracy, but as the ideological gap between the military and society at large grows, it may become more difficult for soldiers to conduct their professional affairs in an ideologically neutral way.

Military sociologists who study organizational culture break military socialization into two dimensions—anticipatory and secondary. Anticipatory military socialization encompasses individuals who self-select to belong to the military, thus implying a fit between organizational and personal worldviews.[42] Secondary military socialization takes into account the organization's role in instilling a worldview in its members.[43]

The effects of a growing perception within society that the military is a haven for politically conservative youths will only skew the self-selection rates among politically conservative youths further. Accession policies should be examined to ensure that measures are in place to recruit a more politically diverse element of the citizenry to serve society as military professionals. Such moves as enhancing Reserve Officer Training Corps (ROTC) programs in Ivy League schools or in traditionally African American colleges may help to balance out anticipatory socialization trends. Within the secondary socialization process, over which the military has more direct control, there should be a return to the fundamental principle that Morris Janowitz espoused in his classic work, *The Professional Soldier.* He advocated imbuing within military professionals an ethic of civil service that would view detachment from partisanship as a critical element for assuring society of the military professional's partisan neutrality.[44] Incorporating the principle of nonpartisanship as part of an officer's professional code (rather than relying on a mere tradition, which is always subject to erosion) could help restore balance in U.S. civil-military relations.

The profession must take note that a long-term erosion of the American military's tradition of political neutrality, which tradition has its roots in the advent of the professionalization of the American military led by Gens. William Tecumseh Sherman and Emory Upton as well as Rear Adm. Stephen Luce in the decades following the Civil War, has since taken place.[45] This now-visible trend has profound implications for the profession and its ability to serve society. Certainly, the icons of the profession noted above would be distressed to learn that many officers today regard membership in a particular political party to be a defining element of officership.

The American military at present needs to reinforce the professional norm that it can serve any political party in a principled fashion. Principled professionalism is not determined by the congruity of a particular party's agenda with an individual soldier's principles, but by adherence to the norms of principled officership. *The United States Military Academy Strategic Vision—2010* lays out eight principles of officership to guide officers in carrying out the responsibility that society entrusts to them. One of these key principles is subordination. The guidelines demand that

> Officers strictly obey the principle that the military is subject to civilian authority and do not involve themselves or their subordinates in domestic politics or policy beyond the exercise of the basic rights of citizenship. Military officers render candid and forthright professional judgments and advice and eschew the public advocate's role.[46]

Soldiers must be equally comfortable serving either party that society chooses to govern.

Advancing civil-military norms related to officers' political activities must balance officers' constitutional rights to participate in the political process with necessary limits stemming from the military profession's unique responsibility to society. Thus far in this chapter I have advocated the development of a collaborative national security policy-making environment that draws upon the strengths of each component of the state's national security community. This is at odds with a vision that some civil-military relations theorists espouse, i.e., those who focus on the primacy of civilian control achieved on the basis of two distinct, non-overlapping civilian and military spheres, with the military sphere clearly subordinate to the civilian sphere.[47]

In contrast, the collaborative vision of national security policy-making rejects the notion that military professionals obediently execute the demands of the civilian government as apolitical beings somehow immune from the politics of conflict and distant from the political process itself. National security professionals from both the civilian and military spheres must be participants in the national security decision-making process if competencies from both spheres are to be sufficiently leveraged.

Indeed, the very concept espoused here of the national security professional as a military officer schooled in the political context of military policy requires the participation of at least some military professionals in civilian graduate programs in order to master many of the competencies shared by civilian national security professionals.[48] There are, however, important distinctions to be made between possessing a certain degree of political expertise in order to serve in a policy collaboration role, on one hand, and using political-military expertise as the basis for abusing the limited responsibility of national security professionals within the military sphere in the decision-making process, on the other.

Perhaps the most contentious area of this debate concerns norms governing policy advocacy that may stem from individual political preferences. The terms of this debate often revolve around whether military members' identities are thought to be rooted in their rights as citizens or in membership in a profession with circumscribed rights stemming from the unique obligation of the profession to society. "The mythic tradition of the citizen-soldier is dead," declared defense specialists Elliott Abrams and Andrew Bacevich.[49] The nature of modern war has spurred the gradual evolution of military forces comprised of volunteers whose military service is not a compulsory, transitory departure from civilian life, but rather a compensated choice to make military service their lives' callings.[50] Today soldiers consciously decide to depart civilian life to serve society within the military profession. In doing so, these all-volunteer force members are more beholden to professional norms than their conscripted forbears.

Those more approving of partisan behavior argue that "soldiers in uniform are, after all, citizens, and so long as they obey orders they retain the rights of expression of their counterparts in the civilian world—and most certainly so the moment they doff the uniform."[51] Though this position is possibly valid from a purely legal perspective, it ignores the impact that such political behavior can have on the profession's ability to perform its primary function in society.[52] Society cannot be adequately served by a military profession that refuses to subordinate its individual and institutional interests to the greater national interest, or that is perceived to take sides in the political process.[53]

There was a general perception in the 2000 presidential election that "the military" supported Republican candidate George W. Bush. This did not stem from any specific mass actions by active duty military personnel or participation in overt campaigning, but resulted primarily from a manifestation of self-identification among military personnel, reported widely in the news media, that the Republican party and its presidential candidate were the "best fit" for the military culture.[54]

There still seems to be a consensus within the U.S. military profession that the "tradition of an apolitical military is critical to our democratic system."[55] According to a Marine brigadier general survey response, "The military is held in high esteem by the American public because we remain neutral and nonpartisan. Once the institution is seen as 'looking out for itself' instead of looking out for the country, we risk losing the trust of the American people."[56]

The problem is defining and clarifying professional norms regarding what acts are unacceptably "political." Involvement in a collaborative policy-making process in accordance with the norms laid out in Issue Area One constitutes engagement in politics, but in a manner that is consistent with the military professional's advisory and expert role. However, expressions of partisan preferences that could be perceived as speaking for the profession itself fall outside professional norms. Both budding professionals and seasoned veterans have difficulty distinguishing between legitimate participation in the national security process and behaviors that are gradually steering the profession away from its apolitical roots toward partisanship.

Recommended Civil-Military Norm: *Principled officership requires adherence to an ethic of nonpartisanship. Military professionals must be comfortable serving any political party that prevails in the democratic political process. The demands of principled officership must carry the day, even if they entail limitations on officers' liberties as citizens. Officers should consider the impact that their public profession of political beliefs has on their subordinates and on the relationship of the profession to society at large. Association of the military profession with any single political party undercuts the legitimacy upon which the military depends to serve society.*

Issue Area Four: Civil-Military Norms And Retired Officers

The development of professional norms governing the political activities of retired officers, especially retired general officers, is intrinsically linked to the goal of keeping partisanship out of the active ranks. As one active duty two-star Marine officer noted, "My concern is the effect of a retired general officer's commitment to a political party immediately after retirement on junior officers. Rather than the junior officer taking time to be fully informed on the current issues, there may be a tendency to blindly follow a senior that they admire for his/her service accomplishments."[57]

Another important consideration is the effect that retired officers' actions have on public perceptions of the military as an institutional actor in the political process. As one active duty four-star officer put it, "Constitutional rights are not the issue. Judgment is. The public doesn't distinguish between the active duty General [officer] and retired General [officer]. As a result, the entire military is politicized. If a retired general officer elects to run for office and enter the process that is fine—but not otherwise."[58]

In a domestic political environment where the civil-military gap is growing, such distinctions will be increasingly difficult to make. Indeed, one could argue that the Bush campaign's attempt to line up retired general officers of celebrity status to support candidate Bush and to participate directly in the campaign through the Republican convention and on the campaign trail was an attempt to create the impression in the public's mind that "the military" was behind Bush. If asked to identity general officers of the U.S. military, probably few Americans could name a general officer on active duty, but many could likely recall the names Colin Powell and H. Norman Schwarzkopf from their association with the successful outcome of the Gulf War.

While these individual retired officers may justify their involvement in the Bush campaign as merely the exercise of their rights as private citizens, their individual decisions may by themselves—or certainly in the aggregate—begin to look like an institutional preference. But is it in the best interests of the profession as a whole to be perceived to hold partisan preferences and to publicly advocate them as a participant in the political process? Such a development necessarily has a negative impact on the profession's retention of its nonpartisan tradition, which has been an essential element of forging effective national security policy.

The profession would be well served if retired officers paused to consider the impact of their individual actions on the profession and on the military's retention of a nonpartisan stance. Professional self-policing in lieu of strict legal restrictions on retired officers' free speech rights is the only feasible course of action in this issue area. Attempts to legally regulate such political activities would be frustrated in the courts and might run counter to the goal of infusing in emergent military professionals a sense of responsibility, even later in retirement, to protect the profession's capacity to serve society. Instead, the preferred method is to turn to the profession's internal capacity to educate officers on their responsibility in this area and to foster professional judgment as well as expected standards of behavior.

Most retired officers, especially retired general officers, expect to be treated as members of the profession in retirement. Although retired officers may eagerly shed the various restrictions that governed their lives on active duty, it is customary to refer to a retired officer by his or her rank. Many retired officers pursue second careers in the defense industry. It is not unusual for these officers, especially those of higher rank, to include their rank on their business cards. Anecdotal testimony of active duty officers in the field as buttressed by my own experience indicates that such retirees as contractors collaborating with active duty personnel are often bestowed great deference in the workplace due to the weight that their active duty rank still wields.[59] In addition, norms of protocol call for affording former senior officers particular courtesies commensurate with their active duty rank.

Andrew Abbott in his seminal book *The System of Professions* singled out the mastery of a body of abstract knowledge as the definitive characteristic of a profession.[60] In the same vein, Huntington characterizes a professional as having acquired a body of expertise through prolonged education enhanced through experience. Once these standards of professional competence are gained, they "are capable of general application irrespective of time and place."[61] In the military profession, rank is a measure of professional competence, with the most expert members at least notionally being awarded the highest ranks. The military profession is unique because it requires its most expert members to retire at the peak of their professional competence to make room for the next generation of senior leaders. It is not surprising, then, that in retirement senior officers are often sought out to share their expertise to shape policy decisions or to advise active officers. As the masters of abstract knowledge, still of great utility to the current generation of active senior leaders and to society in general, military officers are essentially professionals for life.

Corporateness is another essential element of professionalism that carries into retirement. The common bonds of shared professional competence, social associations, and continued interaction with the profession through professional organizations, as well as access to various exclusive amenities (e.g., medical care, Post Exchange shopping, commissary privileges, and access to most post services), connect retired officers with the active force for life.

More uncertain in its applicability to retired officers is responsibility. What ongoing responsibilities do retired officers have to the military profession? Huntington argued that "the principal responsibility of the military officer is to the state."[62] Furthermore, Huntington continues, "His responsibility to the state is the responsibility of the expert adviser. . . . He cannot impose decisions upon his client which have implications beyond his special competence."[63] With regard to the continuing service to society as a military expert, a four-star active duty officer commented, "Upon retirement, general officers have all the rights of citizens not otherwise in government. However, should they endorse a political party or candidate their usefulness to their service or DOD as a military expert becomes compromised!"[64]

Finally, while an explicit legal code[65] may regulate some aspects of this responsibility to society, "to a larger extent, the officer's code is expressed in custom, tradition, and the continuing spirit of the profession."[66] To the extent that a retired officer remains a member of the military profession, commitment to some continuing responsibility to society must remain.

On all three counts, expertise evinced in the mastery of abstract knowledge, corporateness, and responsibility, retired officers retain important aspects of the professionalism ascribed to serving officers. The active forces and society grant retired officers qualified authority to continue as members of the military profession. However, "authority often confers obligation."[67] The question at hand is what are the limits of this professional obligation and who sets them?

Abbott offers a framework to settle jurisdictional conflict across and within professions.[68] One potential application of this framework to the conflict between the professional obligations of active and retired officers is to consider the work of retirees as a profession within a profession. As the current stewards of national security accountable and subordinate to the political leadership, active duty senior leaders set the professional norms. These current senior leaders can establish a constructive requirement for retired officers to reflect upon the adverse impact the unrestrained exercise of their political rights could have on the profession and its ability to participate neutrally in the political process and to serve society.

Such professional regulation would, in effect, subordinate the retiree sphere of military professionals under the active strategic leadership in professional areas that have a continuing impact on the military profession. This will require the renegotiation of jurisdictional boundaries within the profession that both recognizes retirees as members of the profession and obligates them to adhere to a set of professional norms aimed at preserving the military profession's ability to perform its exclusive function within society. A self-regulating profession develops norms in the formative educational and training experiences of its new members and continuously enforces such standards as new members progress to greater positions of responsibility in the profession.

Recommended Civil-Military Norm: *Retired officers have a continuing responsibility to serve the military profession. While such professional responsibilities are more limited than those of the currently serving senior leadership, some degree of professional obligation remains. Active duty senior leaders as the stewards of professional norms should set the expectation that retirees consider the impact of their individual and mass actions on the profession. Useful guidelines in this area include considering the impact particular activities could have on the active serving force, contemplating whether individual choices could be perceived as institutional positions, and finally exercising judgment regarding the degree to which the expression of partisan preferences affects the military institution's capacity to be perceived as nonpartisan in the policy-making process.*

Issue Area Five: Balancing The Functional And Societal Imperatives

In civil-military relations literature, the functional imperative to provide for the national defense, and the societal imperative to preserve and protect the democratic processes of the republic, are often presented as competing and implicitly incompatible obligations.[69] It is important to consider analytical frameworks that do not assume that these imperatives are always in opposition and to reflect on contexts where these imperatives may be mutually reinforcing.[70]

Military professionals in service to democratic states, as noted at the beginning of this chapter, must be as functionally competent as possible to secure the national security interests of the state. Additionally, they are simultaneously charged with performing this function within the societal context of liberal democracy. Their responsibility is to defend the state while concurrently appreciating that the pursuit of national security is affected by national values, national character, and the ideology of the state.

Developing norms in this issue area is inextricably linked to the desired professional judgment associated with the first two issue areas identified in this chapter. These civil-military norms focused on the military professional's role in the policy-making process. Mastering the "first principle" inherent in the responsibilities-competencies match-up is key to grasping the mandate to balance the functional and societal imperatives.

The history of American civil-military relations is replete with examples of the military institution battling with its civilian masters over the incorporation of societal values in the military. The integration of blacks and women both came amidst protests from within the military profession that such changes would harm the institution's ability to perform its function.

In the case of President Harry S. Truman's initiative to integrate the armed forces through Executive Order 8981, Truman used his executive powers to make a first move toward enacting the broader civil rights agenda that he campaigned on in 1948. The desegregation order came the same day that he issued an executive order calling for fair employment within the federal government.[71] Both moves were aimed at correcting what Truman considered to be glaring injustices incompatible with national values.[72] Chief of Staff of the Army, Gen. Omar Bradley, opposed the President's desire for instant integration of the armed forces, arguing it "would be hazardous for [the United States] to employ the Army deliberately as an instrument of social reform."[73] Resistance was also rooted in postwar political realities that made the Army reluctant to alienate powerful Southern legislators and the large portion of its own officer corps that hailed from Southern states.[74] In the end, integration under General Bradley was slow and gradual. Ironically, the functional imperative demanding that sufficient manpower be available to effectively man the units being sent to Korea spurred Bradley's successor, Gen. J. Lawton Collins, to fully comply with the integration order.[75]

A crucial barrier to giving women the same opportunities to advance through the officer ranks as men was removed in 1975 with President Gerald Ford's signing of Public Law 94-106, which called for the admittance of women to the service academies in 1976. Clinging to tradition and citing an inevitable dilution of the nation's ability to produce combat leaders, the Army, Navy, and Air Force stood united in their opposition to the Congressional pressure to open the academies' doors to women.[76] But such a stance was out of sync with various compelling forces at work within the societal imperative. First, indications of society's readiness to embrace the concept of equal opportunities for women was evident in Congress' passage of the Equal Rights Amendment (ERA) in March 1972.[77] Second, Congressional sentiment to admit women to the academies had been brewing for several years due to the frustrations of individual Congressmen from both parties to have their female nominees admitted. Additionally, lawsuits were crawling through the legal system on behalf of women applicants.[78] The Congressional hearings pitted the functional imperative against the societal imperative. The services' arguments against encroachment on its professional autonomy clashed with the Congressional will to integrate. Ultimately women were admitted with the passage of amendments to DOD legislation pending in both the House and the Senate without much Congressional opposition.[79]

Democratic military professionals must develop the judgment required to distinguish between "social engineering" and legitimate evolutions of professional practices reflective of the democratic values of the state. Professional autonomy in a democracy does not mean that the profession can regulate itself in ways that are incompatible with democratic values. Many armies in service to authoritarian societies, such as those that characterized the former Soviet bloc, may have had long traditions of abusing conscripts, keeping their internal operations secret from the public, and excluding undesired elements of the society from its ranks. But when democratic institutions began to take root, the newly empowered democratic civilian leaders—increasingly accountable to the newly empowered democratic citizenry—demanded change. The shift in the political system sparked inevitable changes in the form of military professionalism practiced in the state.[80]

Ideally, military institutions that are attuned to the discrepancies between professional norms and democratic norms will anticipate the inevitability of externally driven change. Such proactive behavior affords the military profession the opportunity to autonomously incorporate changes more on its own terms. The next barrier to fall is likely to be the open integration of gays into the military. Recent surveys show that societal attitudes are well ahead of the military on this issue. The 1999 TISS study reported that while 75 percent of military leaders are opposed to integrating gays, 57 percent of the public and 54 percent of civilian leaders are in favor of doing so.[81] The right configuration of societal forces and the political support of a Democratic Congress and/or a Democratic president may ultimately compel the change. Arguments that such a change will erode the functional imperative will likely be advanced again, but just as with the previous instances of integrating blacks and women the institution will likely adapt without significant damage to its functional capabilities. Such a result, however, depends on the military institution setting the appropriate conditions within the profession for the change to succeed.

Recommended Civil-Military Norm: *Tension between the functional and societal imperatives is a constant feature of civil-military relations in a democracy. Military professionals in service to democratic states must recognize that their jurisdiction over the profession is limited in that it is inherently connected to societal values and the realities of civilian control, leaving ultimate control over the profession in the hands of their constitutional overseers. It is in the best interest of the military to foster the development of professionals who are engaged with their societies. Officers must stay abreast of societal forces and their reflection in the preferences of those empowered with civilian control—the President and Congress. Such engagement* *will allow the profession to better manage civilian-driven change and improve the prospects to retain professional standards of good order and discipline essential to the functional imperative. Military institutions that fail to anticipate societal-driven change will inevitably succumb to the realities of the system of democratic political control, which allows civilians vested with the appropriate authority to demand that democratic values prevail. Military professionals must carefully distinguish between their responsibilities to objectively advise on the potential adverse impact of societal-driven change and their professional inclination to thwart initiatives that threaten to alter the status quo of military culture.*

Issue Area Six: Maintaining Linkages With Democratic Society At Large

Finally, much attention in recent years has been paid to the existence of a "civil-military gap." Tom Ricks set off a debate with his July 1997 cover story in *The Atlantic Monthly* detailing his conclusions based on field interviews of young Marines who felt alienated from their former friends and family upon their return home from boot camp.[82] There was a certain hostility toward civilian society evident in the soldiers' reaction to the undisciplined and perceived immorality of their civilian peers' behaviors and lifestyles. Ricks's book *Making the Corps* (1997) elaborated on his findings, sparking great interest in whether the military was no longer "of" society, but was becoming "separate" from society. Ricks's observations, combined with the concurrent work of Holsti and others documenting the shift in military officers' party identification with conservative Republicanism, has led to concerns that "instead of viewing themselves as the representatives of society . . . . officers believe they are a unique element within society."[83]

Social scientists have been busy at work examining whether the civil-military gap and its various manifestations exist, and determining the consequences and implications of the gap if it does exist. As James Burk has noted, "They [journalists, policymakers, and scholars studying the gap] ask whether there is a fundamental difference between the military and civilian society and reflect concern about the difficulty of establishing an effective working relationship between the two so both may flourish."[84] This observation hearkens back to the principle laid out at the onset of this chapter—that national security depends on the effective collaboration of the entire national security community, which commits its separate competencies to the task, yet recognizes the distinct responsibilities of each within the democratic framework.

Recommended Civil-Military Norm: *Military professionals must recognize the importance of fostering links with society to ensure that the military never becomes an entity separate from society, but remains always of society. Respect for American democratic institutions allowing for various manifestations of behavior in the civilian community that may be incompatible with military service but quite appropriate within a free citizenry should also be fostered. While military professionals properly may not want to exchange their value system with that of some elements of society, they should be taught to respect the process by which such divergences occur. While societal imperatives may at times be viewed as "anti-functional"* *because they emanate from societal forces rather than security needs, armed forces in service to a democracy must reflect to some degree the culture of the society they are sworn to defend.*[85]

Conclusion

This chapter has argued that, at present, the military profession in the United States does not subscribe to a single set of civil-military norms that regulate its participation in the policy-making process, the general political process, and even its relationship with the society it is charged to serve. Maintaining balanced civil-military relations that will best serve a democratic state's national security needs is a high-maintenance proposition, but essential to the achievement of high-quality national security policy outcomes. Infusing enlightened professional norms in this area would be a major contribution toward keeping U.S. civil-military relations on track toward correcting some troubling trends.

Specific recommendations for each issue area have been discussed throughout this chapter. Common principles link the specific recommendations. The adoption of a coherent set of civil-military norms depends on the profession's acceptance of two key principles. The first is that while there may be overlapping competencies regarding the political leadership's and military participants' expert knowledge relevant to national security, there are distinct differences in the responsibility and authority of each within the political system. The second principle evident across the issue areas is the military professional's obligation to balance the functional and societal imperatives. Principled officership must include the realization that the profession's service takes place within the context of a society with a particular set of political and social values and within a specific democratic political system with unique processes of civilian control.

All levels of professional military education must focus on equipping officers at every stage of their professional development with a set of guiding principles that will ultimately result in the establishment of a shared set of civil-military norms within the American military profession. Such measures will enable officers to sort out the ambiguities inherent in collaborating in the national security-making process, participating in a democratic polity, and remaining an entity that is "of" and not "separate from" the society it serves.

Notes

1. The views and opinions expressed in this chapter are those of the author and are not necessarily those of the Department of the Army or any other U.S. government entity.
2. Richard H. Kohn, "The Constitution and National Security: The Intent of the Framers," in *The United States Military Under the Constitution of the United States,* ed. Richard H. Kohn (New York: New York University Press, 1991), 87.
3. George Washington, "George Washington, Excerpt from 'Farewell Address' (1796)," in *100 Key Documents in American Democracy,* ed. Peter B. Levy (Westport, CT: Praeger, 1999), 70.

4. Ibid.
5. James Madison, "The Federalist No. 10," in *The Federalist,* ed. Jacobe E. Cooke (Hanover, NH: Wesleyan University Press, 1961), 62.
6. Ibid.
7. Joseph A. Schumpeter, *Capitalism, Socialism, and Democracy* (New York: Harper & Brothers, 1950), 253.
8. Peter D. Feaver, "The Civil-Military Problematique: Huntington, Janowitz, and the Question of Civilian Control," *Armed Forces & Society* 23, 2 (Winter 1996): 155.
9. Russell F. Weigley, "The American Military and the Principle of Civilian Control from McClellan to Powell," *Journal of Military History,* Special Issue no. 57 (October 1993): 27.
10. Marybeth Peterson Ulrich, *Democratizing Communist Militaries: The Cases of the Czech and Russian Armed Forces* (Ann Arbor: University of Michigan Press, 2000), 10.
11. Samuel P. Huntington, *The Soldier and the State* (Cambridge, MA: Harvard University Press, 1957), 2.
12. Ulrich, 10-11, 116-153.
13. I attribute the idea that a distinct form of professionalism exists in "national security professionalism" to Peter J. Roman and David W. Tarr. These scholars created this concept as a distinct form of military professionalism related to military professionals' expertise in foreign and security policy as it is carried out across the government. I have adapted the concept to apply to both civilian and military participants in the national security process. See Peter J. Roman and David W. Tarr, "Military Professionalism and Policymaking: Is There a Civil-Military Gap at the Top? If So, Does in Matter?" in *Soldiers and Civilians: The Civil-Military Gap and American National Security,* ed. Peter D. Feaver and Richard H. Kohn (Cambridge, MA: Belfer Center for Science and International Affairs, 2001), 409-11.
14. Recognition that civilian and military officials often have overlapping roles in the policy-making process is found in Christopher P. Gibson and Don M. Snider, "Civil-Military Relations and the Potential to Influence: A Look at the National Security Decision-Making Process," *Armed Forces & Society* 25, 2 (Winter 1999): 193-218.
15. Richard A. Chilcoat, "Strategic Art: The New Discipline for 21st Century Leaders," *U.S. Army War College Guide to Strategy* (Carlisle Barracks, PA: U.S. Army War College, 2000), 205.
16. Edward H. Carr, *The Twenty Years' Crisis: 1919-1939* (New York: Harper & Row, 1946), 1.
17. See Don M. Snider, Robert A. Priest, and Felisa Lewis, "Civilian-Military Gap and Professional Military Education at the Pre-commissioning Level," *Armed Forces and Society* 27, 2 (Winter 2001).
18. See Judith Hicks Stiehm, "Civil-Military Relations in War College Curricula," *Armed Forces and Society* 27, 2 (Winter 2001): 284-92. TISS survey data discussed in this article indicate that professional military education curricula need to clarify the distinctions between behavior that is advisory or advocative in nature—that is, merely offering counsel versus recommending in a pleading fashion. Officers' views on the appropriateness of "insisting" were also probed in the study. "Insisting" characterizes behaviors beyond counseling or forcefully recommending to include actions that demand vehemently and persistently. A significant number of officers surveyed replied that insistent behaviors were at times appropriate in civil-military relations. Finally, the term "political" is often unclear to officers. To many officers "political" tends to have a pejorative quality and is applied to anything pertaining to conflict among participants in democratic policy-making processes. A better understanding of the American political process and the legitimate role that officers play as participant-citizens and as contributors in policy-making could improve officers' comfort with processes that form the fabric of American political life.
19. Huntington, 2.
20. I attribute this insight to Lt. Col. John "Paul" Gardner, USA, U.S. Army War College Class of 2001.

21. Russell F. Weigley, "The American Military and the Principle of Civilian Control from McClellan to Powell," 42-46.
22. Forrest C. Pogue, "Marshall on Civil-Military Relationships," in *The United States Military under the Constitution of the United States, 1789-1989,* ed. Richard H. Kohn (New York: New York University Press, 1991), 205-206.
23. Pogue, 206.
24. David McCullough, *Truman* (New York: Simon and Schuster, 1992), 534.
25. Gaurav Kampani, "The Military Coup in Pakistan: Implications for Nuclear Stability in South Asia," *CNS Reports* (Monterey, CA: Center for Nonproliferation Studies, October 1999); available from *http://cns.miis.edu/pubs/reports/gaurav.htm*; Internet; accessed 11 September 2001.
26. Peter D. Feaver, "Crisis as Shirking: An Agency Theory Explanation of the Souring of American Civil-Military Relations," *Armed Forces and Society* 24, 3 (Spring 1998).
27. On resistance to force structure changes, see "Revolt of the Generals," *Inside the Ring,* 10 August 2001.
28. David Halberstam, "Clinton and the Generals," *Vanity Fair,* September 2001, 230-46.
29. Desch, 32.
30. Ibid.
31. A former Clinton administration National Security Council official quoted in Halberstam, 230.
32. H.R. McMasters, *Dereliction of Duty* (New York: Harper, 1997), 327-28.
33. George C. Wilson, *This War Really Matters: Inside the Fight for Defense Dollars* (Washington, DC: Congressional Quarterly Press, 2000), chap. 6.
34. Lt. Col. John W. Peabody, *The "Crisis" in American Civil-Military Relations: A Search for Balance between Military Professionals and Civilian Leaders,* Strategy Research Project (Carlisle Barracks, PA: U.S. Army War College, 2001), 23.
35. Morris Janowitz, *The Professional Soldier* (New York: The Free Press, 1971), p. 236-41. Janowitz's data stem from a 1954 survey of Pentagon staff officers. He readily admits that the survey question posed asked only for a response based on general political orientation—neither party identification nor the specific content of the conservatism was provided.
36. Ibid. Of those surveyed, 45.3 percent reported that they were a little on the conservative side. An almost equal number of respondents, 21.6 percent and 23.1 percent, reported that they were conservative or a little on the liberal side respectively (p. 238).
37. Ole R. Holsti, "A Widening Gap between the U.S. Military and Civilian Society?" *International Security* 23, 2 (Winter 1998/99): 11.
38. Thomas E. Ricks, "Is American Military Professionalism Declining?" *Proceedings* (July 1998); available from http://proquest.umi.com; Internet; accessed 12 September 2001.
39. Ibid.
40. Ole R. Holsti, "A Widening Gap between the U.S. Military and Civilian Society?" 13.
41. Ole R. Holsti, "Of Chasms and Convergences: Attitudes and Beliefs of Civilians and Military Elites at the Start of a New Millennium," in *Soldiers and Civilians: The Civil-Military Gap and American National Security,* ed. Peter D. Feaver and Richard H. Kohn (Cambridge, MA: Belfer Center for Science and International Affairs, 2001), 33.
42. "Worldview" is the term that Winslow used in her study. In this context, worldview refers to political ideology.
43. Donna Winslow, *Army Culture,* Report published in fulfillment of U.S. Army Research Institute Contract No. DASW01-98-M-1868 (2000), 13.
44. Morris Janowitz, *The Professional Soldier* (New York: The Free Press, 1971), 233.
45. Huntington, 230-31.

46. *United States Military Academy Strategic Vision–2010* (West Point, NY: Office of Policy, Planning, and Analysis, 1 July 2000), 7-8.
47. See Huntington and Desch.
48. For an extensive treatment of civilian and military competencies as factors influencing the contrasting power of military professionals and civilian policy-makers in the policy-making process, see Christopher P. Gibson and Don M. Snider "Civil-Military Relations and the Potential to Influence: A Look at the National Security Decision-Making Process," 193-218.
49. Elliott Abrams and Andrew J. Bacevich, "A Symposium on Citizenship and Military Service," *Parameters* 31,2 (Summer 2001), 19.
50. Eliot A. Cohen, "Twilight of the Citizen-Soldier," *Parameters* 31, 2 (Summer 2001); this argument runs throughout the article.
51. Ibid, 28. Cohen presents but vehemently disagrees with this perspective.
52. The courts are increasingly supporting this position. See the two articles by Lloyd J. Matthews in *Army* titled "The Army Officer and the First Amendment," January 1998, 25-34, and "The Voorhees Court-Martial," September 1998, 17-23.
53. Note again the principles of officership required for service in the 21st century in *United States Military Academy Strategic Vision—2010.*
54. See Steven Lee Myers, "The 2000 Campaign: The Convention; Pentagon Taking Opportunity for Show," *New York Times,* 28 July 2000; Richard H. Kohn, "General Elections; The Brass Shouldn't Do Endorsements," *Washington Post,* 19 September 2000, A23; Steven Lee Myers, "The 2000 Campaign: Support of the Military; Military Backs Ex-Guard Pilot Over Pvt. Gore," *New York Times,* 21 September 2000, A1; "Nonpartisan Military Best," *Omaha World Herald,* 16 October 2000, 6; David Wood, *Newhouse News Service,* "E-day Attack; Military Set to Invade the Polls; Observers Worry about Surge in Partisan Politics," *Times Picayune,* 20 October 2000, 5; David Wood, *Newhouse News Service,* "Military Breaks Ranks with Non-Partisan Tradition; Many in Service Turn to Bush, Reject Political Correctness," *Plain Dealer,* 22 October 2000, 16A. See also Ole R. Holsti, "Of Chasms and Convergences: Attitudes and Beliefs of Civilians and Military Elites at the Start of a New Millennium," 31-32.
55. Comment of an active duty Army General Officer in response to the *U.S. Army War College Survey: Retired Generals and Partisan Politics* (October 2000). Survey was conducted by U.S. Army War College student, Lt. Col. William R. Becker, as part of his Strategy Research Project, *Retired Generals and Partisan Politics: Is A Time Out Required?* (Carlisle Barracks, PA: United States Army War College, April 2000), 30.
56. Comment of a Marine one-star general in response to the *U.S. Army War College Survey: Retired Generals and Partisan Politics* (October 2000). See Becker, 37.
57. Comments of a Marine two-star officer in response to the *U.S. Army War College Survey: Retired Generals and Partisan Politics* (October 2000). See Becker, 41.
58. Comments of an active-duty four-star officer in response to the *U.S. Army War College Survey: Retired Generals and Partisan Politics* (October 2000). See Becker, 47.
59. Interviews with Army War College students regarding their experience in the field with retired general officers working with active units as contractors, Carlisle Barracks, PA, August 2001. During my own field research in Central and Eastern Europe (CEE) in July 2001, I encountered several retired U.S. general officers working as contractors for CEE governments or representing a particular defense corporation whose employment or access was based on active duty expertise and/or reputation.
60. Andrew Abbott. *The System of Professions: An Essay on the Division of Expert Labor* (Chicago, IL: University of Chicago Press, 1988), 102.
61. Huntington, 8.
62. Huntington, 16.
63. Ibid.

64. Becker, 47.
65. See *Joint Ethics Regulation (JER)* (November 2, 1994), 2-304, p. 29 for a description of restrictions placed on retirees with regard to capitalizing on their retired military status. Additionally, sections 672, 675, and 688 of *Title 10, United States Code* detail the authority to recall retirees to active duty.
66. Ibid.
67. Abbott, 60.
68. See Abbott, 69-89. Abbott offers five "settlements" that can be applied to jurisdictional disputes: (1) The claim to full and final jurisdiction; (2) The subordination of one profession to another; (3) Division of labor that splits jurisdictional areas into two interdependent parts; (4) Allow one profession an advisory role vis-à-vis another; (5) Divide jurisdictions not according to content of work, but according to the nature of the client.
69. For a more in-depth discussion of this conflict and its impact on military culture, see John Hillen, "Must U.S. Military Culture Reform?" *Orbis* 43, 1 (Winter 1999): 43-57.
70. James Burk, "The Military's Presence in American Society, 1950-2000," in *Soldiers and Civilians,* ed. Peter D. Feaver and Richard H. Kohn, 247-74.
71. Robert H. Ferrell, *Harry S. Truman: A Life* (Columbia, MO: University of Missouri Press, 1994), 298.
72. McCullough, 588-89.
73. Omar N. Bradley, *A General's Life* (New York: Simon and Schuster, 1983), 485.
74. Leo Bogart, *Project Clear: Social Research and the Desegregation of the United States Army* (New Brunswick, NJ: Transaction Publishers, 1992), xxi-xxii.
75. Ferrell, 588-89.
76. Jeanne Holm, *Women in the Military* (Novato, CA: Presidio Press, 1993), 307.
77. Holm, 264.
78. Holm, 305-12.
79. The House amendment passed 303-96 and the Senate amendment was accepted without a roll call vote. Holm, 310.
80. Ulrich, 11.
81. Peter D. Feaver and Richard H. Kohn, *Project on the Gap between the Military and Civilian Society: Digest of Findings and Studies* (Chapel Hill, NC: Triangle Institute of Security Studies, October 1999). These findings are also reported in Feaver and Kohn, *Soldiers and Civilians.*
82. Thomas E. Ricks, "The Widening Gap Between the Military and Society," *The Atlantic Monthly,* July 1997, 66-78.
83. Conclusion of Army Maj. Robert A. Newton who conducted a survey of Marine officers in 1995, as quoted in Ricks, "The Widening Gap Between the Military and Society."
84. James Burk, "The Military's Presence in American Society, 1950-2000," 247.
85. Hillen, 45.

13 Officer Professionalism in the Late Progressive Era[1]

Lance Betros

Many observers of contemporary civil-military relations in the United States claim that the professionalism of today's officer corps has declined, along with its time-honored tradition of political neutrality. The evidence they cite usually includes the high-profile efforts of former Chairman of the Joint Chiefs of Staff Colin Powell to influence political decisions on force structure, gays in the military, and peacekeeping missions.[2] Additionally, they point to the cases of open disrespect shown by military personnel toward President Bill Clinton, the affiliation of the majority of officers with the Republican Party, and the growing propensity of retired officers to endorse political candidates.[3] The bottom line, as they see it, is that the increase in political activity among officers has led to a decline in professionalism and a heightening of civil-military tensions.[4]

The purpose of this chapter is to examine civil-military relations in historical context by focusing on officer professionalism during the late Progressive Era, roughly the years of Woodrow Wilson's first administration. Although there are fundamental differences in the American political and security environments between then and now, there are also intriguing similarities. Like today, the United States was a major economic power enjoying an extended period of relative peace. Public interest in defense issues was low, and there was disagreement about the nature of military threats to the nation. Overseas constabulary missions obscured the focus on fighting wars, yet the officer corps for the most part enjoyed public acceptance of its professional legitimacy.[5] Finally, the political environment was highly partisan, with Republicans attacking Democrats for being weak on defense.

Officers of the late Progressive Era engaged in a wide range of politically oriented activities intended to advance the interests of the Army. Some of the activities were legitimate, such as congressional testimony and writings in professional journals. Others were less so, as they challenged and sometimes exceeded the norms of professional conduct. These latter activities included inappropriate writings in mass-circulation publications; speeches and interviews; informal contacts with members of Congress and the news media; liaison with political, civic, and business leaders; and involvement in interest groups and civic organizations.

Some historians argue that the officer corps had a long history of political activism and that all of the aforesaid activities were within the bounds of accepted behavior. To a degree this is true—throughout the nineteenth century officers routinely engaged in political activities to advance professional interests. These activities were generally indirect, nonpartisan, and therefore non-threatening to those who feared a powerful military establishment.[6] Moreover, the American constitutional system actually promoted such conduct. In separating the powers of the executive and legislative branches, the Founders blurred the channels of civilian control over the military. The result was to encourage political behavior among officers, who sensed the latitude to play one branch against the other without the need to align with a political party.[7]

Like their predecessors, officers of the late Progressive Era were politically active. They differed, however, in their willingness to engage in activities that bordered on partisanship. With surprising frequency, many of Wilson's officers tried to circumvent the channels through which they customarily provided information and advice. Instead, they attempted to mold public opinion, build political constituencies, and pressure politicians to enact legislation favorable to the Army. Their interests, though professionally oriented, put them in close alignment with the Republican Party and, as a consequence, cast a partisan shadow over their activities. Not since Reconstruction had the Army been so in tune with the priorities of a single political party.

Wilson's officer corps, while highly professional in many respects, engaged in partisan activities to an extent that modern officers would consider inappropriate and unprofessional. My intent in pointing this out is not to minimize the ill effects of officer partisanship today; rather, it is to view current civil-military tensions in the context of American history in hopes of offering insight into the nature and scope of the problem.

Officer Partisanship

The modern American officer corps took shape in the aftermath of the Civil War. U.S. military leaders, sensitive to the growing complexity of war in the industrial age, understood that improving officer professionalism was the key to a reliable military establishment. In particular, William Tecumseh Sherman, Commanding General of the Army from 1869 to 1883, emphasized education and training as a means of creating an officer corps capable of serving competently and providing expert counsel to civilian leaders. To that end, he reasoned, "no Army officer should form or express an opinion" on party politics.[8] Recognizing the harmful consequences of partisanship, Sherman and others like him exhibited a growing awareness of the officer corps as a truly professional body, one whose "code of conduct included unprotesting obedience" to civil authorities.[9] Samuel Huntington, in his 1957 classic *The Soldier and the State*, embraced Sherman's ideas and provided a compelling theoretical case against officer partisanship:

> Politics is beyond the scope of military competence, and the participation of military officers in politics undermines their professionalism, curtailing their professional competence, dividing the profession against itself, and substituting extraneous values for professional values. . . . Civilian control exists when there is this proper subordination of an autonomous profession to the ends of policy.[10]

While it is an article of faith that partisanship erodes professionalism, it is no less true that officers have political opinions. They comprise one of the most highly educated professional groups in America, and the variety of their experiences at home and abroad develops an appreciation for the connection between policy and practice.[11] More specifically, they understand that national security decisions made by the President and Congress affect them directly. Given the close relationship between the political decisions of civilian leaders and the day-to-day life of the military, it is no wonder that officers tend to be politically aware.

Woodrow Wilson's officers were no exception. As a group, they were well educated and professionally competent. They traveled extensively, especially to places like the Philippines, the Caribbean, and Mexico where the United States had undertaken military operations. Their social roots connected them with the nation's business and political leaders. The most articulate and ambitious sought to advance policies they believed would benefit the nation's security and, sometimes, themselves. Usually they opted for professionally accepted methods, but they strayed out of bounds with surprising frequency.

One of the most common methods of expressing their views was through the written word. Army officers were prolific writers in the early part of the century. Their work appeared most frequently in the professional journals that flourished in the aftermath of the Civil War. Targeted at a military audience, these writings provided a window into the thinking of officers on a multitude of issues. Most of the topics were mundane—from mess procedures in the field to life on an Army post—but they also included argumentative pieces on the weightier issues of the day. On the pages of publications like the *Army and Navy Journal*, *Journal of the Military Service Institution of the United States*, and the journals of the Infantry, Cavalry, and Artillery Associations, they debated among themselves the issues that would affect their lives and livelihoods.[12]

That officers wrote in professional journals was to be expected. More unusual was the frequency with which they wrote for the general public. Military writers contributed routinely to prominent magazines and newspapers, and published a surprising number of policy-oriented monographs. Some of the writing was of a purely historical, technical, or sentimental nature.[13] Oftentimes, however, it was argumentative, calculated to mold public opinion on issues of national security and influence the policy decisions of national leaders. In discussing a proposed article with the editor of *Collier's Magazine*, for example, Capt. John McAuley Palmer made clear his intent. The article was on a militia pay bill then being drafted in Congress, and he hoped to convince lawmakers to make revisions in the wording of the legislation. He thought it important that his "point of view should be presented to the public . . . to prevent the passage of a bad pay bill."[14] A few years later, Palmer's manuscript for a monograph extolling the virtues of universal military training caught the attention of Tasker Bliss, Assistant Chief of Staff of the Army. Bliss encouraged Palmer to publish the work "*without delay*" because it promised to build congressional support for legislation that would establish a mass army and a federal reserve force—an outcome feared by many Democrats in Congress. He opined that there "are a great many Congressmen who believe in the desirability of . . . national training . . . but hesitate to formally approve and advocate it." Palmer's book, he believed, would help secure a favorable outcome in the debate by mobilizing public opinion in the War Department's favor.[15]

Palmer was one of the more prominent military writers of the era, but certainly not the only one. Many other officers, particularly those chosen for duty on the General Staff, routinely submitted articles and books for publication. In his book *Social Attitudes of American Generals, 1898-1940*, Richard Brown surveyed the writings of the 465 Army officers who held flag-officer rank during that period. His footnotes and bibliography reflect an amazing breadth of writing by military authors in a wide range of publications. Brown found articles in a total of fifty-four different periodicals, *not counting professional journals*. Most popular was *The North American Review* with 68 separate articles, followed by *Scribner's Magazine* (24), *Saturday Evening Post* (22), *National Geographic* (22), *Independent* (20), *Annals of the Academy of Political and Social Science* (19), and *Vital Speeches of the Day* (19).[16] These periodicals catered to a segment of the population most influential in national politics. In many cases, the titles of the articles made clear that the generals' writings were argumentative, not expository, and intended to help steer the debate on defense policy.[17]

In addition to the article-length pieces appearing in periodicals, some officers wrote monographs on defense-related issues. Palmer's *An Army of the People* advocated a citizen-based army created through universal military training.[18] Maj. Gen. Leonard Wood published two books on various aspects of military preparedness, his favorite topic in the years immediately preceding the Great War.[19] Retired Gen. Francis Vinton's *The Present Military Situation in the United States* was yet another voice in the chorus of military preparedness advocates.[20]

Few Army officers wrote more prolifically than General Wood, Chief of Staff of the Army from 1910 to 1914. In addition to the books already cited, he wrote several articles on the need to enhance military preparedness and establish universal military training.[21] His writings appeared in the nation's most popular magazines and thus reached a wide and impressionable audience. He energized a large segment of the population, typified by Theodore Roosevelt, who envisioned a moral renaissance through universal military service. Likewise, he provoked the vehement opposition of members of the peace movement, who viewed the military as wasteful and immoral, and southern Democrats, who still harbored memories of the Union Army's pillaging of the Confederacy.[22] Divided though they were, both groups felt the impress of Wood on the national political debate.

Army officers wrote for public consumption with great frequency despite the fact that they well knew the ethical restrictions on such activity. As professional officers, their duty was to provide unbiased professional advice to civilian leaders, not advocate policy positions or shape public opinion. Yet they intended much of their writing to do just that. Captain Palmer wrote an article for the *Infantry Journal* that "frankly assailed" the National Guard's value as a military reserve force, a sensitive issue in 1915 when preparedness advocates were lobbying for a citizen army based on universal military training. The editors at first refused to publish the article. "They didn't say so," speculated Palmer, "but I think they regarded it as impolitic."[23] The fact that the editors of a professional journal would blanch at publishing Palmer's article reflected the awareness in the professional community that military writers should avoid crossing the line into affairs political.[24]

In case officers had any doubt about the impropriety of their writings, President Wilson clarified the matter. Annoyed with officers' public commentary on the situation in Europe following the outbreak of war in August 1914, he ordered all officers, active and retired, "to refrain from public comment of any kind upon the military or political situation on the other side of the water."[25] To Wilson it seemed "highly unwise and improper" that officers should attempt to publicly advance a political agenda or interfere, even if unintentionally, in the conduct of foreign policy.[26] A few months later, Secretary of War Lindley Garrison broadened the prohibition to include "giving out for publication any interview, statement, discussion, or article" on military policy, whether at home or abroad.[27] Such expressions of political opinion, he explained, would be "prejudicial to the best interests of the service."[28]

Despite the unequivocal guidance, Army officers found ways to continue writing. Palmer lobbied throughout 1915 for permission to publish *An Army of the People*, claiming disingenuously that the book was of a "general nature" and therefore did not violate the spirit of Wilson's and Garrison's orders.[29] His claim was false, but the book's potential benefits—mobilizing the pro-preparedness lobby to pressure Democrats into supporting universal military training—proved irresistible to the Army leaders. With Bliss's help, Palmer finally won approval to publish the monograph from Garrison, himself a supporter of universal military training.[30] Not coincidentally, the book was published about the time Congress began work on the National Defense Bill of 1916, which would address many of the issues raised by Palmer in his book.[31]

If Palmer went through official channels to get approval to publish, other officers were not so conscientious. Capt. Billy Mitchell had found his views on military preparedness to be popular with readers, and the financial rewards were welcome as well; hence, he must have been frustrated by the restrictions on military authors.[32] He confided to his mother the method he used for getting around the restriction: "We have to write under assumed names . . . as a recent order prohibited any public expressions of opinion on our part." As an afterthought he advised, "Do not say anything about it."[33] A few weeks later, after penning an article for *World's Work* that would appear in the February 1915 issue, he told her, "My article . . . probably will be the leading contribution. I will not sign my name to it."[34]

Another way officers circumvented the ban on publication was to act as consultants to sympathetic editors. Considering that the media elite was staunchly behind the idea of universal military training, officers did not have to look far to find an outlet for their views.[35] Capt. Frank McCoy corresponded regularly with the editor of *Outlook* concerning the magazine's editorial content. "If you find things to criticize," wrote the editor about draft editorials on military issues, "I hope you will be quite frank in your advice."[36] Following his assignment to a unit near the Texas-Mexico border, McCoy provided to Herbert Croly, a personal friend and the editor of the *New Republic*, his views on the military situation there. Croly kept McCoy's comments confidential, but he probably used them as the basis of an article that appeared in October 1915.[37] Billy Mitchell likewise maintained his media contacts. He bragged to his mother that the "stuff under the military expert . . . you see in the *Chicago Tribune* is what I give them."[38]

Some officers simply ignored the ban. Gen. John J. Pershing, in a featured article in *The New York Times Magazine*, advocated universal military training because it was "exceedingly improbable" that voluntary recruitment would satisfy the need for trained men in war. Pershing antagonized many Democrats when he dismissed the concerns of those who feared militarism. "No person with a grain of common sense," he argued, "really thinks this nation would be more likely to go to war because it happened to be prepared against aggression. It is only the peace-at-any price man" who thinks so.[39]

In addition to being prolific writers, military officers spoke publicly in a variety of venues. They were no more circumspect in their speaking than military authors were in their writing. Maj. Henry Allen, a nationally known equestrian and popular lecturer, used a Sportsman's Club dinner in 1912 as a forum to criticize proposed legislation that would have reduced the size of the cavalry force by one-third. After apologizing to the audience for discussing a topic that "borders a little on the domain of politics," he went on to encourage their support in defeating the bill.[40] Capt. George Moseley, a General Staff officer who had drafted a bill on compulsory military service, received in early 1916 an invitation from the Academy of Political Science to lecture on the topic.[41] He would have liked to do it; unfortunately, that year he "had to decline a number of invitations" to speak because the Secretary of War's gag order was still in effect.[42] No one was more outspoken than Leonard Wood; despite the Secretary's order, he held informal interviews, passed memos to newsmen, and gave speeches to social and civic groups. He took pains to conceal his activities—"Very frank talk, no reporters," he noted in his diary after one such encounter with the public.[43]

Military officers often used the press to advance their political views, but they did so at their peril. In March 1914, Col. William Gorgas, the Army surgeon who helped eliminate yellow fever in the Panama Canal Zone, went public in advocating the repeal of canal tolls. Gorgas's comments enraged Wilson, who ordered the officer censured. Secretary Garrison wrote Gorgas a "severe letter of condemnation" and reiterated his disapproval of "Army officers indulging in public utterance on political questions."[44] Frank McCoy used more subtle means of reaching the press. Opposed to legislation drafted by Democrats that he believed would have imperiled Army readiness, McCoy wrote to a sympathetic Harrisburg, Pennsylvania, newspaper editor to arouse media interest in the issue. He sent along copies of reports by the Secretary of the Army and Chief of Staff with the "hope that you will do what you feel is consistent" with the effort to defeat the bill.[45] It is unclear how the editor responded, but McCoy's actions left no doubt about his determination to use the press for partisan purposes.

The conservative partisanship of Army officers was attributable partly to their common background as sons of America's social and political elite. Seventy-six percent of general officers from 1898 to 1940 were the sons of businessmen, professionals, public officials, and military officers—in short, the most elite segments of society.[46] Prior to 1920, almost one half of the officer corps was from the Northeast, home to many of the nation's leading businessmen, scholars, and political leaders.[47] The region was strongly Republican at the time, and the party's views on military preparedness mirrored those of most officers.[48] Like their civilian counterparts, officers were predominately Anglo-Saxon, upper-middle class, and old-stock Americans. They were overwhelmingly Protestant, with two-thirds affiliated with the conservative and hierarchical Episcopalian and Presbyterian sects.[49]

Reflective of the common social profile of military and civilian elites was the coterie of men who, at various times, lodged at 1718 H Street in Washington, a short distance from the White House. Frank McCoy and two fellow bachelor officers first rented the apartment in 1907 to be close to their work.[50] They entertained frequently and soon found they had much in common with Washington upper crust—businessmen, journalists, politicians, government officials, and diplomats. During their temporary assignments away from Washington, they invited civilian friends to take up residence at the apartment. By 1918, a remarkable group of young men had resided at 1718 H Street. Besides the three Army officers, the "family" included Willard Straight (founder of the *New Republic* and a J. P. Morgan partner), Benjamin Strong (governor of the New York Federal Reserve Bank), Basil Miles (editor of *Nation's Business*), Arthur Page (editor of *World's Work*), Andrew Peters (Massachusetts congressman and Boston mayor), James Curtis (assistant secretary of the treasury), Henry Fletcher (future ambassador to Mexico), and Joseph Grew (ambassador to Japan). For many years "family" members maintained close friendships and relied on each other for advice and assistance in personal and professional matters. McCoy corresponded frequently with family members on many contemporary issues, and his opinions invariably found their way into articles and editorials in prominent journals. Likewise, his friends helped McCoy secure coveted assignments through their influence in elevated Washington circles.[51]

Capitalizing on the commonality of their backgrounds, professional officers found many venues for interacting with civilian elites. Some of the contacts were purely social—they mingled, for example, in upper-crust clubs and civic organizations like the University Club in New York, the Metropolitan Club in Washington, and the Algonquin Club in Boston.[52] They kept in touch with their Progressive-minded contemporaries, most of whom had similar visions for national security policy. Accordingly, Captain Moseley, still assigned to the General Staff, felt comfortable writing to Dr. Henry Drinker, president of Lehigh University, to commend him for his views on universal military training. Frustrated by Democratic opposition to the program in Congress, Moseley lamented that the Army would have to wait until "a new public sentiment, led by strong, righteous men, demands that our whole military institution be recast along modern lines."[53]

The military preparedness campaign from 1914 to 1917 provided an ideal opportunity for Army officers to rub elbows with the country's social and political elites, Republicans in particular. Despite some opposition from within the ranks, most officers supported universal military training as a means of expanding the size of the Army and improving the quality of the reserve force. Officers of the General Staff testified frequently to that effect in Congress.[54] Some officers, however, went beyond this legitimate form of expression by involving themselves in grass-roots efforts to raise public awareness and support and to influence political leaders. The rolls of the Association for National Service, for example, included many Regular

Army officers; indeed, the organization's letterhead stationery listed four of them as members of the association's National Advisory Council.[55] In another case, Gen. Leonard Wood was so supportive of the pro-military American Legion (not related to the post-1918 organization of the same name) that in February 1915 he allowed its national office to be collocated with his headquarters and detailed his aide as an advisor. The arrangement did not last long; because of the Legion's ties to the Republican Party, Secretary Garrison reprimanded Wood and ordered him to desist from associating with partisan groups in the future.[56] Though Garrison was himself a supporter of military preparedness, he could not condone overtly partisan efforts by the officer corps to achieve it.

Political Patronage

To this point, we have considered how officers of the late Progressive Era engaged in partisan activities to advance their political agendas. Sometimes, however, they dabbled in politics to reap personal, not institutional, gains. Officers in the peacetime army of the early twentieth century had a variety of grievances revolving around such issues as pay, allowances, leave, promotion, and duty station. As dissatisfied as they might be, they understood the constraints of the Army in remedying these problems itself. Some officers therefore took matters into their own hands by seeking the patronage of politicians, government officials, and other influential persons to help them secure preferment of one sort or another.[57]

The stagnation in promotions was a particularly sore point for many ambitious officers. Despite the decision of Presidents William McKinley and Theodore Roosevelt to select some relatively junior officers as brigadiers, the vast majority of officers waited many years for advancement in a promotion system based on seniority.[58] The frustrations associated with the long waits for higher rank prompted some officers to exploit their connections with political leaders in hopes of beating the system. The example of John J. Pershing, a captain selected by Roosevelt for promotion directly to brigadier general, thus bypassing over 700 more senior officers, was instructive. Although Pershing was undeniably an officer of great merit, many of his officer contemporaries and Roosevelt's political enemies attributed the selection to the preferment of his father-in-law, Senator Francis Warren, chairman of the Military Affairs Committee.[59] Another prominent general, Leonard Wood, clearly owed his rank to political connections. Starting out as a contract surgeon, he came to command the famed Rough Riders during the Spanish-American War and formed a close friendship with his second-in-command, then Lt. Col. Theodore Roosevelt. After the war his fortunes rose in tandem with Roosevelt's, and by 1910 he had become the Army's Chief of Staff. Wood had many enemies in the Army and Congress, but his use of political patronage to achieve high rank was viewed with envy.[60]

The examples of Pershing, Wood, and others were not lost on the rank and file. Col. Robert Bullard, an Alabamian who graduated from West Point in 1885, came to Washington in 1911 to attend the Army War College. His study of the art of war soon gave way to more practical pursuits, as he used the opportunity to line up endorsements for promotion to brigadier general. He worked his political contacts, concentrating on southern Democrats in Congress with whom his family had connections and, in at least one case, offering to assist in an upcoming political campaign.[61] The tactic seemed to work: in December 1911, the White House received endorsements for Bullard's promotion from ten members of Congress. Desiring the help of an even higher power, Bullard asked James Gibbons and John Ireland, Catholic bishops with influence in Washington circles, for their endorsements. Bullard, a convert to Catholicism, hoped to convince the prelates that the Church "was entitled to more representation among our generals." Perhaps Gibbons and Ireland were more sensitive than Bullard to maintaining proper civil-military relations, as they opted not to respond.[62]

Henry Allen, an 1882 graduate of West Point, had greater justification than Bullard for seeking political help in getting promoted. In 1912, with thirty years of distinguished service, he was still only a major, despite having served temporarily as a brigadier general as the Chief of Constabulary in the Philippines. Of his old friend Elihu Root, former Secretary of War and current senator from New York, Allen asked "whether you would see fit to assist me" in getting promoted to lieutenant colonel.[63] He very shortly received the higher rank. A few weeks later he was at it again, this time seeking the position of Chief of Insular Affairs, which came with the rank of brigadier general. Allen campaigned hard for the position, contacting several prominent members of Congress and asking bluntly, "Would you not make a personal request of the president in connection with this matter?"[64] Leaving nothing to chance, he penned another request for consideration and took "the liberty of writing this personal letter direct" to President Taft.[65] Despite the entreaties, Taft picked someone else for the job, and Allen would have to wait for the Great War to regain his rank as a general.[66]

The cases of Bullard and Allen were not unusual in the early twentieth century. Many ambitious officers chafed at the limited potential for advancement and resorted to politicking to enhance their chances. In this cutthroat environment, officers sometimes succumbed to the temptation of compromising their professional ethic for personal gain. Col. James Harbord recognized this even as he sought the endorsement of Kansas Senator Charles Curtis for promotion to brigadier general. Harbord explained that he "should not now appeal for your political aid, did I not know that my principal rival . . . is basing his efforts on influence."[67]

An executive order signed by Roosevelt in 1905 had specified that personnel actions within the Army and Navy would henceforth be dictated by "the official records of the War and Navy Departments, respectively, to the exclusion of other sources of influence."[68] Army officers hardly seemed to notice the order, and certainly Roosevelt's precedent of selecting junior officers for promotion emboldened many others to try their hand at winning preferment. Alarmed by the growing frequency of these efforts, the Secretary of War republished the executive order in 1913 and threatened future transgressors with punishment.[69] Amazingly, Brig. Gen. William Carter, one of the Army's most senior and respected officers, ignored the warning. He sent two letters to Secretary Garrison recommending himself for the position of Chief of Staff. Garrison's published response was swift and severe: "I would consider it just as much an impertinence for a man to tell me whom I was to marry . . . as for any one to tell me who should be Chief of Staff."[70] Needless to say, Carter did not get the job.

Securing patronage was not a partisan activity in the sense of aligning with political parties and molding public opinion. Regardless, it demonstrated the audacity of a generation of officers who felt comfortable manipulating the political system for personal gain. Additionally, it reinforced the habit of resorting to politics to achieve short-term objectives, whether personal or professional. The same officers who hobnobbed with the nation's social and political elites in search of patronage did not hesitate to use those contacts to advance the interests of the Army, even when their actions bordered on partisanship.

Measuring Professionalism

The extent to which Army officers of the late Progressive Era engaged in partisan activities was in curious juxtaposition to their reputation for professionalism. By Wilson's time, the officer corps had come a long way from the amateurism and incompetence that had characterized American military operations for much of the nineteenth century. They reflected the improvements in the Army's officer education system both at West Point, where the large majority of officers received their commissions, and at the Army staff schools that were teaching an increasing share of the officer corps.[71] Officers clearly conceived of themselves as professionals, and historians by and large have characterized them as such. How then could they have engaged so frequently and blatantly in partisan activities that were in violation of the military ethic? Answering this question requires consideration of the criteria for measuring officer professionalism.

Functionalist Critique

Over the past half-century, scholars have advanced a variety of models to define and characterize professions. Despite their many differences, the models generally fall into two main categories: "functionalist" and "constructionist." Samuel Huntington typified the former category. He defined a profession as an occupational group organized to meet important social needs through expertise, responsibility, and corporateness. Officers were professional, according to Huntington, to the extent they applied specialized knowledge for socially responsible purposes and adhered to corporate standards of thought and action—that is, the military ethic. An essential element of the ethic was a prohibition against partisan activities, which undermined professional expertise and the ability to render unbiased advice to civilian leaders. Only when officers acted responsibly by shunning politics and focusing on professional matters could they provide the most useful service to society.[72]

Huntington thought highly of the generation of officers that emerged in the late Progressive Era and fought in the Great War, while other scholars, using similar models of professionalism, agreed.[73] Given the evidence of officer partisanship in the late Progressive Era, however, a reassessment of military professionalism may be in order. The officer corps proved to be something less than the apolitical servants of the state envisioned by the functionalists; instead, many of them engaged in partisan behavior to advance Army priorities. There were at least four reasons for this breach of the professional ethic.

First, the partisan activities of senior officers encouraged similar conduct from the rank and file. Junior officers looked admiringly on men like Leonard Wood, who nurtured political constituencies sympathetic to the military. They flattered him by imitation, writing and speaking on a variety of political issues. Wood may have been the most overtly political officer of the age, but many other generals—as well as officers who would become generals in World War I and the interwar years—also expressed their views publicly in an effort to shape public policy. It was only when the Army's senior leaders came out unequivocally against partisan activities that the officer corps began to police itself more effectively; even then, there was a good deal of recalcitrance as old habits proved hard to break.

Second, the conditions that prompted officers to seek patronage likewise encouraged partisanship. During the late Progressive Era, the Army's seniority system left political patronage as the only means of bypassing the bottleneck in promotions. We already have seen how some officers resorted to unseemly tactics to secure political endorsement. Even when successful, they knew that patronage tarnished their professional standing by creating jealousies and ill will among officers and undermining the effort to promote only the most qualified men to positions of higher rank. Perhaps not as obvious, patronage reinforced patterns of thought and action that made engaging in partisanship less odious to officers pursuing favorable policy goals. In both cases—patronage and partisanship—officers overcame their professional scruples to advance short-term priorities at the expense of long-term institutional interests.

The ideological affinity between the officer corps and the Republican Party was a third reason why officers engaged in partisanship. Officers and Republicans shared a conservative-realist outlook on national security issues. They harbored a pessimistic view of human nature and believed in the inevitability of war; the outbreak of World War I in 1914 reinforced their convictions. As a result, both officers and Republicans embraced the cause of military preparedness and looked to each other for support. Wilson and his secretary of state, William Jennings Bryan, on the other hand, were archetypical liberal-idealists when it came to foreign policy, and Wilson's support of the military preparedness movement was belated and tepid at best. The divergence between the two parties on matters of defense was not lost on Army officers—of those who voted in national elections, the majority probably favored Republicans.[74] Capt. Ola Bell, West Point class of 1896, reflected the sentiments of many officers in 1916. "American prestige," he opined, "has sunken to a low ebb under the Wilson administration. We are regarded with contempt by the whole world. Our policies are all weak and vacillating."[75]

Finally, officer partisanship was largely due to the nature of the defense issues under debate in the late Progressive Era. The Army was grappling with a host of divisive issues far removed from the relatively tranquil days when its focus was fighting Indians on the western frontier. Officers could see clearly the need for greater military preparedness, particularly after the outbreak of war in Europe, but the majority of Americans, particularly Democrats, still clung to their traditional isolationism and distaste for a large standing army. Emotions ran high as preparedness advocates, civilian and military, tried to arouse the nation from its slumber. The outcome of the debate would have enduring effects on the size, composition, and orientation of the Army; officers were therefore deeply committed to influencing the debate, despite knowing the proper limits of professional military advice. In the view of many officers, the benefits of partisan action in pursuit of military preparedness outweighed the potential problems. It was a short-term perspective, but highly seductive to men who were committed to advancing the interests of the Army and, in their view, the nation.

Based on the functionalist criteria for measuring professionalism, officer partisanship undermined civil-military relations during the Wilson administration. In particular, the association of officers with civilian proponents of the preparedness movement strained the relationship between the military and some members of Congress, mostly Democrats. Viewing Army officers more as an interest group than as neutral public servants, these congressmen—many of them southerners predisposed to favoring the militia over a standing army—discounted military advice and voted almost reflexively against preparedness legislation.[76] They were wary of military counsel that seemed too closely aligned with the Republican Party, and military policy in the late Progressive Era suffered accordingly.[77] The Army was so clearly in tune with the opposition party that Wilson and other Democratic leaders could never be sure of officers' nonpartisan support of administration policies. This sentiment was at least partially responsible for Wilson's selection of Hugh Scott, a long-time family friend and confidant, as chief of staff over Tasker Bliss, a preparedness advocate considered to be the front-runner for the job.[78]

In addition to straining civil-military relations, partisanship undermined professionalism by eroding military expertise. Officers who immersed themselves in policy issues had less time to study the emerging developments in military science. Between 1914 and 1917, Europeans were learning hard lessons in combat that the U.S. officer corps should have noticed. Instead, the most gifted American officers busied themselves with politics and efforts to mold public opinion and influence defense policy. Had they directed more of their intellectual energies toward analyzing the tactical and operational developments of the Great War, they might have lessened the heavy casualties suffered by the American Expeditionary Force during the campaigns of 1918. Pershing's men fought bravely and with élan, but they entered the conflict with an infantry-oriented tactical doctrine—"open warfare"—that was ill suited to the dominating firepower of machine guns and high-explosive artillery and the defensive strength of enemy entrenchments.[79] The British and French, after millions of casualties of their own, had begun to modify their doctrine to the new conditions of war. The Americans never did so, despite having almost four years from the outbreak of war to the first major engagement of U.S. forces to make the necessary changes. Stubborn reliance on their own outmoded doctrine and a dismissive attitude toward European tactical techniques cost the United States dearly in the blood of its soldiers.[80] These failings were attributable to many factors, but prominent among them was the misplaced focus of the officer corps, whose collective mind had concentrated more on politics than on the art of war prior to 1918. When military leaders finally shifted their focus to the tactical and operational issues of war-fighting, it was too late to avoid the bloody consequences of their earlier neglect.

Constructionist Critique

In contrast to the functionalist critique of military professionalism in the late Progressive Era are the more positive views of the constructionist models. These models began appearing about a decade after publication of *The Soldier and the State* in 1957. They were skeptical of the functionalist assumptions that professions emerged to satisfy important social needs and that professionals acted in the best interest of their clients. Instead, constructionists advanced conflict-oriented theories that viewed the formation of professions as the outcome of social competition for control over abstract knowledge (i.e., expertise) within a field of work. Occupational groups become professional to the extent they are able to carve out and defend a "jurisdiction"—the domain within which their abstract knowledge is applied.[81] Professionals compete for jurisdiction by a variety of means, most effectively by demonstrating the ability to solve problems related to the occupation. If successful, they may convince society of the relevancy of their work and thereby solidify their professional status. These three elements—expertise, jurisdiction, and legitimacy—comprise the foundation of the constructionist model of professionalism.[82]

Officers of the late Progressive Era worked hard to acquire abstract knowledge related to their calling. They attended the service schools and joined the professional organizations that had emerged in the waning decades of the nineteenth century. As we have seen, they showcased their budding intellectualism in many articles, speeches, interviews, and other expressions of professional opinion. These efforts seemed to pay off—by the time of the Great War, Army officers had earned a reputation for professional expertise, which in turn heightened their legitimacy in the eyes of the public. It did not matter that they focused their intellectual efforts more on political than professional concerns; this shortcoming would not be evident until the entry of the United States in World War I.

In addition to their expertise and legitimacy, officers of the late Progressive Era successfully delineated and defended their jurisdiction. They saw it as their duty to educate America's intellectual, business, and political elites on military matters, and they used their social contacts with these groups to good effect. The General Staff, newly created in 1903, was the focal point of intellectual ferment among officers and the lead agent in the Army's outreach activities.[83] It occasionally released official reports to the press, universities, business and civic groups, boards of trade, and prominent citizens in an effort to influence public opinion and advance Army interests. Similarly, officers enlisted the help of sympathetic lobbying groups, such as preparedness organizations, to advance their cause.[84] These efforts, along with those of like-minded civilians, succeeded in consolidating the Army's jurisdiction over the land defense of the nation. When the United States entered the Great War, no organization competed with the American Expeditionary Force or challenged its authority to wage war on behalf of the nation. General Pershing enjoyed exclusive jurisdiction over the conduct of operations within his sector to an extent that would have made his counterparts in the latter half of the century envious.

Although the constructionist assessment of the officer corps accords with the laudatory opinions of most historians, the model itself is flawed. The constructionist model discounts the importance of self-sacrifice and social altruism in defining professions; it emphasizes, in place of these virtues, market-based measures of success. In the competitive, conflict-oriented environment assumed by constructionists, society bestows legitimacy on the professions that dispense their services most economically and efficiently. As sociologist James Burk points out, however, "Relying on forms of discourse borrowed from the marketplace, one cannot explain or justify self-sacrifice for the public good that military . . . service often requires."[85] Market-based legitimization is appropriate when there are questions about the value or allocation of professional services. In the case of national defense, however, moral considerations affecting such intangibles as leadership, unit cohesion, and combat effectiveness supersede purely economic considerations. Military professionals must think primarily in terms of *effectiveness* rather than *efficiency* even though doing so might put them at a disadvantage in the competition for legitimacy. This is the primary reason why constructionist models—currently in vogue in academic circles—have limited utility in measuring the professionalism of the officer corps. The officer corps should not ignore the effect of market forces on achieving professional legitimacy, but it must never be lured into changing its values-based professional ethic in response to political or social forces.[86] Were it to do so, the officer corps might win legitimacy as a profession but squander its effectiveness as a fighting force.

Nurturing Officer Professionalism

Given the political activism of officers during the late Progressive Era, we must disabuse ourselves of the notion of a mythic professionalism of an earlier age to which contemporary critics sometimes allude in debates over the current state of officer professionalism. The officers who wrote and spoke publicly on political matters, mingled with the nation's social and political elites, and sought political patronage were the same men who rose to high positions during and after World War I. Some were contemporaries of George Marshall, the patron saint of officer professionalism and the exemplar of apolitical service to the state. Many officers, however, were not "saints," insofar as they held strong opinions about political issues and acted on them in ways that today would be condemned as serious breaches of professionalism.

Additionally, assertions that the increasing partisanship of the modern officer corps has eroded professionalism must be balanced by the knowledge that officers of an earlier generation contended with related problems of their own. In many ways, today's officers adhere to higher professional standards than their forebears of the late Progressive Era. They generally shun overtly partisan activities, confine their writing and speaking to professional and scholarly forums, and adhere to published guidance on involvement in political activities.[87] Moreover, reforms in the Army personnel management system have virtually eliminated patronage as a factor in promotions, except at the very highest levels. All in all, the breaches of professional decorum that characterized the officers of the late Progressive Era are far less evident today.

There are many other ways, of course, that modern officers can transgress the boundaries of proper professional conduct, and the observers of civil-military relations have been thorough in exposing them. The sources of recent tensions are many, but dominant among them is the uncertainty concerning the nation's security requirements in the post-Cold War world. The stakes are high—they include a definition of the roles and missions of each branch of service and billions of dollars for corresponding weapon systems, force structure, and personnel. Interest in how these issues are resolved has led some officers to go beyond their advisory role and into the realm of advocacy and partisanship. Like the officers of the late Progressive Era, they sometimes overlook the consequences of partisan activities and focus instead on the potentially favorable outcomes of their actions.[88] This short-term outlook undermines officer professionalism and civil-military relations as much today as it did in Wilson's time.

The most effective way of encouraging professional behavior among officers is to buttress the institutional values of the military. In particular, senior leaders must continuously reinforce the military ethic, which embraces the notion of apolitical service to the nation and rejects partisan efforts to mold public opinion, build political constituencies, and influence policy decisions. In a hierarchical organization such as the military, the example of senior leaders in adhering to professional values is critical to the acceptance of those values throughout the institution. This is especially true when defense issues become the focus of partisan debate and officers find themselves, willingly or not, drawn into it. Senior leaders, through their example, mentorship, and authority, are in the best position to prepare the officer corps to overcome these challenges to ethical behavior.

While avoiding partisanship can be difficult for senior officers who work near the apex of national government, it is not impossible. The weighty responsibilities of these officers include advising civilian leaders, protecting service interests, and executing the national security policy of the nation. Routinely they are buffeted by powerful influences—party politics, competition between the executive and legislative branches, interservice rivalry, and lobbying by interest groups—that complicate their job and blur the line separating ethical from unethical behavior. Regardless, the line is still discernible and, with good judgment, they can stay on the proper side of it. Officers can render honest advice or testimony without being partisan so long as their comments go no farther than the venue in which the advice was sought. Moreover, they can provide opinions in the most partisan environments without appealing directly to the public or attempting to build political constituencies in pursuance of a favored policy. Engaging in such activities would cast them as politicians, a role military officers have no business playing.

In addition to the example of senior leaders, professional education is an important factor in encouraging ethical behavior. Army educational institutions, especially those at the pre-commissioning and junior-officer levels, should sharpen their focus on the historical study of the military as a profession. Officers should enter the Army understanding and accepting the values of the professional military ethic, and subsequent education should emphasize the subject. An important part of that understanding is the concept of responsible behavior in service to society, which implies the rejection of partisan activities. The sooner military officers embrace the professional ethic, the fewer problems they will encounter in dealing with the temptations to engage in partisanship during their careers.

Personal example and professional education are the methods of choice for encouraging proper behavior, but punishing transgressions is equally important. Making an example out of the relatively few officers who exhibit partisan behavior should dramatize to the rest the consequences of stepping outside the bounds of ethical conduct. The sanctions must be severe and applied consistently; otherwise, their utility will diminish and officers will view them as arbitrary. As evidence from the late Progressive Era made clear, officers flouted Army policies on partisan behavior when they sensed that the leadership vacillated in enforcing them or ignored the policies themselves. Not until senior leaders levied stiff sanctions against such behavior did the officer corps begin to pay attention.

Officer professionalism is the key ingredient in effective civil-military relations. It takes years to develop and must be nurtured continuously by the senior leaders of the Army. Military professionalism declines when the officer corps engages in partisanship and loses its focus on fighting and winning the nation's wars. This is something the critics of civil-military relations well understand, and they provide a valuable corrective to the bureaucratic complacency that may blind military and civilian leaders to underlying trends. Given the experience of the officer corps of the late Progressive Era, today's Army leaders would be wise to stay vigilant.

Notes

1. The views and opinions expressed in this chapter are those of the author and are not necessarily those of the Department of the Army or any other U.S. government entity.
2. Colin Powell, "U.S. Forces: Challenges Ahead," *Foreign Affairs* 72 (Winter 1992/93): 32-42; idem, "Why Generals Get Nervous," *New York Times*, 8 October 1992.
3. Peter D. Feaver and Richard H. Kohn, "Digest of Findings and Studies: Project on the Gap Between the Military and Civilian Society" (paper presented at the Conference on the Military and Civilian Society, Cantigny Conference Center, Wheaton, IL, 28-29 October 1999), 2; Richard H. Kohn, "General Elections," *Washington Post*, 19 September 2000.
4. Thomas E. Ricks, "The Widening Gap between the Military and Society," *Atlantic Monthly*, July 1997, 66-78; Russell F. Weigley, "The American Military and the Principle of Civilian Control from McClellan to Powell," *Journal of Military History* 57 (October 1993): 27-58; Charles Lane, "TRB from Washington: Green Line," *New Republic*, 15 November 1999, 6; Ole R. Holsti, "A Widening Gap between the U.S. Military and Civilian Society? Some Evidence, 1976-1996," *International Security* 23 (Winter 1998/99): 5-42.
5. Many Americans, in particular southern Democrats, still had a reflexive suspicion and dislike of the Regular Army dating back to the Civil War. Regardless, the officer corps, like many other emerging occupational groups in the Progressive Era, was gradually winning acceptance as a legitimate profession within American society. Allan R. Millett, "Military Officership and Professionalism in America," Mershon Center of the Ohio State University, Columbus, OH, 1977.

6. William B. Skelton, "Officers and Politicians: The Origins of Army Politics in the United States before the Civil War," in *The Military in America*, ed. Peter Karsten (New York: Free Press, 1980), 90.
7. Samuel P. Huntington, *The Soldier and the State: The Theory and Politics of Civil-Military Relations* (Cambridge, MA: Harvard University Press, 1957), 163-92.
8. Ibid., 232.
9. Weigley, "The American Military," 39.
10. Huntington, *The Soldier and the State*, 71-72.
11. The overall level of officer education has risen and fallen over the years, but it always has been relatively high. In 1988, for example, ninety-seven percent of Army brigadier generals possessed a post-graduate degree, compared to only thirty percent of top business executives. Michael Satchell, "The Military's New Stars," *U.S. News and World Report*, 18 April 1988, 33-40.
12. A survey of the professional writing of the early twentieth century by Army and Navy officers is in James L. Abrahamson, *American Arms for a New Century: The Making of a Great Military Power* (New York: The Free Press, 1981). See in particular the notes for chapters 5-8 and page 239 of his excellent bibliographical essay. See also the footnotes and bibliography of Richard C. Brown, *Social Attitudes of American Generals, 1898-1940* (New York: Arno Press, 1979).
13. An example of historical writing was the series of articles by Col. George Goethals on the building of the Panama Canal, *Scribner's Magazine*, March-June 1915. An example of technical writing was that by the Army's Chief of Ordnance, Brig. Gen. William Crozier, "International Settlements," *North American Review* 199 (June 1914): 857-65. An example of sentimental writing was the sympathetic article by Felix Frankfurter and Capt. George H. Shelton, "The 'War Record' of Stimson," *Boston Evening Transcript*, 12 March 1913, which appeared at the time of Stimson's departure from the War Department.
14. John McAuley Palmer to Arthur Ruhl, 26 December 1912, John McAuley Palmer Papers, Box 1, Manuscript Division, Library of Congress, Washington, DC.
15. Italicized words are underlined in the original document. Tasker Bliss to Palmer, 9 December 1915, Folder for "1915," Box 2, Palmer Papers.
16. Richard C. Brown, *Social Attitudes*, 411-14.
17. For example, Gen. William H. Carter left little doubt about his preference for a federal military reserve force when he penned "The Militia not a National Force," *North American Review* 196 (July 1912): 129-135.
18. John McAuley Palmer, *An Army of the People* (New York: The Knickerbocker Press, 1916).
19. Leonard Wood, *The Military Obligation of Citizenship* (Princeton, NJ: Princeton University Press, 1915); idem, *Our Military History: Its Facts and Fallacies* (Chicago, IL: Reilly and Britton, 1916).
20. Francis Vinton Greene, *The Present Military Situation in the United States* (New York: Charles Scribner's Sons, 1915).
21. Some of Wood's more prominent writings, in addition to those already cited, are "Training for War in Time of Peace," *Outlook* 90 (1909): 976-89; "Why We Have No Army," *McClure's* 38 (1912): 976-89; and "The Army's New and Bigger Job," *World's Work* 28 (1914): 75-84.
22. Charles Chatfield, *The American Peace Movement: Ideals and Activism* (New York: Twayne Publishers, 1992), 35-39.
23. Palmer to Capt. Frank S. Cocheu, 3 June 1915, Folder for "1915," Box 2, Palmer Papers.
24. The article in question eventually was published as "The Militia Pay Bill," *Infantry Journal* 11 (November-December 1914): 336-52.
25. War Department, General Order No. 60, 8 August 1914.
26. Ibid.
27. War Department, General Order No. 10, 23 February 1915.

28. Ibid.
29. Palmer to Cocheu, 3 June 1915, Folder for "1915," Box 2, Palmer Papers.
30. Telegram from Bliss to Palmer, 9 December 1915, Folder for "1915," Box 2, Palmer Papers.
31. The National Defense Act of 1916 turned out to be a victory for anti-preparedness activists, primarily because the bill retained the National Guard as the nation's first line of reserve forces. Arthur S. Link, *Woodrow Wilson and the Progressive Era, 1910-1917* (Norwalk, CT: Easton Press, 1988), 187-189.
32. William Mitchell to his mother, 1 or 8 September 1914, Box 19b, William Mitchell Papers, Manuscript Division, Library of Congress, Washington, DC. Mitchell noted that some publishers "pay for what I have to say, and while the opportunity offers I shall do all I can."
33. Ibid.
34. Ibid., 21 December 1914.
35. *World's Work* surveyed newspaper editors for their views on the nation's military preparedness. Of the 261 respondents, only six "showed any doubt of a need for stronger national defense." "The Press on Preparedness," *World's Work* 31 (November 1915): 26-31.
36. Editor of *Outlook* to Frank R. McCoy, 20 November 1914, Box 13, Frank R. McCoy Papers, Manuscript Division, Library of Congress, Washington, DC.
37. "On the Mexican Border," *New Republic*, 9 October 1915, 256-57; Andrew J. Bacevich, Jr., "American Military Diplomacy, 1898-1949: The Role of Frank Ross McCoy" (Ph.D. diss., Princeton University, 1981), 134.
38. Mitchell to his mother, 1 or 8 September 1914, Box 19b, Mitchell Papers.
39. John J. Pershing, "General Pershing Wants Every Man a Soldier," *New York Times Magazine*, 30 January 1916, 3.
40. Speech delivered by Allen to Sportsmen's Club, 19 February 1912, Folder for "1911-1919," Box 40, Henry T. Allen Papers, Manuscript Division, Library of Congress, Washington, DC.
41. Moseley drafted the bill (S.1695) for Senator George Chamberlain in December 1915. He also worked on subsequent revisions. George Van Horn Moseley Papers, Box 10, Folder for "1916," Manuscript Division, Library of Congress, Washington, DC.
42. Moseley to the Academy of Political Science, 1 May 1916, Box 47, Mosely Papers.
43. James W. Pohl, "The General Staff and American Military Policy: The Formative Period, 1898-1917" (Ph.D. diss., University of Texas-Austin, 1967), 255.
44. Ibid., 238, n. 53.
45. McCoy to Vance McCormick, 27 December 1911, Box 13, McCoy Papers.
46. Brown, *Social Attitudes*, Table III, 8.
47. Ibid., Table I, 3.
48. Pohl, "The General Staff," 274; Robert D. Ward, "Stanley Hubert Dent and American Military Policy, 1916-1920," *Alabama Historical Quarterly* 33 (1971): 180.
49. Ibid., 16.
50. At the time, the War, Navy, and State Departments were housed in what is now known as the Old Executive Office Building, immediately west of the White House. McCoy and his apartment mates worked in the War Department headquarters.
51. Andrew J. Bacevich, Jr., "Family Matters: American Civilian and Military Elites in the Progressive Era," *Armed Forces and Society* 8 (Spring 1982): 413.
52. T. Bentley Mott, *Twenty Years as Military Attaché* (New York: Oxford Press, 1937), 49.
53. Moseley to Drinker, 25 March 1916, Box 3, Moseley Papers.
54. See for example the hearing on universal military training, Senate Committee on Military Affairs, 64th Congress, 26 January 1916. During the hearing, three General Staff officers—Brig. Gen. William Crozier, Brig. Gen. Erasmus Weaver, and Capt. George Van Horn

Moseley—testified in favor of universal military training. Copy of hearing record found in Box 10, Moseley Papers.

55. Several examples of the letterhead stationery are in Folder "A," Box 10, S. B. M. Young Papers, Manuscript Division, Military History Institute, Carlisle Barracks, PA. The four active-duty officers were Brig. Gen. William Crozier, Col. G. W. Read, Maj. Charles Howland, and Maj. James A. Moss.
56. John Garry Clifford, *The Citizen Soldiers: The Plattsburg Training Camp Movement, 1913-1920* (Lexington, KY: The University of Kentucky Press of Kentucky, 1972), 51.
57. Marcus Cunliffe, *Soldiers and Civilians: The Martial Spirit in America, 1775-1865* (Boston, MA: Little, Brown, 1968), 289.
58. Weigley, *History of the United States Army* (New York: Macmillan, 1967), 168-69, 291.
59. See *Army and Navy Register* 51 (22 June 1912): 6, for a report of a particularly strident denunciation of the politics behind Pershing's elevation to brigadier general.
60. Ironically, Pershing was one of Wood's staunchest critics. During World War I, he accused Wood of being "insubordinate" and a "meddling political general"; for this reason, Pershing refused to allow Wood to serve in the American Expeditionary Force. "Pershing, a Trail Breaker, Always Stirred Controversy," *Washington Post*, 16 July 1948, 4. Found in folder on "General Pershing," Hagood Family Papers, Johnson Hagood II, Military History Institute, Carlisle Barracks, PA.
61. Allan R. Millett, *The General: Robert L. Bullard and Officership in the United States Army, 1881-1925* (Westport, CT: Greenwood Press, 1975), 229.
62. Ibid.
63. Allen to Elihu Root, 14 February 1912, Folder for "1912," Box 10, Henry T. Allen Papers.
64. See letters from Allen to Senator William O. Bradley, 8 May 1912; from Allen to Senator Murphy J. Foster, 3 December 1912; from Representative Oscar W. Underwood to President Taft, 17 and 18 April 1912, Folder for "1912," Box 1, Allen Papers.
65. Letter from Allen to Taft, undated, Box 10, Folder for "1912," Allen Papers.
66. Allen eventually attained the rank of lieutenant general and commanded a corps during World War I. Heath Twichell, Jr., *Allen: The Biography of an Army Officer* (New Brunswick, NJ: Rutgers University Press, 1974).
67. Harbord to Curtis, 26 January 1909, Folder for "Private Letters," Box 3, James G. Harbord Papers, Manuscript Division, Library of Congress, Washington, DC.
68. Theodore Roosevelt, Executive Order, 7 July 1905, published as War Department General Order No. 112, 13 July 1905.
69. War Department, General Order No. 31, 24 April 1913.
70. Pohl, "The General Staff", 227; "To Change Chief of Staff," *New York Times*, 12 December 1913, 1.
71. During World War I, ninety officers were high-level commanders (commander-in-chief; army, corps, and division commanders; commander of Services of Supply). Of these, 90.2 percent were West Pointers, and 55.5 percent had graduated from at least one of the Army's staff schools. Army War College Curricular Archive, File 68-58, "Memorandum to the Commandant," 31 October 1923, United States Army Military History Institute, Carlisle Barracks, PA.
72. Huntington, *The Soldier and the State*, 70-74.
73. Other functionalist models of officer professionalism are in Morris Janowitz, *The Professional Soldier: A Social and Political Portrait* (New York: The Free Press, 1974); and Allan R. Millett, "The American Political System and Civilian Control of the Military," Mershon Center of the Ohio State University, Columbus, OH, 1979. Both Janowitz and Millett characterize the officer corps of the late Progressive Era as being highly professional.

74. It is well known that few officers voted in national elections in the early twentieth century. Specific information on the percentage that voted and their party affiliation, however, is virtually unrecoverable prior to the 1960s. Local election laws governed voting procedures, and there were no federal statutes safeguarding the rights of military personnel to vote by absentee ballot. The conclusion that most officers favored Republicans is based on my examination of the personal papers of many officers of the late Progressive Era.
75. Diary entry, 8 November 1916, Box 3, Ola W. Bell Papers, Archives Division, United States Military Academy Library, West Point, NY.
76. Ward, 180; Link, *Woodrow Wilson and the Progressive Era*, 180-84; James Middleton, "Congressional Leaders and Our Preparedness for War," *World's Work* 31 (November 1915): 22-25.
77. John Patrick Finnegan, *Against the Specter of the Dragon: The Campaign for Military Preparedness, 1914-1917* (Westport, CT: Greenwood Press, 1974), 3-4. Finnegan argues that the preparedness movement "did not get the United States ready to intervene in World War I," despite the efforts of civilian and military leaders in advocating the cause.
78. "General Scott Is Close Friend of the President," undated newspaper clipping from the [New York?] *Herald*, Folder for "1912-13," Box 4, Hugh L. Scott Papers, Manuscript Division, Library of Congress, Washington, DC .
79. Rod Paschall, *The Defeat of Imperial Germany* (Chapel Hill, NC: Algonquin Books of Chapel Hill, 1989), 167-169
80. Ibid., 191-192; "The High Tide and the Turn: The Western Front Battles of 1918," *British Army Review* 119 (August 1998): 83-92; Holger H. Herwig, *The First World War: Germany and Austria-Hungary, 1914-1918* (London: Arnold, 1997), 416-28.
81. Andrew Abbott, *The System of Professions: An Essay on the Division of Expert Labor* (Chicago, IL: University of Chicago Press, 1988).
82. For an excellent summary of functionalist and constructionist models of professions, see James Burk's chapter titled "Expertise, Jurisdiction, and Legitimacy of the Military Profession" in the present book (Chapter 2). Burk prefers the constructionist model, although he criticizes it for emphasizing only expert knowledge and jurisdiction. Consequently, he refines the model by specifically adding "legitimacy" to the elements defining a profession.
83. See the adulatory article by L. Ames Brown, "The General Staff," *North American Review* 206 (August 1917): 229-40.
84. Abrahamson, *American Arms for a New Century*, 147-50; Mary T. Reynolds, "The General Staff as a Propaganda Agency, 1908-1914," *Public Opinion Quarterly* 3 (July 1939): 407.
85. Burk, "Expertise, Jurisdiction, and Legitimacy."
86. Ibid.
87. Department of Defense Directive 1344.10, "Political Activities by Members of the Armed Forces on Active Duty," 15 June 1990.
88. A counterargument to this critique is that senior military officers, conditioned by the experience of the Vietnam War, have learned to be forceful and unequivocal in presenting their opinions to civilian leaders. Such dialogue may be construed as advocacy or insistence when made public, when in fact it is within the officers' conception of responsible and ethical behavior. Former Chairman of the Joint Chiefs of Staff, John M. Shalikashvili, noted that the Joint Chiefs during Vietnam "fell too far under [Defense Secretary Robert] McNamara's control, rather than make the tough stands and let the President, Congress, and civilians in the Defense Department know the military's viewpoint. He then added, "A lot of us did take that lesson out of Vietnam." Quoted in James Kitfield, "Standing Apart," *National Journal*, 13 June 1998.

IV

Ethics and the Army Profession[1]

Any discussion of the state of the Army's expert knowledge and practice of its own ethic is usually very intense, particularly within the officer corps. The traditional view of military professions, certainly the Army profession, focuses sharply on the institution's ethos and the responsibility of soldiers to adopt and police it.[2] These three chapters tackle this very subject, presenting broadly different perspectives ranging from a straightforward derivation of the moral obligations of Army officers, to a Catholic-informed theology of military ethics, to an investigation of the pragmatic, moral obligations of professionals to preserve and defend their expert knowledge within the nation's political structures.

However, even from these different perspectives, each chapter reviews the current state of the Army's professional ethic and its influence, particularly as regards a "culture of candor," the norm of truth-telling by all members of the profession at all times. As we consider the well-documented gap in trust between junior and senior officers, we must ask how it is that professionals who risked their lives in combat in the junior grades are subsequently conformed all too often to risk-averse decision-making as they rise within their profession. In the field, after-action reviews are generally conducted with a healthy respect for candor. But when dealing with the Army's management systems and political leaders, candor is often sacrificed for the sake of various needs: those of the system (e.g., for high readiness or fuller expenditures); those requiring "fairness" (e.g., inflating evaluations); and those supposedly supporting a "better" Army (e.g., "can do" attitude and micromanagement to zero defects).

This section's authors also propose a return to first principles: the nature of the profession is such that only moral soldiers can discharge their professional duties, and the Army's strategic leaders are morally obligated to the client to maintain a profession of both competence and character. America's soldiers must have, and indeed deserve, moral leaders who are technically able. Leaders not thus qualified are unfit for command. But how do we develop moral soldiers and leaders? Morality cannot be taught via rote rules, clever acronyms, or other such centralized bureaucratic approaches. An essential component of the Army's transformation must thus be an enlightened program for character renewal among Army professionals at all levels, including the reestablishment of a culture of candor, one that will conform Army management systems and external relationships to the profession's essential nature.

Notes

1. The authors express their appreciation to Maj. Douglas V. Ollivant of the USMA Department of Social Sciences, who served as Panel Rapporteur for this section's topic during the USMA Senior Conference, June 2001.
2. See Field Manual 1, *The Army* (Washington, DC: Department of the Army, 14 June 2001), 6-10.

14 The Ties That Bind: The Army Officer's Moral Obligations[1]

John Mark Mattox

Introduction

A recent study by respected historians Donald and Frederick Kagan summarizes the effects of United States defense policy in the 1990s as follows: "Beginning in 1994, reports began to surface that the readiness of the armed forces to fight a war on short notice was eroding—training could not be paid for, equipment was breaking down, overworked people were burning out and leaving the services. The House and Senate armed forces committees began taking testimony and uncovered many problems. By 1998 it was widely recognized that the readiness of the armed forces had eroded seriously."[2]

In geopolitical terms, these historians conclude that "the international situation has already begun to slip from our control. The [1991Gulf War] coalition has shattered; NATO and the United States risk drifting apart. Challengers to the status quo proliferate, along with weapons of mass destruction and the means to deliver them."[3] Moreover, they believe that such measures as are now being taken to remedy the situation are far from adequate: "The military deficit is probably already too great for any 'reasonable' politician to contemplate, and it will only get worse. Not only have readiness and 'quality of life' issues (which affect recruiting and retention of qualified people) suffered badly, but the experiences of even the past few years show that the armed forces are too small. Worse than that, since at least 1995, the Department of Defense has been forced to reprogram money earmarked for modernization toward readiness and current operations. The systems that America will need to fight the major war of 2010 or 2020 are not now in place and are being developed too slowly or not at all. America is not ready now to face a major challenge, and current plans, even in light of current proposals to increase the defense budget, will not make it ready to face a major challenger in the future."[4]

Indeed, since the beginning of the post-Cold War era, the civilian masters of armies throughout the Western world—not alone the United States—seem to have found themselves largely at a loss as to what to do with their sizeable standing armies. For example, they have wondered what their armies' proper jurisdiction should be, what place their armies should occupy in their societies, and what kind of moral standards should be expected of the members of those armies. Evidence of this general air of uncertainty has typically manifested itself in three ways:

1. Drastic defense budget reductions based on the assumption that armies of Cold War proportions are no longer needed;

2. Assignment to those armies of missions well outside the scope of the labor traditionally reserved for armies alone; and
3. Formulation of foreign policies that seek full economic engagement with the world—an engagement totally dependent on international tranquillity—without accounting adequately for the self-serving agendas of power-hungry despots ready to expand their influence as soon as they realize that they will be unopposed militarily.

The answer to questions about what to do with or expect of armies in the twenty-first century may, in fact, be knowable only as more obvious threats to world peace emerge than now seem to exist. However, recognition of this uncertainty does little to assuage the concerns of reflective U.S. Army officers interested not only in the day-to-day performance of individual military duty, but also interested in the long-term welfare of the nation and its Army. Such reflection is healthy and productive for leaders committed to the proposition that the institution entrusted to their care by the nation must survive the test of time. Moreover, the act of questioning what the Army's role should be and how it should set out to accomplish the tasks that lie ahead is certain to produce, as it already measurably has, a desire on the part of its leadership to transform the Army as necessary to ensure that it is relevant to the times. As the 1999 Fletcher Conference observed, "The Army has proclaimed that 'Everything is on the Table' as it pursues transformation."[5] Indeed, *everything should* be on the table if the Army is to gain an honest assessment of its current status and make the changes necessary to maintain its "non-negotiable contract with the American people to fight and win our Nation's wars"[6] in a rapidly changing world.

However, the fact that the Army has placed everything on the table as it undergoes its transformation does not mean, and cannot mean, that everything necessarily must change. Reflective Army leaders certainly realize this as they confront questions of the kind that the leaders of any enduring profession must answer:

1. What does it mean to belong to a profession? That is to say, what obligations does membership in the profession entail?
2. Are there any "fixed points," i.e., professional commitments that never change, or does everything that it means to be a member of the profession hinge ultimately on factors external to the institution (in the case of armies, for example, the international situation or the perceived threat, or perhaps on less glamorous notions such as supply and demand, budget cycles, and the like)?
3. What enduring moral obligations are entailed by membership in the profession? In the case of the U.S. Army, the question becomes, "What is the nature of the moral bond of obligation—if, indeed, such a bond actually exists or should exist—between the U.S. Army officer corps and the nation it serves and the soldiers it leads?"

This chapter will argue that, no matter how extensively the Army must change, adapt, or transform in order to be prepared to meet the uncertain challenges of the future, the answers to the moral-philosophical dimension of these questions *must not* change. Furthermore, the U.S. Army officer corps must feel safe in the assurance that the answers *will not* change.[7] Not only must the answers to these questions remain unchanging in the face of the institution's grand transformation, but the answers given to these questions all must point to the following conclusion: *By the nature of the profession of arms, only officers of firm moral character can discharge adequately their professional obligations to the nation and to their subordinates that they are called to lead.*

While it is gratifying that some (hopefully most) Army officers may find this claim to be self-evident and thus not one requiring a rigorously argued defense, it is at the same time true that potent cultural forces are at work that could have the effect of forcing a radical reinterpretation of the nature of the officer corps' bond of moral obligation. With this in mind, let us examine the philosophical grounds for moral obligation within the officer corps.

The Grounds of Moral Obligation

The officer corps' moral bond of obligation to the nation and to the Army's soldiers ultimately derives from two sources: (1) the essential and distinguishing characteristics of the profession of arms, and (2) the obligations freely incurred through the oath of office by which officers bind themselves to the Constitution. These will be considered in turn.

The Place of the Profession of Arms among the Professions

A well-established corpus of literature exists on the subject of what constitutes a profession and which human labors properly can be called professions.[8] In this context, the question periodically arises as to whether, properly speaking, the "profession of arms" may, in fact, be considered a profession. However, in order to answer that question, there first must be some agreement as to the nature of professions in general.

For the purposes of this chapter, it is not necessary to establish the status of every human endeavor that aspires to be labelled a profession. It merely is necessary to demonstrate that the profession of arms falls completely within that set of human endeavors that may be labelled "the set of all professions" (see Figure 14-1). Given the premise that any human endeavor that falls completely within **B**, the set of all professions, is itself a profession, then it follows that any human endeavor that falls completely within a subset of **B** is likewise a profession. In this case, the subset of **B** at issue is **C**, the set of those segments of society that are recognized widely (if not universally) as legitimately performing functions directly associated with the preservation or termination of human life or freedom in ways accepted by just societies. The qualifying phrase "in ways accepted by just societies" must be added to exclude organizations like the Mafia that engage in the termination of human life but do so illicitly. (For example, the Mafia might claim that, within the context of Mafia subculture, the terminating of human life is legal and otherwise socially acceptable.

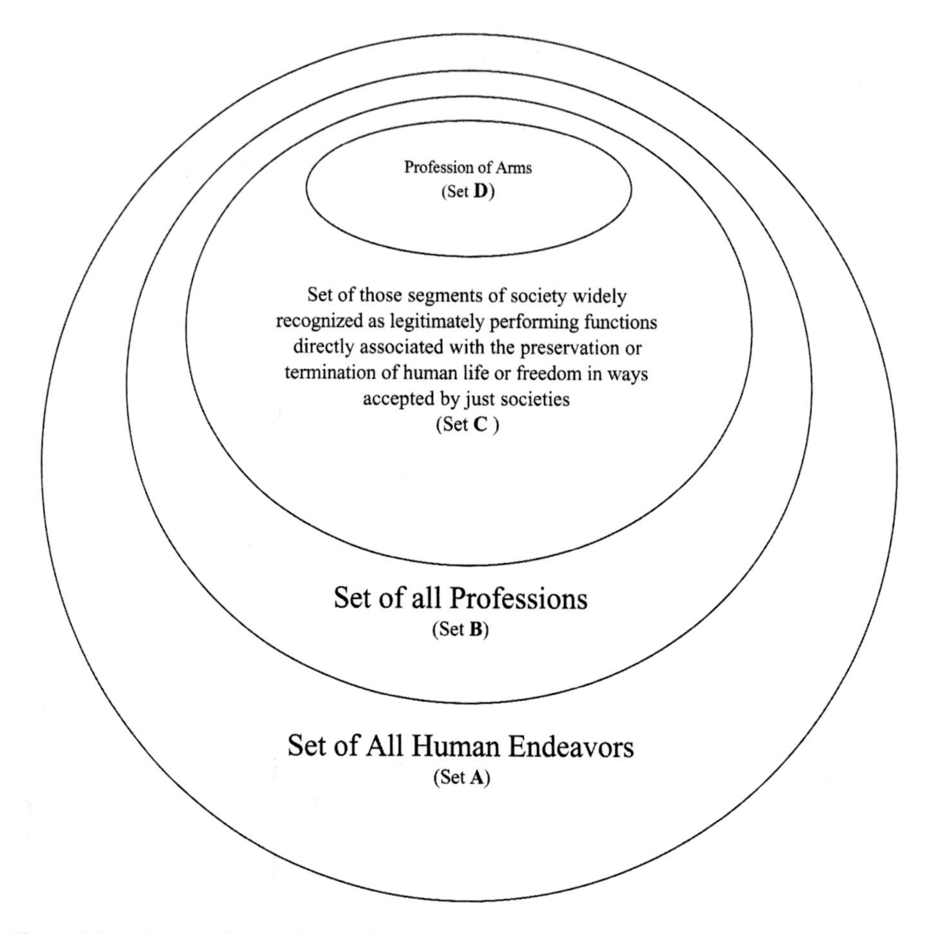

Figure 14-1. *Sets Enclosing the Profession of Arms.*

To this we may reply, without any harm to the argument, that the killing of Jews may have been legal and otherwise socially acceptable in the context of Nazi culture, but that any reasonable, morally sensitive person might be expected to conclude that the Mafia, the Nazi Party, and other organizations like them can be excluded from **C** because of their inherently evil nature.)

This argument embraces the premise that all members of **C** are themselves professions. In this connection, it is worth noting that at least some of the confusion over what counts as a profession exists for semantic reasons rather than logical ones. This is so because of the unfortunate practice in popular culture to elevate the social status of almost any human endeavor by labelling it "professional." Thus, one commonly hears of "professional" hair stylists, "professional" air conditioner installers, and even "professional" golfers and bodybuilders, by which is meant that these pursuits are undertaken on a more or less full-time basis as a source of income.

The confusion could be avoided altogether by applying to these pursuits the traditional vocational titles of "apprentice," "journeyman," and "master" that have been used by tradesmen over the centuries. For example, one can readily imagine an apprentice hair stylist, a journeyman plumber, or even a master golfer or bodybuilder. However, to apply in common, non-technical discourse the term "profession" to a pursuit that has little or nothing to do with the central questions of human existence does a double disservice. First, it artificially elevates the status of some pursuits whose existence is not truly fundamental to a well-ordered society (many societies have functioned without hair stylists, plumbers, and golfers, but few if any have functioned without some form of medical care, legal order, military protection, etc.). Second, it obscures the crucial point that some professions exist in the first place because society has recognized them as essential for the promotion of the public good.

Accordingly, within **C** one finds, among others, the medical profession and the legal profession because both are recognized as having the responsibility to adjudicate issues involving the preservation or termination of life or freedom. However, since **D**, the "profession" of arms, also has as its principal function the preservation or termination of human life or freedom in ways accepted by just societies, one may conclude that **D** is a subset of **C** and therefore also a subset of **B**. Accordingly, given these assumptions, the human endeavor generally denominated the "profession of arms" is, in fact, a profession.

Notice that the logical claim is not that all professions (i.e., the members of **B**) are members of **C**, nor is it necessary for them to be. In fact, it probably is not the case that all things that properly can be called "professions" have as their central function the preservation or termination of human life or freedom. There may be many human endeavors that do not have the central identifying characteristic of **C** but that may be called professions on other grounds. However, it seems reasonable that those segments of society that generally are regarded as having a legitimate claim over matters of life and death, of bondage and freedom, appropriately may be called professions. Accordingly, this argument does not purport to establish the identity of all members of **B**, but merely to establish that among its members, one necessarily finds **D**, the profession of arms.

In a way that is particularly important in light of the aims of the present chapter, it should be noted that regardless of what other sets **D** may intersect or be a subset of, **D** definitely is a member of **C**. This is significant because, as even minimal reflection will reveal, the professions that belong to **C** have *moral* obligations, i.e., obligations derived from the fact that the principal functions of these professions affect the happiness, well-being, flourishing, and, potentially, the life or death of human agents. Indeed, nothing could be more central to the moral enterprise than those matters most directly affecting human existence. Hence, by virtue of its membership in **C**, it must be concluded that **D** is not only a bona fide profession, but also a profession with moral obligations. Thus can be established the claim that Army officers have moral obligations that derive, in part at least, from the fact that they are members of a profession entrusted by society with making decisions that concern the preservation or termination of human life. This is not to say that the profession of arms has *only* moral obligations, nor is it logically necessary to insist that such is the case. It merely is necessary to establish that, among the professional obligations that Army officers have, some of those obligations are properly denominated *moral* ones.

The Place of the Profession of Arms in Society

Since ancient times, the Western intellectual tradition has held that the highest quality of human life could best be realized within the context of the state—the highest form of society. This view was common among the greatest philosophers of ancient Greece and Rome. For example, Cicero regarded the preservation of the state as essential to the moral and physical well-being of the human race. Should this highest form of society cease to exist, man simply would become unable to flourish to the degree otherwise possible, or perhaps at all. Hence, by Cicero's account, the death of any man is less to be lamented than the extinction of the state.[9] In Augustine's paraphrase of Cicero, "Death often rescues individual men from pain, instead of being a disaster to them; but the death of a whole community is always a disaster."[10] Thus, in order to ensure the preservation of this highest form of society, it would seem that, as long as the present bellicose state of human affairs remains, armies will be a sine qua non for any stable state. Similarly, Plato, writing in the *Republic*, pointedly argues that even the ideal state must be prepared to secure itself by having "a special class of guardians"[11] exclusively charged with conducting the "business of fighting"[12] (an enterprise which, one might note with interest, he regards as both an "art and a profession"[13]).

Even Augustine, who takes a radically different view of the state than that found in Plato or Cicero, argues that armies are understandable, if not essential, features of the best of states. The "earthly city," as he refers to human society, never can become an ideal state; that eventuality is reserved for the "city of God" that cannot have its complete realization among fallible human beings. Nevertheless, the earthly city properly has an army to preserve order within, to defend it from without, and even by Augustine's account to serve as an instrument of divine justice. If nothing else, the army properly exists to enable the righteous citizens of the earthly city to cope with their imperfect circumstances until such time as they can partake of perfect justice in the city of God. Whether or not one embraces Augustine's transcendental view (a view which underwrote much of the social and political thought in the West for over one thousand years between Augustine and the Peace of Westphalia), it is nonetheless interesting that even Augustine's worldview includes a privileged place for the profession of arms.

This view of the centrality of the state and the necessity to defend it is by no means restricted to the ancients alone. In fact, no place is this idea plainer than in the writings of the English philosopher Thomas Hobbes (d. 1679). For Hobbes, all human beings live in a constant state of fear of death because of the unremitting forces of the "state of nature," where all humans find themselves as combatants in "the war of all against all": "During the time when men live without a common Power to keep them all in awe they are in that condition called Warre; and such a warre, as is of every man against every other man."[14] According to Hobbes, human beings find refuge from this chaotic world only as they bind together in societies in which they submit their agency to the will of a sovereign (be that one person, such as a monarch, or an assembly of persons, such as in a republic). In exchange for the promise to obey the sovereign, the sovereign "guarantees" (to the extent that it is humanly possible) the security of the members of society by protecting them from the encroaching forces of the state of nature. The sovereign accomplishes this by raising an army. Without such a mechanism to guarantee the security of the state, life is bound to become, in the famous words of Hobbes, "solitary, poore, nasty, brutish, and short."[15]

Armies and Moral Obligation

If the Hobbesean argument is extrapolated one step further, a particularly striking conclusion arises: Since, according to Hobbes, ethical notions such as right and justice cannot exist except within the protected confines of the state,[16] and the state cannot maintain its existence without the guarantee of safety afforded it by an army, the army becomes for Hobbes not only the guarantor of the state's existence but also *the guarantor of morality* (or at least the guarantor of the *potential* for the existence of morality). It may, in fact, be unnecessary to insist upon this latter point. Nevertheless, it is worth noting that in social contractarian moral philosophy of the kind advocated by Hobbes, one can argue that the sovereign is able to "guarantee" the security of the state (i.e., a place where morality can exist—for it cannot exist in the chaotic Hobbesean state of nature) only because he has at his disposal an army. At the very least, and in a similar vein, we can fairly conclude that, to the extent that an army serves as the guarantor of the state's safety, and to the extent that the preservation of the state represents a moral good, the army has a *moral obligation* to do all within its power to ensure the preservation of the state.

One can arrive at the same conclusion by any number of other avenues. For example, the Kantian moral philosopher might argue that if, as the ancients held, an orderly society is necessary to human flourishing (in which, among other things, the society's members enjoy maximum liberty to discharge their moral obligations), there exists a moral duty to defend such a society's existence. Those officers appointed to lead the state's army, then, might properly be said to have a moral duty to ensure, to the best of their abilities, the preservation of the state from outside danger. On this premise, the state has an army both as a matter of practical necessity and as a matter of moral imperative, and the officers who lead the army are morally obligated to lead it well.

Still another school, the utilitarian moral philosophers, might argue that the interests of the greatest number of society's members are best served when a segment of society devotes itself to the defense of the whole. If this be the case, then the utilitarian could likewise conclude that the army exists as a matter of moral imperative, and that its officers are morally bound by utilitarian principle to lead it well.

In sum, whether one elects to embrace Hobbesean principles, or whether one applies Kantian, utilitarian, or similar tests, all permit the conclusion that armies—and therefore most certainly the officers who lead them—are bound by moral obligation to the societies whose responsibility it is theirs to defend.

The Constitution as a Source of Professional Moral Obligation

The idea of a constitution exists inseparably from the idea of a nation-state. It is one that far transcends the sum total of the meanings of the words contained in a written document, for indeed a constitution need not be written (as in the case of the English constitution). Broadly speaking, each state has a "constitution" in that it exhibits some kind of broadly sanctioned political structure. According to Herbert Spiro, "Aristotle used the Greek word for 'constitution' (*politeia*) in several different senses. The simplest and most neutral of these was 'the arrangement of the offices in a *polis*' (state). In this purely descriptive sense of the word, every state has a constitution, no matter how badly or erratically governed it may be."[17]

In the generally accepted sense of the word today, a "constitutional" state is one in which the institution of government is bound, in some meaningful way, by *contract* with at least some of those it governs.[18] Of all the constitutions in history, none illustrates this idea of "contract" better than the Constitution of the United States. (Inasmuch as the idea of "constitution" transcends the written document itself, one can properly speak of the U.S. Constitution as including not only the governmental structure outlined therein, but also the ideals associated with representative and limited democratic government. It also includes the ideals enshrined in allied documents like the Declaration of Independence and the Bill of Rights, as well as laws made under the authority of the Constitution.)

Given the philosophical, theological, and experimental soil from which sprang the Constitution of the United States, it seems clear that the ultimate aim of the social contract it codifies is not merely law and order, or even merely the preservation of liberty, but rather the promotion of human flourishing—to protect all that is virtuous, lovely, of good report, or praiseworthy[19] while at the same time to serve as a hedge against any malicious influence subversive of that aim. In short, it has the highest of human goods as its aim[20]—an aim that it shares with the moral enterprise.

Against this background, the officer's oath of office assumes a much deeper moral significance than some may suppose. Properly understood, the oath taken by Army officers to "support and defend the Constitution of the United States against all enemies foreign and domestic" comprehends not only the governmental structure itself but also all the *ideals* of government and the way of life (including the protection of all those beliefs and practices that make human flourishing—morality—possible) implicit in them. Furthermore, the oath requires that the gravity of the obligations it entails be acknowledged "without mental reservation or purpose of evasion." It binds the officer not merely to "do his or her job," but to "well and faithfully discharge the duties of the office." It concludes with an invocation of divine assistance in fulfilling the duties of this public trust. If there lingers any doubt as to whether the profession of arms is indeed a profession, let us note that no baker or butcher or hair stylist or golfer in *any* society—as important as these undertakings might be—is ever required to incur such a weighty set of obligations!

It is equally important to note that Army officers do not swear allegiance to any particular person or even to all of the people of the United States in the aggregate. Rather, they swear allegiance to the *ideals* embodied in the Constitution. Hence, the obligation of Army officers to defend the Constitution is a *direct* obligation, whereas the obligation to defend the American people is a *derivative* obligation. It derives from the common allegiance to the Constitution acknowledged both by the American people and by the officers of their Army.

The Army Officer as Moral Exemplar

Of course, the claim, however well established, that the Army has a moral obligation to defend society does not, in and of itself, imply that the Army's officers themselves have an obligation to be moral persons. However, this logical consideration does not fully account for the complexity of the issue. Logic deals only in what is possible; but when the realities of human nature are superimposed upon the demands of logic, one must be willing to include in his or her calculation the timeless truth that *as a man thinketh in his heart so is he*.[21] This chapter takes as axiomatic that men and women such as Army officers, faced with tasks that require the most exacting capacity for moral reasoning, can best accomplish those tasks if they are themselves morally well-founded persons.

If the Army's officers are themselves bereft of sensitivity for what is moral—a sensitivity which, as Aristotle would tell us, is cultivated by practice and dulled through disuse—then they are no better than the amoral forces of the state of nature. As a practical matter, "if the Army merely replicates the state of nature, as in societies run by warlords, it ceases to be useful to society."[22] Beelzebub cannot cast out Beelzebub, a house divided against itself cannot stand,[23] and the state of nature cannot defend against the state of nature. Hence, the defenders against the state of nature must themselves be of a different—and one might argue better—nature: one that not only recognizes the demands of morality but also one that consistently makes choices that reflect that recognition.

As a result of Army officers' solemn Constitutional obligation to ensure the defense of principle, it is altogether reasonable to expect them to embody the moral virtues enshrined in the Army's own rendition of soldierly values: loyalty to the ideals embodied in the Constitution (as evidenced, among other things, by their strict personal obedience to the laws established under the authority of the Constitution—"the supreme law of the land"[24]), an unwavering commitment to the performance of duty, a spirit of selfless service, and both moral and physical courage.

Even if moral obligations were not inherent in the professional status of Army officers, the oath they take is sufficient to establish de facto that Army officers are themselves bound to live in accordance with a definite and clearly discernible set of moral obligations. More to the point, they are bound to *embody* certain specific moral virtues. This is so because, without these virtues, officers would be unable to fulfil their derivative obligations to the American people. If officers fail to defend the Constitution by being unwilling to perform their duty without regard for the personal risks involved, or to face those risks courageously, selflessly, or even sacrificially, then the American people could not be guaranteed that Army members will do all in their power to defend the people's safety and way of life.

Hence, if soldiers—and thus all the more the officers who lead them—historically occupy a position of unique and recognized status in society (i.e., as instruments appointed by the state to preserve or terminate human life), it seems altogether reasonable to expect that they should be men and women of character who can be counted upon to conduct their task with due sobriety and moral awareness.

The Army Officer as a Moral Exemplar in War

The claim that Army officers have moral obligations in time of war is well established in the history of the profession of arms. For example, according to Augustine, army officers are to be persons of such honor and integrity that they can be counted upon to deal justly even with their enemies. Augustine urged Boniface, a Roman general, that "when faith is pledged, it is to be kept even with the enemy against whom the war is waged."[25]

Likewise, it long has been understood that, inasmuch as Army officers are entrusted by the state with the power to preserve or terminate lives, they must be individuals of character who are committed to the properly constrained use of that power, both in its legal and moral dimensions. Moreover, officers were not to be bound merely by the external constraint of law, but even more so by the internal constraint of conscience. Augustine urges that officers be guided during war by the desire to achieve a just and lasting peace and not by a lust for blood or a desire to harm: "Even in waging war, cherish the spirit of a peacemaker, that, by conquering those whom you attack, you may lead them back to the advantages of peace."[26] To another officer he writes, "Let necessity, therefore, and not your will, slay the enemy who fights against you."[27] Thus, he urges that the taking of lives in war ought to be minimized to the greatest extent possible. Says Augustine, "For he whose aim is to kill is not careful how he wounds, but he whose aim it is to cure is cautious with his lancet; for the one seeks to destroy what is sound, the other that which is decaying. . . . In all . . . cases, what is important to attend to but this: who were on the side of truth, and who were on the side of iniquity; who acted from a desire to injure, and who from a desire to correct what was amiss?"[28] Augustine unambiguously advocates that a spirit of mercy and forbearance be displayed toward all those who fall into the power of their enemies: "As violence is used towards him who rebels and resists, so mercy is due to the vanquished or the captive, especially in the case in which future troubling of the peace is not to be feared."[29] In the light, then, of such demanding moral imperatives as these, it follows that only men and women of the deepest compassion, clearest sense of justice, and highest integrity would be both able *and willing* in time of war to distinguish between justified and unjustified applications of violence.

The just war tradition has long acknowledged the possibility that the cause for which the state summons its military to war may be less than perfectly just (a morally generous description of at least one side of every war that ever has been fought). On the one hand, therefore, Army officers must be men and women of integrity who, confronted with dark realities, still are willing to do their professional duty, unpleasant though that duty may be, as honorably as they can.

Of equal importance, Army officers are not, and indeed cannot be, automatons. They are moral agents who must recognize their responsibility (1) never to issue an immoral order, and (2) to refuse an order—or even a suggestion—to undertake military operations inconsistent with the ideals they are sworn to defend. Moreover, when an informed judgment leads them to the conviction that an order is morally wrong, they also must possess the maturity to recognize, and the fortitude to accept, the personal consequences of refusal.

The Army Officer as a Moral Exemplar in Peace

That Army officers have fundamental moral obligations in time of peace as well as in war also is well established. For example, according to Plato, those chosen to defend the state "are not to be savage animals, preying on those beneath them, but even if stronger than their fellow-citizens, they will be their friendly allies, and so it is necessary to ensure that they should have the right education and mode of life. . . . [G]old and silver they should neither handle nor touch. And this will be their salvation and the salvation of the State." [30] More recently, the Joint Chiefs of Staff have declared: "Our military service is based on values—those standards that the American military experience has proven to be the bedrock of combat success. These values are common to all the Services and represent the essence of our professionalism."[31]

Unfortunately, there are those in influential positions both in and out of uniform whose conduct suggests a less-than-enthusiastic endorsement of this ideal. For example, given the appalling reality that the citizens of the United States recently were informed from the highest political level that what a public official does in private is not a public concern and, therefore, is no one else's business,[32] there doubtless are those in contemporary society who would shy away from the claim that Army officers have an obligation, by virtue of their privileged position in society, to be moral exemplars not only in war, but also in the conduct of their private, peacetime lives. One version of the argument takes the following form:

> Premise 1. In a democratic republic, those who occupy positions of public trust are expected to act in the public interests.
>
> Premise 2. If such a person functions successfully in that public trust (i.e., "does his or her job"), that person deserves approbation for his or her service, and no other demands can be made on him or her with respect to matters of moral character.
>
> Premise 3. Person X, who holds a public trust, has functioned successfully (as defined above) in his or her office, but has failed miserably in terms of living a virtuous private life.

> Conclusion: Therefore, person X deserves the approbation of the American people for a job well done, regardless of how deep may be the moral flaws evident in person X's private life.

The position of this chapter is that the foregoing argument is unsound by reason of the falsity of Premise 2. This and similar arguments hold a seductive appeal for many members of liberal democratic society. Nevertheless, such arguments are as pernicious as they are seductive. This is particularly so as applied to Army officers. As has already been established, Army officers belong not only to a profession but also to that significant subset of professions whose primary function pertains to the preserving or terminating of human life (recall Figure 14-1, Set C). The professional status conferred by society upon members of this subgroup makes of them the moral demand not only that they *know* how to perform their professional tasks but also that they *be* persons of a high moral quality and that they maintain the highest moral standards at all times.

The same is true for other professions in Set C. For example, the physician rightly cannot claim moral license to use cocaine in the privacy of his or her off-duty time and then come to the hospital to perform brain surgery. The judge rightly cannot associate with criminals and accept bribes from them during his or her off-duty time and then expect to enjoy the public's confidence that he or she will deal justice to those same criminals facing racketeering charges in court. Likewise, because Army officers are charged to preserve or terminate life in the most untidy of circumstances, they cannot afford to present themselves as persons of whom others might wonder whether their conduct would be cavalier, flippant, or oblivious to the demands of morality in the most demanding of situations. Thus, the Army in transformation must take as its inalterable stance that, even in (what may be an increasingly) liberal democracy, where in many circles recognition of moral verities may be classified politely as postmodern, the Army still will hold itself firmly committed to its traditional ideals of military virtue.

Some may argue that idealized thinking of this kind is fine for ideal societies, but that since American society is itself demonstrably less than ideal, any such talk is out of place in the real world. However, those who argue in this way altogether miss the point because they fail to understand the essential meaning of "America." America is special *not* because it is the realization of any social ideal; it never has been ideal, and perhaps it never will be. America is special because it *aspires* to be ideal; it has always so aspired. America's early pioneers—the Pilgrims, the Founding Fathers, and others—conceived of themselves as men and women with a divine mission. The life they built in the New World was to constitute an ideal—"a city on a hill."[33] As poet Stephen Vincent Benét reflects,

> They were founding Zion, not the United States
> —And the seed is sown, and it grows in the deep earth,
> And from it comes what the sower never dreamed.[34]

Indeed, as long as the current state of human nature prevails, the realization of the perfect society must remain beyond our grasp. Nevertheless, societies, like individuals, that strive for the ideal are bound to come closer to realizing it than those that do not.[35] Likewise, it seems reasonable that any approach to the ideal in the present world would include a professional military led by officers who cling tenaciously to the highest moral standards.

Some may argue that an Army that espouses—or even claims to espouse—moral virtues that are "higher" than those of contemporary society at large cannot but be "out of step." They may claim that such an Army is certain to adopt a sanctimonious, "holier than thou" attitude about the way it regards the society it is sworn to defend. "How could such an Army," they may wonder, "truly possess the motivation necessary to fight the wars of a society it regards as its moral inferior?" However, such questioning also misses the point. The Army never has itself been an ideal organization; it is composed of men and women taken from American society and who possess the same foibles as Americans not in uniform. The taking of the oath of Army officership does not automatically transform, as it were, a "sinner" into a "saint." The point is that by the nature of the demands of professional Army officership, those who would serve the state as Army officers must commit to *strive to transform themselves* into men and women who can live up to and preserve the high ideals of the profession of arms. Officers are not expected to be perfect, but they are expected to live, in the words of the West Point prayer, "above the common level of life."[36]

The suggestion that the nation does not want to be defended by men and women who hold themselves to a higher moral standard than that found in society at large is pure nonsense. No reasonable person who considers the matter would prefer that those charged by the state with the responsibility to make decisions relative to the preserving or terminating of human life be men or women of deficient moral character. On the contrary, such a reasonable person would desire that those who undertake these and similar tasks be among the most virtuous persons that the society can produce. Even the most morally depraved person suffering from a terrible disease would rather be operated upon—all other things being equal—by a medical professional of high moral character who could be counted upon to act in the patient's best interest than by a member of the profession who, although as technically qualified, was a moral derelict. Surely a helpless widow who had lost her life's saving through the actions of an unscrupulous swindler would want the lawyer she engages to be one she could depend on to act vigorously to protect her interests and to right the wrongs done to her—not to compound them for his own selfish purposes. Likewise, any society that entrusts its warring powers to military professionals would want, and indeed *expect*, those professionals to apply judiciously and honorably the powers entrusted to them. (Else, why would a nation full of admittedly less-than-morally-ideal citizens find itself enraged and appalled by the actions of *decidedly* less-than-morally-ideal Army officers at My Lai?) Even in time of peace, those same citizens can be expected to hold those who lead their Army to standards of moral conduct above and beyond what they might find themselves willing to tolerate in society at large. (Else, why the national outrage at neglect of sexual abuse by those in authority at Aberdeen Proving Grounds?)

In short, there is no warrant whatsoever for the claim that the ancient wisdom which for millennia has served as the touchstone for professional military conduct in peace and war suddenly has become irrelevant. Thus, while it may be true that professions in jurisdictional competition that do not adapt are destined to die, it is not clear that the profession of arms can meaningfully exist in a democratic society apart from its foundational moral commitments.

Professions exist because, among other reasons, they lay proper claim to special expertise within the purview of their jurisdiction. Thus, when persons outside the profession of arms foist upon the profession a perspective or a practice that is foreign to and resisted by traditional military culture, there is reason to wonder whether those persons (even if they happen to be political masters) actually possess the professional competence necessary to insist on having their way. It is fully conceded that policy-making authority rests ultimately with civilians outside (and authoritatively above) the profession. Surely, no reflective Army officer questions this arrangement that has served the republic well for over two centuries. By the same token, there is great danger in implementing in the Army politically motivated policies against the considered opinion of the uniformed professionals who have devoted their lives to the *effective* military service of the nation. It is incumbent upon the senior members of the officer corps to articulate this reality to the Army's political masters.

In sum, thus far this chapter has sought to establish the following claims with respect to Army officers and moral obligation:

1. Service as an Army officer is indeed a profession;
2. The profession of arms has moral obligations to society;
3. Constitutional government imposes both explicit and implicit obligations on those who live under the order that government is designed to ensure;
4. Army officers incur a direct moral obligation by binding themselves to defend the Constitution and a derivative moral obligation to defend the people of the United States; and
5. As such, Army officers have a solemn obligation to serve as exemplars of morally well-founded conduct in peace and in war, in public and in private.

With these claims in mind, we now can consider, in moral terms, the kind of professional officer corps that will best enable America's Army to accomplish effectively its aim of winning the nation's wars as that Army marches into the twenty-first century.

Toward a Morally Renewed Officer Corps

The value of any philosophical reflection for policy-makers is conditioned upon the degree to which philosophical promptings are capable of practical implementation. In the critique that follows, we shall examine two matters of ethical import that, if not dealt with, may effectively erode the bond of moral obligation between the officer corps and the nation, and between the officer corps and the soldiers it leads. The issues discussed are merely representative, but they merit serious consideration by thoughtful persons who have an interest in the Army's long-term well-being.

The Army as a Business

During the past decade, much has been done to make the Army a more efficient organization (that is, efficient in the commercial sense of generating greater output at less cost). At the same time, some have astutely noted that, in the name of efficiency, the Army has become a less *effective* organization in terms of its raison d'être—its ability to fight and win the nation's wars.

Logic does not dictate any necessary relationship between efficiency and effectiveness. Take, for example, the case of a company that produces replacement parts for tanks and that is under contract with the Army. Since the raison d'être for parts manufacturing companies—like all companies—is to make a profit, the goal the company probably will strive for is to produce the best tank parts (hence, maximum effectiveness) at the least possible costs (hence, maximum efficiency). Both concepts are important, because both help the company to achieve its ultimate aim of making money. Efficiency ensures the minimum expenditure of resources on the front end of the company's investment, while effectiveness ensures that those who buy the company's product will return to buy the product again, thus ensuring continued revenue. Again, however, this end state is merely an ideal, not a logical necessity. For example, if the company undertakes to develop, say, a new kind of replacement part for a tank, one of four possible cases, or outcomes, will obtain:

1. The part will be cheaper to produce (more efficient) and more reliable (more effective); or
2. It will be cheaper to produce (more efficient) but less reliable (less effective); or
3. It will be more expensive to produce (less efficient) but more reliable (more effective); or
4. It will be more expensive to produce (less efficient) and less reliable (less effective).

In Case 1, the project manager overseeing the production of the part gets a promotion. In Cases 2 and 3, the project manager may manage simply to keep his or her job. In Case 4, the project manager gets fired! In any event, an increase or decrease in efficiency does not necessarily imply an increase or decrease in effectiveness—the two variables are logically independent.

While this independence is fairly easy to recognize in the world of business, it may be somewhat more difficult to see when the notions of efficiency and effectiveness are applied to an organization like the Army that does not have the generation of profits as its purpose. Nevertheless, it is equally true that an Army that costs less to operate (or at least is provided with less resources with which to operate) will not necessarily be more effective at its task of winning the nation's wars.

Since businesses have as their ultimate aim the generation of profits, businesses—even those that supply the Army—may from time to time find it prudent to sacrifice effectiveness in order to maximize efficiency. However, no such luxury accrues to the Army; the Army can *never* afford to sacrifice effectiveness in the name of efficiency. To do so means that the Army is willing to risk being unable to accomplish its most basic and critical function of defending the nation. Hence those who insist that greater efficiency will necessarily result in a more effective Army ignore the realities of logic. What is worse, they place the Army at a crossroads not unlike the one at which Alice finds herself in Wonderland when she encounters the Cheshire Cat. Alice asks the Cat:

> "Would you tell me, please, which way I ought to go from here?"
> "That depends a good deal on where you want to get to," said the Cat.
> "I don't much care where—" said Alice.
> "Then it doesn't matter which way you go," said the Cat.
> "—so long as I get *somewhere*," Alice added as an explanation.
> "Oh, you're sure to do that," said the Cat, "if you only walk long enough."

So it is when the crucial distinction between efficiency and effectiveness is lost. Pursuing the path to efficiency does not necessarily ensure arrival at the aim of effectiveness; it merely ensures arrival *somewhere.* If the Army chooses to understand efficiency as the necessary precursor to effectiveness, or worse, if it chooses to become more efficient at the expense of becoming more effective, the whole nature of the institution's moral commitments will require revision. This is so because the business world is, by nature, mercenary in character. People rarely work for businesses because of ideals; they work for businesses out of profit motives. The claim here is not that there is anything wrong with the profit motives of business mercenaries per se, but only that such thinking inappropriate in many military settings. In practical terms, whenever the claim is made that a policy or program will increase efficiency, the Army's leadership must require the claimant to demonstrate how the increase in efficiency will produce a commensurate increase in the overall *effectiveness* of the institution or at least avoid producing a decrease in effectiveness. In fact, it may well be that a justification of this kind should accompany all program proposals that require strategic-level approval.

Risk Aversion

Central to the traditional Army ethic is the idea that Army officers are the defenders of the defenseless; officers use the power and the authority of their office *not* to defend themselves, but rather to defend from injustice those that cannot defend themselves. The profession of arms is truly to be a selfless undertaking in which risk to self is understood to be the price of the privilege of being entrusted with the life-taking and life-preserving power of the state.

This point figures into Hobbesean social contract theory. According to Hobbes, the sovereign is able to "guarantee" the security of all members of society *except* those in the sovereign's army, for they must be sacrificed if necessary in order to preserve all else. Hence, if Hobbes conceives morality to mean anything at all, he must conceive that the sovereign's army is *morally* bound to protect society, if necessary at the cost of the complete sacrifice of its members' personal interests.

However, one need not adopt Hobbes' concept of morality in order to conclude, with Hobbes, that to suppose that society can be defended without personal risk to those in the army is patently absurd. An army exists to assume risks on behalf of those who rightly lay claim to the army's protection. It exists as a corps of trained professionals who are better equipped to perform the task of defense than any layman would be. Risks to the army are not to be mitigated by never putting it in harm's way, for in harm's way is the place where the army, better than any other organization, is suited to be. Risks to the army are mitigated through the thorough training of officers and soldiers; and, indeed, they require thorough training. As Plato asks:

"[A]re we to believe that a man who takes in hand a shield or any other instrument of war springs up on that very day a competent combatant in heavy armor or in any other form of warfare—though no other tool will make a man be an artist or an athlete by his taking it in hand, nor will it be of any service to those who have neither acquired the science of it nor sufficiently practiced themselves in its use?"[37]

Risk assessment as it is currently practiced in contingency operations merits thorough review. The standard applied in training that the death or serious injury of a soldier is unconscionable cannot be conceived as transferring directly to contingency operations. Soldiers must be trained to assume that the environment into which they are being sent is hostile and that they are there to *act* and not to be acted upon. If a task regarded as vital to the national interests requires Americans to go to a place with instructions to "hunker down" and neither get hurt nor allow anyone in their charge to get hurt, the task so described arguably does not belong to the Army. Indeed, the idea that the Army (i.e., the institution, and not its soldiers individually considered) will go anytime, anywhere for *any reason* is a romantic notion that requires thorough reassessment.

While it is understood that the ultimate decision on how to use the Army is one reserved for the highest political level, the Army leadership must articulate this concern vigorously and repeatedly to the Army's controlling but non-professional masters so that the institution's position relative to risk assessment is well understood: If the Army's political masters wish to reduce the risk of death or injury to America's sons and daughters in uniform, the answer is not to avoid sending them to places where they legitimately should be sent or to discipline their officers because someone in the officers' charge gets injured in the line of duty. Rather, the way to reduce risk is to provide the Army with adequate funding and other resources so that its officers can adequately train and equip the soldiers placed in their care and thus prepare them to the greatest possible degree to encounter and overcome risks successfully.

Undoubtedly, there are certain things about the Army that need to transform. As pertaining to those things of moral import that must change, the Army's task is clear:

1. It must examine itself for moral shortcomings, with special attention to those shortcomings that are of its own making (or own allowance). This is an urgent and continuous need because, while the Army cannot cure many of society's ills, it *can* act to remedy its own shortcomings, and then hold itself accountable to fulfil its moral obligations.
2. It must communicate its moral commitments in ways that will appeal to the moral sensitivities of reasonable men and women in a liberal democratic society who may not embrace for themselves the same moral imperatives as does the officer corps. In this regard, even if the American people were either to expect or to require less in terms of moral commitment from its Army officer corps, the officer corps can neither expect nor allow less of itself.

Although the officer corps does, and should, seek to refine its understanding of its own profession and to adapt to the exigencies of the new century, it does not need a new set of moral commitments. It simply must face up squarely to the moral commitments it already has.

The good news is that the task of living in a way that acknowledges the institution's deeply rooted moral commitments is accomplishable by good men and women who are willing to put service before self. Virtue *can* be taught; people *can* be transformed tomorrow into something better than they are today. Moreover, some of the most morally committed men and women to be found anywhere in the world are among the ranks of the U.S. Army—its officers, noncommissioned officers, and soldiers.

However, as the Army transforms, its professional officer corps must take great care to distinguish those things that may or must change from those that must not. Whatever transformation the Army and its officer corps undergo, that transformation must not assume that the work of the officer corps can be divorced from moral considerations—considerations grounded in history and in tradition, as well as in the nature of professions and constitutional government.

Notes

1. The views and opinions expressed in this chapter are those of the author and are not necessarily those of the Department of the Army or any other U.S. government entity.
2. Donald Kagan and Frederick W. Kagan, *While America Sleeps* (New York: St. Martin's Press, 2000), 432.
3. Ibid., 434.
4. Ibid., 433.
5. 1999 Fletcher Conference—Findings and Recommendations; available from *http://www.army.mil/cmh-pg/documents/fletcher/fletcher-99/F99-F&R.html*; Internet; accessed 20 November 2000.
6. "On the Army Transformation," Congressional Testimony delivered by General Eric K. Shinseki, Chief of Staff, United States Army, before the Airland Subcommittee on Armed Services, United States Senate, Second Session, 106th Congress, 8 March 2000.
7. Naturally, it is hoped that claims of the kind advanced here would find appropriate application to the Army as a whole, the Army's sister services, and, in principle at least, the services of many of America's allies. However, in keeping with the purpose of the present anthology on Army professionalism, the arguments that follow will be tailored to apply directly to the United States Army Officer Corps.
8. Of particular note in the context of the larger study on the Army Profession of which this chapter is a part, an important representative of this corpus of literature on professions is the work of Andrew Abbott entitled *The System of Professions: An Essay on the Division of Expert Labor* (Chicago, IL: University of Chicago Press, 1988).
9. As Cicero argues, "But private citizens often escape those punishments which even the most stupid can feel—poverty, exile, imprisonment and stripes—by taking refuge in a swift death. But in the case of a State, death itself is a punishment, though it seems to offer individuals an escape from punishment; for a State ought to be so firmly founded that it will live forever." Hence, Cicero argues, "There is some similarity, if we may compare small things with great, between the overthrow, destruction, and extinction of a State, and the decay and dissolution of the whole universe"(Marcus Tullius Cicero, *De Re Publica* 3.23, in *De Re Publica* and *De Legibus,* trans. C. W. Keyes [Cambridge, MA: Harvard University Press, 1928], 211-13).
10. Augustine, *City of God*, 22.6, trans. Henry Bettenson (London: Penguin Books, 1984), 1032.

11. Frederick Copleston, *A History of Philosophy* (New York: Bantam Doubleday Dell Publishing Group, Inc., 1985), 1:226.
12. Plato, *The Republic* 2, 374 b, in *The Collected Dialogues*, ed. Edith Hamilton and Huntington Cairns (Princeton, NJ: Princeton University Press, 1961), 620.
13. Ibid.
14. Thomas Hobbes, *Leviathan* (London: Penguin Books, 1985), 185.
15. Ibid., 186.
16. Ibid., 188.
17. *Encyclopaedia Britannica*, 15th ed., s. v. "Constitution and Constitutional Government," by Herbert John Spiro.
18. U.S. Army Col. (Retired) Alexander P. Shine, personal correspondence with the author, 11 March 2001.
19. Philippians 4:8.
20. Shine, 11 March 2001.
21. Proverbs 23:7; see also Isaiah 10:7.
22. Shine, 11 March 2001.
23. Luke 11:17, 18.
24. U.S. Constitution, art. 6.
25. Augustine, *Letters* 189.6, trans. J.G. Cunningham, in *The Nicene and Post-Nicene Fathers*, ed. Philip Schaff, First Series (Grand Rapids, MI: Eerdmans Publishing Company, 1956), 1:554.
26. Ibid.
27. Ibid.
28. Augustine, *Letters* 93.8, 1:385.
29. Augustine, *Letters* 189.6.
30. Copleston 1:228; Plato, *The Republic* 3.417 a 5-6
31. U. S. Department of Defense, Joint Publication 1, *Joint Warfare of the US Armed Forces* (Washington, DC, 11 November 1991), 7.
32. William Jefferson Clinton, nationally televised address, 17 August 1998; available from *http://www.cnn.com/ALLPOLITICS/1998/08/17/speech/transcript.html;* Internet; accessed 13 March 2001.
33. Matthew 5:14.
34. Stephen Vincent Benét, "Pilgrims' Passage."
35. Shine, 11 March 2001.
36. Ibid.
37. Plato, *The Republic* 2.374c, d.

15 A Message to Garcia: Leading Soldiers in Moral Mayhem[1]

James H. Toner

Note from author: In the account that follows, Col. Pedro Garcia, his wife Maria, his sons Dom and Josh, and Colonel Garcia's Air War College classmate, Lieutenant Colonel Morgan, are entirely fictional, serving merely as literary devices to assist me in drawing certain moral lessons for professional military officers. Any parallels between these fictional characters and actual persons are unintended. The author's dependence upon one faith tradition is a natural consequence of the "shared religious convictions" between the author and Colonel Garcia. The "letter" following the italicized prelude below is rooted in those convictions.

When Lt. Col. Pedro Garcia, United States Army, arrived as a student at the Air War College in 1999, he and I soon discovered that we had a great deal in common. Our friendship began by accident. Colonel Garcia was speaking in the corridor to another War College student, Lt. Col. Claude Morgan, when Colonel Morgan saw me nearby, mistaking me as an old friend of his. As we sorted that out, I learned that the two colonels were discussing one of my favorite topics, baseball, and that Colonel Garcia was seeking advice about batting on behalf of his two sons, Dom, 15, and Josh, 13. Dom couldn't catch up to a fastball and Josh had just met a new enemy, the curveball. Before long, Colonel. Garcia—now "Pete" to me—and I were sitting at a table discussing hitting strategies and drills his sons could use. It was the start of a lot of batting practice I was to give the Garcia boys during their time in Montgomery, Alabama, the location of the Air War College.

It was the beginning, too, of a year of lengthy serious conversations between Pete and me on a wide variety of subjects: the state of our country and of the world, the United States Army and his flourishing military career, his reservations about what he perceived as increased tension—Pete called it a "chasm"—between the profession of arms and the American society he loves, the challenges to his faith presented by contemporary military operations, and the ethical cases—Oh! those cases—which we thought through and discussed and debated all year long. My wife Rebecca and Pete's wife Maria became good friends, and Dom and Josh became, if not my "nephews," at least my batting disciples! Pete is a graduate of Notre Dame, where he earned his Army commission through the Reserve Officers' Training Corps. I earned my doctorate at Notre Dame, and my oldest son Chris earned his degree and Army commission at Notre Dame. We often attended mass together, and Pete is an excellent example of commitment to Christian faith. That commitment—although one I shared—led to a number of arguments, one of them very heated, in fact threatening the foundations of our friendship. But I get ahead of myself. I, too, had been an Army infantry officer for one tour of active duty, 1968-1972, though I did not serve in Vietnam. But teaching at a military college (Norwich) and more than ten years of teaching at a War College had kept me at least somewhat current with military matters.

Pete's ten-month course of study at Maxwell Air Force Base zipped by, the highlight of the year proving to be Pete's promotion to colonel, along with the knowledge that he was off to brigade command soon after War College graduation. Much of what we discussed was never settled, so a couple of weeks before Pete left Maxwell, he extorted a promise from me. I remember the conversation:

"Jim, you guaranteed me several months ago that I would ultimately see things your way about ethics and the profession of arms. As it is, the only thing I agree with you about is improving Dom's hand speed to catch up with fastballs."

"That's not fair, Pete. You also agreed with me about teaching Josh to keep his front shoulder in so he could hit the curveball! But as for the rest—well, I wish we could continue our conversations. You've given me a lot to think about, Pete. And I'm sorry—no kidding—that I haven't resolved all your problems. But I should leave something for you to do, after all. Isn't that what we teachers are supposed to do?"

"Of course. But I was thinking . . ."

". . . not again!"

"No, seriously: I was thinking that since Maria and the boys are going to remain here during my tour, I'll have at least some time to continue our conversations. Would you do me a favor? Take a while and write me a letter about some of the problems we've discussed. But no reading assignments and absolutely no quizzes!"

"All right, Pete. I'll do what I can. And maybe I can go you one better. I know your thinking on so many of these subjects that I can anticipate some of your reactions—and objections. I'll try to take these into account as I write."

"That will be great, Jim. Would you object if I shared your correspondence with others?"

"Not at all. But keep in mind that I'll be writing to you from a religious perspective we largely share. I'll be assuming an agreement between us which in most cases clearly exists."

"But, Jim, is it really possible any more, given what we agree is the decline in morality both nationally and internationally, to have a truly ethical American armed force? I could be on the verge of making military decisions that could result in the deaths of scores of people. I want some answers before I make those decisions!"

Department of Leadership and Ethics
Air War College
Maxwell AFB, AL 36112-6427

Hi Pete:

This is the letter I promised you I'd write. I'm properly going to leave the family talk to Maria and the boys, but I thought I'd assure you that Maria is doing great taking classes at the college, and the boys are really in their element at Catholic High

School. I miss batting practice with them, but I've promised to get back out on the diamond with them once basketball season is over. So Dom accuses me of pestering them to read, eh? Guilty! Every time they come over for hamburgers, I hit them with history, politics, and geography. And they get their share of Bible questions—especially from Rebecca! But they're doing great, Pete. Super young men!

Postmodern Moral "Standards"

It's kids like Dom and Josh we need, Pete. I am, as you know, profoundly upset by events in our country over these last thirty years or so. Although Neil Howe and William Strauss assure us that the "Millennial Generation" is bound to be great,[2] I frankly don't think it's getting better. Too often, the examples all around us are of moral squalor. I think we can see that in the person of the ubiquitous cartoon character who, as a decal on the back windows of cars and trucks, urinates on people and things. There have been some great movies recently—I think of *Saving Private Ryan*, *Schindler's List*, and *Glory*—but so much of what appears is trash, or worse. The entertainment industry seems to thrive on sex, violence, brutality of all kinds, and outright hostility towards the virtues and values you and Maria want your kids to grow up with.

You know, I constantly hear that when I was growing up in the late 1950s or when my Mom and Dad were growing up in the 1930s, there were plenty of problems then, too. And that's right. All I have to do is ask African-Americans here in Alabama about that. When I speak favorably of military virtues, I am often greeted with cloaked derision from people who mistakenly think I'm suggesting that the Army dictate values to American civilians. But that's hardly my position. I do think that the armed forces must practice the virtues—and to the extent that they exemplify the foundational virtues of wisdom; justice or truthfulness; moral and physical courage; and temperance, modesty, or self-control—they can serve as a source of inspiration for all of us.

The armed forces led the way in making integration work. Sometimes I think you guys in uniform don't get enough credit for that. We all know racism hasn't been completely cured in the military, but you've done a great job of making "equal opportunity" a reality and not just a slogan.

So black Alabamians can point with some pride to the fact that, largely because of their courage (and that of one of their own, Dr. Martin Luther King, Jr.), progress in race relations has been made. Things are better now than they were in the segregated 1950s. Certainly, improvements have been made—and in a number of fields. For example, Dom and Josh aren't doing "duck and cover" drills today, getting under their desks at Catholic High, to "protect themselves" in case of nuclear attack. We have made, or so it seems, some progress on the Chemical-Biological-Radiological front. But progress even there can be as illusory, as the 11 September 2001 terrorists attacks on the World Trade Center and Pentagon show.

I grew up loving the Boston Red Sox. Today we have world federation wrestling and "extreme fighting." TV and the movies were bad enough in the 1950s, but *The Lone Ranger* and *Howdy Doody* can't be confused with *Jerry Springer*. The language that kids use routinely today in public I did not hear even in locker rooms years ago. How, for instance, do you compare Eisenhower and Clinton, morally speaking? What passes for comedy today would probably have horrified even teenagers years ago. And even in the 1960s, there was not acceptance of abortion, euthanasia, and infanticide (partial-birth abortion). On the last night of one of the political conventions recently, a local Eagle Scout group was invited to lead the Pledge of Allegiance. Did you read this, Pete? They were met by boos and catcalls from adults in the audience because of the Scouts' position on homosexuals.[3]

I've read numerous books which contain statistics about crime. It's up; it's down; it's more white collar; it's more blue collar; it's more—or less—urban. But use instead as your standard of measurement your old high school. Have you visited your alma mater recently, Pete? It must have been pretty good since your education there got you into Notre Dame. How is it doing today? How are the teachers? Will they patrol the boys' and girls' rooms? How is the level of respect for teaching, for learning, for the school itself? What do the kids talk about? Whom do they admire and seek to emulate? Did you know, Pete, that when I was growing up I didn't even know players on the Red Sox *got paid*! It seems as though money, income, and finances dominate us today as never before. If the love of money really is the root of all evil (1 Tim 6.10), we have much to worry about.

Every kid I knew in the 1950s—including me (if you can believe it!)—was interested in sex. But the picture has changed. When I was in high school, we wouldn't have *publicly* dated girls in which we might have been interested for only one reason. Remember the word *risqué*? It makes no sense any more. Neither does the word *promiscuous*. But when kids are saturated with sex everywhere they watch and read and listen and see, what are they to think of as "right"?

Oops . . . slipped again, didn't I? Talking about "right" as if there were such a thing. Almost all the kids I personally know are cultural or ethical relativists today. What is "right" or "wrong" depends solely upon customs or circumstances, upon friends or gangs, or upon momentary urges or appetites. Glands and gonads dictate, too often I think, what is "right" and what is "true."

I know . . . I can be accused of implicitly talking about a golden age. You and I know such a golden age never existed. We frequently knew as kids, however, that when we misbehaved (another word that is gone by the boards today), we were acting (as teenagers so often do) contrary to a standard. What standard? The church standard. The customary standard. The standard of ladies and gentlemen. So often today, actions are justified—if that's the term to use—by the argument that *there is no standard*. Who are *you* to impose *your* values on *me*?

Well, let me retreat. Maybe there is a standard, after all: the "me" standard. Around us swirl the eddies of moral nihilism. When we have no idea about what truth is, why and how should we concern ourselves about what taste ought to be? So how can we criticize kids—or older people—for bullying, brutal, or bestial behavior (booing kids trying to say the Pledge of Allegiance, for example!) when "everything goes" as long as the cops don't spot it? Parents too frequently on their way to the well-stocked liquor cabinet tell their kids not to use pot. Priests and ministers fall into pederasty or profit-making. Teachers don't read widely or don't care much, going on strike when they so decide, often despite laws forbidding it. The idea of "student-athlete" is burlesque. Entertainment caters to the basest instincts. Corruption festers in the professions. And our young people see . . . what? At all times and in all professions there have been departures from the standard of right and wrong. But today, we deny the standard itself. There is nothing discourteous, for there is no courtesy. There is nothing indecent, for there is no decency. There is nothing profane, for there is no holiness. What is good or true or beautiful depends only upon "me": my emotions, my desires, my pleasures.

But there's another perspective, too. You know how much I believe in the military ethos—and how firmly I defend it. But I bristle when I hear soldiers cavalierly (and arrogantly) talk about how much better they are than the civilian world. My Lai, Okinawa, Aberdeen, Tailhook. What about the soldiers who have brought shame and infamy to the profession of arms? *The military ethos is morally defensible because—and to the extent that—it is rooted in transcendent truth.* The same is true with the political order in which we find ourselves. Our obedience to our military services and to our political society thus depends upon their fidelity, in turn, to the permanent ideals from which spring our faith. That one adjective, *permanent*, speaks volumes, for it tells us that there are enduring and exceptionless norms (e.g., Ps 119.89-90; Heb 13.8), and that there is a truth to which we owe first and full allegiance (Jn 14.6). That is *really* a counter-cultural view! The idea that some obligations transcend our political-military duties is not a notion that subverts our professionalism, but rather one that endows it with direction and dimension. You are a professional, Pete, because you understand yourself within the settings of both the sacred and the profane; that is the tradition of the virtues, one expression of which is the profession of arms at its best. There are, I fear, too few people today whose education is rooted in the classical virtues.

At Notre Dame we can find students who are committed Catholics, young men and women deeply devoted to the Christian faith. You can find the same kinds of young Protestants at Baylor, say, or at Wheaton—deeply committed to Biblical values. Similarly, you can find many fine young people, committed to Judaism at, for example, Brandeis. Now how many of these intelligent and ethically concerned students are ethical absolutists? How many will say that there are clear and wholly compelling ethical standards which may not be transgressed? I don't have an answer. I have an intuition based upon a number of anecdotal sources telling me that many of these students deny the possibility of an objective moral standard. If they deny that, they do not understand what being Catholic or Protestant or Jewish means. How can one say he believes in God and then deny an absolute standard of morality? I understand cultural and ethical relativism, but I do not for a moment grasp how ostensibly traditional Christians and Jews can accept God on the one hand and implicitly deny the scriptures—Christian and Hebrew—on the other.

The highest good on many campuses is toleration. Remember G. K. Chesterton's thought about that? He believed that toleration is the virtue of people who don't believe in anything. Of course, we should and must practice toleration when it comes to respecting people of different races and religions. But there must be a time, too, to be intolerant. As William Bennett has put it, "A defining mark of a good republic is precisely the willingness of its citizens to make judgments about things that matter." Without being "judgmental," he points out, "Americans would never have put an end to slavery, outlawed child labor, emancipated women, or ushered in the civil rights movement. Nor would we have prevailed against Nazism and Soviet communism, or known how to explain our opposition."[4]

I can tell you only what I've heard from a number of people speaking under the promise of non-attribution: Many, perhaps most, of the young people entering officer accession programs today are, at heart, ethical egoists whose main value is reflected in their mirror. If that's true, then many of the company-grade officers in your command right now are primarily interested in themselves. When you talk about "Army values" of loyalty, duty, respect, *selfless service*, honor, integrity, and personal courage; or when the Air Force talks about integrity, excellence, and *service before self*; or when the Navy and Marine Corps talk about honor, courage, and *commitment*—you're all speaking a foreign language. I don't mean to suggest that these officers are "bad" or that they'll necessarily fail you or their troops in a crisis. On the contrary, they are bright and well schooled in numerous practical skills. But often their ethical frame of reference is decidedly different from that which is at the heart of the traditional military ethic.

Consider this. The heart of all ethics consists in the ability morally to transcend the group. If the group—whatever it is—dictates right and wrong, then there is no Ultimate Standard to which to appeal. It is precisely ultimate standards which are ignored or denied by today's secular culture. We must not think that the privates or second lieutenants in today's Army are importing this ultimate standard with them when they walk into the orderly room to report for duty. The chances are, they do not have this standard to appeal to. For you and me, the Ultimate Standard is God, the Church, the Bible. We have a way of judging the country itself—as well as, of course, the Army and any of its procedures and practices. But you and I also know very good people *without* religious conviction whose actions are wholly consistent with orthodox religious teaching. I would hold, however, that, as Robert Bork has written, they "are living on the moral capital of prior religious generations,"[5] or, as Frank Sheed put it, "we still tend to live by what remains of the Christian ethic, though there is no longer any clear grasp of its foundation; and concessions to human weakness have left it pretty tattered."[6]

But suppose we deny the sacred standard. Then we are left, not with truth, but with, at best, only loyalty. Have I told you what I've heard about the late U.S. Air Force Col. John Boyd? When introducing himself to his commanders, he supposedly said: "Ask for my loyalty and I'll give you my honesty. Ask for my honesty and you'll have my loyalty."[7] But what is honesty or truth if . . . if there is no honesty or truth but only my conception of it? Then I'd better be loyal. But to what? How tough it is for all of us to understand that our first commitment must be to truth, not to society or to country, not to service, not to branch, not to self.

So, when you testify before Congress someday after your command time is done and some two-star tells you to "Get on the team, Colonel Garcia" and reminds you about who writes your OERs and the color of your uniform, then ask yourself what comes first, loyalty or truth? *I know your answer, Pete, but I also know your frame of reference.* Remember the time we boiled it down to this: Soldiers must be prepared to give life or limb for their country—*but not their soul*—for no country can properly ask that of its soldiers. But, you and I, as Catholics, agree about the existence of the soul, with all that implies. What if one denies that existence, *with all that implies*? I can hear you now! "So, Jim, what does that denial mean to me as a commander? What do I do about it *right now*?"

One thing you can't do is to set up command-sponsored religion classes, although the old notion of "Character Guidance" (which we used to have in the Army more than thirty years ago) is a serious and solid idea, as long as it is nondenominational instruction (and best taught by line officers). A couple of years ago, I assisted an Air Force line officer and student from Air Command and Staff College, a Protestant fundamentalist whom I respected for his commitment to his religious values, with some difficulties he was having on a course paper. He told me that in his next assignment, he wanted to use Commander's Time to "win converts for Christ." I told him that he was free at appropriate (read: off-duty) time to proselytize, but that he was legally (and quite correctly) forbidden to use "Air Force time" for such overt religious activity. I insisted that he get prior counseling from the Judge Advocate General, which he did. But what you can do is talk about the Army values: what they mean; what kinds of character they foster; and how incompatible they are with egoism or with ethical relativism. Then you "walk the talk"; you practice what you preach.

Although you and I would contend that the values we cherish derive from religious precept and a treasury of spiritual values, the guidance you provide soldiers must not be rooted in a given religion, for soldiers are free to be either religious or nonreligious. But they are not free to refuse to abide by the warrior ethos. As one of the Army's publications tells us, "The warrior ethos concerns character, shaping who you are and what you do. In that sense, it's clearly linked to Army values such as *personal courage, loyalty to comrades, and dedication to duty*. Both loyalty and duty involve putting your life on the line, even when there's little chance of survival, for the good of a cause larger than yourself. That's the clearest example of *selfless service*."[8] We know that ethical egoism doesn't always clash with military service, but it's very likely to, especially when the chips are down. So I think you can start soldiers on the road to character—on the road to professionalism—by teaching them that loyalty builds organizations and can save lives.

But remember that however necessary loyalty is, it isn't sufficient for the good person or the good soldier. Neither is utilitarianism—or its alter ego, proportionalism. But we'll get to that later.

Postmodern Military Challenges

At the same time that recruits to the profession of arms enter, much too frequently, from a background of religious indifferentism, moral confusion, and values anarchy, they are confronted after training with a frenetic deployment schedule and with military-moral situations that would try the wisdom of Solomon and the patience of Job. Have you read Martin Van Creveld's *Transformation of War*? There he talks about how war is becoming the struggles of primitive tribes.[9] Navy Capt. Larry

Seaquist uses the term "community wars" to describe the rogue politicos who use fear and brutality to create "soldiers" with no inkling of the law or morality of the West: "The new faces of war are a ragtag collection of ne'er-do-wells, teenagers, and ordinary citizens temporarily dragooned into service by the local thug-in-chief." But, he says, "the rabble regularly outgun us," and they can be viciously efficient, such as when rampaging Hutus slaughtered—with handguns and machetes—some 800,000 of their neighbors in about a hundred days. This is a more lethal record, Seaquist tells us, than that of Hitler's murderers.[10] You read much of Robert Kaplan's stuff when you were at the War College. Sure, he can be incredibly self-promotional, but his admonitions about anarchy on the horizon seem altogether too prescient to ignore. The list is interminable: Rwanda, Haiti, Kosovo, Sierra Leone, Chechnya, Colombia, Bosnia, Liberia, Afghanistan, and on and on. As Army Col. A.M. Coroalles has said about training junior officers in field operations: "You've got to educate folks in [solving] problems they've never seen before."[11]

I am reminded of a passage in a book published in 1989 by Paul Seabury and Angelo Codevilla. The book appeared as the Cold War was ending and the Berlin Wall was coming down. It was a time of great euphoria, remember? Soon we were discussing "peace dividends," and we began to dismantle—to "down-size" and to "right-size"—our armed forces. "Today's world," they wrote, "is filled with more people animated by greater hatred and possessed of more means to make war than ever before."[12] Then came the Gulf War—and a reminder that there really are bad guys out there. In Army circles, T.R. Fehrenbach's book about Korea used to be very popular; it's time to re-read it, for it tells us about the need to fight, and perhaps die, in the mud.[13] And that is not a message that plays well in a postmodern society such as ours, though there seem to be intimations of resolve thus far in American public attitudes toward our military operations in Afghanistan. Although we want to believe well and to act wisely, we know that we live in times of great political and ethical challenge. We know that evil exists and, in a word, we know that life is *tragic*.

When Samuel Huntington's book about the clash of civilizations was published, a number of people criticized it—and occasionally on solid ground. I remember, however, how much you said you got out of it after a careful reading. Remember this point? "The West won the world not by the superiority of its ideas or value or religion . . . but rather by its superiority in applying organized violence. Westerners often forget this fact; non-Westerners never do."[14] Air Force Col. Charles Dunlap's 1996 article "How We Lost the High-Tech War of 2007" put it starkly: the bad guys' strategy was and is "to make warfare so psychologically costly that the Americans would lose their 'will to win.'" Dunlap even theorizes that an enemy state might drop a nuclear weapon—perhaps purloined—on one of its own cities, blaming us. Dunlap quotes Ralph Peters who told us that we will face warriors who have acquired a "taste for killing," who are irrational by our standards, who commit atrocities "that challenge the descriptive powers of language," and who "will sacrifice their own kind in order to survive."[15]

Remember the story I told you about my lecture at the Marshall Center in Germany where I spoke about ethics to a number of officers from eastern European and Soviet successor states? One of the officers, utterly perplexed by my talk, told me essentially that I was "clueless," for it was obvious that I failed to understand that success in warfare meant (1) winning and (2) returning home, regardless of what you had to do or whom you had to kill to accomplish those objectives. When I pressed my points about how it is wrong to violate moral or ethical standards or the Bible or the Koran or . . . whatever, he gave me the same kind of look you might give an otherwise intelligent, cosmopolitan man who expressed serious belief in Santa Claus. The more I talked to him about ethics, the more I felt like Pollyanna or Micawber or some other ultra-idealist from the world of fiction. The only rule that made any sense at all to him was that there are no rules in warfare. I was there that day to teach him; I fear he taught me—and I didn't like the lesson.

Remember a few years ago when North Korea threatened to attack South Korea if international sanctions were imposed upon them because of their continued noncompliance with the nuclear nonproliferation treaty? The Clinton administration responded by saying that North Korea would receive international assistance in developing peaceful nuclear power, plus hundreds of thousands of barrels of oil. The inspections did not take place. As Donald and Frederick Kagan have pointed out, the Clinton administration "declared victory, but even at the time it was laughably clear that all we had accomplished was to bribe the North Koreans to back away from their threat." Their conclusion is that we have learned to buy off the bad guys. But how long will it last? "Will we," they ask, "like England [in the 1930s], one day find ourselves facing both an urgent need to rearm and insuperable political and economic pressure not to?"[16]

Throughout our history until the Korean War, our national enemies were, at least in terms of military science, often similar to us—the British, the Mexicans, the Spanish, and the Germans of World War I. There is even some such similarity with regard to the Germans and Japanese of World War II in terms of "centers of gravity" and the like. But it's altogether likely that the road-warrior societies of the future—filled as they are with boy-soldiers or with criminals or with approaches to waging war that we would properly regard as ethically insane—will challenge us militarily and morally as never before. The Korean War, of course, is still not legally resolved—and we are still debating today what happened at No Gun Ri more than fifty years ago. The Vietnam War serves in many ways as an example of the kind of "community war" you'll have to wage in the future.[17] And even as a full colonel with more than two decades of Army experience and education, in a sense what you know of Vietnam is what you know of, say, World War I—it is all book learning to you.

Remember that even the Gulf War, which seems to fit into the mold of World War II as a "good war," was unpopular on many campuses and that Congress came very close to denying President George Bush the latitude he needed to prepare for likely combat there. We remember the dissent against the Vietnam War, but we have largely forgotten how unpopular the war in Korea had become by the fall of 1952, when Eisenhower and Stevenson ran for the presidency. To most Americans, the best part of the Gulf War was that it was very short, although we know certainly now that it was too short—and that Saddam's military might was not destroyed to the extent that it should have been. There most certainly is an American character, and, with all its good and bad, it's manifested in all of us who are or who have been in uniform.

In no way do I mean to challenge or to question your own commitment and courage as a soldier during the Gulf War—or to impugn the professionalism or the patriotism of the thousands of other GIs who faced Iraqi soldiers—but what do you suppose might have happened had Saddam mustered the military might to hold out for many months or even many years (as some, recall, were predicting on the eve of hostilities)? I thought that the *moral* reasons for that war were compelling. Many disagreed, of course, contending that we were trading blood for oil. One of the authors of *Just War and the Gulf War*, George Weigel, has pointed out that much of mainline Protestantism and many in the Catholic Church have abandoned the Christian realism of Reinhold Niebuhr, Paul Ramsey, and John Courtney Murray, substituting in its place "psychologized and quasi-utopian understandings of international public life, which suggest the possibility of a world without conflict."[18] Remember the priest at Notre Dame who told you that you could not be at the same time a soldier and a Catholic? I have run into that (mistaken) sentiment many times in my career, although I think it is profoundly unbiblical—and even heretical, a re-appearance of Pelagianism.

But I guess that gets us where, really, you wanted me to go. Given the realities of a postmodern United States and the increasing geopolitical ugliness of the world which you have to deal with, what is military ethics today all about? How do you maintain true professionalism in the face of the kinds of missions—like your present one—you find yourself committed to? As you once beseeched me, "Tell me how I can deal with ugliness and how I can keep my soldiers from becoming what they're resisting." The new Army Leadership manual touts values, attributes, and character which are utterly at odds with the kind of savagery (think of Bosnia and Kosovo) our soldiers often have to deal with *as peacekeepers*. All soldiers see their own clash of civilizations when they deal with the kinds of military and moral problems beyond their ken. A high-ranking soldier-peacekeeper, speaking at the Air War College under the promise of non-attribution a couple of years before you were here, told us that as "negotiations" began between two warring gangs in his area of operations, one gang sent its agenda to its rival gang in a large plastic hatbox. Why not in an envelope? Because an ordinary envelope could not have contained the message plus the bloody head of a child from the other gang. Of course that other gang then replied with its own message, together with the head of a child from the first gang.

Someone has said that certain matters in ethics should cause decent people to shudder. Imagine using the head of a murdered child as a macabre visual aid to make a point about your agenda for upcoming negotiations. What kind of savage could do such a thing? And how does it brutalize the mind and soul of the disinterested onlooker? If you can find the time, read Anthony Loyd's book, *My War Gone By, I Miss It So*. It is filled with butchery and barbarism as reported by a journalist who has written from Chechnya, Afghanistan, Sierra Leone, and Kosovo. I could quote at length from some of the stomach-turning scenes painted so powerfully by Loyd, and others like Mark Bowden in *Black Hawk Down*.[19] But you know the stories, Pete. How is it possible for anyone to commit the kinds of atrocities described there? But, similarly, how is it possible for anyone to stick scissors into the head of a baby emerging from the womb into daylight and, through the hole thus created, suck out the baby's brains?

Remember Senator Harry Truman's comment during World War II that we should just sit back and let the German Nazis and the Russian Communists destroy each other? You can read that opinion today about many of the world's conflicts. Just sit back and let the brutal bastards kill one another. But isn't that an indication of how far down the slippery slope of life we've already come? Don't we have a moral—and, arguably, military—duty to intervene to prevent slaughter when and where we can? There is a powerful news film now available that shows how a contingent of international peacekeepers were ordered out of a place in Rwanda in which hundreds of refugees had gathered, pleading with and begging the peacekeepers not to depart, for the refugees knew that when the soldiers left, they would be hacked to death. They even asked the soldiers to shoot them, preferring to die by bullets rather than by machetes. So, what do you do when your commander tells you to leave the helpless, knowing that they are about to be slashed and hacked and murdered, every one? And how long do you stay there? And how many do you evacuate and to where?

Moral Decision-Making

In the often brutal arena of world politics, how do you find the right course and keep to it? I remember some of our long conversations about this subject, and I can summarize what we agreed upon—en route to what we disputed. In my War College classes and lectures, I try to give students "tools" to use in ethical decision-making. I argue that there are three pieces to the ethical puzzle:[20] customs, rules, and outcomes. *Customs*—the mores, the usual expectations of one's culture—can, to a certain degree, guide us. In Shirley Jackson's short story, "The Lottery," a village held a lottery each year to choose a family; the "winning" family participated in a follow-up lottery to choose one of its members—who was then stoned to death to ensure good crops. Lousy meteorology! But if that is what the society's customs dictate, who are we to challenge them? After all, that would put us in the ranks of those who challenge cultural relativism. But of course we must challenge cultural and ethical relativism (cf. 2 Cor 10.12), provided we see events through the prism of a higher understanding. In this case, of course, empirical science testifies to the sheer stupidity of murdering someone to "ensure" good crops. Nonetheless, customs *are* important, and one should not lightly dismiss tradition. It's part of the picture, but often not the entire thing.

It is certainly critical to soldiers that they not "go native." If in the course of accomplishing their mission, they adopt the customs of savage people—some of whom decapitate little children in order to make "negotiating points"—they will have thereby betrayed not just Army values and American norms, but standards (the permanent things) available to anyone of "ordinary sense and understanding." These standards (so often ignored today as we earlier discussed) are best seen in the Sermon on the Mount, which is, as Father Servais Pinckaers, O.P., has told us in his books, "a model of the ancient moral catechesis" and a "charter of the Christian life."[21] Some things, truly, are written on our hearts (Bar 3.7, Ez 37.26, Jer 31.33, Heb 10.16, Rom 2.15); there are some things we cannot not know, although, as J. Budziszewski has written, we can easily deceive ourselves about what we know.[22] Army values are meritorious, secular reflections of transcendent, sacred virtues; such military core values, understood and properly applied, can help us remain constant to truths "written on our hearts."

In addition to customs, there are *rules* which we should and usually can follow in deciding upon a proper ethical course of action. The rules found in Exodus 20—the Ten Commandments—are not a bad place to start, and one does not have to be Jewish or Christian to accept the spirit of those commandments. We also have the positive or man-made law known as the Uniform Code of Military Justice, whose rules are pretty clear. Beyond that we have Rules of Engagement that can guide soldiers in specific tactical circumstances. When I was in Officer Candidate School at Fort Benning (a long time ago, I admit), we learned the five-paragraph field order which, at that time, did not provide for "Commander's Intent." That is an important concept, proving that there is unlikely to be a single rule which clearly governs soldierly conduct in every circumstance. It's important for subordinates to understand, not just their superiors' literal orders, but also the pattern and purpose of their thinking. I don't mean to sound patronizing about this, and I'm sure you would take great pains in explaining this to your own soldiers, but there are surely times when rules and orders must be broken. One hopes, however, that soldiers who ignore or contravene orders will, when time and circumstances permit, make a clean breast of affairs with their superiors, providing the rationale for their judgment—and not seek to cover up their decisions. Remember Ambrose Bierce's civil war short story, "One Kind of Officer"?[23]

In addition to customs and rules, we must involve ourselves with *outcomes* or *results*. A part of any good ethical judgment is concern with circumstances, including consequences. If we act, for example, out of a fear of death, this can diminish our responsibility for the possible evil of our act. But the teaching we accept is that circumstances "can make neither good nor right an action that is in itself evil" (*Catechism of the Catholic Church*, #1754). The critical point is that good actions are not determined *only* by situation or by circumstance. We begin to approach our major area of disagreement, don't we?

As you used to tell me when you were here at the War College, a soldier may not have time to sit in a library carrel and casually ponder the "tools" of ethical judgment. When the chips are down, you said, soldiers sometimes have to act *now*, and when that is true the sole reliable guide to action is proportionalism. I know you like the idea of "proportionalism"[24] because, as a Christian, you think it's more morally acceptable than utilitarianism or consequentialism, which judges actions only on the basis of outcome or probable outcome. Proportionalism, by contrast, suggests a more measured, more theologically "decent," weighing of alternatives—not excluding the best projected final outcome—a kind of teleological ethics at least partially in keeping with the Church's ancient teaching. My rejoinder that proportionalism is mere crass utilitarianism wearing a miter did not impress you.

Remember the case we used to fight about—Bernard Williams's story?[25] You're on vacation in X-land and, while on a hike, are seized by revolutionaries who also have taken twelve X-lander police captive. They plan to let you go regardless, but tell you that they will shoot all the X-lander police unless you shoot one of the police, which will then prompt them to release unharmed the other eleven. The police tell you that they have reason to believe that the revolutionaries will keep their word and the deal seems rational to them: eleven will live if one dies; all twelve will die if one doesn't. To help you, they're willing to hold a lottery to choose the unlucky sacrificial victim. You will be given a small pistol with one round only. Do you shoot one of the hostages?

Your argument was "yes," citing proportionalism. "The good of the many outweighs the good of the one," you said, prompted by some movie you'd seen. Never take your ethics from movies, Pete, unless maybe you've watched *The Ten Commandments*. Not to act in this circumstance would be sheer cowardice, you told me; not to act would preserve your "purity" but only at the price of twelve deaths. Surely, the greater good would be to shoot one of the hostages.

You know my argument: To shoot one of the hostages is not only doing the terrorists' work for them but it is purely and simply an act of murder. You are implicitly telling me that the life of the hostage you shoot has value only insofar as it is related to others' lives; it has in and of itself no value and is therefore dispensable. But life is sacred—I know you believe that as a Christian.

I well recall your rather heated comeback: that because you believed life is sacred then as a Christian you *must* act—not to take a life but to save eleven lives! I countered with the evidence from a TV show (your movie vs. my TV show—quite an intellectual slugfest!). On a M*A*S*H episode Hawkeye Pierce, some soldiers, and a number of South Koreans are on a stalled bus behind enemy lines in Korea. The North Koreans are sending patrols out into the area, but the bus is camouflaged, waiting for the patrols to depart, allowing the Americans to jumpstart the bus and head back to friendlies. But a baby is crying and may give away the position. Is it all right to smother the child to protect the others from being noticed by the North Koreans? You hesitated. Then I unloaded. It's your wife Maria and Josh as a baby, crying. *Now* is it OK, Pete? *Now does the good of the many outweigh the good of the one?* I know that made you angry—the only time we actually *quarreled* during an "argument."

A few days later, I recall, you told me that I had won only by a debater's tactic—and that it was unfair to personalize the argument. Then you hit me with another example to try to defend proportionalism. A U.S. Army infantry platoon is in prepared positions when armed enemy soldiers approach on line, each soldier carrying a child. You remember I called this example impossibly conjectural (or dumb), and I thought it was—until I read a similar story in Mark Bowden's *Black Hawk Down*.[26] We are back to ugliness, aren't we? So, is it all right to shoot the children in order to retain the positions and, most likely, to keep your soldiers from being shot if they try to retreat to supplementary positions? And how far can those soldiers retreat? Twenty yards? Twenty miles? You were consistent. You told me that, although you'd do everything in your power not to hurt the children, you would not surrender your position or withhold protective covering fire in the event of a pull-back. Your argument is that the likely greater good, in both the short term and long term, is the defense of the positions. Your "clinching" argument is that the injuries to or the deaths of the children, however regrettable, might best be understood as "collateral damage."

Given the ubiquitous evil I talked about early in this letter, I think that, tragically, this kind of situation may well become more common than we care to think about. I never expressed an opinion as to the morally dictated course in this last situation because, as you know, I rather cavalierly dismissed it—until encountering *Black Hawk Down*. Pete, I think . . . well, I think you're right; but I also think you draw the wrong conclusion. There is a parallel, but a slim one, between shooting the hostage policeman and the hostage children. You will have to murder that policeman in cold blood. You intend his death. But when firing at the advancing enemy soldiers, you most certainly do not intend harm to the children they have as hostage. In fact, you'll search for a way out—flanking fire? fire at the exposed parts of the enemy soldiers? getting to them for hand-to-hand combat? Of course, your soldiers will do everything possible, if they fire at the advancing soldiers, not to hit the children. But if I were the lieutenant there, I too would defend my position—even if it meant endangering the children. My small arms fire would, I am afraid, have two results: defense of my position, which is good; and the deaths of one or more of the children, which is bad. We do not seek the bad, either itself or in the course of seeking another end. The good effect is not achieved directly through, or as a consequence of, the bad effect which happens together with the good and which is not wanted or willed by any of us defending our position and ourselves. And the good effect, arguably, outweighs the bad.

Physicians and medical ethicists often use this concept (known as the "Principle of Double Effect") to think through the problem of giving heavy doses of painkiller, which might be life threatening, to terminal patients. If we believe that life is sacred, then we do not wish to hasten death, but we do wish to mitigate pain. A doctor can ethically give the painkiller because, although it's dangerous in this scenario, the bad effect (danger to life) is not intended, is outweighed by the good effect (relief from pain), and does not cause the good effect but is simultaneous with it.

As Catholics, you and I are both absolutists and believe in an objective order of morality. We know that the first and fundamental task—the first absolute of any legitimate warrior ethic and of all genuine soldiers—*is to protect innocent life*. As Gen. Douglas MacArthur said in 1946 when he reviewed the case of Japanese General Yamashita, who had been convicted of war crimes: "The soldier, be he friend or foe, is charged with the protection of the weak and innocent. It is the very essence and reason for his being." But we also know that, in this vale of tears in which we live, such is not always possible or practicable. We know that the innocent can suffer—and have suffered, sometimes terribly. But it may be the Protestant ethicist, Norman Geisler, who helps us the most with the question of whether absolutes can be "compromised." After all, if something is eternally true and perpetually binding, there are no righteous exceptions to it. You and I have just said that we'd shoot toward—although hoping to God we wouldn't hit—babies. And we are prepared to kill our soldier enemies. How do we square that with the Commandment? Geisler refers to "graded absolutism" (also called "qualified absolutism" or "contextual absolutism").[27] I've used the term Universal Ethical Obligation (UEO). A UEO is a working absolute, but it admits of competing claims to conscience; more precisely, it can give way to another UEO that supersedes it in certain circumstances. For example, I have the obligation to protect my men and my position; I also have the obligation to protect the lives of the children (and, insofar as possible, even the lives of the advancing enemy soldiers). I now have to make a reasoned, prudential, virtuous choice between competing UEOs—what might be thought of as "dueling duties."

We thus come full circle, back to the means by which we make ethical choices in the midst of trial and turbulence, of tragedy and trauma. We explore local and military *customs*; we educate ourselves about that treasury of moral doctrine and dogma which we humans have learned over the centuries—*rules*. We think hard about *outcomes*, both certain and almost certain, realizing that the outcomes must be broadly construed to embrace prudential consideration of attendant circumstances. Then we make a decision, re-adjust as necessary, and report our reactions and results to appropriate authorities as time and events allow. But there is nothing mechanical or formulaic about this decision-making, for the key ingredient is *you* and the *virtues* which are your character (cf. 2 Pt 1.5-9, Gal 5.16-26).

I come back to the proposition that proportionalism is wrong because, no matter how you dress it up, it is sheer utilitarianism. Again, as Catholics, we would agree with Pope Paul VI, quoted by John Paul II in *The Splendor of Truth* (or *Veritatis Splendor*): "Though it is true that sometimes it is lawful to tolerate a lesser moral evil in order to avoid a greater evil or in order to promote a greater good, it is never lawful, even for the gravest reasons, to do evil that good may come of it (cf. Rom 3:8)—in other words, to intend directly something which of its very nature contradicts the moral order. . . ." John Paul adds that "circumstances or intentions can never transform an act intrinsically evil by virtue of its object into an act 'subjectively' good or defensible as a choice."[28]

"Dirty Hands"

Thus, there are some wrongs we must never do. Does that imply that there are wrongs you *can* do? Perhaps. So now I will tell you what I did not when you were at the War College: there may be *some* wrongs, in *some* cases, which we must not shrink from doing. And I greatly fear that the moral evil of the world—and your position in it—may put you in circumstances requiring *dirty hands*.

We have fallen a long way here, from absolutes through UEOs and the Principle of Double Effect to Dirty Hands, by which I mean the deliberate commission of a wrong. There is no way to justify this, no way satisfactorily to explain it; doing Dirty Hands business means that you have fashioned your own morality in place of God's, however much you may not want to. The only standard you may be able to apply here is your judgment at the time of crisis; it is the ultimate kind of "situational ethics." Michael Walzer might *disagree* with me when I say that Dirty Hands means his concept of "supreme emergency" at a personal level; but it is as good an explanation as I can make. Trying to use Double Effect in these circumstances is specious, I believe. As the late Elizabeth Anscombe once cogently put it: "It is nonsense to pretend that you do not intend to do what is the means you take to your chosen end. Otherwise there is absolutely no substance to the Pauline teaching [Rom 3.8] that we may not do evil that good may come."[29] There are some things I must never do because they are "always gravely illicit" (CCC #1756); in other words, there are things I *absolutely* must not do—e.g., murder—because it is everywhere and always wrong. As one man, I know that I cannot do evil that good might result, regardless of possible or probable consequence.

As a steward of the public good or as a military officer, however, are there times and places for me to do what is evil so that justice might reign? I know you'll agree with me that we have a solemn responsibility to preserve our integrity, our personal moral purity. But as a public steward—as a military officer—you are not free so easily to indulge your conscience, I'll admit. If the Secretary of the Treasury today decides to give all of his personal wealth to a worthy charity, we might well cheer his great generosity. But he is not free to give away *public moneys*, of which he is the custodian or steward. Suppose President Franklin Roosevelt in 1941 had been a pacifist; he might have chosen not to prosecute the war. He could not have ethically surrendered America to Japan, however, so even as a pacifist he might have had to compromise his conscience by leading our country into the just war of 1941.

Are there times when, in your capacity as an infantry officer, you may or must use Dirty Hands—that is, willingly do an evil act? Isn't this only ultimate utilitarianism, which I have just repudiated? Consider: You are, say, a battalion intelligence officer and you have virtually certain information that two enemy soldiers in your custody know the location of units about to ambush your battalion. You're in a helicopter with the prisoners. Can you push one into a net beneath the helicopter, deceiving the other about the "death" of his buddy, probably inducing him to talk when he thinks he's next to go? We know that this is wrong; that it is torture; that it is a war crime. But we know, too, that the pressures of command are compelling; the idea that if you torture—"just a little"—you can gain precious information and perhaps save your men's lives is almost irresistible. So it proved to be recently for an Army officer in Kosovo who held an unloaded weapon to the head of a Kosovo Albanian during interrogation, leading, in time, to his being charged with conduct unbecoming an officer and communicating a threat. The officer later admitted, "I was totally wrong." But isn't that what soldiers do, "communicate threats"? Sure: but not by becoming vigilantes or thugs. One soldier had it right in saying that the unit involved in the "interrogation incident" was "a street gang in Kosovo—an embarrassment to the United States Army."[30] When you torture you destroy, first, your integrity; and, second, your career, for it will follow in flames, as probably it should, for no institution can knowingly tolerate or accept such practice by its members.

When do we cross the line separating tough, hard-nosed, aggressive military operations from intimidation tactics inconsistent with Army values? To lament the difficulties of peacekeeping duties is justifiable, but such duties do not excuse improper behavior. Remember when Maria, Rebecca, you, and I watched the movie *Dirty Harry*? As I recall the plot, policeman Dirty Harry captures a bad guy who, trying to extort money, has buried alive a child whose oxygen supply will shortly run out. Can and should Dirty Harry torture the bad guy to find out where the child is? This is not something he wishes to do *for himself*; but it's not something, either, that can be accepted as police practice. It is the wrong thing to do; but is there, in this single and unrepeatable case, a "*wronger*" thing to do—that is, not to torture the criminal? If this were Dom or Josh, wouldn't you torture that kidnapper-killer in a second? But isn't that why doctors don't operate on their own family members? And isn't that a key reason that we have commissioned officers, whose terrible task it can be to insist upon the dry logic of law and high purpose even in the heat of passion and the desire for information at any cost, for vengeance at any price?

Pete, what I am saying here can be reduced to a simple, yet very harsh, sentence: As a soldier you are prepared to risk your life for your country; but, as a soldier, you may also be risking your *soul* for your country—but *risk* does not mean *relinquish*. You know that Catholic theology tells us to avoid the occasion of sin.[31] One could argue that Christians ought not to involve themselves in either political or military affairs, for, if my reasoning is right, you may be involved in situations, especially in the moral squalor of contemporary world affairs, that could lead to your making decisions that might be deeply troubling from an ethical standpoint. You may be risking your soul for your country. And that is a risk that no country has a right to ask of its citizens or of its soldiers, for we know we must always obey God before men (Acts 5.29). Still, what are we to do? Can we withdraw into a politically quiet shell or enclave, awaiting the Second Coming and disdaining affairs of state? Have we no obligation to love God and neighbor in the realm of politics? We do, of course; and we are obliged to act the very best we can, torn as we will sometimes be between competing obligations. In our country, with its eternal optimism, we forget that life is, at its heart, tragic. I suspect that the years ahead will re-teach us the lessons of life and politics which informed classical political realism. All around us are military situations in which, as Seabury and Codevilla put it, "The only rule is that there is no rule."[32] How do you conduct yourself as a man of character and principle, as a gentleman, and, most critically, as a Christian, in circumstances of moral nihilism? We know that the heart of *jus in bello* is discrimination and proportionality. But maybe Charles Dunlap is right in saying that "when societies propagate evil, democracies must be prepared to visit upon them force so staggering it will produce fundamental change."[33]

Clean Hearts—and Conscientious Leaders

Of President Harry Truman, for example, what are we to think? Did he perform his trust wisely and well in the terrible circumstances of August 1945 when he twice ordered atomic bombs to be used? As Christians, we ask mercy on his soul; as Americans we ask whether he acted and reasoned rightly. I believe he did. Far from contending that the ends always justify the means, I argue only that there can be cases of such extraordinary moral and political import—almost bizarre, very rare, probably unique—that even as there is no precedent *for* them, neither should there be precedent *from* them. *This is why leadership is so critical*, for I am "betting" in these cases, not upon customs or rules or even outcomes (for no one is clairvoyant), but upon men and women of great—one wishes to write "noble"—character.

In situations of momentous military, moral, or political pressure, it is imperative that we have good men—people—as leaders. I do not say here "good leaders" but good men as good leaders. "Army leadership begins with what the leader must BE, the values and attributes that shape a leader's character. It may be helpful to think of these as internal qualities: you possess them all the time, alone and with others. They define who you are; they give you a solid footing. . . ."[34] Retired Air Force Brig. Gen. Malham Wakin once gave us the same advice: "Fully accepting the

Aristotelian wisdom that moral character develops out of repeating good actions, that it cannot be ordered but can be exemplified and imitated, our advice to those who aspire to become worthy military leaders can be none other than that of the ancient Chinese sage: 'The way to do is to be.'"[35]

What I am saying is that the good man as a leader knows when to follow orders and when not to. Moreover, he also knows *when to make exceptions*. When a good man is confronted with a horrifying situation such as the one in which enemy soldiers with children as shields are advancing on his troops' perimeter, or when a good man must decide whether to get into the *Enola Gay* and drop an atomic bomb on an enemy, what specific chapter and verse, what precedent, what rule can adequately inform him? In such appalling circumstances, whom does he consult? Here, he needs not information but formation. He must have developed, through continual practice (including reading, study, and prayer or meditation), the strength of character to make, if entirely necessary, Dirty Hands decisions (cf. 2 Chr 30.18-19; Ps 51.12). We are, as Aristotle tried once to tell us, what we continually do. We might add the corollary that we continually do what we fundamentally are.

Now consider our current plight. As Barbara Tuchman has put it, government remains "the paramount field of unwisdom because it is there that men seek power over others—only to lose it over themselves." And: "Although professionalism can help, I tend to think that fitness of character is what government chiefly requires."[36] Well, do the past few years make you optimistic about our likelihood of developing "fitness of character" in our leaders? If our postmodern society is corrosive in its effect on character-building in homes and schools and churches—as I think it is—then where will our genuine leaders come from? How can we trust them?

The military tries to answer these questions as far as officers are concerned. Read your commission again. It is a statement of trust and confidence in your patriotism, valor, fidelity, and abilities. That is the teaching, also, of the Church: "Those too who devote themselves to the military service of their country should regard themselves as the agents of security and freedom of peoples. *As long as they fulfill this role properly*, they are making a genuine contribution to the establishment of peace"[37] Most particularly and most urgently, I as a citizen trust and confidently expect that you as a soldier will make *proper* decisions as a public steward of character and competence. And I believe that you will carry out your duties with honor, not only in routine matters, but in the gravest circumstances, where, ultimately, the decisions you make flow from all you are and all you believe—even to the point of Dirty Hands.

If you must make that kind of decision, I have to trust in *you*. This is not an attempt to deify you or to exalt subjectivism or ethical egotism. All I am saying here is that there may be times when your life and even your soul may be on the line. I believe in "virtue ethics" (or aretaic ethics)[38] to the point that I concur with Anton Myrer, author of *Once an Eagle*, which all Army officers ought to read: In it, the protagonist, Col. Sam Damon, tells his son (and us): "If it comes to a choice between being a good soldier and a good human being—try to be a good human being."[39] No one has ever put it more concisely, more cogently.

I want you as a commander to have a well-formed conscience. I hope you will follow the dictates of that conscience, *despite* the rules we know and accept, if exigent circumstances so demand. And I hope you then face squarely and manfully the consequences of your decision and your action. But what your fellow citizens expect of you as a soldier—and the report of your stewardship that you must render—is, in fact, not much different from our higher obligation (as in Rom 2.6-11 or 1 Pt 4.10).

Remember the early 1960s film *Fail-Safe*? The United States has accidentally dropped a nuclear weapon on Moscow due, principally, to technological failure. The Soviets are about to retaliate when the U.S. president, trying to avert World War III, orders a U.S. Air Force officer to drop a nuclear weapon on New York City—to placate the Soviets. Talk about Dirty Hands! Can the President order that? Can he, given the scenario, *not* order that? Can the officer fly the plane and drop the bomb? (By the way, in the film, the officer's wife and children are in New York City that day, as is the President's wife.) The officer flies the plane, drops the bomb, and then kills himself.

How do you read me, Pete? I should decide that dropping the bomb is a terrible necessity. Right? My Dirty Hands moral theory must come into play here, mustn't it? No. I cannot drop that bomb, nor can I applaud the President for ordering it or the officer for doing it. I understand the President's decision, but I cannot condone it. These are wholly innocent American citizens; international nuclear war is not a certainty (although as painted for us in the film and book such war is admittedly very likely); and the President and the Air Force officer, finally, cannot kill their own countrymen in the name of national security. I'm saying that the ends here do not justify the means.

I know: I just finished saying that it's possible, *but extremely rare*, that good ends can justify bad means. Here, in the *Fail-Safe* scenario, I cannot accept that rationale. Upon reflection, neither would you. For we can never justifiably call "collateral damage" what is, in fact, "the indiscriminate destruction of whole cities or vast areas with their inhabitants" (CCC # 2314). To do such a thing would not merely dirty our hands; it would poison our hearts and pollute our souls.

The Things We Cannot Not Know

In 1972, Barbara Tuchman, whom I mentioned in a different context above, spoke at the Army War College on the subject of "Generalship." She told the audience that the West Point motto of "Duty, Honor, Country" may no longer suffice. "Country," she said, "is clear enough, but what is Duty in a wrong war? What is Honor when fighting is reduced to 'wasting' the living space—not to mention the lives—of a people that never did us any harm? The simple West Point answer," she continued, "is that Duty and Honor consist in carrying out the orders of the government. That is what the Nazis said in their defense, and we tried them for war crimes nevertheless." Tuchman appealed to the soldiers assembled there at Carlisle Barracks *to think*.[40]

But Tuchman actually did not have to urge the soldiers to think. Thinking *should* be regarded as essential to professionalism. The *Manual for Courts-Martial*, for example, tells military personnel that "it is a defense to any offense that the accused was acting pursuant to orders unless the accused knew the orders to be unlawful or *a person of ordinary sense and understanding* would have known the orders to be unlawful."[41]

We may understand the phrase *ordinary sense and understanding* to refer to conscience. The central idea, after all, is that if you receive an order from your two-star, you are to obey it unless you know the order is unlawful or unless *you should know* (because it's "written on your heart") it to be unlawful. In other words, the *Manual for Courts-Martial* insists that there is a standard of general knowledge and of ordinary morality according to which our actions may be judged. The ordinary private in Lt. William Calley's platoon at My Lai, if he actually did not know that the order given there to murder Vietnamese was wrong, should have known. All decent people know that! It is no defense to say, "I didn't know such-and-such was wrong," for the rejoinder is that you most certainly should have known. We are back to an earlier point: *There are some things that we can't not know.* That is very, very important. It takes us, if we let it, into theological waters, such as those found in 2 Cor 3.3, in Heb 8.10, and in CCC # 1958.

Some would argue that, in a multicultural society, we can do little in the way of discussing ends or purposes or substance. That is why the military oath of office is sworn in support of the Constitution. At the same time, however, Pete, you and I realize as Christians that our first and foremost obligation is to God. Remember Sacred Heart Basilica at Notre Dame, where under the archway of one of the doors are chiseled the words *God, Country, Notre Dame*? To us, God must always come before country. We have what we might call a *perfect duty* of obedience to God through our formed conscience and a compelling but still *imperfect duty* to the state and its representatives and officers (as in Mt 22.21, Rom 13.1-7, 1 Pt 2.13-17, Ti 3.1, CCC #1880).

You've heard my argument on this before. We live in a society in which we often hear about *rights* (and perhaps not enough about duties). But we very rarely hear about *Right*. The various core values of the armed services are pathways to *Right*, by which I mean virtuous ends. If all of what our services do to maintain our national security is done without concern for—or in violation of—standards of good and evil, of honor and shame, that are obvious to anyone of "ordinary sense and understanding," then we have paid too high a price, and we have destroyed what we ostensibly sought to preserve. If we are going to become like the barbarians who routinely murder and commit other savage practices (such as allowing rape by soldiers as a means of terrorizing the populace, thus facilitating military control over the defeated country), then why not simply join the barbarians?

The object of military ethics, after all, is to help soldiers understand that there are things they cannot do, lines they cannot cross, means they cannot employ—even if they thereby sacrifice certain military advantages—because, by so doing, they destroy, not just *rights* but *Right* itself. *Right* refers to those things we cannot not know; it is prima facie virtue, a perception of what is true and transcendent, despite what divergent (and occasionally defective) cultures might say; it is an understanding of the mind and heart about what makes life itself holy. As John Paul II has written, people exist in particular cultures but are *not* "exhaustively defined by that same culture." And: "If there is no transcendent truth, in obedience to which man achieves his full identity, then there is no such principle for guaranteeing just relations between people. Their self-interest as a class, group, or nation would inevitably set them in opposition to one another. If one does not acknowledge transcendent truth, then the force of power takes over, and each person tends to make full use of the means at his disposal in order to impose his own interests or his own opinion."[42]

Isn't that, essentially, what it's all about? Isn't that, finally, why the military ethic is *good*? To the extent to which the warrior ethos, the military ethic, or the profession of arms participates in what is good and true (cf. Phil 4.8), you as a soldier *owe*—a very important verb—allegiance to that ethic, and those you command in the name of that ethic similarly owe obedience. But when countries or commanders or specific orders depart from or, worse, violate a transcendent standard, then citizens and soldiers alike—people of ordinary sense and understanding—must very carefully consider whether the country has forfeited its claim upon them (as in the case of Germany in World War II) or whether the commander's orders have become immoral and therefore illegal.

But there must be a standard—and that, I fear, is what the new barbarism denies and what our postmodern society increasingly scorns. Remember how impressed you were with José Ortega y Gasset's argument that "barbarism is the absence of standards to which appeal can be made"?[43] His book first appeared about seventy years ago; what do you think he'd say of *today's* moral nihilism? Probably much the same thing that Alasdair MacIntyre might say: that "modern politics itself expresses in its institutional forms a systematic rejection of [the tradition of the virtues]," for contemporary government "does not express or represent the moral community of the citizens, but is instead a set of institutional arrangements for imposing a bureaucratized unity on a society which lacks genuine moral consensus." The problem is that "the nature of political obligation becomes systematically unclear."[44]

Teaching What We Are

So, finally, after all this, you'd want me to say how all this can be taught! The services have different words for their core values, even though the meanings are similar. They are necessary, I think, but not sufficient. Philosophy and theology do offer us four virtues (not just "values") which now, in turn, I offer to you. The first is *wisdom,* for without it we can hardly expect to differentiate good from evil. I define it as the ability reasonably to evaluate (and to bring order to) the temporary in light of the eternal and the changeable in light of the changeless. Human nature doesn't change. If it does, how can we expect to read Sophocles or Euripedes or Herodotus or Homer and truly appreciate the emotions expressed there? Some of the lessons, for example, of the Peloponnesian War are timeless. We spend much time seeking information superiority when, so often, the problem we have is wisdom deficiency. And, no, I am not trying to make either you or your men philosophers! But you have a *legal obligation* to judge immoral orders. How can you hope to do that unless there is some serious and substantial education given to soldiers along with solid, hard-nosed training? The cardinal virtues are sometimes referred to as *hinge virtues* because all the others depend upon these four, which in effect are one virtue.

Justice (or truthfulness) is the second cardinal virtue. It refers to habitual right thinking and speaking and worthiness of conduct. *Fortitude* or courage is the third cardinal virtue, meaning firmness in difficulty and constancy of purpose toward the good. *Temperance* or moderation or self-control is the fourth, and it refers to direction and discretion with respect to appetites, instincts, and urges. I regard these four virtues as a composite; you can't have one without the others. What we have built here is a four-component image of the good man or woman. These are not debatable "values"; they are essential virtues, and virtue is conformity to standards of Right (cf. Wis 8.7, Rom 12.2; Col 1.9-12; *CCC* #1804-1809).[45]

So where does the "standard of Right" come from in the U.S. Army, which is a secular armed service of a secular state? It comes from you and from the culture and confraternity of your fellow leaders at all levels, insofar as it conforms to what is the eternal standard (cf. Jn 5.30, 1 Cor 2.14-16, 1 Thess 4.1; *CCC* # 215). And you already know the standard because it is found and exemplified daily by those you respect and admire (although certainly not in all of us and never perfectly in any of us). Every commander—every good teacher—knows that, in the end, he or she will be followed, not because of rank or special privilege or particular position, but because he and she are people worthy of trust and confidence. Those words appearing on your commission mean so much because they, not power or rank, are ultimately what your soldiers will follow when the chips are down. They will follow you, not just because you are *Colonel Garcia* (although that of course is an important part of it), but because you are *Pedro Garcia*, who has the competence of the good soldier and the character of the good man.

You will teach the standard of Right to your soldiers, not just because you *do* it but because you *are* it. When FM 22-100 says "Be, Know, and Do," it has leadership exactly right. Every time you make a serious moral judgment, you become that judgment; every time you issue a command, you not only tell your subordinates what to do but what to be. That is why, in the horrible circumstances in which you or your soldiers might find yourselves in the months ahead in a world seemingly gone morally mad, I trust in you because of the moral compass which is yours from your education, your experience, your expertise. You *do* on the basis of your information; you *are* on the basis of your formation. Ethics, in the final analysis, is caught, not just taught. Your command will do right because of the example you'll set, helping its members to be right.

Well, Pete, it's time to quit. I've tried to write from the heart as well as from the head. You know I like Anton Myrer's novel, which I've already quoted to you once here. I can't resist closing with another excerpt. Colonel Damon tells his son: "That's the whole challenge of life—to act with honor and hope and generosity, no matter what you've drawn. You can't help when or what you were born, you may not be able to help how you die; but you can—and you should—try to pass the days between as a good man."[46] I believe, Pete, that if we try sincerely to know and to do what is good, we can become, if not good, then at least better men. In that effort—toward which the cardinal virtues compel us—lies, not just character, but the fulfillment of doing a noble job nobly.

All the best, Pete!
—Jim

Notes

1. Although I am solely responsible for any errors of fact, interpretation, or expression in this chapter, I wish to thank Col. Alexander Shine (USA, Ret.) for his keen interest and significant contributions to this effort. The opinions expressed in this chapter, of course, do not imply the endorsement of the Department of Defense, the Department of the Army, the Department of the Air Force, Air University, or the Air War College.
2. Neil Howe and William Strauss, *Millennials Rising: The Next Great Generation* (New York: Vintage, 2000).
3. *Montgomery Advertiser*, 9 September 2000.
4. William Bennett, *The Death of Outrage* (New York: The Free Press, 1998), 121.
5. Robert Bork, *Slouching Toward Gomorrah* (New York: Regan, 1996), 275.
6. Frank Sheed, *Theology and Sanity* (San Francisco, CA: Ignatius, 1978), 394.
7. Quoted by Grant T. Hammond, *The Mind of War* (Washington, DC: Smithsonian Institution, 2001), 211.
8. Headquarters, Department of the Army, FM 22-100, *Army Leadership: Be, Know, Do* (Washington, DC: U.S. Government Printing Office, 1999), 2-21.
9. Martin Van Creveld, *The Transformation of War* (New York: Free Press, 1991).
10. Larry Seaquist, "Community War," *U.S. Naval Institute Proceedings* 126 (August 2000): 56-59.
11. *Baltimore Sun*, 4 September 2000.
12. Paul Seabury and Angelo Codevilla, *War: Ends and Means* (New York: Basic Books, 1989), 14.
13. T.R. Fehrenbach, *This Kind of War* (New York: Macmillan, 1963).
14. Samuel Huntington, *The Clash of Civilizations and the Remaking of World Order* (New York: Simon and Schuster, 1996), 51.
15. Charles J. Dunlap, Jr., "How We Lost the High-Tech War of 2007," *Weekly Standard*, 29 January 1996, 22-28.
16. Donald Kagan and Frederick Kagan, "Peace for Our Time?" *Commentary* 110 (September 2000): 45-46; see also their book *While America Sleeps* (New York: St. Martin's Press, 2000), 400-35.
17. On this theme, see David Donovan [Terry Turner], *Once a Warrior King* (New York: McGraw-Hill, 1985). This is an excellent study of use to company- and field-grade officers.
18. James Turner Johnson and George Weigel, *Just War and the Gulf War* (Washington, DC: Ethics and Public Policy Center, 1991), 85.
19. Anthony Loyd, *My War Gone By, I Miss It So* (New York: Atlantic Monthly Press, 1999); and Mark Bowden, *Black Hawk Down* (New York: Atlantic Monthly Press, 1999).
20. Theologians often contend that there are three sources of a morally good act: the object chosen, the end or the intention, and the circumstances. See *Catechism of the Catholic Church* [hereafter *CCC*] (New York: Doubleday, 1994), #1750.
21. Servais Pinckaers, O.P., *Morality: The Catholic View*, trans. Michael Sherwin, O.P. (South Bend, IN: St. Augustine's Press, 2001), 8.
22. J. Budziszewski, *The Revenge of Conscience* (Dallas, TX: Spence, 1999), ch. 2.
23. This story, from Bierce's collection *Can Such Things Be?* (1893), tells of Captain Ransome's fire on his own troops because of the faulty orders of his general. The general is killed in battle, leaving Ransome to be punished for his general's error.
24. For a study of proportionalism, see Sean Fagan, S.M., *Does Morality Change?* (Dublin: Gill and Macmillan, 1997). There is much in this book that is mistaken, however. Cf. the encyclical letter of Pope John Paul II, *The Splendor of Truth* (Boston, MA: St. Paul [1993]), #32.

25. See E. L. Miller, *Questions That Matter* (Boston, MA: McGraw-Hill, 1998), 395.
26. Bowden, *Black Hawk Down*, 106.
27. Norman Geisler, *Christian Ethics* (Grand Rapids, MI: Baker House, 1989), 97-132.
28. *The Splendor of Truth*, sections 80 and 81.
29. Michael Walzer, *Just and Unjust Wars*, 2d ed. (New York: Basic, 1977), ch. 16; Elizabeth Anscombe, "War and Murder," in *War, Morality, and the Military Profession*, 2d ed., ed. Malham M. Wakin (Boulder, CO: Westview Press, 1986), 294.
30. *Washington Post*, 19 October 2000.
31. *CCC, #408, 1768, 1865-1869.*
32. Seabury and Codevilla, *War: Ends and Means*, 10.
33. Charles Dunlap, Jr., "Rethinking Noncombatancy in the Post-Kosovo Era: The End of Innocence," *Strategic Review* 28 (Summer 2000): 17.
34. FM 22-100, *Army Leadership*, 1-4.
35. *War, Morality, and the Military Profession*, 197.
36. Barbara W. Tuchman, "An Inquiry into the Persistence of Unwisdom in Government." *Parameters* 10 (March 1980): 8-9.
37. Vatican Council II, *Gaudium et Spes* (7 Dec 1965), no. 79; my emphasis.
38. Aretaics is the "science of virtue—contrasted with *eudaemonics"(Webster's Third New International Dictionary of the English Language Unabridged,* 1968*).*
39. Anton Myrer, *Once an Eagle* (New York: Berkley, 1968), 913; in the HarperCollins edition of 1997, see 936.
40. Barbara W. Tuchman, "Generalship," *Parameters* 2 (1972): 11.
41. Joint Service Committee on Military Justice, *Manual for Courts-Martial, United States,* 1995 Edition (Washington, DC: U.S. Government Printing Office, 1995), rule 916 (II-109), my emphasis.
42. *The Splendor of Truth*, sections 53 and 99.
43. José Ortega y Gasset, *The Revolt of the Masses* (New York: W.W. Norton, 1957), 72.
44. Alasdair MacIntyre, *After Virtue*, 2d ed. (Notre Dame, IN: University of Notre Dame Press, 1984), 254-55. For MacIntyre's comments about the extent to which modern barbarianism has infected us, see 263.
45. In *Morality*, Father Pinckaers writes that "the cardinal virtues are simply four forms of charity" (21; cf. 87).
46. Myrer, *Once an Eagle*, 660; in the HarperCollins edition of 1997, see 677-678.

16 Army Professionalism: Service to What Ends?[1]

Martin L. Cook

Introduction

Much has been written in recent years about the state of professionalism in the contemporary U.S. Army officer corps. This chapter explores the concept of professionalism itself as a normative term. From an ethical perspective, the question is how to understand professionalism so that two equal values, somewhat in tension with one another, are preserved: the unquestioned subordination of military officers to Constitutionally legitimate civilian leadership, on the one hand, and the equally important role of professional expertise of the officer corps in providing professional military advice, unalloyed with extraneous political or cultural considerations.

I will argue that a good deal of the contemporary confusion in the profession results from two equally distorted understandings of these issues: some emphasize subordination, to the virtual neglect of legitimate professional expertise; others emphasize professional expertise, to the point of threatening illegitimate insubordination of the military to political leadership. Proper conceptual clarification of the terms will substantially reduce those misunderstandings.

"Professionalism" as a Normative Term

Typical Army officers have a fairly limited moral vocabulary. The central terms are "integrity" and "professionalism," and those terms are rarely examined critically. One praises an officer by saying he or she has "integrity," and one criticizes by saying he or she lacks it. One criticizes a wide range of behaviors, from how an individual dresses to the orderliness of an office or the individual's ethical character by saying they show a lack of professionalism, and one praises an equally wide range of conduct by saying the individual "is a real professional."

A professional group can successfully manage to maintain a widely shared standard of conduct and professional self-understanding, even with such unexamined vocabulary, during times of relative stability in the profession and its environment. When members of the profession are engaged in common activities; when the environment and culture in which they do their professional work are stable; and when the assumptions about the nature and purpose of their work are widely shared, critical examination of that vocabulary is perhaps unnecessary.

But during times of great change in the nature, function, and security of the profession, such unexamined unanimity will be found inadequate to articulate the nature of the challenge. If, for example, a profession has become accustomed to acting in a given sphere with complete autonomy, great tensions will arise when societal or legal changes challenge or limit that autonomy. Consider, for example, the enormous tensions managed care has precipitated in the medical profession. Increasingly, medical professionals question whether physicians can be "true professionals" as their autonomy in medical practice is increasingly circumscribed by case managers and preferred-provider limitations. Was the way physicians experienced the conditions of their work before managed care a necessary condition of their professionalism? Or are the stresses now being experienced simply the kind of stress any change inevitably brings with it?

By any measure, the officer corps of the Army is moving through a period of enormous change in the nature, purpose, and meaning of the profession. In recent years, it has been deprived of its central defining threats—threats that have shaped the nature and practices of the military culture for several generations. It has experienced a wrenching reduction in the size of the force, which in turn has had extensive effects on the normal officer career progression and prospects for professional advancement. Its equipment is aging and, it increasingly appears, not highly relevant to the kinds of military engagements it is being asked to perform. Among the military services, the Army in particular is undergoing a process of very rapid transformation in force structure, equipment, and doctrine that many do not fully understand and, in some cases, endorse.

Not surprisingly with so many shifts occurring, Army officers in their professional role are somewhat unsure of themselves, a typical sign of a profession in transition and change. Its junior members (captains) are leaving the profession at higher rates than expected. Significant numbers of the most successful colonels are declining brigade command, thus effectively refusing professional advancement. Distrust between levels of the profession is extremely high, and morale is low.[2]

As is to be expected, the moral vocabulary of the profession is invoked to "explain" what's occurring among Army officers. When officers don't behave in ways associated with earlier, more settled times, they are seen as "lacking integrity" and manifesting a diminished sense of "professionalism." When they rank force protection as their primary mission in far-flung peacekeeping operations, they manifest already declining "professionalism."[3] When they fail to embrace the profession as essential to their personal sense of identity and leave military service after short careers, those who remain are inclined to blame the departing individuals for their failure to share the "professional" ethos.

In fact, however, the departing individuals may be barometers indicating that the shifts in the role and mission of the profession have been so great as to change their relation to the profession dramatically. Further, they may reflect shifts in broader societal attitudes toward work, family, and career loyalty that merely play out in military culture in distinctive ways, but which have significant parallels throughout the society.[4]

Before too quickly deploying the standard vocabulary of professionalism, one should pause and inquire whether the vocabulary is indeed adequate to the task. Unless the central terms are conceptually precise, one runs the risk of invoking the military's standard limited moral vocabulary in criticism of individuals and classes of officers without accurately grasping the phenomenon one wishes to describe.

Eliot Friedson shrewdly raises a caution about too uncritical an embrace of the language of "professionalism" to exhort and criticize. He notes the ideological sense of "profession" which allows the term to be used manipulatively:

> The ideology may be used by political, managerial, and professional authorities to distract workers from their objective lack of control over their work, to lead them to do the work assigned to them as well as possible, and to commit them to means and ends others have chosen for them.[5]

Friedson is suggesting that in circumstances in which workers have lost control over their work or become disillusioned with the resources and support available to them, one would expect their superiors to employ the language of professionalism. The function of that language is to motivate workers to labor on, oblivious to the objective shifts in the character and meaning of their work. In such a circumstance, professional workers are no longer able to define the scope and nature of their core professional competencies, and are being tasked to accomplish missions for which they feel ill prepared by their professional competence and experience.

Before embracing the conclusion that the root problem with the contemporary officer corps is a declining sense of professionalism, one should step back and determine the nature and purposes of the military officer corps as professionals. What is the fundamental nature and purpose of the profession? What are the timeless attributes of the profession that must be maintained in all circumstances, and what elements are contingent on particular historical and political circumstances? Are there elements that have historically been foundations of the self-understanding of the profession that are in flux?

Necessarily, one cannot address the issue of Army professionalism ahistorically. Besides the obvious changes in the size and mission of the Army and the generational shifts in attitudes among younger officers, there are still larger shifts in the society from which and within which the military recruits and functions. Is the "postmodern" military a reality?[6] Certainly the internal and external environments within which the military functions are changing dramatically. Charles Moskos suggests eleven critical dimensions in which the military of the future will be fundamentally different from the military of the Cold War—dimensions as diverse as the military's media relations, force structure, perceived threat, and the roles of women and spouses in military culture.

If such fundamental changes are, indeed, in progress, it seems unlikely that the professional ethos of the profession will remain constant. Clearly in a time of enormous change in the nature, scope, purpose, and size of the military, some confusion and reconsideration of purpose are inescapable and appropriate. In this chapter, I will examine the question of the nature of the military as a profession in the contemporary context. I will offer some reflections on continuity and change of the professional military ethic as it emerges from this period of transition and uncertainty. And I will offer some normative suggestions regarding the shape military professionalism should take as the dust settles and we emerge from this time of transition and uncertainty.

The Military Profession in a System of Professions

The conceptual framework of the larger Army Professionalism study to which this chapter is a contribution is derived from Andrew Abbott's book, *The System of Professions: An Essay on the Division of Expert Labor.*[7] Fundamental to Abbott's conception of professions is the idea that professions apply bodies of abstract knowledge to a sphere of professional tasks. Societal recognition of the appropriateness of a profession's sphere constitutes its professional "legitimacy." Perhaps most important to Abbott's view is the idea that legitimacy is not a fixed, immutable fact. Rather, professions jockey for position within a given society as different professions and non-professional groups attempt to gain jurisdiction over spheres of human labor that previously have belonged to other professional groups, or alternatively as professions attempt to protect their historical jurisdiction from encroachments by competitor professions.

Within this framework, one may ask whether the current uncertainty in the Army officer corps is a manifestation of precisely this kind of jurisdictional conflict in progress. If the answer is affirmative, then the changed character of the geopolitical situation (an "external system change" in Abbott's vocabulary) has required the Army (and the other military services) to engage in a wide range of activities that differ from the classic Cold War war-fighting mission. Those new demands and tasks differ considerably from the activities, behaviors, and expectations that defined the profession for fifty years.

The jurisdiction of the Army, for example, has been large-scale armored and combined arms land warfare throughout the Cold War. While it could be assigned many other tasks, the core jurisdiction remained fixed, and the culture, values, and professional standards of the Army remained linked to that core jurisdictional function. Continuing this line of analysis, one sees that individuals raised in that professional culture may well experience professional confusion and uncertainty in the face of new and different tasks. So much of their professional self-understanding is bound to those tasks. So much of their body of professional knowledge, expertise, and agreed-upon ways of functioning are tied to that context.

It is to be expected that, in the face of quite different expectations, they question whether the new demands are consistent with their understanding of the profession. It should not be a surprise that professionals reared and rewarded within a relatively fixed set of assumptions about the nature and purpose of the profession experience discomfort and professional uncertainty as the verities of that context begin to dissolve.

If this is the correct view of what's occurring,[8] then the question before the profession (in terms of Abbott's framework) is whether to compete in a jurisdictional struggle for new roles and missions, or to cede the territory to other professional contenders. In other words, given that political leadership and American society generally are choosing to engage in peacekeeping and peace-enforcing missions around the globe, the Army must either adapt to them or prepare for considerable professional stress.

In this evolutionary way of viewing the matter, the Army is analogous to a species that may or may not successfully adapt to a changing environment. In

Abbott's view, professions live and die in the face of the evolutionary pressure of adaptation. Those that cling to forms superseded by external change face the prospect of joining astrologers and phrenologists in the dustbin of cultural history.

One common view—perhaps the majority view within the profession—takes for granted the idea that the Army should and must embrace all tasks assigned by "society," and views the reluctance to do so on the part of some officers as evidence per se of a declining sense of professionalism in the officer corps. As Don Snider, John Nagl, and Tony Pfaff state the matter:

> In a democracy, an Army does not get to choose the missions it accepts—at least, no professional army does. The hesitancy of the U.S. Army to accept wholeheartedly the missions it is currently being given strikes the authors of this paper as cause for concern in the context of military professionalism. *We believe that means defining the Army's organizational purpose, its essence, simply as serving the American society, and fighting the conflicts they approve, when they approve them.* Any other essence or purpose statement places the institution in the illegitimate and unprofessional position of declaring its intellectual independence from the society it was formed to serve.[9]

A number of points are striking about this analysis. First, it asserts that "intellectual independence from the society" is, per se, unprofessional. I will argue that the question of intellectual independence is considerably more complex than this formulation suggests. Second, it presupposes that the "hesitancy" of the Army to accept current missions is itself "cause for concern" regarding the Army's professionalism. Third, it offers a definition of the core professional function "simply as serving the American society." On each of these points, I will argue that the correct analysis is considerably more complex, and complex in ways that bear directly on a proper understanding of professionalism for the future U.S. military.

What Are Professional Obligations?

A few preliminary remarks are in order to avoid a misunderstanding of my view. There is no question that military members are obligated to follow legal orders of their superiors and to serve American society as the society's civilian leaders see fit. Self-evidently, civilian control over the military means that the military does not pick and choose its own missions. The essence of the soldier's commitment to service entails the unlimited liability clause that he or she may be required to sacrifice life and limb in following those orders and striving to complete legally assigned missions.

In all those ways, the commitment (especially the voluntary commitment) of the soldier to selfless service of the society and dutiful obedience to Constitutionally valid authority is the root of the nobility of the profession, and the source of American society's trust in and respect for its profession of arms. Indeed, it is for these very reasons that any overt affiliation of the vast majority of the military with a single political party is cause for concern. It is why endorsement of political candidates and causes by prominent retired officers using their military reputations for political purposes is potentially damaging to the public's perception of the professionalism of its officer corps.

But having said all those things, I would still argue that one cannot take the view expressed in the quotation above and retain any meaningful sense in which military officership truly is a profession. Obedient service is a crucial element of military professionalism, granted. Further, there can be no question that at the end of the day the soldier's obligation comes down to that: to follow legal orders, even if he or she believes the orders to be misguided, foolish, and likely to cause his or her death. But, I will argue, leaping directly to this "bottom line" reality of the military life not only fails to articulate what is important about the military as a profession, but also actually eliminates the very space in which the exercise of professionalism can and does occur.

Intellectual independence from society is the essence of professionalism, not at all evidence of a lack of it. It is the essence of a profession that members possess unique knowledge and the skill to apply that knowledge to a given range or sphere of service. A formulation such as that above, with its exclusive emphasis on obedient service, undercuts the centrality of the "abstract knowledge" aspect of Abbott's (or any other) model of professionalism. As Friedson expresses the point, "Professionalism entails commitment to a particular body of knowledge and skill both for its own sake and for the use to which it is put."[10]

In another essay, Don Snider and Gayle Watkins acknowledge the importance of professional expertise. Discussing the possibility that "the Army should be allowed simply to deprofessionalize, becoming an obedient but nonprofessional military bureaucracy," they list as one of the "two key benefits" of a professional military "the development and adaptation of military expertise."[11]

I believe the tension between this formulation and the one cited by Snider et al. above is one of the central issues. How does one correctly capture the inherent intellectual component of professionalism and professional expert judgment without overstating it and suggesting insubordinate or disobedient behavior based on that level of expert knowledge, which is presumably superior to that of the military's civilian masters?

If military officers are not, in fact, in possession of a body of knowledge that constitutes the intellectual component of their profession, the idea of "professional military advice" become meaningless. On that view, the skill and wisdom about how to employ military forces reside not among military officers at all, but among those who use the military as obedient tools, employed in ways and for ends beyond the purview of military officers themselves.[12]

But clearly this is not the case. Military personnel, especially officers, undergo extensive periodic education and training designed to impart professional knowledge and expertise, and they practice the application of that knowledge at all levels of their work—from field training exercises, through National Training Center rotations, to strategic-level exercises and war games. The result of that extensive professional development is, indeed, a degree of intellectual independence from the society officers serve, grounded in their unique professional focus.

The question of Army professionalism is wrongly framed, I would argue, by the suggestion that intellectual independence is somehow inherently unprofessional or even insubordinate to civilian control in a democracy. The proper question is not whether there is professional intellectual independence, but rather how the intellectual independence inherent in the very concept of professionalism is properly exercised. But this is just to pose the issue, not to resolve it.

Consider a circumstance in which professional military advice has been rendered but not accepted by the professional's civilian superiors. Here the officer must decide whether this is the occasion for obedient service and deference to civilian leadership (the normal case) or whether the course of action chosen by political leadership is so at variance with sound professional judgment that conscientious resignation should be entertained as a possibility.

Unwillingness to render obedient service to policies an officer considers deeply flawed and utterly at variance with sound professional judgment is not, I submit, evidence of a *lack* of professionalism but rather a high manifestation of it. Indeed, the very essence of a profession (as distinct from other occupations) is that the professional does possess and exercise independent judgment.

To cite Friedson once again:

> [A]nother critical element of work that nourishes professionalism lies in the nature of the relationship between client and professional. . . . In the case of employment in large-scale organizations we can put it crudely as whether or not policy is that customers are always right and that the organization and its members exist solely to serve, even cater, or their desires as long as they are willing to pay. Is the policy to provide whatever customers or clients desire, even if their capacity to evaluate the service or product is seriously limited and what they desire contradicts the better judgment of the professional?[13]

One of the most widely read and valued cautionary tales in contemporary U.S. military literature is H. R. McMaster's *Dereliction of Duty*, a historical analysis of the relation of the Joint Chiefs of Staff with the Johnson administration during the Vietnam War.[14] The main point of McMaster's book is that the Joint Chiefs failed to deliver professional military advice to President Johnson regarding the conduct of the war in Vietnam. The book is not widely read among the officer corps primarily because of its historical value, but rather as a morality tale on the lack of military professionalism on the part of the Joint Chiefs. According to McMaster's account, the primary moral failure of the senior officers is that they did not effectively exercise their intellectual independence and insist, to the point of resignation if necessary, that their professional military judgement be heard and accorded due weight by political leadership.

Of course it goes without saying that it would be unprofessional in the extreme to appear to accept the guidance of political superiors and then to subvert it through less than enthusiastic implementation or downright evasion of the spirit and letter of the order. But of more practical concern, it is the attempt to "have it both ways"—maintaining one's status as a serving military officer while simultaneously being insubordinate to duly constituted leadership—that makes conduct unprofessional for our purposes here—not (I repeat) the intellectual independence of the professional.

It is this aspect of the issue that the Snider et al. quotation above correctly captures, that is, the observation that "hesitancy" to accept missions assigned by civilian authorities is a cause for concern regarding Army professionalism. This, I believe, is correct. But in order for the analysis to portray the problem accurately, we must assess the causes of the hesitancy. Insofar as hesitancy derives from application of the profession's body of knowledge, the root problem may not be solely the willing obedience to authority that is part of the essence of the military profession. In addition, it may derive from the changing character of the missions being demanded and, arguably, from the profession's legitimate concern that it is being tasked beyond existing resources and professional competency.

When professionals are asked to perform tasks that lie beyond their scope of training and competency, they may hesitate for good and professional reasons as well. A medical analogy would be the demand that a general practitioner of medicine perform coronary bypass surgery; a legal analogy, that a personal injury attorney is pressured to take a bank fraud case. Thus, before we draw the simple deduction that hesitancy implies lack of professionalism, we should explore the possibility that prudent disinclination, like intellectual independence, is a manifestation of professional seriousness, not a lack of it. We would not think well of the professionalism of the GP who attempted the heart bypass; perhaps we should at least pause to reflect before we assume that the combat arms officer reluctant to take on police missions is culpably resistant to authority.

These observations point to the problematic nature of the third point raised in the Snider et al. quotation: that is, the proposed definition of the Army professional's core and defining obligation. The authors suggest simply that service to society is an adequate description of that obligation. But as we have seen, this cannot be an adequate definition of any profession's obligation. It stresses only the social dimension of professional obligation without giving equal weight to the aspect of professionals having to do with defining their own body of knowledge, having their own highly developed internal sense of the proper application of that professional knowledge, etc.

To say that physicians, or lawyers, or ministers exist to serve society is, of course, correct. But to infer therefrom that they have an obligation *as professionals* to collect trash or serve as school crossing guards, should society ask that of them, is ludicrous.[15] The point of these examples is that simply serving society's requests can never be an adequate definition of the obligation of any profession (although one might imagine a nonprofessional defined as a universal factotum of the society).

The reason it cannot be adequate is that it neglects the profession's own internal dynamic in assessing its body of knowledge, determining its relevance to a given identified social need, and negotiating jurisdictional questions with the larger society. The essential point here is that it is a multilateral negotiation, not a unilateral one. In an occupation, the hiring party can define every detail of the nature, scope, and method of the work, and then hire individuals willing to perform that work on those terms. Garbage collectors do not arrive on the job with a body of professional knowledge about the manner and scope of their work. In a profession, however, the case is different. The profession brings its own superior sense of the scope and limits of its expertise to the negotiation. If professionals do not possess that sense, then they "are no longer able to exercise the authoritative discretion, guided by their independent perspective on what work is appropriate to their craft, that is supposed to distinguish them."[16]

Of course the society may confer legitimacy on a profession or remove it. It may, for example, decide that chiropractors or Christian Science practitioners or acupuncturists are, or are not, "health care providers" for the purpose of insurance coverage. It may decide that ancient and well-established bodies of "professional knowledge" confer no legitimacy on the supposed professionals (e.g., astrologers) even though some individuals and groups may continue to treat them as if they were professionals.

Having cleared some ground and (I believe) misunderstanding of the issues, I will now turn to the central questions: What is the essential nature of professionalism in the U.S. Army? How, in the present, ought that profession to be reconceived to adapt to the new social, cultural and geopolitical context within which the contemporary Army functions?

Toward a Normative Account of Army Professionalism

I propose organizing a normative account of professional officership in terms of the following major elements: professional knowledge, professional cohesion, and professional motivation and identity.

Professional Knowledge

Entry into any profession is a kind of initiation into a body of knowledge primarily, if not exclusively, generated, transmitted, and built upon by fellow members of the profession. New members become familiar with that body of knowledge, learn where the major energy of the field is currently going, and, ideally, aspire to contribute to it as they join the ranks. As an individual comes to think of herself as a member of the profession, she increasingly acquires a familiarity with its technical vocabulary and knowledge, and acquires the ability to speak it technical language with facility. He learns the identity of a pantheon of archetypal members of the profession and stories of their contributions to the profession. She becomes familiar with a set of institutions, awards, honors, and so forth that members of the profession know and value (and that generally individuals outside the profession do not). He picks up, almost unconsciously, the small signals in dress, attitude, speech, etc. that members use to signal to each other that they are members of the same professional group.

It is also important to note that, in most professions, there is a core of the "real" members of the profession who have fully imbibed this knowledge. But there is also a large penumbra of para-professionals, knowledgeable in some areas essential to the profession's jurisdiction, but weakly if at all encompassed in the profession as profession. To use medicine as an example once again, phlebotomists and x-ray technicians are essential to the core functions of medical care, but clearly are not members of the medical profession.

In this light, one is tempted to suggest that the kinds of missions being assigned to the post-Cold War Army may seem to many in the profession as not calling upon and exercising core professional competencies. For example, the kinds of units that are required for effective operations in Bosnia and Kosovo are civil affairs and military police units, which are in short supply and largely in the reserves. The logic of "task effectiveness"[17] suggests that the active-duty Army ought either to convert a significant number of is combat units to civil affairs and military police functions or to rather explicitly renegotiate jurisdiction issues regarding those functions with the society the Army serves.

Resistance to doing so would be considerable, and for understandable reasons. In a particular health crisis one might find that the demonstrated priority need is for low-technology public health workers and perhaps family practitioners. No matter how true and strong that objective demand, however, one could predict that surgeons (for example) would resist being tasked to fulfill those roles. It's important to understand the nature of that resistance.

The nature of professional commitment is complex. True, surgeons have a shared commitment to patient care and welfare. But they have invested years of training, reading, and practice in acquiring extremely technical knowledge and in learning its application. While from a certain level of generality all this is in the service of patient health and welfare, one simply cannot leap from that level of description to generate the professional obligation for surgeons to retool their professional work. To understand the group and individual psychology of the profession, one must come to grips with the tremendous personal and societal investment in their identity and expertise, not merely as health care worker, or even as generic physicians, but *as surgeons*.

It is not adequate in describing this phenomenon to focus only on the grand organizing purpose of the activity in patient welfare. One must also see the degree to which mastery of complex skills, equipment, and abstract knowledge has become embedded in the very sense of self of the surgeon. One can predict with certainty that surgeons asked to care for patients in ways that do not draw on those skills (no matter how effectively they treat the patients) will experience professional disillusionment, lowered morale, and diminished sense of commitment to the profession. The point would be equally valid for the "strategic leaders" of surgery as a profession in the face of a societal demand that their profession make such an adaptation.

One might be tempted to call such a reaction "unprofessional." After all, is not the essence of the medical profession a willingness to serve health and the welfare of patients? At some level, that is correct—and in the circumstance I describe, if the objective need were for the kinds of professional skill I have outlined, the result would indeed be what Abbott calls a jurisdictional dispute. Further, should surgeons persist for a long period of time in their own internally defined sense of professional identity, their skills would become less and less relevant to the health needs of their society, and their profession would wither and perhaps even die for failure to adapt.

But clearly it is reasonable for at least a period of adjustment to occur. During that period, the profession collectively and surgeons individually would have to engage in some serious self-reflection and examination of their motives. Is their commitment to patients and health indeed their fundamental motivation? Or is their love of their specific set of skills and knowledge as surgeons in fact the foundation of their identity?

Before their profession was undergoing challenge, of course, that was not a forced choice: they could have both, in the confidence that the skills they took professional pride and identity in mastering did indeed serve the well-being of patients. But now, in changed circumstances, what had been a happy convergence now becomes a hard choice.

Furthermore, the profession may be so conservative, so wedded to its earlier self-understanding, that the client comes to lose faith and patience, and simply withdraws jurisdictions from the profession, awarding them to more flexible or newly minted groups, willing to meet the demands of the changed situation.

I would argue that an analogous phenomenon may be occurring in the contemporary Army officer corps. Officers who have invested their lives and careers in mastering complex combat arms skills in particular branches of the Army (Infantry, Armor, etc.) experience an analogous professional disorientation when they are consistently tasked to execute missions which draw little if at all on their body of knowledge and its application. Yes, of course, they understand their function is to serve the nation selflessly. But in addition their entire professional careers have been guided by acquiring ever more sophisticated skills and abilities. Their entire sense of professional identity has been formed by stories of exemplars of the profession: great combat leaders, heroic figures who sacrificed themselves in battle for the good of country and comrades.

Other communities and professions have different symbols, different archetypal stories, of course. But for professional soldiers, newly tasked with peacekeeping missions and humanitarian aid, it is understandable if they react by thinking, "This is not what I signed up for." It is understandable (even if not finally defensible) if their traditional symbol system and narrative of heroic self-sacrifice cannot justify death and injury on such missions—and that force protection thus emerges as a high value.

Of course, like the surgeons in my example above, it may turn out that the objective need of U.S. society is for continuous capability to engage in military operations other than war (MOOTW) and rarely for the skills associated with more traditional combat arms. If that is the case, and if the Army is unwilling to embrace the missions and come to reverence the professional knowledge required to execute them as it does for combat skills, the profession will fail in the competition for jurisdiction.

It is far from self-evident what should happen at this historical juncture. Just as surgeons, individually and collectively, would face some hard choices in the scenario I suggested, so would the officer corps. Assuming that some effective combat arms capability is a perpetual societal need, the Army may choose to make the case that the preservation of such capability is the essence of its understanding of military professionalism.

If it takes that position and upholds it successfully, the Army should be prepared to be radically downsized, unless and until it is necessary to build up a significant combat force when large-scale combat looms again on the horizon. This pattern is, of course, traditional in American military matters—the period of the Cold War being quite anomalous viewed through the lens of American history and our customary suspicion of standing armies. Those civil affairs, policing, and other functions necessary to MOOTW could be transferred to another group or profession ("peacekeepers") who would enthusiastically embrace the culture, symbols, knowledge, and skills necessary to carry out those missions with effectiveness.

One should be leery of dismissing this possibility out of hand. Surgeons might well decide it was ultimately more important for them to maintain the purity of surgery as a profession than to maintain "market share." Some surgical skills will remain necessary, they might argue, and leave it to others to provide public health and general practice medicine (or in Abbott's terms, cede the jurisdiction).

So too Army officers might conclude that they are willing to pay the price of reduced force structure and funding in the name of preserving the knowledge and skills central to the core functions of the combat arms. In that circumstance, if society remains committed to conducting other-than-war operations, it will necessarily constitute a force for that purpose. Inevitably, some subset of military skills (organization, logistics, etc.) will be necessary to carry them out efficiently. But the agents who conduct them may wear different uniforms, share different symbols, venerate different heroes, and understand their professional motivations in fundamentally different ways than combat professionals do.

In other words, the issue is wrongly framed as "professionalism" (which would counsel universal obedience to "the client's" demands) versus "unprofessionalism" (which resists some demands as beyond the sphere of professional expectations). Rather, it is an intellectual and leadership challenge *within* the profession. Is the profession of arms, having gained, transmitted, built upon, and capitalized upon a specific body of knowledge and its application in a particular form of expertise, flexible enough to choose to remake itself and compete for a new jurisdiction that requires different skills and bodies of expert knowledge?[18] This is the fundamental question before the profession, I would argue, at this historical juncture.

Professional Cohesion

Another central feature of a profession, as distinct from a mere occupation, is the ways in which individuals identify with the profession and with fellow members of the profession. As Friedson expresses the point:

> [T]he social organization of professions constitutes circumstances that encourage the development in its members of several kinds of commitment. First, since an organized occupation provides its members with the prospect of a relatively secure and life-long career, it is reasonable to expect them to develop a commitment to and identification with the occupation and its fortunes. . . . Second, since an organized occupation by definition controls recruitment, training, and job characteristics, its members will have many more common occupational experiences in training, job career, and work than is the case for members of a general skill class. . . . Such shared experience . . . may be seen to encourage commitment to colleagues, or collegiality.[19]

There is considerable evidence that the contemporary Army officer corps is experiencing difficulty in this dimension of the evolving profession. Study after study shows a deep distrust between levels of rank and status in the profession. Mentoring of junior officers by their superiors, even the most minimal sort such as discussion of OERs, is lacking. Any suggestion that colonels and captains are in any sense of the word "colleagues" would be met with derision in the contemporary officer corps.

One might say that formal aloofness between ranks is inevitable in a rigidly hierarchical profession such as the military. But this would be false. The relation between senior attending physicians in teaching hospitals and their interns is every bit as hierarchical as the military; senior partners of major law firms are without doubt as superior to their junior associates as senior military officers are to their subordinates. In both cases, the professional fate of junior members of the profession are very much dependent on the assessment and evaluation they receive from their seniors.

Yet clearly law and medicine are more effective at instilling a sense of shared membership in a single profession than the contemporary military is. It seems equally clear that most junior physicians and lawyers identify senior members of the profession whom they admire and strive to emulate, and in some cases look to for mentoring. Other professions find mechanisms (e.g., Medical Society committee work, formal dinners, and law firm social events) that allow and encourage engagement between junior and senior members on grounds of collegiality in addition to hierarchy.

At their best, professions instill and express the sense that all members of the profession are part of a continuous professional chain, linking the developing body of knowledge, expertise, and institutions through time. For the Army, too, the "long gray line" has at times been a felt reality of the profession, providing an emotional and psychological sense of continuity and connection between the generations of officers. While many contemporary officers do feel that continuity, it seems clear that something has diminished it in the contemporary state of the profession.[20]

There is much anxiety and not a little confusion in the contemporary Army about the professional connection between senior and junior officers. Sometimes it is expressed in terms of generational differences, sometimes in terms of respect and trust between age cohorts, sometimes in terms of how mentoring can be improved within the profession.[21]

Each of these perspectives is relevant and important to understanding the challenges to professional cohesion and retention of officers. But I suspect the root problem lies in something more fundamental and attitudinal. There is no substitute for the fundamental mind-set that members of the profession, regardless of rank, are colleagues, engaged in a common enterprise that matters deeply to them. If that mind-set is present, then each member feels a loyalty to the other, grounded in his or her common professional identity. If each thinks of professional identity in this way, each takes pride and responsibility in preserving, developing and transmitting the body of knowledge that resides at the core of the profession.

A powerful example of exactly this sort of engagement in a junior's professional development is apparent in Dwight Eisenhower's early career. During a posting as brigade adjutant in Panama, Eisenhower engaged in regular and extensive reading and discussion of professional literature with his commander, Fox Conner. One biographer, Geoffrey Perret, remarks, "Ike had always been exceptionally intelligent. What had been lacking until now was the time to read and the chance to talk over ideas with someone who was both a true intellectual and an accomplished soldier."[22] Perret attributes Eisenhower's intellectual awakening and deepened sense of the intellectual component of the profession to the fact that a senior officer and intellectual was willing and able make it part of his responsibility to engage in professional development and dialogue with his subordinate.

Eisenhower's experience leads us to inquire about the presence of Fox Conner types in today's Army. To what degree do senior officers in today's officer corps both model the profession as soldier-scholar and aggressively seek out promising younger officers to assist in the junior's professional development? Only the soldier-*scholar* preserves both aspects of truly professional officership: excellence in performance of military skills as currently understood by the profession *and* the contribution as a scholar to the continuing evolution of the body of professional knowledge that advances the profession through time. The United States Military Academy's strategic vision document well captures this essential dual requirement of the military professional. It states:

> While change is expected in any era of strategic transition, it nevertheless will severely test Army leaders. In order to serve the nation properly, the Army is pursuing an ambitious transformation agenda that will enable it to be strategically responsive and dominant at every point on the spectrum of operations—from actual combat to peacekeeping missions to humanitarian assistance. Forces will be more mobile, lethal, and agile, and better able to address the needs of our national security strategy. Leadership will also need to become more creative, able to cope with ambiguity, more knowledgeable about the intricacies of other world-views and cultures, more aware of emerging technologies, and capable of adapting rapidly to the changing contextual realities of evolving missions. In preparing its graduates for this service, West Point will continue to assess and update its developmental programs, as it has done in the past.[23]

Anecdotal evidence suggests that professional mentoring such as Ike received from Conner is a rarity in today's Army. Were it less rare, both the intellectual component of the professional and the bonds of trust between superiors and subordinates would in all likelihood be less problematic.

Professional Motivation and Identity

The fundamental questions of Army professionalism turn on the questions of the precise nature of professional motivation within the officer corps and its flexibility in adjusting the body of professional knowledge to the new kinds of missions now being assigned to the military.

I have argued elsewhere that the relation between the contemporary volunteer military and American society can usefully be construed as an implicit contract.[24] Military personnel volunteer to serve the society in the application of coercive and lethal force. They serve on terms of unlimited liability, in which they follow lawful orders in full recognition that they may die or be severely injured in fulfillment of those orders. The terms of the contract may appear somewhat different, depending on whether it is viewed legally or morally. From the legal perspective, military personnel are obligated to follow all legally valid orders of their superiors. Morally, however, I have argued that political leaders have an implicit contract with their volunteer military personnel.

The terms of the "contract" are that the military officer agrees to serve the government and people of the United States. He accepts the reality that military service may, under some circumstances, entail risk or loss of life in that service. This contract is justified in the mind of the officer because of the moral commitment to the welfare of the United States and its citizens.

Such considerations are critical in our assessment of professionalism because they go to the heart of professional motivation and self-understanding. It is because of the ends to which their expertise and commitment will be devoted that officers are able to justify to themselves the grounds of their service. It is because they see themselves as engaged in defense of the values, security, and prosperity of their family and nation that their service has moral meaning.

It is not surprising, therefore, that in a time of greatly diminished threat following the demise of the Soviet Union, some of that moral self-understanding is undergoing challenge. Further, deployments in Kosovo and Bosnia raise additional challenges to professional self-understanding and motivation.

On the one hand, soldiers engaged in those operations generally report that they enjoy the opportunity to exercise their professional skills and abilities. On the other hand, precisely because the reasons for those deployments do not clearly link up to the moral core of professional self-understanding, that is, defense of the vital interests and survival of the American people and state, many officers worry that they are eroding their core war-fighting competencies and wonder whether such operations are "what they signed up for."

As other chapters in this book demonstrate, this particular understanding of the nature and purpose of the profession is not timeless. Rather, it has been forged in the remarkable and historically unique furnace of the Cold War, especially the latter stages of the Cold War when the military transitioned to an all-volunteer or all-recruited force.

For those who chose to make the military their profession during this period, the threat was clearly identifiable and palpable, and the connection between their professional expertise and the defense of vital national interests was clearly visible. The Persian Gulf War provided a manifest demonstration of the professionalism and competence of that Cold War army and to a large degree reinforced the professional values and assumptions that had shaped the reform of the Army after Vietnam.[25] Having "grown up" in an Army that took quite justifiable pride in that far-reaching reform and enhanced sense of professionalism, it is not surprising in the least that many officers experience hesitation and reluctance to refocus their professional activities and self-understanding on activities ancillary to the war-fighting prowess that enabled that reform.

Perhaps the clearest expression of the connection between the war-fighting focus and the self-understanding of the profession is to be found in the Powell-Weinberger Doctrine, which, to many officers, seems to state immutable truth. But it does not, of course. Rather, as Suzanne Nielsen's chapter in this book demonstrates (Chapter 10), it is an attempt by the military to stipulate its own contract with society, the terms under which it is willing to be used, all in the guise of objective professional military advice.

At root, however, the military does not set the terms of its social contract—certainly not unilaterally, and essentially (although it is a negotiation to some degree between the profession and the society) not at all. As the times change and as the strategic needs of the nation change, the contract changes as well. As is made clear in the historical review of the profession by Leonard Wong and Douglas Johnson in the present book (Chapter 4), the only real constant in military professionalism is that the military constitutes a "disciplined, trained, manpower capable of deploying to a possibly dangerous environment to accomplish a mission."

The contemporary challenge of Army professionalism is for the profession itself to engage intellectually with the changing nature of the environment and to embrace fully the need to lead in its own adaptation to that environment. The strategic reality is that large-scale land warfare is not on the immediate horizon of probability but that many other uses of disciplined forces the Army can provide are. Given this reality, it makes little sense to cling desperately to a self-understanding of the profession that began to lose relevance after the Cold War. Neither can the Army afford to accept other types of missions only grudgingly, while funneling as much energy and as many resources as possible into husbanding the capability for the preferred professional activity of large-scale combat.

What is required is for the Army's officer corps to enthusiastically embrace the reality that the nation requires a different and more complex set of skills of its Army today than defined it following the reforms of the 1970s. As with any profession in a time of change, it is psychologically understandable that a period of transition is likely as the profession mentally assimilates the changed circumstances.

But finally the profession must come to recognize that the wider and more complex range of missions the contemporary strategic context necessitates falls within the profession's jurisdiction. In practical terms, that means the profession itself must apply its own body of abstract professional knowledge to that range of problems and, when necessary, devote the intellectual work required to expand and enhance that knowledge. If new kinds of units, equipment, and training are necessary to the effective prosecution of such missions, the profession itself should spend the political and intellectual capital to develop and acquire them.

The contemporary Army transformation plan, in crafting a lighter and more deployable force, is a step in the right direction. But even that plan, still focused on large-scale land combat, will not be sufficient to address the widening range of missions the contemporary environment makes likely. Admittedly, the contemporary scene is confusing, and it is difficult to foresee with accuracy every kind of new mission the Army may be tasked to perform.

But some are clear: peacekeeping and peace-enforcement missions will require larger numbers of civil affairs and psychological operations units. The assumption that units trained for high-end combat are really the most effective units to perform those missions is certainly debatable. Urban warfare and Special Forces operations against non-state actors in urban environments are likely missions. Countering an enemy's use of various kinds of weapons of mass destruction looms on the horizon. It is an essential obligation of the profession itself to engage intellectually to determine how best to structure, train, and equip itself to perform those missions with the greatest degree of effectiveness.

Clearly, the Army is aware of such likely missions and is preparing, to some degree, to perform them. But equally clearly, they are viewed as secondary to the core function, which is large-scale combat (even with the future Objective Force which will be lighter, more deployable, and more lethal).

The benefit of viewing professions through Abbott's lens is that it avoids viewing the profession statically and ahistorically. Rather, it sees the profession as evolving through time in interaction with its environment and with other claimants to the profession's jurisdiction. At the root of the challenge to Army professionalism is the necessity to create and sustain the intellectual creativity to get ahead of environmental changes, to embrace them, and to demonstrate the intellectual flexibility to inspire the nation's confidence that it can meet the demands of the changing security environment with enthusiasm. Such a profession transmits and extends its corporate culture and its developing intellectual engagement with a body of expert knowledge into the future environment.

Notes

1. The views and opinions expressed in this chapter are those of the author and are not necessarily those of the Department of the Army or any other U.S. government entity.
2. For a representative sampling among the sea of articles making these points, see Justin Brown, "Low Morale Saps U.S. Military Might," *Christian Science Monitor*, 8 September, 2000, 3; and Fred Reed, "Military Service Warning Labels," *Washington Times*, 10 September, 2000.
3. Don M. Snider, John A. Nagl, and Tony Pfaff, *Army Professionalism, The Military Ethics, and Officership in the 21st Century* (Carlisle Barracks, PA: Army War College, Strategic Studies Institute, December 1999).
4. See Leonard Wong, *Generations Apart: Xers and Boomers in the Officer Corps* (Carlisle Barracks, PA: Army War College, Strategic Studies Institute, October, 2000).
5. Eliot Friedson, *Professionalism Reborn: Theory, Policy and Prophecy* (Chicago, IL: University of Chicago Press, 1994), 124. It is interesting to note that exactly the sort of institutional context Friedson's analysis offers is also part of Snider and Watkin's analysis of the Army's present situation. They write, "[T]here is a gross mismatch between institutional capabilities and national needs." "The Future of Army Professionalism: A Need for Renewal and Redefinition," *Parameters* 30, 3 (Autumn, 2000): 8.
6. Charles C. Moskos, "Toward a Postmodern Military: The United States as a Paradigm," in *The Postmodern Military: Armed Forces after the Cold War,* eds. Charles C. Moskos, John Allen Williams, and David R. Segal (New York: Oxford University Press, 2000), 14-31.
7. Andrew Abbott, *The System of Professions: An Essay on the Division of Expert Labor* (Chicago, IL: University of Chicago Press, 1988).
8. This is, of course, the basic thrust of the analysis offered in Don M. Snider, John A. Nagl and Tony Pfaff.
9. Ibid., p. 21. Emphasis in original.
10. Friedson, *Professionalism Reborn,* 210.
11. Snider and Watkins, 7.
12. To further illustrate this misunderstanding, suppose one were to say that because physicians serve patients and the health of the society more generally, it would be unprofessional of them to assert intellectual independence and superior knowledge of medicine. Clearly, it is only insofar as physicians do possess intellectual independence and superior knowledge that they are of any use to the society. To say that, however, is not to say that society does not have the right to limit the discretion of physicians to use that expertise as they see fit. Society may well (for example) rule out medical procedures that are medically indicated if they are too expensive from the perspective of public policy. In that case, the physician would have to decide whether to provide a less expensive alternative procedure, even though it was less desirable from a narrowly

medical perspective. Presumably, there might be some point where resources are so constrained that the physician would conclude that it would be "unprofessional" to continue to practice medicine in such an environment.

13. Friedson, 211.
14. H. R. McMaster, *Dereliction of Duty: Johnson, McNamara, the Joint Chiefs of Staff and the Lies That Led to Vietnam* (New York: HarperCollins, 1997).
15. Needless to say, one can imagine some scenario in which trash collection or guarding school crossings becomes a critical social need. In that situation, one might well appeal to physicians to assist in those activities. But one would be appealing to them as citizens, or as benevolent individuals, not because of any obligation inherent in their professional status. If they are employees of an organization, one might even require such service of them in the exigency of the moment. But even in that case, they would not be rendering service *as professionals*, but rather merely as able-bodied employees.
16. Friedson, 211.
17. The term is from Snider and Watkins, op. cit., 7.
18. An interesting historical question arises which I am not competent to address. To what extent did experienced combat leaders following the Civil War experience a similar professional disorientation when they were tasked with garrisoning the West and with national development missions in the late 19th century?
19. Friedson, 122-123.
20. Edwin Dorn et al., *American Military Culture in the Twenty-First Century: A Report of the CSIS International Security Program* (Washington, DC: Center for Strategic and International Studies, February 2000).
21. See Wong, *Generations Apart*, for a full discussion of these issues.
22. Geoffrey Perret, *Eisenhower* (New York: Random House, 1999), 88.
23. U.S. Military Academy, *United States Military Academy Strategic Vision—2020* (West Point, NY: U.S. Military Academy, 1 July 2000), 4.
24. Martin L. Cook, "'Immaculate War': Constraints on Humanitarian Intervention," *Ethics and International Affairs* 14 (2000): 55-65.
25. See James Kitfield, *Prodigal Soldiers* (New York: Simon and Schuster, 1995) for a detailed account of the much eroded state of military professionalism in the aftermath of the Vietnam War and of the reforms then instituted to increase the effectiveness and professionalism of the Army. This is significant, I would argue, because it is precisely the culture and institutions created by those reforms that are now threatened with the challenges of the current situation.

V

Professional Leadership[1]

We turn now from the external focus of the previous three sections to a more introspective perspective as we consider the Army profession's internal jurisdictions: creating, developing, and maintaining expert knowledge, and embedding that knowledge in members of the profession. A profession's ability to compete in external jurisdictions—for the Army, conventional and unconventional war, operations other than war, and homeland security—is influenced by how well it accomplishes essential professional tasks in its internal jurisdictions. Clients will reject, or competitors will overtake, professions whose practice in their external jurisdictions is based on inadequate accomplishments in their internal jurisdictions owing to outdated knowledge or weak expertise. Surprisingly, the Army's internal jurisdictions are under more competitive pressure than one might suspect. The recent pressure to outsource many functions has opened these jurisdictional doors to private corporations who are becoming responsible for more and more of the Army's internal work, ranging from educating ROTC cadets to designing the curriculum for the Army's mid-level officers.

In this section, researchers address these critical internal jurisdictions of the Army from several perspectives, in particular the leadership perspective, since how to lead soldiers is one of the foundations of the Army's expert knowledge. In fact, the first two of the four chapters address the development of officer leaders. Chapter 17 presents insights from ongoing research within the Army officer corps on measures of psychological maturity. Given the new nature of Army operations in the 21st century, it raises serious questions about whether the Army's traditional rank structure is still compatible with the level of maturity needed to handle the new tasks. Chapter 18 by Anna Simons presents an anthropological perspective on the potential use of moral and ethical principles to better develop officer-leaders. Principled leadership, she finds, is the sine qua non for inducing soldiers to accomplish their missions.

The final chapters address two leadership issues little discussed within the Army during the past decade: the spiritual needs of soldiers in combat and their leaders' responsibility to provide for those needs (Chapter 19); and the tasks and self-concept of the strategic leader of a profession (Chapter 20). The conclusions reached in these chapters are straightforward and, in both cases, seriously challenge current Army doctrine and practices. Doubtless they will appear provocative to much of the conventional wisdom among Army professionals, but professions can renew and redefine themselves only when the prevailing orthodoxies are challenged.

Note

1. The authors express their appreciation to Maj. Christine Sandoval of the USMA Department of Social Sciences, who served as Panel Rapporteur for this section's topic during the USMA Senior Conference, June 2001.

17 Making Sense of Officership: Developing a Professional Identity for 21st Century Army Officers[1]

George B. Forsythe, Scott Snook, Philip Lewis, and Paul T. Bartone

Since the end of the Cold War, in a very fundamental way the U.S. Army has been under attack. Setting aside the Pentagon strike on 11 September 2001, this onslaught has come not from terrorists or rogue states, and the target has not been our military formations in the field or our nation's borders. Since 1989, what's been at risk has been our Army's professional center of gravity, its sense of self. One way to conceptualize this assault is to think of it as an "identity crisis," a social and psychological battle fought at both individual and institutional levels of analysis, with each dynamic reinforcing the other in a self-fueling vicious cycle. It is difficult to overemphasize the magnitude of the changes that have led to this crisis. The Berlin Wall came down; the Soviet Union collapsed; the Cold War ended. Gone was the relatively simple bipolar globe that had dominated our geopolitical landscape for almost forty-five years. What would such monumental changes mean for the U.S. Army? What would this loss of the existing world order mean for a generation of Army officers whose extant professional identity was defined largely in opposition to an enemy now defeated? What would such global upheaval mean for how our Army thinks about itself? In the present chapter we provide a way of considering professional identity as part of the fundamental process by which we, as human beings, construct and periodically reconstruct a meaningful view of the world and our place in that world.

Identity, whether we are talking about the institutional identity of the Army or the professional identity of Army officers, does not occur in a vacuum. To a significant extent our identity as officers and as an institution is a reflection of how we are expected to function by the larger society of which we are a part. "Victory" in the Cold War resulted in significant change in both the quality and quantity of demands levied on the U.S. Army. With the Soviet Union gone, American society demanded more than simply trained and ready war-fighters. Over the past decade, our Army as an institution has been asked to execute a broad range of missions across the entire spectrum of operations. Such changes at the institutional level have required individual Army officers to adopt a wide variety of roles and operate in increasingly complex and ambiguous environments. Not surprisingly, such dramatic changes in the substance of what we do as an Army have sparked serious debate about fundamental issues of identity—who we are, what we're about, and, perhaps most fundamentally, *how* we make sense of what it means to be a military officer of the United States in the early twenty-first century.

Figure 17-1 illustrates the multidimensional nature of this struggle. We will argue that central to understanding the issue of identity and identity development is the distinction between "content" and "structure." Identities at both institutional and individual levels of analysis consist of both content and structure. Consider for a moment the two types of individual identity depicted in the bottom row of Figure 17-1. In the left column is the content or "what" of our individual identities as Army officers. Content is the surface material—our behaviors, ideas, preferences, values, aspirations, and leadership style. The content of an officer's identity might include the seven Army values, or Duty-Honor-Country, or the principles of officership, or an amalgam of all of them. Indeed, the debate within the profession about roles and missions, and the associated questions about an officer's identity—am I a warrior or more?—is fundamentally about the content of an officer's identity. How we answer this important "content" question will have significant implications for such issues as how we train and allocate our resources, and whether we, as individuals, choose to remain in the Army. This is the way we customarily think about professional identity. Our focus in the present chapter, however, is not primarily on this aspect of identity. Instead, we will focus mainly on the bottom right quadrant of Figure 17-1, the "structure" of individual identity. Focusing only on issues of identity content will lead to an incomplete understanding of professional identity issues and more importantly to a lack of understanding of the *process* of identity formation and transformation. What, then, are these other identity issues? What is the "structure" of identity and identity development?

The structure of identity refers to the way in which any particular identity, be it warrior or peacekeeper, or something else, is constructed. As Figure 17-1 suggests, it is the "how" of identity formation. For example, some officers will construct their understanding of the peacekeeper identity as a set of skills and behavioral roles that

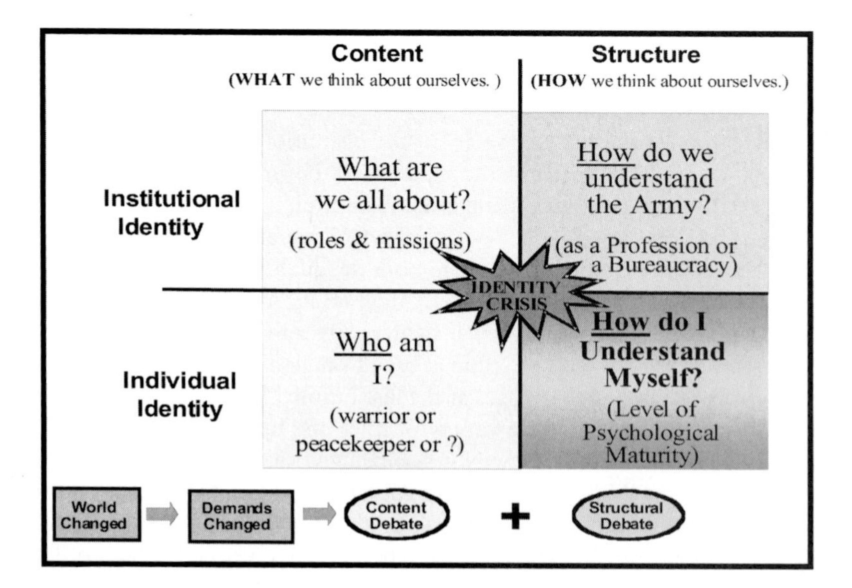

Figure 17-1. *Identity Matrix.*

must be mastered to fulfill that identity. For these officers, if one has the requisite skills and can employ them effectively, then one is a peacekeeper. Other officers will construct the peacekeeper identity in a more complicated fashion. To them being a peacekeeper may mean more than merely meeting a set of role expectations. It may also entail embodying a set of societal values about the proper place of a superpower in a unipolar world. For these officers, the fixed roles and narrow competencies of a peacekeeper are embedded in a broader context of societal values about the rights of ethnic minorities and the centrality of democratic social and political institutions to the creation of just societies. The "how" of the skills-roles identity entails enacting certain behaviors. The "how" of the contextual identity entails locating oneself within a broad set of societal views about the place of the military in a democratic, multi-ethnic society. We will argue that these contrasting identities are not merely different in content. We will argue that the latter identity requires and reflects a higher level of psychological maturity than the former. It entails a more sophisticated and psychologically mature way of structuring one's identity than does the former. Failure to appreciate the fact that identities vary by structure (level of maturity and sophistication) as well as by content will yield an incomplete understanding of our ongoing struggle with issues of professionalism in the Army.

Figure 17-1 maps how the content-structure distinction and the institutional-individual level of analysis interact to provide a more complete framework for our current debate. The most familiar arguments are found in column one. For the past ten years, the Army has helped fight a war on drugs, drilled wells in Africa, aided the victims of natural disasters around the globe, fought a successful conventional war in the Gulf, battled warlords in Somalia, and stemmed the tide of ethnic cleansing in the Balkans. Such a significant shift in roles and missions was bound to have an impact on the Army's identity, its organization, its officers, and its profession. The resulting debate is about the content of our identity. What is the Army really about? Fighting wars, keeping peace, building nations, or something else?

Debate over these content issues at the institutional level has generated corresponding angst at the individual level. How should soldiers think of themselves? As warriors, peacekeepers, policemen, statesmen, or something else? Arriving on the heels of a relatively narrow professional self-concept adopted during the Cold War—that of warrior—this decade's increase in scope of institutional missions and individual roles has left the Army and its soldiers searching for a coherent new identity. Such uncertainty at the very heart of our profession can manifest itself in many dysfunctional ways. Alarming trends in recruitment, morale, readiness, leadership, and retention can all be viewed as symptoms of a deeper malaise, one rooted in fundamental questions about the very essence of who we are and what we do.

As noted above, however, identity is not only about content; there is an equally important structural dimension to our sense of self, both for the Army as an institution and for its members as individuals. While column one issues (Figure 17-1) have captured current headlines, we may find that column two holds greater promise for illuminating our current crisis and for generating new and better approaches to lasting change. While eventually we must resolve questions of content, perhaps a deeper understanding of *how* we think about ourselves and our experiences, the *structure* of our identities, will lead to more robust insights into these fundamental issues of self. We leave it to others to directly address issues of content; our work here is situated within the broader debate over *how* we think of ourselves, both collectively as an institution and individually as psychologically maturing adults.

In contrast to issues of identity content, issues of identity structure are universal. They pertain to all professional identities. The content of identities varies by profession. Questions like, "What are we all about?" and "Who am I?" are boundary issues. Answers to such content questions are both vocation specific and time dependent; they serve to differentiate one profession from another. Structural issues, on the other hand, are more general; answers to these broader questions provide the framework within which issues of content are addressed. The two types of identity issues are related. Structure influences content; *how* we think about ourselves influences *what* we think about ourselves. In the example presented earlier, the peacekeeper who structured his identity in terms of roles will be limited to seeing himself in terms of specific peacekeeping behaviors. In contrast, the officer with a broader peacekeeper identity will construct his identity using shared, societal values and ideals. What each considers his professional identity to be will be the same (both embrace a peacekeeper identity). But it is important to note that the breadth and complexity of the two identities are very different. It is this structural component of identity that provides the lens through which we view ourselves, either collectively or as individuals. In this sense, structure constrains what level of content can be seen. Being clear about how an identity is structured will provide us with significant leverage in understanding all issues of identity, particularly in terms of how well we understand the issue of identity development.

At the institutional level, structural issues of identity (upper right quadrant of Figure 17-1) address questions of organization. How do we think of ourselves as an institution? What is the structure of our *collective* consciousness? Is it organized primarily as a large bureaucracy or as a time-honored profession? How we resolve such issues at the institutional level will not only shape how we collectively think about whether we should be war-fighters or peacekeepers, it will also frame individual-level issues of identity as well. Fundamental questions of individual member identity are colored by one's sense of being either a professional or a bureaucrat. Once again, however, we'll leave this question of structure at the institutional level to others. To repeat, our focus in this chapter is on the lower right-hand quadrant of Figure 17-1—the structure of individual identity.

Given that the lower right quadrant of Figure 17-1 describes the individual component of structural identity, we are taking a decidedly psychological approach. To understand officer development from this perspective, we have turned to the literature on adolescent and adult identity development for insights. Our work is grounded in the constructive-developmental theory of Harvard psychologist Robert Kegan[2]. We have found his approach provocative and illuminating because it allows us to focus on *how* officers understand themselves rather than on *what* they understand. In the remainder of this chapter we extend Kegan's general theory on adolescent and adult identity development to the specific context of officer development in the U.S. Army. We will not explain Kegan's theory in detail; rather we extrapolate his work to the subject of the present volume—officer professionalism. We look first at the basic ideas of Kegan's theory as they apply to officer development. Next, we present the findings of our research within the officer corps. Our intent is to explain what we currently know about officer development. We conclude with a discussion of the implications of our work for re-professionalizing the officer corps.

Kegan's Theory of Identity Development

Robert Kegan is primarily interested in *how* people make sense of themselves and their world, what we have referred to above as the "structure" of identity. In Kegan's view, the level of sophistication of how one structures an understanding of oneself and one's experiences lies at the heart of self-concept or identity. Two key ideas underpin Kegan's view of identity. First, Kegan believes that we actively *construct* our understandings; we don't simply receive them from others. We build our understandings of ourselves from our experiences. It should be noted that two individuals may share a common experience but understand it (construct it) very differently. Second, Kegan believes that we progress through a finite series of universal and progressively more complex stages in *how* we construct our understanding. It is the foregoing two ideas in combination that make Kegan's approach to identity and identity development a powerful explanatory system. We shall extend these ideas to the officer corps, specifically to the proposition that how officers understand themselves and their experiences in the Army lies at the core of their professional identity—who they are as professionals. Note the use of the word "how" rather than "what" in the preceding sentences. Although the content of an officer's identity is important, we believe that how officers make meaning is more fundamental to understanding professional identity. In this regard, we offer an approach to this topic that differs from that of other authors in this anthology. Nevertheless, this approach shouldn't be unfamiliar to anyone who has ever heard a service school instructor say, "We are not trying to teach you what to think as much as we are teaching you how to think." In terms familiar to Army officers, this approach offers a way of thinking about the BE in the Army's BE, KNOW, and DO framework for leader development.

Within Kegan's framework, development involves major qualitative shifts in how people construct their understanding of themselves and their world. Kegan's theory is about changes in meaning-making, where development involves progressively more complex ways of constructing this understanding—complex in terms of the breadth of perspective and the organization of internal and external experiences. From this perspective, as our understanding becomes more complex, it also becomes more encompassing—it's a broader world view. We see more, we can step back and be objective about more things, we are more mature, and we are better able to deal with the psychological and interpersonal demands of life. Because we "see" more, we are able to establish deeper connections to others. Within this framework, officer development involves qualitative shifts in how officers make sense of themselves and their experiences. Each shift leads to a progressively broader perspective toward oneself as a professional and one's relationships to others within and outside the profession. These are precisely the sorts of changes advocated by the Army's literature on leader development—developing broad, encompassing, and abstract perspectives that enable leaders to deal with the mental and emotional challenges of operating in complex and ambiguous situations. The present analysis suggests that such changes will occur only if officers have reached a certain level of structural development.

Kegan explains the structure of the self-concept in terms of what he calls "subject-object relations." "Subject" refers to the psychological lens through which we construct meaning—the perspective from which we understand our selves and our experiences. "Object," on the other hand, is that which can be can be understood psychologically; it is what we can be "objective" about. Subject-object relations, then, is an internal psychological process involving the interplay between *what* we can understand and *how* we understand it—between what we can take a perspective on ("object") and the perspective we bring to bear ("subject"). In general terms, then, development entails the shifting of subject to object, making it possible to take a perspective on and view more broadly what was previously the way in which we organized our understanding of the world. In this manner, each step in development leads to an expanded world view, which gradually becomes broader and more encompassing as we gain the ability to step back and "see" ourselves and our experiences in ways that we couldn't before. We move from being embedded in a point of view to being able to step back and assess it. Table 17-1 shows Kegan's stages in terms of the subject-object distinction to illustrate this point.

Stage	Subject	Object	Features
0	Reflexes (sensing, moving)	None	Incorporation
1	Impulses, perceptions, feelings	Reflexes	Single perspective; can't take other's point of view
2	Needs, interests, wishes	Impulses, perceptions, feelings	Can take multiple perspectives, one at a time.
3	Mutuality, shared meaning	Needs, interests, wishes	Takes multiple perspectives simultaneously
4	Self-authored system of values	Mutuality, shared meaning	Has own personal perspective on relationships and societal ideals
5	Universality	Self-authored system of values	Recognizes that own perspective on experience is a created convenience

Table 17-1. *Kegan's Developmental Stages.*[3]

Stages of Officer Identity Development

For purposes of understanding officer identity, we will focus on Kegan's post-childhood stages, 2, 3, and 4, because it is these stages and their development that are most relevant to our understanding of officer identity and officer identity development. Stage 5, reserved for the rare and exceptional individuals among us, will not be treated. Although the ages at which the various stages emerge vary considerably, stage 2 begins to emerge at about age 5 for most children and is the typical stage of most early teenagers. There are, however, a significant number of successful adults who have not yet moved beyond this relatively early developmental stage.[4] Stage 3 thinking begins to emerge in adolescence for most individuals and is the modal stage of American adults. Stage 4, when it develops at all, appears to be a feature of middle adulthood.

Kegan's Stage 2

As shown in Table 17-1, stage 2 individuals have moved their immediate impulses and embeddedness in a single perspective from subject to object. As a result, individuals who are functioning at stage 2 are able to play social roles, delay gratification, follow rules, develop and pursue enduring interests, construct a stable sense of self, and participate in close relationships based on social exchange and reciprocity. Unlike individuals at stage 1, these individuals are not impulsive; they possess the ability to postpone short-term indulgences in favor of long-term goals, and they have a sense of agency. Newly able to reflect upon their immediate desires, they can take responsibility for their actions. Stage 2 individuals' sense of who they are is tied up in their ability to effectively pursue their enduring interests, including enacting the behaviors associated with desired roles. As leaders, stage 2 individuals focus on following the rules and being responsive to the needs and interests of their subordinates and superiors. They believe that one usually has to "go along to get along." However, stage 2 officers tend to view others from the point of view of their own agendas (note that in Table 17-1 they are shown as being subject to their needs, interests, and wishes). Stage 2 individuals understand others in terms of how these others can help or hinder them in satisfying their own personal interests.

As officers, they tend to view professional values and standards as rules to be complied with, and do so in order to accomplish something personally important to them or to avoid negative consequences. They are ultimately motivated by the desire to meet their own needs. While they may be concerned about what others will *do* to or for them, how others *feel* about them does not become a part of how they feel about themselves. There is no shared perspective, because a shared perspective is a structural achievement of stage 3. It is certainly true that in many cases the content of a stage 2 officer's agendas is professionally acceptable—getting more training or a more challenging job, for example. But because of structural limitations, the underlying motivation of a stage 2 individual is always to satisfy his or her personal needs. There is an absence of a shared reality—how others will be affected by what I do is often relevant, but it is not a part of my sense of my stage 2 self. Thus, interpersonal mutuality and a sense of corporateness are not possible for a stage 2 officer. In a fundamental way, any professional identity adopted by a stage 2 officer is always an individual identity. The identity of peacekeeper, for example, could be enacted as a set of desired role behaviors. But the stage 2 officer would lack the structural capacity to be identified with or have a sense of community with others who share that identity.

To illustrate, consider the following excerpt from a twenty-year-old West Point cadet in his sophomore year who is describing what it means to him to be a leader. This cadet is describing why he feels more motivated when he has a leader who "puts out a hundred and ten percent." This excerpt, and the others that follow, are drawn from research we have been doing with West Point cadets and Army officers, research that we will describe in more detail later in the chapter.

> The leaders that I've had in the past, the ones that are really good, I perform well in there. If they are putting out a hundred and ten percent, and they are doing their best, you can visually see it; their uniform looks really cool. You know, if their shoes are really shined, you know, mine should get that good too. And if I see them putting a hundred and ten percent out for us, they're doing it for us—their own troops—and they really care about them, then I'm going to work that much harder for that person.

What sounds stage 2 about this is that the relationship between the cadet and his leader (platoon sergeant) is construed as a sort of quid pro quo exchange—he puts out for me, so I'll put out for him. This view can be accommodated by the stage 2 capacity to take two perspectives, the cadet's and the platoon sergeant's, one at a time. Still, it's possible that this cadet's orientation to the platoon sergeant's "caring" is more mature than a stage 2 perspective. The interviewer seeks to clarify this issue (interviewer's comments and questions are in bold type):

Why Is It Important to You That They Care?

> I guess for me it's because of the tangible rewards of what will happen if somebody does really care about you. Maybe, like for instance, if they are putting a hundred and ten percent out, then you're going to put a hundred and ten percent out there. And then you are all going to feel better about everything else.

Because You Think You'll Be Able, Yourself, to Succeed More?

> Yeah, yeah, definitely.

So we see that the interviewer's questions elicit the notion that the cadet will get more of his own needs met if he has a motivated leader (the "tangible rewards"). And presumably he "feels better about everything else" because he, individually, is being more successful. This is clearly a stage 2 perspective on leadership. Note that the interviewer got beyond the content of good leadership (putting out for your subordinates) to the underlying structure of how this cadet understands this particular leadership principle (balancing reciprocal needs and interests so you can get what you want).

Although stage 2 understandings are essential to effective performance as an officer, embeddedness in self-interest is both practically limiting and normatively anathema to officer professionalism. There is nothing wrong with having selfish interests. But military officers (and most other professionals) are expected to routinely subordinate self-interest to the greater good. In the terms of Kegan's theory (see Table 17-1), this is the shift from being "subject to" my needs and interests to having ("making object") needs and interests. It is this developmental shift that signals the emergence of stage 3.

Kegan's Stage 3

Stage 3 officers have developed the ability to hold multiple perspectives simultaneously, which enables them to regard themselves and others in terms of abstract, inner qualities and to be directly affected by what they think others think about them. Now that they can hold another's perspective simultaneously with their own perspective, that other perspective can be used to *co*-construct or make sense of the self. How others view them becomes a part of how they view themselves. This is a major psychological accomplishment. Unlike stage 2 individuals, those in stage 3 can be "team players," able to subordinate their self-interests to professional ideals.

We might think of the transition from stage 2 to stage 3 in terms of professional socialization—the shift from understanding the self in terms of individual interests to an understanding of self as a part of something larger than the self, as a member of a profession. It is important to note that this is a change of identity structure, not merely content. Because of this structural change, stage 3 officers have a much more sophisticated view of officer professionalism than those at stage 2. They are able to *internalize* the profession's expectations, and they strive to live by these expectations because of their shared bond with other professionals. They also are able to understand themselves in terms of abstract internal qualities such as honesty, duty, and responsibility. Stage 3 officers understand the seven Army values as more than simply rules to be followed; rather, they see these values as internalized qualities that the Army expects will characterize all soldiers.

To illustrate, consider the following excerpt found in an interview of a cadet scored at stage 3. Like the previous excerpt, the cadet is talking about good leadership. The content is similar, but the structure of his understanding is very different.

> I think the best cadet is the one who really understands that his real mission in life—in the Army—is being there for his troops. I mean he's got to have a commitment to the Army's goals overall.

How Do You Know That's the Right Thing?

> I guess it's one of those gut feelings that you have. And I'm of the firm belief that if you don't want that, if you're here just for yourself, you know, get the heck out of here. And, you know, the Academy kind of makes you realize that taking care of your troops is the most important thing.

So You Have That Responsibility. Where Does That Responsibility Come From?

> I think the responsibility as a leader to take care of your troops comes from the fact that you're the lieutenant. In addition to the fact that you're a human being. You know, as a human being you have a responsibility, I think, to take care of everybody else.

What makes this a stage 3 response is the cadet's embeddedness within other points of view (e.g., the Academy's) as the source of his own point of view. Those points of view include the Academy's expectations of an Army officer, the responsibilities of having a professional position ("you're the lieutenant"), and an idealistic societal view of the responsibilities of membership in the human race, all of which are part of this cadet's point of view. Missing is an embeddedness in his individual needs. Indeed, he critiques self-interest (a stage 2 way of constructing the self) when he admonishes those who are "here just for yourself" to "get the heck out." Critique of a lower stage (in this case identification with a personal agenda) indicates the ability to take a broader perspective and signals a more advanced stage.

The transition to stage 3 is just the first step in the development of a professional identity. There are limitations to stage 3 capacities that lead to an incomplete understanding of what it means to be a professional. Because we are subject to shared meaning at stage 3, in many ways we are not psychologically distinct from our relationships. Others' feelings about us are a part of how we experience ourselves. As a result, we often expend far too much energy avoiding hurting others' feelings. And because we assume that others, like us, make our feelings about them a part of *their* identities, we tend to feel responsible for how others experience us. In the terms of Table 17-1, at stage 3 we are subject to mutuality. In a psychological sense, there is no independent "self" that we bring *to* our relationships, because there is no "self" apart from our relationships. We have a shared identity at stage 3 rather than a personal identity. With regard to professional identity, stage 3 professionals "are the profession"; they have fully "bought in" and take on all the expectations of the profession. As a result, stage 3 professionals are unable to reconcile the competing expectations that are a natural part of professional practice, because they have not yet "integrated" the expectations of the profession into an autonomous self system. Lacking such a system of self-authored values, the stage 3 professional has no basis for resolving competing expectations.

Kegan's Stage 4

Kegan asserts that psychological maturity—what he terms "psychological autonomy"—is not possible until stage 4 is achieved. Stage 4 officers are able to organize their stage 3 social and interpersonal identifications using individually constructed systems for organizing values and standards. At stage 4, one has developed the capacity to use self-authored principles and personal standards to reflect upon one's personal and professional experiences. At stage 4, officers' *shared* ideals and professional expectations shift from subject to object; stage 4 officers are capable of taking a perspective on their shared ideals and assessing them in terms of self-authored principles and standards. Officers with a stage 4 understanding of the profession now "own" the values of their profession; they have internalized these values in a fundamentally different way than have stage 3 officers. They can now take an entirely personal perspective on the profession's expectations, critique them, assess them, and organize them in more complex and sophisticated ways. And, importantly, stage 4 officers have developed a system of personal values and principles that permits them to reconcile professional and personal value conflicts.

The transition from stage 3 to stage 4 involves the transformation from being one's relationships to having relationships. For officers, the transformation involves a shift from being a member of a profession to being a professional. Stage 4 is concerned with psychological independence. Self-authored values and principles provide a perspective on one's relationships and socially defined identities. While both stage 3 and stage 4 officers can articulate broad values and ideals (content), only stage 4 officers can justify their values in terms of what they personally believe rather than appealing to a shared external standard. We believe that the transition from stage 3 to stage 4 is a critical transformation that must be made if officers are to *lead* and *shape* the profession as it faces the challenges of the new era. Such leadership requires officers who are not psychologically embedded in the profession, but who are professionals capable of stepping back and viewing the profession objectively. As we have said, this capability is characteristic of a stage 4 orientation toward the profession.

The following excerpt illustrates a self-authored stage 4 perspective. It comes from an interview with an officer at the Army War College (from research conducted by Philip Lewis and Owen Jacobs[5]). This individual is in his early 40s and has about 20 years of commissioned service. The officer is talking about an experience in which he felt particularly successful.

> I also feel to be successful you have to undergo hardships. You know, units and organizations, whether it's a staff organization—coming together not during good times, but overcoming bad times—inspections, field exercises, and stuff—that's where they come together. That's where you become successful. But my personal success is geared not by the jobs I have, by the ranks or medals I obtain, but by the accomplishments of the units and the people below me.

What's the Most Valuable Kind of Information That You Get for Yourself in Terms of How You Are Doing?

> It's my opinion, my assessment. I have never really doubted—even though OERs and awards from your superiors are nice pats on the back—probably my greatest satisfaction has always come from me just going out and watching somebody do something they could not do. Units who had problems before are now successful.

So It's Your Own Judgment about These Things?

> If I think that's the best we can do, then I feel good. That's my philosophy. There have been times where I thought we did a good job and my boss has not. But, once again, I felt good and at peace with myself.

This officer has a clear sense for how he evaluates his own success. It is a stage 4 understanding because it is based on an internal standard; it is not dependent on either tangible rewards (stage 2) or how others feel about or judge him (stage 3). What is most valuable about how this stage 4 officer constructs his understanding of his performance as a military officer is that he has his own internal compass for proceeding even when he is receiving ambiguous or conflicting signals from others. In short, he can function as an independent decision-maker, one who shapes the environment in which he operates instead of merely reacting to it.

Transition Between Stages

To this point in our discussion we have focused on Kegan's stages as if most individuals function at one of three adult stages. In fact, however, individuals spend a substantial part of their lives in transition between stages, as will be seen in the developmental data presented below. Committed as we are to the idea that it is desirable to attempt to facilitate psychological growth, it makes sense to try to understand the nature of developmental transitions. In general, individuals in transition demonstrate an ongoing involvement in both the conceptually simpler constructions of the earlier stage and the conceptually more complex constructions of the next stage. Consider the following excerpt from an interview with a West Point cadet:

> Like say you set a goal for, you know, I'm going to do a hundred push-ups. And you work real hard on it, and then one day you just get down—things are happening right—and you do a hundred. And you feel good about yourself, because you met the standard. And then you hand in your card and someone says, "Wow, a hundred push ups." Something like that, it boosts your confidence if you can do something. But I think it also boosts it a little bit if somebody would tell you that. "You did that well."

Tell Me Why It Seems to Add to That.

> I know it does add to it, but, ah, I'm not really sure why. . . . When people know that you can do something, well, then I think that's important.

So There's Something about Someone Else Knowing You Did Well That Strengthens Your Confidence at this Point?

> And that's going back to it—if people have a good image of you then that boosts your confidence. You know, people will be more likely to put responsibility on your shoulders. People don't have to worry about you messing things up. And that is why, at times, I consider it a success. . . . But I think it's important to me to have a good image. And to have other people have a good image of you. And I think that, you know, that can only help in a whole bunch of different areas. I mean, it's hard to explain as to why I think a good image is important . . . . just don't think it's easy to explain.

From this excerpt and other portions of the interview, it was clear that this cadet was in the early part of a transition from Kegan's stage 2 to Kegan's stage 3. This is revealed in several ways. First, a whole new way of attaching meaning to this cadet's experience is beginning to emerge. It is so new that he is having difficulty explaining it to the interviewer. In addition, it appears that rather than merely determining his self-worth by meeting certain personal goals (the chief source of meaning at stage 2), he is beginning to be able to make another's image of him a part of his experience of himself (evidence of stage 3 simultaneous perspective-taking). So both a stage 2 and a stage 3 source of identity are operating. It is early in the stage 2 to stage 3 transition, because the ultimate goal (and source of meaning) is his "getting things done," and the value of an internal sense of confidence, derived from how others see him, is that it helps him get things done.

A second example, illustrating early signs of the transition from stage 3 to stage 4, comes from an interview of an officer who is a student at the U.S. Army Command and General Staff College.[6] The officer is talking about how he has changed during his time at the Command and General Staff College and how he will enter his unit on taking up his next assignment.

> I can take a look at myself. What do I really stand for? Because when I leave here I don't want to be a product that is blown by the wind, you know, left and right. I

> want to enter an organization, introduce myself, interact, and then be known as, this is what _____ stands for. And I want to have a strong conviction that that's what's really important. And walk forward with that. And that they can count on that. That's where they hang their hat and say, okay, that's what he's all about. And to do that, I think again I've become more introspective.

In this excerpt, we see evidence of the beginning of the transition beyond stage 3, manifested in this officer's desire to have his own internal compass that guides his actions. He doesn't want to be blown about by external perceptions of who he is and what he should do. But his perspective is not yet a fully self-authoring (stage 4) perspective; it is still part of a shared experience—he wants to "be known as" an independent thinker. He is not yet centered in his *own* view of himself. We will have more to say about the types of experiences that promote developmental transitions at the end of the present chapter.

Research on Officer Identity Development

For the past decade, we have been involved in a research effort to understand Army officer development using Kegan's framework. Our goal has been to describe developmental shifts across an Army officer's career and to investigate the relationship between identity development and indicators of professional competence. In this chapter, we report our findings from a cross-sectional study of officer development from precommissioning through battalion command. Some of our data are taken from a much larger study of leadership development at the United States Military Academy, part of the Baseline Officer Longitudinal Data Set (BOLDS) Project. We have combined these longitudinal data with cross-sectional data gathered from majors and lieutenant colonels to produce an overall picture of officer identity development. One question we have addressed is whether there is evidence that cadets, who are just entering our profession, show lower levels of psychological maturity—narrower perspectives in Kegan's terms—than do mid-career and late career officers.

Research Participants

Cadet participants were randomly selected from the entire entering Class of 1998. Data from 38 cadets were collected during the first six weeks (Cadet Basic Training) of their first year (Time 1). Time 2 interviews were conducted with 31 of the original 38 participants during spring semester of the second year. Additionally, 24 new participants were randomly selected and interviewed for the first time, providing a total Time 2 sample of 55. During the senior year, 35 of those from the Time 2 group were re-interviewed (Time 3). Of these, 21 were part of the original Time 1 group, 34 had been interviewed at Time 2, and 20 were interviewed at all three points in time. Attrition was a function of a variety of factors, including resignation, involuntary separation, and refusal to participate. While these samples may appear small in size, they are actually quite large for this type of research—indeed, to our knowledge this is the largest longitudinal test of Kegan's theory, and it is the only one involving traditional college-age students in a professional school context. Officer participants represent convenience samples from two grade levels: majors and lieutenant colonels. Fourteen majors attending the U.S. Army Command and General Staff College in the Class of 1996,[7] and 28 officers attending the Industrial College of the Armed Forces and the Army War College in the Class of 1991, were interviewed.[8]

For all three groups of participants, developmental assessments involved semi-structured interviews using the method detailed by Lahey, Souvaine, Kegan, Goodman, and Felix.[9] Following these procedures, trained interviewers prompted respondents to recall their own recent experiences in response to stimulus words printed on index cards. The stimulus words, reflecting emotion-laden themes, included *sad*, *success*, *torn*, *anxious and nervous*, and *important to me*. After thinking about their experiences and making notes on the cards, the respondents were invited to begin by describing one of the events that came to mind in response to a stimulus card. The interviewer then asked a series of follow-on questions to elicit the underlying perspective the respondent was using to understand the event. When the interviewer was able to elicit the respondent's underlying and most encompassing perspective, the interviewer invited the respondent to move on to describe an event elicited in response to another card. This second event was then explored to help the interviewer discover the most encompassing perspective the respondent was using to make sense of the event. An assessment interview normally lasted between sixty and ninety minutes during which three to four events were explored.

All interviews were audiotaped and transcribed into hard copy for scoring. Four psychologists trained in the Kegan interview and scoring procedures scored the cadet interviews at Time 1. Average inter-rater agreement across these 4 scorers was 63% (to within one-fifth of a stage) and 83% (to within two-fifths of a stage).[10] A single individual, a Kegan-trained expert with demonstrated expertise in administering and scoring constructive-developmental interviews, scored all the later (Time 2 and Time 3) cadet interviews, the CGSC interviews (majors), and the senior service college interviews (lieutenant colonels).

Cross-sectional Data on Developmental Levels

Table 17-2 summarizes the developmental level scores for senior service college students, CGSC students, and USMA senior (Time 3) cadets. As can be seen, there is a general tendency for the senior service college students to have the highest developmental scores. Fully half of this sample of late career officers were demonstrating full stage 4 levels. More than three-quarters of the rest were in the stage 3 to stage 4 transition, and none were scored below a full stage 3 level. The USMA seniors showed the lowest developmental level scores. None were shown to be at Kegan's stage 4, and fully a third of this group had not yet reached stage 3. As expected, the mid-career CGSC officers were intermediate between the USMA cadets and the senior service college students in their developmental levels. Like the senior service college students, a substantial proportion (43%) of the CGSC officers had reached Kegan's stage 4. Interestingly, the same proportion of CGSC students scored in the stage 2 to stage 3 transition as did USMA seniors (28% in both groups).

In short, there was a much broader range of developmental level scores among the CGSC students than might be expected. It is not clear what to make of this finding. First, it should be acknowledged that the sample of CGSC students was quite small (N = 14). It is also not clear how this sample was selected. The CGSC interviewer was a CGSC student at the time of the interviews, and it appears that he recruited interviewees from those students he knew personally. If this is true, then the sample may not have been representative of the entire class of CGSC officers. In contrast, the senior service college students were nearly all of the students in three seminar groups. The senior service colleges (in this instance the U.S. Army War College and the Industrial College of the Armed Forces) attempt to assign students to seminar groups in a manner that insures that all the seminar groups are roughly comparable.

Kegan Developmental Level Scores[11]	USMA Seniors (N=32)	CGSC Students (N=14)	Senior Service College Students (N=28)
2	2 (06%)		
2(3)	3 (09%)		
2/3	5 (16%)	3 (21%)	
3/2	1(03%)	1 (07%)	
3(2)	0 (00%)	0 (00%)	
3	15 (47%)	3 (21%)	3 (11%)
3(4)	4 (13%)	1 (07%)	3 (11%)
3/4	2 (06%)	0 (00%)	3 (11%)
4/3		0 (00%)	3 (11%)
4(3)		0 (00%)	2 (07%)
4		5 (43%)	14 (50%)

Table 17-2. *Kegan Developmental Level Scores for Military Leaders at Three Different Career Points.*

The other major difference between CGSC officers and senior service college officers is that the latter colleges are more selective. At the time the Kegan interviews were conducted, approximately 65% of the Army's majors were selected for the residential CGSC program, whereas only 10% to 12% of the Army's lieutenant colonels were selected for a residential senior service college program. In short, the senior service college students are a more highly selected sample than are CGSC students. It is possible, therefore, that the apparent increase in developmental levels between mid-career and late career suggested by these data are actually only a reflection of the selection effect. In other words, it is possible that more of the officers who are already at Kegan's stage 4 when they are majors get promoted and sent to a senior service college than do majors who are at stage 3 or lower.

It is harder to argue that the differences between the senior cadet developmental level scores and the other two groups are a function of differential selection. None of the 32 USMA seniors who were interviewed had yet achieved Kegan's stage 4. Indeed, only two were well into the stage 3 to stage 4 transition. It seems likely, therefore, that the stage 3 to stage 4 developmental transition, if it takes place at all, is traversed only after officers have embarked on their Army career. The other cross-sectional data bearing on the question of whether there are progressive changes in Kegan developmental levels are the USMA Class of 1998 interview data collected during the freshman, sophomore, and senior years (the senior year data were also presented above). Recall that not all cadets in the USMA sample were interviewed at all three time points. Indeed, a new sample of cadets was randomly selected and then added to the interview study during the sophomore year. These cadet data, therefore, can be considered cross sectional; they are presented in Table 17-3. The longitudinal data for USMA cadets interviewed both sophomore and senior year are presented later in the present chapter.

In general, these cross-sectional cadet data are consistent with the proposition that Kegan interview scores reflect a progressive developmental process. The most striking finding shown in Table 17-3 is the dramatic increase in the percent of cadets scored as functioning at Kegan's stage 3 and above in the senior year. At the beginning of the freshman year only 16% were at stage 3. By the middle of the sophomore year this percentage had risen to 27%. By the senior year, these percentages had more than doubled to 66%. These data suggest that there is, for many cadets, significant developmental progression between the freshman and senior year.

Kegan Developmental Level Scores[12]	Freshmen (N=38)	Sophomores (N=52)	Seniors (N=32)
2	8 (21%)	12 (23%)	2 (06%)
2(3)	3 (08%)	12 (23%)	3 (09%)
2/3	16 (42%)	11 (21%)	5 (16%)
3/2	4 (10%)	2 (04%)	1(03%)
3(2)	1 (03%)	1 (02%)	0 (00%)
3	6 (16%)	11 (21%)	15 (47%)
3(4)		3 (06%)	4 (13%)
3/4			2 (06%)

Table 17-3. *Cross Sectional Kegan Developmental Level Scores for USMA Class of 1998 as Freshmen, Sophomores, and Seniors.*

Longitudinal Data On Developmental Levels

A more direct assessment of developmental change can be gained by inspecting the Kegan stage scores of those USMA cadets who were interviewed on two occasions. Figure 17-2 presents those data for subjects interviewed during both their sophomore and senior years. Data from the freshman year interviews are not presented because the cross-sectional data suggest that overall there was little developmental change between the freshman and sophomore years. Figure 17-2 provides direct support for the proposition that for many West Point cadets the college years are a time of significant developmental change. Sixteen of the 28 subjects (57%) showed developmental progression. Only 2 subjects showed regression. Of the 10 subjects who showed no change from their sophomore to their senior year, half were already at stage 3 during their sophomore year, a stage beyond which many college students are not expected to move until somewhat later in adulthood, if at all. With the exception of the two instances of regression, which could be a function of the difficulty of obtaining a reliable developmental score, the longitudinal data presented in Figure 17-2 suggest that Kegan developmental level scores are tapping a variable that shows progressive developmental change for many cadets during the college years.

Subject	Sophomore Year	Time 2 of Change	Direction Senior Year	Time 3
	A05	2	→	3/2 –3
	A08	3-3(4)		3(4)
	A09	2		2
	A10	2	→	2/3
	A15	2-2(3)		2(3)
	A17	2	→	3/2
	A19	3		3
	A21	3		3
	A23	2/3	→	3/4
	A24	2/3	→	3
	A25	2(3)	→	3
	A26	2(3)	→	3(4)
	A27	2(3)		2(3)
	A28	2		3/4
	A31	2/3		2/3
	A32	3		3
	A36	3	←	2/3
	A37	2-2(3)	→	3
	B11	2(3)	→	3
	B12	3/2	←	2(3)
	B15	3		3
	B17	2	→	3
	B23	3/2-3	→	3(4)
	B24	3/2	→	3(4)
	B25	2(3)	→	2/3
	B32	2/3	→	3
	B34	2/3		2/3
	B35	2(3)	→	3

Figure 17-2. *Individual Stage Scores of USMA Cadets with Scorable Interviews at Both Time 2 and Time 3*[13].

Previous descriptions of the stage 2-to-3 transition have suggested that this developmental shift typically takes place during the pre-teen or early teen years.[14]

However, these descriptions have not been based on actual data. The study we report here, the only one with a college-age cohort, suggests that this key transition takes place somewhat later, probably during the late teens and early 20s. We speculate that this finding also holds true for cadets in ROTC programs as well.

Implications For Officer Professional Identity Development

Preoccupied with their own needs and interests, those cadets who are still functioning at Kegan's stage 2 cannot be expected to have the broad, internalized understanding of and commitment to ethical codes and other professional standards that we ask in our officer development programs. Rather, stage 2 cadets will likely see ethical codes and professional standards solely as guides for behavior. They are likely to conform their actions to these guides in order to garner rewards and avoid negative consequences. Cadet development programs will not be successful in instilling desired values in these less mature cadets unless the broad educational environment in which they operate promotes identity development toward a shared perspective on officer professionalism (i.e., to Kegan's stage 3). Consequently, cadet development programs must be tailored with this in mind. Programs must encourage cadets to behave in ways that are consistent with the professional values while at the same time promoting development of a shared perspective for considering those values that is grounded in a sense of what it means to be a member of a profession. This is indeed what happens at the United States Military Academy, and, as our data suggest, many cadets appear to make progress toward a stage 3 understanding during their cadet years. Precomissioning programs at West Point appear to be successful in socializing cadets into the profession—cadets are "joining the profession" and becoming members of the officer corps. However, it is also clear that very few cadets at West Point are likely to have developed to the point where they are self-authoring professionals—stage 4—by the time they are commissioned.

Turning now to the cross-sectional findings for the officer samples, we see what appears to be progress from stage 3 to stage 4 between the end of precommissioning and mid career, at least for most officers. It seems highly likely that most lieutenants and captains function with at least a stage 3 perspective on themselves and their experiences.[15] Hence, they see themselves as members of the profession. From a stage 3 point of view, they understand themselves in terms of the expectations that the profession has of them, and these expectations are organized around qualities that characterize an officer. But the stage 3 officer is also subject to the expectations of others outside the profession—spouses, children, parents, and friends. Because stage 3 individuals internalize others' expectations, their primary developmental challenge comes about when the expectations of these different sources are in conflict—such as family and profession. We expect that stage 3 officers will experience considerable tension when trying to resolve these competing demands. Indeed, an inability to do so may be an important source of the current retention problem with Army captains.

However, if properly managed these challenges can set the stage for the transition from stage 3 to stage 4. Our sample of majors showed considerable variability.

A few are still completing the stage 2-to-3 transition. We were surprised to find evidence of stage 2 functioning (21% were still early in the stage 2-to-3 transition)—officers still subject to personal agendas—in this select group of officers. On the other hand, many officers in the sample of CGSC majors (43%) had completed the transition from stage 3 to stage 4, manifesting the internalization of personal value systems with which they should be able to resolve conflicting demands. An even larger proportion of our sample of senior service college lieutenant colonels (50%) had completed the transition from stage 3 to stage 4. The remaining officers are still in transition from Kegan's stage 3 to stage 4.

In the aggregate, these findings indicate a pattern of continued development toward a principle-centered, self-authored understanding of officer professionalism as one advances in rank. In the absence of complete longitudinal data, it is difficult to know for certain whether these findings actually represent development or whether they show the effects of selection of higher functioning officers across the career span. However, they are at least suggestive of developmental possibilities. Furthermore, the results suggest that experiences during the second decade of a career—roughly the period from the staff college to the senior service college—may influence the transition from stage 3 to stage 4 perspectives.

To most Army officers, the term "officer professional development," or OPD, summons an image of dayroom classes, professional reading lists, and staff rides. Thus far, the Army's approach to professional development has primarily emphasized improving tactical and technical expertise by filling officers' professional kitbags with the latest knowledge and skills, the latest training and doctrine. While such an approach works well to keep officers "informed," how effective is it at influencing professional identity development? If redefining and renewing the Army profession requires fundamentally changing officers' self-concepts, can we rely on our traditional training model to help us "*trans*form" the officer corps? Will officers who are simply better "*in*formed" be able to meet the increasingly complex demands of the twenty-first century?

The present chapter offers an alternative way to think about these questions and officer professional development, both in terms of what is being developed and also how we might attempt to influence it. In terms of the Army's doctrinal BE, KNOW, DO leadership framework, the task of reprofessionalizing the Army is primarily one aimed at the BE component. Clearly, "developing" professional knowledge and skills is important to "being" a professional; but when we talk about "development" in this chapter, we are talking about influencing officers' self-concepts, their sense of who they are, their worldview, how they know what they know, their character, their identity.

To be successful in this new era, today's officers not only have to be better informed, they also have to be more mature psychologically. Increasingly, modern officers are required to make sense of unstructured tasks, exercise judgment under conditions of ambiguity and uncertainty, and do so with little or no direction from higher headquarters. Leading under such conditions, we argue, requires not only a tactically and technically proficient expert, but also officers who operate from a broader perspective, officers who "see more of the world"—in short, more psychologically mature adults. Developing enough officers equipped with such self-concepts or identities is an essential task if we are to succeed at redefining and renewing the Army as a profession. Our research has implications for how the Army approaches officer education (Officer Education System), career management (Officer Personnel Management System), and the challenge of Army reprofessionalization. To tip the scales in favor of professionalization as opposed to bureaucratization, we will have to emphasize education (long-term, deep understanding) in addition to training (short-term, efficient performance) in our approach to officer development.

Many of the demands that we currently place on officers throughout their careers require levels of psychological maturity beyond that which many have achieved. Structural limitations imposed by each stage of adult development have significant implications for how officers make sense of their professional experiences. Knowing where officers are in their structural development should inform what we can reasonably expect from them, how we educate them, how we assign them, and ultimately how we approach their development.

Recommendations

As a profession, we must come to grips with the fundamental differences between *training*, *education*, and *development*. Each intervention process has its place; each has its limitations. Redefining and renewing our profession are primarily an educational and developmental challenge. Our ingrained tendency will be to *train* our way out of this. But an approach based primarily on a traditional training paradigm will not achieve the types of outcomes we require to remain a vibrant profession. Our current organizational structure and culture do not support officer *development*; they support *training*. Where do *education* and *development* fit in the Army's Training and Doctrine Command (TRADOC)?

We propose that the Army reframe TRADOC to include *education* and *training* as subsets of the broader process of officer *development*. To the extent possible, the Army should focus on skill acquisition (training) in units and organize to educate and develop in the schoolhouse. As recommended in the *Army Training and Leader Development Panel Officer Study Report to the Army*, "Leaders should focus on *developing* the 'enduring competencies' of self-awareness and adaptability."[16] To develop these enduring competencies, we must realize the limitations of training and better organize to support *education* and *development*.

If many officers are in over their heads from a psychological development perspective, then perhaps it's time to reconsider our entire approach to how we select, evaluate, promote, and assign officers within our profession. If the trend continues towards a more distributed battlefield, characterized by increased volatility, uncertainty, complexity, and ambiguity; toward smaller units having increasingly lethal weapons and mobility; toward command by officers who have to react faster to more significant challenges at lower and lower levels with greater and greater autonomy—then perhaps the British have it right. Perhaps we should consider 40-year-old company commanders.

We are proposing that promotions and assignments be based on a broader definition of officer development. Our post-Cold War Army requires officers that are more "psychologically mature" at more junior levels. We should reconsider our current approach to officer assignments, which is based primarily on the size of the unit, and instead align officers and positions based on the level of perspective-taking required by the position and the developmental level of the officer. We should consider, for example, assigning much more senior (psychologically mature) officers to much smaller units. The Army should also consider using complete 360-degree feedback and self-assessment mechanisms "with teeth" tied into OERs for all officer operational assignments to better support the development of our two "enduring competencies"—self-awareness and adaptability.

The Army profession is currently facing significant boundary/jurisdictional issues—an identity crisis with both institutional and individual implications. This chapter adds an additional dimension to this debate—the psychological-structural dimension. Understanding officer development from this perspective raises several profound issues. First, it is very difficult to understand, let alone thoughtfully change, that in which you are psychologically embedded. Therefore, it may take officers with at least a stage 4 perspective (independent of the existing institution) to effectively resolve fundamental issues of professional identity. Second, one of the primary characteristics of a profession is its ability to "self-regulate." Our research suggests that perhaps it also takes someone with a stage 4 (self-authoring) perspective to truly "lead" a profession (as opposed to merely being a "member of" a profession). Ultimately it is up to commissioned officers to redefine and renew the Army as a profession. Such a daunting task will require an officer corps made up not only of technically and tactically proficient bureaucrats, but also of psychologically mature professionals. We believe that understanding officer development from a psychological perspective will help us significantly in our quest to develop effective professional identities for twenty-first century Army officers.

Notes

1. Research for this chapter was supported in part by the Army Research Institute for the Behavioral and Social Sciences. The views and opinions expressed in this chapter are those of the authors and are not necessarily those of the Department of the Army or any other U.S. government entity.
2. Robert Kegan, *The Evolving Self: Problem and Process in Human Development* (Cambridge, MA: Harvard University Press, 1982); Robert Kegan, *In Over Our Heads: The Mental Demands of Modern Life* (Cambridge, MA: Harvard University Press, 1994).
3. Derived from Kegan, *The Evolving Self,* 86.
4. Kegan, *In Over Our Heads*, 192-195.
5. Philip Lewis and T. O. Jacobs, "Individual Differences in Strategic Leadership Capacity: A Constructive/Developmental View," in *Strategic Leadership: A Multiorganizational-Level Perspective*, eds. R. L. Phillips and J. G. Hunt (Westport, CT: Quorum Books, 1992), 121-137.

6. We are indebted to Lt. Col. Patrick Sweeney for conducting interviews with officer students at the U. S. Army Command and General Staff College. The data from the interview sample were gathered by Lt. Col. Sweeney while he was a student at the staff college in 1995-1996.
7. Interviews were conducted by Lt. Col. Patrick Sweeney.
8. Interviews were conducted by Dr. Philip Lewis as part of another study.
9. Lisa Lahey, Emily Souvaine, Robert Kegan, Robert Goodman, and Sally Felix, *A Guide to the Subject Object Interview: Its Administration and Interpretation* (Cambridge, MA: Harvard University Graduate School of Education, Subject-Object Research Group, 1988).
10. Interview scores may represent a full stage (e.g., stage 2 or 3) or a transition between stages (somewhere between two stages). For example, transitional scores between stages 2 and 3 are 2(3), 2/3, 3/2, and 3(2). These scores represent differences in the degree to which a stage 2 or a stage 3 perspective dominates. Hence agreement between scorers within one-fifth of a stage might involve scores of 2/3 and scores of 3/2. Continuing with this example, agreement within two-fifths of a stage involves scores between 2/3 and 3(2).
11. Scores of 2(3), 2/3, 3/2, and 3(2) represent progressive scorable distinctions in the transition between a full stage 2 and a full stage 3; scores of 3(4), 3/4, 4/3, and 4(3) represent progressive scorable distinctions in the transition between a full stage 3 and a full stage 4.
12. Ibid.
13. "A" subjects entered the study as freshmen; "B" subjects entered as sophomores.
14. Kegan, *The Evolving Self*, 167; Lahey et al., *Guide to the Subject-Object Interview,* 94-131.
15. Preliminary data from a sample of Army captains enrolled in a master's degree program in preparation for the role of West Point Tactical Officers for the Corps of Cadets suggest that the vast majority of these officers are functioning at stage 3.
16. Department of the Army, "The Army Training and Leader Development Panel Officer Study Report to the Army," Department of the Army (May 2001): OS-3; available from *http://www.army.mil/atld/report/pdf;* Internet; accessed 16 August 2001.

18 Backbone vs. Box: The Choice between Principled and Prescriptive Leadership[1]

Anna Simons

Officers belong to what is often described as the world's oldest profession. Do soldiers really predate priests or chiefs? We'll never know for sure. But officership has to count as one of humanity's most enduring social inventions. Without question, significant shifts have occurred in terms of whether officers command from the front or the rear, compel or impel, transform or transact, and manage or lead. However, this chapter will contend that the services which officers and only officers can provide haven't really changed. Nor can they. In addition to fulfilling organizational and functional requirements, officership meets bio-political as well as sociological demands—demands which will remain the same no matter how dramatically future battlescapes alter.

Though the Army must always worry about what looms ahead, I want to suggest that it is equally essential to consider what still lurks within, namely, the all-too-human desire to be directed, if not led. This chapter will suggest a way in which it may actually be possible to address both concerns via the same approach—principled leadership—thereby taking advantage of what we do know in order to address what we can't know.

My general thesis is this: not only does operating from a set of core convictions or principles grant leaders at all levels maximum flexibility while minimizing a range of different frictions, but having principled leaders satisfies soldiers' demands. Absent the flexibility and adaptability afforded by principled leadership, the Army's future success would depend on its present ability to foresee accurately tomorrow's operational concepts and scenarios. That is a tall order. No one could have forecast U.S. involvement in Somalia twelve or, arguably, even two months before the fact. Were we the least bit prepared for an exodus of over 630,000 refugees from Kosovo or for their later return? Who would have suspected in August 2001 that we'd be fighting a war in Afghanistan in the fall? As these or any number of other recent military interventions would suggest, the only thing we can say with assurance today is that we will be surprised again tomorrow. In terms of how to react, meanwhile, we can no more count on controlling the what, where, and when of future confrontations than we can command the weather; we don't always understand others' motivations or their methods. What we should be able to rely on, though, is the extent to which we know ourselves. And where our own strengths lie *as* Americans *is* in principles, and not just in tactics, techniques, and procedures, vital as these are. This is yet another reason to (re)turn to principles, though the gist of my argument is more elemental still: in conditions of flux humans need to be able to count on something fixed, while what is fixed has to be flexible too, paradoxical as that may sound.

In what follows I contend that humans need a hierarchy of values even when, and perhaps especially when, circumstances call for maximum behavioral flexibility. From an anthropological perspective, I demonstrate the extent to which officers often resemble Big Men more than they do Chiefs (terms I'll return to below), which means that they impel rather than compel and employ carrots as well as sticks. I make the point that an officer's most effective tool is to engage in principled behavior. I then go on to examine some of the principles of officership which have withstood the test of time, as well as those which seem to be challenged by postmodern change. I use Special Forces teams and two superlative military advisors—T.E. Lawrence and Edward Lansdale—to illustrate that principled officership remains integral even in decentralized situations. These are exactly the kinds of situations which soldiers confront. The world, after all, lacks geopolitical balance, units increasingly experience turbulence, and combat always causes turmoil. In the face of this, principles, like a backbone, offer consistency, resilience, and strength.

The Hierarchical Imperative

As humans, we are, to borrow from Lionel Tiger and Robin Fox, "imperial animals."[2] We invariably form hierarchies. Even in what anthropologists call "stateless" societies, which by definition lack formal government—where there are no permanent government structures, police forces, or officials of any kind—leaders arise. In foraging societies, for instance, one man may be a better hunter. Someone else might be a better healer. In societies where elders, as a group, deserve respect some individuals will always be listened to more closely; inevitably there are a few who command more attention than their peers. Why? Because they strive for this? Or because others won't, don't want to, or can't measure up?

Interestingly, in much of the leadership literature little is said about the competition involved in rising to the top. Anthropologists' accounts, in contrast, describe strategic struggles among New Guinea Highlanders, Swat Pathans, Kwakiutl, with the list stretching on and on. In both sets of literature the focus tends to be on what makes leaders effective and what effects they then have on their followers. Occasionally there are discussions about followers' desires: what they want of, from, and in their leaders. But not even in leadership manuals is it explained why followers might want to be followers and not leaders themselves.[3]

The literature never explicitly labels leaders winners and followers losers, but reading between the lines this is what it strongly suggests. If leaders have such qualities as ambition, character, drive, and ideas, then followers must lack them. Or they lack something else, perhaps self-esteem, self-confidence, skills, or smarts. Yet, what if many simply prefer to not compete? What if, instead, they'd rather belong? Or what if they like being directed by someone else, and would rather work for someone they admire, and alongside others, rather than for (or by) themselves? To anyone with drive or ambition this must sound unreal, impossible, or false.[4]

Indeed, to most Americans "moving up" must seem to be a universal goal. However, hierarchicized divisions of labor that have existed for generations generally persist not only because those at the top manage to successfully constrain those below, but because there are costs associated with exerting control. Or, to turn this around, so long as the price paid by subordinates is not too high (or can be rationalized away), the contest to become superior (or even autonomous) may not be considered worthwhile. Concomitant with different values (and rewards) attached to different kinds of work are different expected behaviors. Generally speaking, individuals lower down the ladder are granted (or grant themselves) more latitude to express themselves freely, display emotions, and act out in public.[5] The obverse is true for those in control; they must act controlled. Ergo the stiff upper lip (at least in public).[6] In other words, regardless of—and maybe even in psychic proportion to—the benefits leaders accrue, leadership is presumed to exact what many regard as too burdensome or too difficult a toll (though not a toll those who become leaders must mind).

Despite our "take-charge" attitudes, American ambivalence toward leadership—and even toward leaders themselves—runs deep. At the moment, for instance, we are consumed with weighing, justifying, and bemoaning what might be lost as we choose greater responsibilities for higher pay but longer hours. Attaining power costs time; exerting power consumes energy. We constantly weigh the trade-offs. No less than the former Secretary of Labor himself, Robert Reich, has recently questioned the sacrifices he made during his own rise to the top. Yet, no matter how predisposed we might be to think in terms of quid pro quos, it is unclear whether rewards and sacrifices are ever directly proportional.

For instance, in the all-volunteer force the quid pro quo we might expect for leaders vis-à-vis followers, in terms of risks and rewards, is turned on its head. The playing field is even when it comes to who makes the ultimate sacrifice: everyone commits to it. Yet, having said this, do senior officers work harder than juniors? Or put in longer hours? Lieutenants and captains would probably scream No. But the seniors earn more money and receive innumerable additional benefits. How do they justify this? How would we?

The clear answer is that those higher up the chain of command are responsible for more lives, more equipment, and more weighty decisions.[7] The more heretical answer would be that the Army can't operate without chains of command and has to have something to make its leadership division of labor as perceptible as is the division of labor between, say, armor, artillery, and infantry. The irony is that the further away from low-level, hands-on troop command that one moves, the harder it may be to clearly define precise leadership responsibilities. All the more reason, then, to make the hierarchy look as strong and well-delineated as possible, though in significant ways the more removed from soldiers officers are, the more control over them they can have.

Officers: Big Men or Chiefs?

Anthropologists often draw a distinction between two types of leaders: Big Men and Chiefs. Big Men have prestige. Chiefs hold power. Prestige is evanescent; it can't be consolidated over time or inherited without continual expenditure of effort. In the hunt for more prestige, Big Men strive to maintain and attract more and more followers from whom they solicit the makings of grand feasts where everything collected will then be consumed or given away. The point is to *seem* big, a point they must make over and over again in the face of never-ending, often-escalating competition. Reputation, in such a contest, is paramount. Thus, Big Men devote inordinate amounts of time to impression and image management, planning, plotting, and worrying. And for what ends? At best all they can do is dominate the social scene. They can't dominate anything else.[8]

Chiefs, in contrast, not only dominate, they domineer.[9] Unlike Big Men, Chiefs have coercive authority. Either they have armed forces at their disposal, who assist them in making others do what they want, and/or they possess divine powers. As a consequence, Chiefs have the capacity to inspire fear as well as command respect. They *are* in control. Or, to redraw the contrast, Chiefs can make people do things; at best, Big Men can only get them to want to help.

So which, would we say, describes officers? In today's Army, would we characterize officers as Big Men or as Chiefs? The academic answer is, of course, that they represent a bit of both, though the ratio of Big Manship to Chiefliness seems to increase as individuals rise through the ranks. All officers are Chieflike by virtue of being vested with command authority; those who command soldiers can not only give direct lawful orders but can also mete out nonjudicial punishment. And like leaders who wield coercive authority elsewhere, officers of all types and in all positions seek to maintain distance between themselves and their subordinates.[10] There are legions of ways in which they do this—physically, materially, ideologically, and/or symbolically.[11] The officer-enlisted divide itself exemplifies this, though as officers advance they make use of all sorts of distancing techniques to preserve separations even among themselves. This is written all over uniforms. But it is also made evident in, for instance, the size of higher-ups' staffs and how difficult gaining access to those higher-ups may be—or is made to be.

Of course, there are practical reasons for limited access, large staffs, and other perquisites of power which wind up reinforcing distance. But we also shouldn't forget that distinction is integral to hierarchy—something which is much easier to maintain (and cultivate) in garrison, during peacetime, and in higher headquarters, than in the field, during war, and in operational units. Without question, the hierarchical hope is that patterns of deference, obedience, respect, and order established during peacetime and training will carry over into war. Or, alternatively, the goal is that no one should think that such patterns don't carry over. But consider: when a unit is on its own, confronting discomfort, danger, or disaster, its officers actually live no differently than their soldiers and wield no more power than a Big Man could.[12] Though officers may be in a position to tell subordinates what they want them to do, they can't force them to do those things, let alone make them do them well.[13] Nor can they prevent their soldiers from acting like the fictional Sergeant Svejk, an exemplary goldbricker.[14]

Everyone has heard stories about soldiers who have found ingenious ways to thumb their noses at commanders by following the letter rather than the spirit of the law (even in combat). Thus, regardless of the fact that the stick looms larger than the carrot in the sense that those who choose not to follow orders *will* be punished (Big Men, remember, can't punish), officers who hope to succeed in battle, to accomplish their missions, and/or to rise in the chain of command are far better off using incentives than threats. Field Manual 22-100, the Army's leadership bible, itself concentrates on carrots: how to treat and counsel subordinates in order to get the most out of them. Presumably this wouldn't be doctrinal advice if followers' enthusiasm and support wasn't understood to be essential to mission accomplishment and leaders' success.

Interestingly, when we examine how Chiefs (and kings, emperors, rulers, and dictators) tend to treat their armed forces—as opposed to their subjects—we see something very similar. In societies where a national leader depends directly on wielders of force for his authority he will dangle all sorts of incentives in front of them in order to keep them eager to work on his behalf, with him, and for him. Tellingly, in this country the higher up the chain one looks the less officers have to depend on subordinates to help secure their reputation or assure their longevity as leaders. Instead, where subordinates' impressions matter most is at the tip of the spear, in units. There, most officers realize, at least in theory (and certainly this is what they are taught), that how well they do depends on what they can get, *not force*, their soldiers to do, while to accomplish this—as all the manuals make clear—they must lead.

In classic Big Man societies, where the societal ideal is to be generous, Big Men lead by being big, with the selfish aim of proving themselves *bigger than*. In the Army, according to the most recent edition of FM 22-100, "lead" means "be, know, do."[15] Does this mean, as with Big Men, "be, know, do" *better than*?

Time-Tested Principles

According to Raimondo Montecuccoli, a preeminent seventeenth-century commander and strategist, "The ideal commander was warlike, in good health, and of martial stature. He should possess moral strength, prudence, and above all have 'force,' a quality embracing courage, fortitude, energy, and determination, similar to the *virtu* demanded by Machiavelli and the *constantia* praised by Lipsius."[16] Some three centuries later, U.S. Army Chief of Staff Edward Meyer emphasized the importance of character: "Character is ingrained principle expressed consciously and unconsciously to subordinates, superiors, and peers alike—honesty, loyalty, self-confidence, humility, and self-sacrifice. Its expression to all audiences must ring with authenticity."[17] Tim O'Brien, a writer and veteran of Vietnam, describes a platoon leader he admired this way: "He was insanely calm. He never showed fear. He was a professional soldier, an ideal leader of men in the field."[18]

If we consider the continuities, two things stand out. First, the inner character and inner convictions leaders are said to have must be made manifest somehow. And second, the ideals they embody have to resonate. Is it mere coincidence that the traits Edgar Puryear finds in most top military leaders—integrity, humility, selflessness, concern for others, reverence, and showmanship—can all be outwardly displayed?[19] Hardly. Leaders have to give evidence, somehow, that they are worth following, as well as being personally worthy of the sacrifices they might (and likely will) demand. Those who would be leaders will therefore find it advisable to come across as confident, credible, courageous, decisive, just, honest, loyal, and selfless. Moreover, they should set the example and manifest self-control.

Leadership manuals and guides draw from at least two broad sources.[20] One is clearly the past: what character traits, attitudes, and behaviors have characterized exemplary military leaders through the ages? The second is human nature—or, to be more precise, the unchanging nature of social relations. At any rate, the profitable use of examples from the past implies that the nature of social relations is unchanging. For instance, even as FM 22-100 warns aspiring leaders about "the stress of change" ahead, it refers them to Field Manual 100-5, *Operations*, because:

> FM 100-5 provides a doctrinal framework for coping with these challenges while executing operations. It gives Army leaders clues as to what they will face and what will be required of them, but *as COL Chamberlain found on Little Round Top*, no manual can cover all the possibilities. The essence of leadership remains the same: Army leaders create a vision of what's necessary, communicate it in a way that makes their intent clear, and vigorously execute it to achieve success."[21]

I quote the passage above at such length because its claim—that the essence of leadership hasn't changed since the unexpected will always occur—tells only half of what we already know. It's really the essence of human relations that hasn't changed, but recurs in familiar ways. Otherwise, it would make little sense for the manual to use Colonel Chamberlain at Gettysburg to illustrate exemplary leadership. Gettysburg was fought 138 years ago. Presumably everything about war has changed since then. Except, of course, the human element.

Not surprisingly, John Mattox is able to trace "principles of moral military leadership" back to Augustine's day,[22] while, as Lloyd Matthews might add, it is not so much the "broad ethical ideals" themselves which have proliferated over the centuries but rather the literature about them.[23]

Yet, no matter how carefully the historical record has been plumbed to yield lists which double as guides, the catch for every officer is that he or she has to recreate this ideal past all over again in the present. And though the Army may offer all sorts of methods for assisting officers to do this, it can't control for all of the dynamics they will encounter. There is interpersonal chemistry to take into account. Also, individual officers may be no more adept at fieldcraft (or soldiering) than are their soldiers. Nevertheless, what soldiers still want officers to be *is* "better than."[24] And they want this without officers implying that they think they are "better than." Thus, the only realm in which officers can't be bested turns out to be the overarching realm of principled, moral, and ethical behavior.

Since there is no better-than-principled behavior, *being* principled amounts to a virtue in and of itself. Moreover, the fact that it is never easy to remain principled, especially in war, abroad, under stress, during crises, *and* when presented with contrary temptation, helps render the foregoing proposition that much more significant and lends incalculable worth to officers who are principled.

Soldiers need principled leaders for both noble and prosaic reasons. A leader who acts in a principled way represents order and inspires cohesiveness as well as confidence, which are the antitheses of what war promises. If we rethink the attributes desired in leaders—fortitude, strength, composure—they represent fixity in conditions of extreme flux. One can think of a range of related traits—dependability, trustworthiness, reliability, consistency, good judgment—to describe someone who is principled.

There is probably more tacit agreement in the combat literature about the significance of being principled than there is regarding which principles officers should adhere to. Some principles, such as those brought forward by Don Snider, John Nagl, and Tony Pfaff, are suggested as those from which officers "should draw both their vision and their motivation."[25] Other lists of principles describe how officers should act and what they should do to be considered leaders. All, again, have been carefully distilled from what has worked through time and across space. Significantly, though, not a single one of these tells officers what to do in specific situations;[26] they can't. Given the messiness of real world encounters, officers have to be able to apply situational ethics to whatever might confront them. But if we think about it, this means that the thing officers most need—which is to know *what* to do—no one else can provide, except possibly those subordinates who are there with them (on the spot, in the field, in combat, in the unit). Yet *to lead* means never having to be told by subordinates what is the "right" or "moral" thing to do.

This, I would argue, is the source of an officer's power. Whether he or she can *consistently* make the "right" decisions determines whether or not others will consistently follow. Soldiers expect officers to be the authority on what is the right thing to do—tactically, doctrinally, legally, morally. Whenever officers can prove soldiers correct about this, they lead.

One could almost argue that this renders officership a profession within a profession: officers' work is leadership. Their survival in the system depends on their moral expertise, their maintaining control over the jurisdiction of deciding what is or is not the appropriate thing to do, and their being able to prove the legitimacy of their decision-making to subordinates (and civilian authorities) through journals, special schools, courses, and academies. However, there is at least one problem with such a comparison. Historically, professions never have formal, permanently fixed, and clearly ranked hierarchies. At most they display informal and fiercely contested pecking orders in which seniority matters, but in which senior partners do not *command* junior partners. Just consider: generals and their aides will never share the same status while still in uniform. This alone renders the kind of hierarchy the Army has—and requires—unique. At least thus far.

Postmodern Challenges

However, hierarchy as an organizing principle, at the small unit level especially, may be in jeopardy. In 1974, Morris Janowitz was already noting that "the basis of authority has shifted from that of an authoritarian domination to manipulation, persuasion, and group consensus."[27] He attributes this shift to "changing societal values" and the "impact of technology." Writing more than a decade later, Daryl

Henderson likewise comments on "a significant historical shift downwards in the locus of control."[28]

Even more recently, in describing postmodern leadership challenges, French sociologist Bernard Boëne and British sociologist Christopher Dandeker acknowledge remarkable congruence between their conclusions and those reached by two Israelis, Boas Shamir and Eyal Ben-Ari.[29] The latter's prediction is that:

> The military organization of the future is likely to be much more "organic" in nature. Organic organizations are characterized by a more flexible division of labor, decentralization of decision-making, low reliance on formal authority and hierarchy and on rules and regulations to coordinate work, and greater reliance on nonrestricted, two-way, informal communication and coordination systems.[30]

Along the same lines, but in an even more pointed way, David Ronfeldt and John Arquilla, among other war futurists, advocate a new and hybrid form of networked organization "in which 'all-channel' networks are fitted to flattened hierarchies."[31]

Those who take such a tack, predicting or promoting the (d)evolution of hierarchy (or its transformation into something else, or less), do so out of the belief that something fundamental has changed in the world at large. They cite young peoples' expectations and attitudes, the technological and organizational capabilities of our opponents (and ourselves), and/or the increasingly diffuse nature of conflict, which may involve state, parastate, and substate actors anywhere in the world. They argue that so long as potential missions run the gamut from coalitional war-making to coalitional peacekeeping, there is a "greater need for versatility, flexibility, and adaptability" in size, design, responsiveness, recruits, in—essentially—everything.[32]

To recast what they are saying in only slightly different terms: basically, as the world's youth become better-informed, more nonconformist, and more self-absorbed,[33] as our enemies become more politically, economically, and socially but not conventionally militarily disruptive—the more complex and multinational our missions will become, and the harder it will be to exert control. Experientially, militaries have long had to cope with challenges posed by entropy, the intrinsic tendency toward disorder. As Martin van Creveld points out, there have generally been two alternative responses when the ground shifts tectonically: tighten "command *and* control," or decentralize and delegate in turn.[34] The problem with the latter approach, though, is that to decentralize, and still win, vertical integration must somehow be maintained. At the very least, everyone has to be singing from the same sheet of music—something far easier to achieve when everyone already shares internalized "values and norms," and when there is already "unity of thought."

As an organizational principle, unity of thought has few parallels. It facilitates unity of effort among separated forces in fluid environments, and is liberating in at least two senses. Any part of the organization should know how to replicate (or repair) the whole. This has to offer psychic comfort. At the same time, with less structure to worry about, it should be easier for units to reconfigure in response to novel situations. The staff of Moltke the Elder, for instance, shared "ethical and even religious ideals which served as the basis for calm confidence."[35] John Gates credits the success of American Army volunteers in the Philippine insurrection of

1899-1901 to their being a "self-assured group with a self-conscious progressive orientation and a commitment to such traditional values as Duty-Honor-Country."[36]

Since the days of Napoleon, armies have worked hard to instill common values and norms when these were not already present. However, as Shamir and Ben-Ari caution, "achieving coordination through culture" may be increasingly difficult today.[37] They make this comment in light of "the frequently changing composition of the organizational and interorganizational frameworks that bring together members of different organizational and sometimes national cultures."[38] However, their observation may be equally apt when applied to advice offered today's leaders: namely, to be sensitive to, and even appreciative of, subcultural, gender, and other differences, as well as differing opinions.

This cannot but complicate a leader's role, which among other things is to build consensus *for* (or consensus from) without soldiers discovering, in a Heisenbergian twist, that the act of trying to reach a consensus among themselves prevents its achievement. At the same time, the need to soften (never mind flatten) hierarchy to foster strength through diversity reveals the Army's default in what may well be its chief responsibility to those it commissions, which is to place them in charge of soldiers who, as individuals, should all regard themselves as members of the Army—their *national* Army—first, and members of other groups or subcultures a clear and distant second.[39]

At a minimum, soldiers should come to units already regarding themselves *as* soldiers.[40] Otherwise, unity of effort—which requires that individuals think in terms of the good-of-the-group and not themselves—won't always (and may seldom) be achieved. Under conditions of confused (or even multiple) allegiances, organized decentralization will never work. In his book, *Command in War,* van Creveld analyzes four armies which purposely disaggregated their units and succeeded.[41] Worth noting, however, is that authority was purposely diffused between units, not diluted within them.

Organized Decentralization: A Contradiction in Terms?

If organized decentralization (or some approximation thereof) is to be introduced, there are in the meantime at least two sets of examples worth paying attention to: that of the Special Forces, and two different "armies of one" symbolized by T.E. Lawrence and Edward Lansdale.

Curiously, despite all the discussion in the military leadership literature about teams and teamwork, and what may or may not be applicable from the realms of business and management, little is said about the fact that teams already exist in the Army, and have for decades. More to the point still, many do exactly what it is said conventional units will do in the future. For example, Special Forces operational detachments—the term "operational detachment" itself speaks volumes—are not only designed to "deal in the gray" but to be left on their own for long periods of time, as is presumably the case in Afghanistan today. Until the end of the Cold War, Special Forces (SF) could be considered distinctive thanks to its unconventional warfare, counterinsurgency, and foreign internal defense missions. These required that SF soldiers be adept at winning hearts and minds and building rapport. SF may no longer be quite so alone when it comes to plying these particular skills, since peacekeeping operations and humanitarian interventions make similar demands. But the label "unconventional" unquestionably holds when applied to the teams' organization and the officer-enlisted relationship.

The individual who is officially in charge of an "A" Detachment—the captain—is often only nominally in charge of the warrant officer and ten noncommissioned officers (NCOs) on his team. Usually, the team sergeant has the more commanding presence. In part, this is a function of the team's division of labor: the team sergeant (an E8) is responsible for the world inside the team room while the officer interfaces with the world beyond. But also, the captain is not just chronologically younger than many of the men who serve under him (sometimes by as much as a decade), he is more inexperienced than most. Thus, they are unlikely to automatically look up to him, nor can he often present himself as an authority on anything having to do with their job(s), though he can know more about their next mission, about the country where they might deploy, the people there, etc.

Special Forces NCOs joke that the definition of a good captain is someone who knows when to defer to them. And, indeed, smart captains seem to understand exactly how to do this to earn respect: they don't initiate competition they can't win and, ideally, they don't engage in competition at all. In other words, they don't act like experts, while when they defer to others *as* experts it is in only certain domains.

This is not quite what the Army's leadership manuals advise, or what these officers would have done previously when commanding soldiers younger and more inexperienced than themselves. So how do they know, when new to SF, to do this? How do they know *how* to do this? The short answer is, they don't. Their success is simply an outgrowth of principled behavior.

Perhaps the best way to illustrate what I mean by behaving in a principled way is to compare two of the unconventional world's most prominent and successful figures, T.E. Lawrence and Edward Lansdale, though in the course of achieving their successes (Lawrence in Arabia, Lansdale in the Philippines) neither led a team of compatriots.[42] Instead, both men were military advisors: they could not command, nor could they control. All they could reasonably do was exemplify, suggest, project, and, to an uncanny degree, fit FM 22-100's prescriptions for how to lead.

Each did so, though, using radically different techniques. Lawrence was adept at everything Bedouin; Lansdale never tried to live or speak like a Filipino. Lawrence was an expert about the Bedouins long before he began advising them; Lansdale was interested in reading people, not about them.[43] Each operated at what amounted to opposite ends of the advisory spectrum. Nevertheless, both knew their own strengths and weaknesses (again, as FM 22-100 advises), and both fully appreciated what they could and could not be expert in *from their advisees' point of view*.

Nothing spells this out better than Lawrence's "Twenty-seven Articles," his own lessons learned as distilled into principles for others to use. For instance, here is Article 18:

> Disguise is not advisable. Except in special areas let it be clearly known that you are a British officer and a Christian. At the same time if you can wear Arab kit when with the tribes you will acquire their trust and intimacy to a degree impossible in uniform. It is however dangerous and difficult. They make no special allowances for you when you dress like them. Breaches of etiquette not charged against a foreigner are not condoned to you in Arab clothes.[44]

In other words, Lawrence was not just savvy enough to know how to dress like a Bedouin (Articles 17, 19, and 20 offer more specific advice about what was best to wear), but he also understood what Bedouins would and wouldn't let him get away with as either a Briton (which he was) or a Bedu (which he knew he could never be).

In contrast, Lansdale never developed anything like Lawrence's expertise regarding what Filipinos would have allowed him to do or not do as an intimate. But, then, he never had to. He found that he didn't need to operate from a set of principles related to them (let alone a set of principles built around *how* to relate to them). He excelled instead by operating from his own core convictions about what it meant to be an American, and only had to understand what Filipinos expected *this* to mean. In other words—while Lawrence proved consistent regarding Bedouins' expectations of him as someone who knew what it meant to be Bedouin, and thus knew in which areas it was acceptable to best them (e.g., endurance, marksmanship, etc.), and when to not even try (e.g., in discussions about religion and women)—Lansdale's expertise was in democracy and in being American.[45]

Not surprisingly, Lansdale's approach turns out to be in keeping with how most Americans behave abroad. We invariably project American values and ideals. The fact that these include (if they don't revolve around) being and acting the same, no matter where or among whom we find ourselves, may itself be our defining principle, while the fact that we judge one another accordingly only reaffirms our commitment to that principle.[46]

From American Principles to Principled Americans?

Although every society can be said to have a set of principles by which its members are expected to live, we regard ours as universal human truths. We not only place emphasis on our absolute equality as individuals, both before the law and before one another, but our expectation is to always be treated fairly, as anyone should be, regardless of position, status, or rank.

At first glance this might seem to make hierarchy unAmerican. However, as Gerald Linderman points out in a book chapter titled "Discipline: Not the American Way" (in his seminal work on combat experience during World War II), Americans will accede to hierarchy so long as it's equitable. What American combat soldiers in World War II desired, for instance, "was not a standing equivalent to that of officers but acknowledgment of an equivalent worth."[47] As he explains elsewhere, what American soldiers have historically wanted (and needed) from officers is to know where officers stand, what they stand for, what they won't stand for, and that they will always be fair.[48] This was true during the Civil War, when officers were often elected, and still held a century later, with appointees in place.[49]

Why would soldiers be so consistent and adamant about this? No doubt because, as Lloyd Matthews has written,

> Such revered documents as the Declaration of Independence, the Preamble to the Constitution, and the Bill of Rights have given rise to an American political tradition in which liberty and equality remain vibrant touchstone ideals among U.S. citizens. Though these values obviously cannot receive full or even substantial expression in military service, they do instill boundary expectations in the minds of service members that military leaders ignore at their peril.[50]

Not all of this, of course, is particularly or peculiarly American. For instance, Linderman quotes a British soldier who says, "Trust depends on a man's knowing that his commander thinks of him as a person and therefore treats him fairly, and looks after him."[51] But we, being a longer-lived, more deeply-rooted meritocracy, may push this even further. Meritocracies work only when status is achieved, not ascribed. And the Army is a classic meritocracy. Therefore, the presumption among soldiers is that those above them must have, somehow, demonstrably proven that they deserve to be where they are. And like Big Men, officers must keep proving and demonstrating this worth. Yet, to whom? What superiors need from officers seems vastly different from what subordinates do, in part perhaps because what merits worth for superiors is not necessarily what merits worth for subordinates. Often, though, this leads to wildly different judgments by subordinates and superiors about the same officer. Must this be? Should this be?

If principled behavior binds us together as Americans, and particular principles (such as equity) define us as a nation, then surely being able to be counted on *to be* principled is the one standard by which the Army could—and should—measure fitness, up and down the chain of command. Arguably, this is how we all judge one another anyway—rank, status, position be damned.

If the Army doesn't do this more consistently—or explicitly—it runs two risks: first, that informal pecking orders will undermine its formal order; and second, that the trappings of hierarchy are shown to be just that—devices. Neither is something the Army can long afford, since without differentiating people—and making subordinates *want* those differences to matter—control won't stick.[52]

Backbone vs. Box

Again, it seems important to bear in mind that the Army does not need to maintain control merely for its own sake, but for the sake of those it seeks to control. Not everyone wants to lead. Either people recognize their own limitations, or realize they can't (or don't want to) set limits for others. Limits is a shorthand way to think about officers' domain: they remind soldiers of and about limits, set limits, interpret limits, and help subordinates stay within or push beyond accepted limits.

Limits have always been critical, but today—given real-time scrutiny by the news media, international agencies, and concerned citizens—they may be more important than ever before, especially since the Army has never been confronted by so many culturally alien, morally messy, and potentially victoryless missions. All the more reason to grant more, not less, control to those on site, those who know what's happening on the ground and can better read the local situation. To be able to grant leaders at the tip of the spear more control, though, demands and requires trust. Trust, meanwhile, comes from knowing subordinates can be counted on. This is less tautological than it is self-reinforcing: if everyone operates from the same set of principles—and proves principled—there would be no need for doubt or second-guessing.

Also, principles, once internalized, are always there; doctrinal manuals and models may well not be. Not only does operating from a strong principled core/corps offer maximum flexibility, but it avoids artificially constraining (as may a model) or delimiting (as can doctrine). Perhaps the best way to conceptualize this is graphically. Think of principles as a Backbone. With deeply embedded core convictions, officers should be able to operate anywhere; they don't have to think about what *not* to do; they know from within. But lacking such core convictions they would have to be told. Because command has to make sure that everyone understands where the limits are to what's acceptable, constraints would be set externally and generated from without, not within. Ergo the Box.

Without question this is what some people would prefer—the comfort of being in a Box. Boxes offer protection, among other things. Small wonder it's then difficult to pry people out of them. Or that the Army finds itself *needing* to tell leaders to "think outside the box."

Of course, if there were no Box, there'd be no reason to tell leaders to think outside it. If, instead, the Army trusted—and entrusted—officers to be principled it could think, talk, and operate in terms of Backbone instead.

Which would build and bolster confidence? Which represents the more positive approach? Surely having Backbone is the means test the Army needs to apply in peopling its hierarchy: only those with inner conviction have the strength, resiliency, wisdom, and courage to set limits for those who prefer that limits be set. By switching metaphors—from Box to Backbone—the Army would simply be acknowledging this and using to maximum advantage our human penchant to self-select the extent to which we'd rather lead or follow.

To summarize. What followers want is to be able to count on their leaders, while what the Army should ask of its leaders is that they be able to be counted on. At the very least, they must be able to be counted on to know what *not* to do—tactically, ethically, strategically, politically.[53]

If equity, consistency, and acting in a principled way are inherent to being American, then it only makes sense to apply this standard—of officers being able to be counted on *for* principled behavior—up and down the line.

So long as superiors no less than subordinates can count on officers for principled behavior, they can relax top-down control (though not monitoring). By granting units more autonomy, they then grant themselves more flexibility.

In a messy world, few things are more important than flexibility at the strategic level, where national interests are at stake, or at the operational level, where the outcomes of campaigns are at stake, or at the tactical level, where lives are at stake.

But at the tactical level, down where the bullet meets the bone, soldiers have another need even more important than *flexibility*: they need something fixed, something they can trust and rely on no matter what. That something is their officer-leaders, officer-leaders who invariably are fair, morally courageous, and principled.

Notes

1. Thanks to Reginald Davis for an important off-hand remark; to Joe Andrade, Nick Mullen, and Ben Higginbotham for insightful comments on my initial draft; to Leland Young for prompting me to rethink and rephrase any number of critical points; and to Gayle Watkins for generously offering better ways to think about more things than I've managed to discuss. I'm (alas) responsible for all advice not taken, and any errors.
2. Lionel Tiger and Robin Fox, *The Imperial Animal,* 2nd edition (New Brunswick, NJ: Transaction Publishers, 1998).
3. Followers of specific leaders, yes—there are ample testimonials to what first attracted people to famous and yet-to-be famous men (and women) and then what kept them enthralled. Charisma is prominently mentioned. But not even Max Weber, who introduced the concept, believes charisma can be defined.
4. Also, consider: as Americans we are socialized not only to compete—for grades in school, for attention from our parents, teachers, and one another—but we are taught from an early age that the harder we choose to work the more this will propel us ahead. Not everyone learns this lesson soon enough. Not everyone accepts it. Meanwhile, keeping up with the Joneses when the Joneses are a moving target creates a vastly different set of societal ideals and standards than keeping up with the Joneses would if the Joneses just stood still.
5. See Paul Riesman, *First Find Your Child a Good Mother: The Construction of Self in Two African Communities* (New Brunswick, NJ: Rutgers University Press, 1992).
6. See V.G. Kiernan, *The Lords of Human Kind: Black Man, Yellow Man, and White Man in an Age of Empire* (New York: Columbia University Press, 1969; reprint 1986).
7. As one infantry captain and former general's aide has pointed out: the higher up the chain one moves, the better prepared and not just compensated officers are, though (as he also notes) no one puts in longer hours than do generals.
8. Sociobiologists, evolutionary psychologists, and some biological anthropologists would argue that the more successful a Big Man is/remains, the more wives he would likely have, the more offspring, and the higher his inclusive fitness or reproductive success. However, inclusive fitness (as they point out) is not a conscious strategy. Instead it is the genes' way of disseminating themselves. Unfortunately, this explains everything but nothing at the same time. As much pleasure as a Big Man in the New Guinea Highlands may take in procreating (and recreating), it is surely not because he wants to make his genes happy. At most he wants as many children as possible to spread his name.
9. Most of those we think of, and still refer to, as chiefs in this country—e.g., Chief Joseph, Sitting Bull, Crazy Horse, Osceola, etc.—were Big Men. They had no coercive authority. They could never compel followers to stay with them. Nor could they order or command men in battle. This granted an immense advantage to the U.S. Army.
10. Some classic techniques of doing this: people might be forbidden to set eyes on a ruler, except from a distance. Or he might never be seen to eat, let alone engage in any bodily function, by anyone outside his retinue.
11. For examples, see Abner Cohen, *Two Dimensional Man: An Essay on the Anthropology of Power and Symbolism in Complex Society* (Berkeley: University of California Press, 1974). Also, David Kertzer, *Ritual, Politics, and Power* (New Haven, CT: Yale University Press, 1988).

12. To take an extreme example: an officer might threaten to shoot a soldier for disobeying a direct order in combat, but once dead the soldier will never be able to comply with that order. If the officer needs the soldier, can he afford to threaten to shoot him? When officers need coercion most, they can afford only to use it carefully, strategically, and sparingly.

    The comparison between officers and Big Men actually holds in at least one other regard. In Big Man societies there may be status differentials, but these don't translate materially. Big Men eat the same, dress the same, and live the same as their dependents. More to the point still, there is a one-for-all, all-for-one ethos in Big Man societies (which is the ethos skillful Big Men work hard to reaffirm). Can't the same be said of officers at platoon and company levels, especially when they are in the field?
13. Soldiers themselves represent their only immediate means of force.
14. For more on this, see F.G. Bailey, *The Kingdom of Individuals: An Essay on Self-Respect and Social Obligation* (Ithaca, NY: Cornell University Press, 1993).
15. Though "exemplify, suggest, project" might be no less accurate, and seems to be what the manual itself suggests.
16. Gunther Rothenberg, "Maurice of Nassau, Gustavus Adolphus, Raimondo Montecuccoli, and the 'Military Revolution' of the Seventeenth Century," in *Makers of Modern Strategy: from Machiavelli to the Nuclear Age*, ed. Peter Paret (Princeton, NJ: Princeton University Press, 1986), 61-62.
17. Edgar Meyer, "Leadership: A Return to Basics," *Military Review:* 4 [journal online]; available from *http://www-cgsc.army.mil/milrev/english/janfeb97/meyer.html;* Internet; accessed 18 January 2001.
18. Richard Holmes, *Acts of War: The Behavior of Men in Battle* (New York: The Free Press, 1985), 347.
19. Edgar Puryear, *19 Stars: A Study in Military Character and Leadership* (Novato, CA: Presidio, 1994/1971), 396.
20. For examples of the literature I refer to, beyond sources already cited, see: Dandridge Malone, *Small Unit Leadership: A Commonsense Approach* (Novato, CA: Presidio Press, 1983); Lloyd J. Matthews and Dale Brown, eds., *The Challenge of Military Leadership* (Washington, DC: Pergamon-Brassey's, 1989); Robert Taylor and William Rosenbach, eds., *Military Leadership: In Pursuit of Excellence,* 2nd edition (Boulder, CO: Westview Press, 1992), in addition to FM 22-100.
21. Headquarters, Department of the Army, Field Manual 22-100, *Army Leadership: Be, Know, Do* (Washington, DC: Department of the Army, 1999), 3-51; emphasis mine.
22. John Mattox, "Fifth-century Advice for 21st Century Leaders," *Military Review* [journal online]; available from *http://www-cgsc.army.mil/milrev/English/MayJune98/mat.html;* Internet; accessed 18 January 2001.
23. Lloyd J. Matthews, "The Evolution of American Military Ideals," *Military Review* 78 (January-February 1998): 51-61.
24. As Bernard Boëne points out: "Of the rank and file combatant, a minimum amount of courage or valiancy is expected—the capacity to control fear in the face of danger, discomfort, pain or misfortune, and to transgress the social taboos of civilian life when ordered to do so—together with physical agility and stamina, loyalty, and good will. Mastery of fairly simple tactical and technical skills is required.

    Of the leader, something more is expected: competence, composure, and self-control, inspiration by example, comprehension, and manipulation of human relations in the exercise of authority, formal (discipline) and informal (personality)." See "How 'Unique' Should the Military Be? A Review of Representative Literature and Outline of a Synthetic Formulation," *European Journal of Sociology* 31 (1990): 30; emphasis mine.
25. Don M. Snider, John A. Nagl, and Tony Pfaff, *Army Professionalism, the Military Ethic, and Officership in the 21st Century* (Carlisle Barracks, PA: U.S. Army War College, Strategic Studies Institute, December 1999), 36.

26. For instance, under the subheading "Integrity" FM 22-100 counsels, "People of integrity consistently act according to principles. . . . People of integrity do the right thing not because it's convenient or because they have no choice. They choose the right thing because their character permits no less. Conducting yourself with integrity has three parts:
    - Separating what's right from what's wrong.
    - Always acting according to what you know to be right, even at personal cost.
    - Saying openly that you're acting on your understanding of right versus wrong."

    However, nowhere is what *is* right described, delineated, or defined. See FM 22-100, 2-31.
27. Daryl Henderson, *Cohesion: The Human Element in Battle* (Washington, DC: National Defense University, 1985), 44.
28. John Johns, *Cohesion in the U.S. Military* (Washington, DC: National Defense University Press, 1984), 22.
29. Bernard Boëne and Christopher Dandeker, "Post-Cold War Challenges and Leadership Strategies in West European Military Institutions," in *Leadership for Change*, ed. Gwyn Harries-Jenkins (London/ Hull: European Research Office of the U.S. Army/ University of Hull, 1999), 11-29; Boas Shamir and Eyal Ben-Ari, "Leadership in an Open Army? Civilian Connections, Interorganizational Frameworks, and Changes in Military Leadership," in *Out of the Box Leadership: Transforming the Twenty-First-Century Army and Other Top-Performing Organizations,* eds. James G. Hunt, George E. Dodge, and Leonard Wong (Stamford, CT: JAI Press, 1999), 27.
30. Shamir and Ben-Ari, 27.
31. John Arquilla and David Ronfeldt, "Looking Ahead: Preparing for Information-age Conflict," in *In Athena's Camp: Preparing for Information Age Conflict,* eds. John Arquilla and David Ronfeldt (Santa Monica, CA: Rand, 1997), 440.
32. Shamir and Ben-Ari, 27.
33. See Boëne and Dandeker. Also, see Shamir and Ben-Ari. This is yet another reason to reinforce principles. Everyone, ideally, should be taught principled behavior—and particularly since it can't be taken for granted that this has been taught in schools or in homes. Also, principled, good-of-the-group, selfless behavior, as Snider, Nagl, and Pfaff point out, differs significantly from what is or would be acceptable principled behavior among civilians. This, too, has to be emphasized.
34. Martin van Creveld, *Command in War* (Cambridge, MA: Harvard University Press, 1985).
35. Van Creveld, 149.
36. John M. Gates, *The U.S. Army and Irregular Warfare* (1998), chapter 5, 5 [book on-line]; available from *http://www.wooster.edu/jgates/book-contents.html;* Internet; accessed 20 November 2000.
37. Shamir and Ben-Ari, 27.
38. Ibid, 28.
39. The Yugoslav army is an unfortunate example of a nationalist institution which wasn't quite nationalist enough. Although numbers of Yugoslav soldiers and officers regarded themselves as Yugoslavs first, and Croat, Bosniac, or Serb second, and many deserted and refused to fight for what rapidly became the Serb army, not enough either quit or refused to quit. Slobodan Milosevic was thus able to use paramilitaries to make up for his military shortfall, alongside Serb forces who remained in uniform.
40. As numerous people continually note, this is something the Marine Corps does extraordinarily well. The Army, interestingly enough, seems able to do this only in *selective* units, but even in these (e.g., the 82nd Airborne, Rangers, Special Forces) there is a self-selection process which is likely to attract only individuals who already think in these terms.
41. The four armies which purposely disaggregated with successful results were the French, Israeli, German, and Roman.

42. On Lawrence see John Mack, *A Prince of Our Disorder: The Life of T.E. Lawrence* (Cambridge, MA: Harvard University Press, 1976; reprint 1998). Also, T.E. Lawrence, *Seven Pillars of Wisdom* (New York: Dell Publishing, 1926; reprint 1963). On Lansdale see Cecil Currey, *Edward Lansdale: The Unquiet American* (Washington, DC: Brassey's, 1998). Also, Edward Lansdale, *In the Midst of Wars: An American's Mission to Southeast Asia* (New York: Fordham University Press, 1991).
43. The contrast could be summarized this way: Lawrence did because he could, while Lansdale didn't because he couldn't. Lansdale couldn't learn other languages. He was also a miserable shot. But he clearly excelled at communicating with people of all walks of life—in large part by always treating everyone as though they were of worth.
44. Mack, *A Prince of Our Disorder*, 465.
45. This is not to suggest that promoting democracy as vigorously as Lansdale did is what the U.S. should always do. Far from it. But this is what his mission called for; this is also what local leaders sought. Also, though Lansdale was not a Lawrence in his approach to local culture, he showed a keen interest in whatever Filipinos were interested in. His natural curiosity, empathy, and the extent to which he treated everyone with respect won him more friends than any amount of prior scholarship could have. In an ideal world, of course, advisors should have both.
46. This also explains, in part, Lansdale's failure in Vietnam. Though some critics claim Lansdale mistakenly tried to apply the same cookie-cutter programs in Vietnam which had worked so well in the Philippines, the real problem was that he wouldn't compromise his beliefs about how to achieve democracy and security, and Diem wasn't sufficiently receptive. There is actually every indication that had he been able to convince Diem to do what had been done in the Philippines—namely, secure free and fair elections, professionalize the military, initiate land reform, and root out corruption—things might have turned out very differently.
47. Gerald Linderman, *The War Within War: America's Combat Experience in World War II* (New York: The Free Press, 1987), 199.
48. Gerald Linderman, *Embattled Courage: The Experience of Combat in the American Civil War* (New York: The Free Press, 1987).
49. Johns, 33, reiterates the importance of this: "A sense of fairness (not equality) is critical. All members must perceive . . . that they are being treated fairly in terms of rights and obligations."
50. Matthews, 59.
51. Linderman, *The World Within War*, 227.
52. Of course, the more obvious—and significant—the division of labor is, the less noticeable hierarchy needs to be. Although this would seem to fly in the face of business models for teamwork, in which individuals are allowed (and even encouraged) to define their own roles, this describes SF teams. The weapons sergeants are weapons specialists, the communications sergeants are communications specialists. Everyone but the team captain plays a clear and significant role, which is not to say that captains can't or don't. But the fact that some NCOs consider officers superfluous, extraneous, and unnecessary on teams reveals how much is expected of them. To lead, officers had better add something. To convert potential critics into followers they must find a way to contribute to the good of the group, and they must do so even while proving themselves first among equals. Outside of combat, the fastest way to do this is by *being principled* and recognizing that it is a leader's duty to ensure principled behavior by, for, and from everyone.

19 Reality Check: The Human and Spiritual Needs of Soldiers and How to Prepare Them for Combat[1]

John W. Brinsfield

> If I learned nothing else from the war, it taught me the falseness of the belief that wealth, material resources, and industrial genius are the real sources of a nation's military power. These are but the stage setting . . . national strength lies only in the hearts and spirits of men.
>
> S.L.A. MARSHALL[2]

In the quest to reexamine and possibly redefine the Army profession, the key roles, skills, and knowledge required of military leaders are indispensable elements for analysis. No profession can compete with competent outsiders without defining itself, its special expertise requirements, and its solutions to the problems of transformation and change in perceived influence, power allocation, internal organization, and organization of knowledge to support its special claims to jurisdiction.[3]

The historical mission and jurisdiction of the joint military services are to win the nation's wars. All other missions are secondary to this national security responsibility. Yet at the beginning of the twenty-first century, our conception of America's security umbrella has been broadened to include domestic police, fire, and drug enforcement activities as well as international humanitarian and peacekeeping missions—to the detriment, some would say, of the Army's main war-fighting role. In fact, the system of professions within which the Army competes is crowded with American government entities such as the other military services, the State Department, Border Patrol, Drug Enforcement Administration, Federal Bureau of Investigation, Central Intelligence Agency, and Federal Emergency Management Agency. The nation's formerly well-integrated system of professions addressing security has mushroomed "without a commensurate expansion in the legal, cultural, or workplace mechanisms that legitimate each profession's jurisdiction."[4] Mission creep, in other words, has challenged the Army's traditional understanding of its role in the nation's defense.

Moreover, in the quest to establish its professional boundaries, the Army has had to rely on civilian leadership, often with little or no experience in the military, for its mission definition and resources, all the while competing commercially with academia and the marketplace for the hearts and minds of the recruits who may become its future leaders.[5] These challenges, among many others, seem to require a redefinition of and reemphasis on the components of military professionalism and leadership for the future.

Soldiers are the Army's heart, life force, and strength no matter what their mission may be. They define the Army's effectiveness, success, or possible failure. They must respond to the unique demands of the profession of arms: total commitment, unlimited liability, possible lengthy separations from family, community and civilian primary support systems, and total loyalty to a values-based and service-based organization. In time of war, they may be asked to sacrifice themselves for the nation and for one another as guardians of the republic. Any internal analysis and definition of the profession of arms must include, therefore, an appraisal of the soldier's human dimension, lest, to paraphrase the words of one Civil War general, they be asked for more than they could possibly be expected to give.[6]

As part of such an effort, this chapter seeks to analyze the human and spiritual needs of soldiers as part of the special knowledge required by Army leaders to motivate, train, and command their personnel and their units in peace and war. It also suggests some considerations for preparing soldiers psychologically and spiritually for future combat operations.

The working hypothesis is that all soldiers have human needs and most have spiritual needs broadly defined, and that converting these needs into strengths of will and character is an important part of combat leadership—and therefore of Army professionalism. The chapter is composed of three major parts: (1) a definition and discussion of human and spiritual needs, including an analysis of the theory of needs as applied to soldiers; (2) a description of some of the past efforts to capitalize on human and spiritual needs so as to achieve confidence, cohesion, and courage; and (3) a discussion of proposed combat training considerations as related to the human dimension of soldiers in the future. Because certain aspects of human nature cannot be directly observed but must be inferred from observed behavior, the data for analysis rely on multidisciplinary sources which include the humanities as well as the social sciences.

Assumptions

Since the subject matter of this analysis deals with the needs of the soldier, a review of sources relating to the individual will be useful before we move to the organizational or profession level. Much of the research data involve individual responses from soldiers in small units rather than Army-wide studies. In taking this approach, it may be assumed that military leaders do and will recognize their dual obligations to complete their missions successfully and to take effective measures to ensure the health and welfare of as many personnel as possible within their commands. This is an ancient canon of the military art, explained by Sun Tzu's *The Art of War* in the early part of the fourth century B.C.:

> And therefore the general who in advancing does not seek personal fame but whose only purpose is to protect the people and promote the best interests of his sovereign, is the precious jewel of the state. Because such a general regards his men as his own sons they will march with him into the deepest valleys. He treats them as his own beloved sons, and they will die with him. If he cherishes his men is this way, he will gain their utmost strength. Therefore The Military Code says: "The general must be the first in the toils and fatigues of the army." [7]

There are, of course, many other authoritative utterances regarding the commander's duty to care for soldiers, but few of such established antiquity. In present literature, if one were to seek guidance for a career in the Army, Field Manual 22-100,

Army Leadership: Be, Know, Do, states simply, "Accomplish the mission *and* take care of your soldiers." [8]

The second assumption is that a holistic knowledge of the human and spiritual needs of soldiers, yet to be defined, will be of value to the military leader in providing support and resources for meeting these needs, thereby strengthening the capacity of the fighting force to complete its missions successfully. In war, soldiers' comfort, inasmuch as comfort is possible, affects morale and thus combat effectiveness.[9] The Creed of the Noncommissioned Officer includes this concept in the brief declaration that " all soldiers are entitled to outstanding leadership; I will provide that leadership. I know my soldiers and I will always place their needs above my own."[10] Gen. Creighton Abrams, former Army Chief of Staff, goes to the heart of the matter:

> The Army is not made up of people; the Army is people. They have needs and interests and desires. They have spirit and will, strengths and abilities. They are the heart of our preparedness and this preparedness—as a nation and as an Army—depends upon the spirit of our soldiers. It is the spirit which gives the Army life. Without it we cannot succeed. [11]

If leadership means gaining the willing obedience of subordinates who understand and believe in the mission's purpose, who value their team and their place in it, who trust their leaders and have the will to see the mission through, then leaders must understand two key elements: leadership itself and the people they lead. [12]

Definitions and Discussion: Religion, Spirituality, and Human and Spiritual Needs

On the last page of his classic study of the psychology of soldiers, *The Anatomy of Courage* (1967), Lord Charles Moran approached the subject of religion and spiritual power:

> I have said nothing of religion, though at no time has it been far from my thoughts. General Paget asked me once to talk to officers commanding divisions and corps and armies in the Home Forces. When I had done, they broke up and came to me, one or two at a time, questioning. Often that night I was asked about the importance of religion. Speaking as if they did not know how to put it, they separately told me how faith had come into the lives of many of their men. Is it so strange? Is it not natural that they are fumbling for another way of living, less material, less sterile, than that which has brought them to this pass? What are they seeking? [13]

Lord Moran's questions are well posed, for the separate disciplines of psychology and religion often look to separate sources of authority, separate methodologies, and different language to describe human behavior. Nevertheless, many psychologists, sociologists, anthropologists, and physicians recognize the phenomenology of religion as abstracted from any claims concerning its essence. In other words, religion may be studied and respected as an element of culture without subscription to its content. W.I. Thomas, one of the classic sociologists of the past century, explained that "if a culture believes something to be real, we must respect that belief in dealing with that culture."[14]

Many soldiers in the American Army culture do identify with a specific religious faith—some 299,958 or 64% of active duty soldiers in April 2001—but many are also reluctant to define too closely what they mean by religion, faith, and especially spirituality.[15] Even though spiritual strength is mentioned in many Army publications, including the 1999 edition of the U.S. Army Training and Doctrine Command's *TLS Strategy: Change, Readiness, and the Human Dimension of Training, Leader Development and Soldiers* and the 2001 edition of the Department of the Army's Well-Being Campaign Plan, there are comparatively few useful definitions that have been published. [16]

Part of the reason why soldiers are reluctant to discuss religion openly is their perception that religion is a very personal subject. Professor Morris Janowitz found two generations ago "a tendency among leaders in a political democracy, and especially among the military, to resent being questioned about their religious background." [17] A strong adherence to a particular religious point of view can be perceived as politically divisive and detrimental to unit cohesion. More commonly, religious language itself is not well understood, for the same terms may have different meanings in different faith groups. Military leaders like to have a clear idea of what they are saying and supporting, as do most people.

At the same time, many educational institutions, including the United States Military Academy, have recognized a spiritual domain in their philosophies of comprehensive education. The Cadet Leader Development System, a strategy for total commissioned leader development at West Point, links the spiritual domain to a common quest for meaning in life:

> This [spiritual] domain explicitly recognizes that character is rooted in the very essence of who we are as individuals, and discerning "who we are" is a lifelong search for meaning. Cadet years are a time of yearning, a time to be hungry for personal meaning and to engage in a search for ultimate meaning in life. Formally recognizing this fundamental aspect of human development is not unique to West Point; educators have long held that individual moral search is an inherent, even vital, component of any robust undergraduate education. In other words, cadets' search for meaning is natural, it will occur, whether or not we explicitly recognize and support it as an institution or not.[18]

For some, the quest for meaning will lead to questions of religion. For others, meaning is found through spirituality, a broader and possibly less distinct category than institutional religion. Is there a useful taxonomy for terms such as religion, spirituality, identity, ultimate meaning, and self-actualization in individual development?

Dr. Jeff Levin, Senior Research Fellow at the National Institute for Healthcare Research and a scholar of religion and medicine, tackles the problem of defining religion and spirituality as follows:

> Historically, "religion" has denoted three things: particular churches or organized religious institutions; a scholarly field of study; and the domain of life that deals with things of the spirit and matters of "ultimate concern." To talk of practicing religion

> or being religious refers to behaviors, attitudes, beliefs, experiences, and so on, that involve this domain of life. This is so whether one takes part in organized activities of an established religious institution or one has an inner life of the spirit apart from organized religions.
>
> "Spirituality," as the term traditionally has been used, refers to a state of being that is acquired through religious devotion, piety, and observance. Attaining spirituality—union or connection with God or the divine—is the ultimate goal of religion, and is a state not everyone reaches. According to this usage, spirituality is a subset of a larger phenomenon, religion, and by definition is sought through religious participation. [19]

Dr. Levin goes on to observe, however, that in the last thirty years the word "spirituality" has taken on a wider meaning. New Age authors and some news media have limited "religion" to those behaviors and beliefs that occur in the context of organized religious institutions. All other religious expression, particularly private meditation and secular transcendent experiences including feelings of awe and oneness with nature, are now encompassed by the term "spirituality." This wider definition reverses the relationship between religion and spirituality to make the former now the subset of the latter. [20]

Many scholars of world religions agree that Levin's wider definition of spirituality seems to fit the beliefs of many faith groups, even those with non-theistic views. Although the majority of the world's religions do claim to be the vehicles for a personal experience with God, Allah, Brahman, or one of the other of the world's named deities, there are others for whom spirituality is a non-theistic pilgrimage to individual enlightenment, wisdom, and transcendence. For example, in Zen or Ch'an Buddhism, "the highest truth or first principle is inexpressible," that is, the divine is so remote from human perception as to make its essence indescribable, thereby rendering an organized, doctrinal religion impossible; however, a mind-expanding, experiential awakening called "satori" is still available through meditation, mentoring by masters, and self-discipline. [21] In the Shinto religion of Japan, the perception of "kami" may be simply the reverence one has for the awesome power and beauty of nature even though gifts are frequently left at Shinto shrines for the spirits that inhabit such places. [22] The spiritual goal of reaching Nirvana is found in both theistic Hinduism and non-theistic Theravada Buddhism. The Falun Gong meditation which began in China in 1992 consists of spiritual exercises to promote health, cure illnesses, and allow the practitioner to absorb energy from the universe in order to ascend to a higher plane of human existence, but there are no named deities. [23]

Thus, to summarize the period since the 1970s, the context of religious institutions and spiritual practices in America has become enormously more diverse. Although in 1998 approximately 90% of the American people professed to be religious and 63% (169 million) identified themselves as affiliated with a *specific* religious group, the number of separate religious denominations has grown in a sixty-year span from about forty-five in 1940 to more than 2,000 at present.[24] This enlargement of religious and spiritual options suggests that Levin and others are correct to identify spirituality with the individual quest for greater insight, enlightenment, wisdom, meaning, and experience with the numinous or divine.

Religion does refer in most current literature to the institutionalization of symbols, rites, practices, education, and other elements necessary to transmit the specifics of religious culture to the next generation.

However, there is no evidence that the world's major religions are in decline. Indeed, as Samuel Huntington has argued, there is a worldwide revival of interest in traditional faiths, including Christianity in Russia, Buddhism in Japan, and Islam in Central Asia, faiths which offer meaning, stability, identity, assurance, and fixed points of reference in the face of the "clash of civilizations and the remaking of world order." [25] Moreover, in a recent poll taken by Blum and Weprin Associates of New York, which surveyed adults across America, 59% of those polled said that they were *both* religious and spiritual. Only 20% identified themselves as "only spiritual," and only 9% viewed religion in a negative way.[26] Nevertheless, with the growth of communication technologies and the availability of knowledge at the individual level, the spiritual quest for future generations may depend on some traditional religious institutions but will certainly be directed toward meeting individual needs.

The Theory of Human Needs Applied to Soldiers

The psychological study of soldiers is a relatively new academic endeavor. In the preface to his book, *The Anatomy of Courage*, Lord Moran, who had served as a medical officer in France and Flanders during World War I, explained that "there was no book in the English language on the psychology of the soldier before 1945."[27] *The Anatomy of Courage* was designed to fill that gap. It was an attempt to answer questions about what was happening in men's minds during combat and how they overcame fear. Since 1967, when Moran's book was published, there have been numerous studies on military psychology, military psychiatry, and combat motivation.[28] It was from an analysis of motivation and behavioral theory that the theory of needs as applied to soldiers found its most eloquent proponents.

The nature of the relationship between motivation and human behavior has been a subject of philosophical and psychological interest for centuries. There are multiple modern formulations which seek to explain motivation in general, including hedonistic, cognitive, drive reduction, and needs theories, to mention a few.[29] For more than forty years, a popular theory in U.S. Army literature which outlines both a description of human behavior and of motivation was Dr. Abraham Maslow's concept of *self-actualization* as the driving force of human personality, as set forth in his 1954 book *Motivation and Personality*.[30] Dr. Maslow was associated with the humanistic movement in psychology. Humanistic psychologists emphasize the person and his or her psychological growth.[31] Maslow described self-actualization as the need "to become more and more what one is, to become everything that one is capable of becoming," or, in other words, to be all one can be.[32]

According to Maslow's self-actualizing theory, the components of identity arise from two sources: the individual's unique potential and the different ways the individual copes with impediments placed in the way. Maslow identified two kinds of needs: basic needs arranged in a hierarchy which included physiological safety and security, love and belongingness, and esteem or recognition needs; and metaneeds which included spiritual qualities or metaphysical values such as order, goodness, and unity.[33] Basic needs are deficiency needs and must be fulfilled before a person can turn attention to the metaneeds. Metaneeds are growth needs; if properly satisfied, a person will grow into a completely developed human being—physically, emotionally, and spiritually—and have the potential to become a self-actualized person.[34]

Maslow recognized a spiritual component in the human personality, but argued that it was a natural component which sought meaning in a cause outside oneself and bigger than oneself, something not merely self-centered, something impersonal.[35] Moreover, the spiritual need impelled persons toward vocations, callings, and missions which they described with passionate, selfless, and profound feelings.[36]

Maslow believed that metaneeds or spiritual needs are universal, but that only self-actualized people usually attempted to meet them. "This is to say," he wrote, "that the most highly developed persons we know are metamotivated to a much higher degree, and are basic-need-motivated to a lesser degree than average or diminished people are. The full definition of the person or human nature must then include intrinsic values as part of human nature. These intrinsic values are instinctoid in nature, i.e., they are needed (a) to avoid illness and (b) to achieve fullest humanness or growth. The highest values, the spiritual life, the highest aspirations of mankind are therefore proper subjects for scientific study and research."[37]

Finally, Maslow argued that the spiritual aspirations of the human personality are a natural phenomenon, not a theological construct nor limited to the domain of religious institutions. In his book *Religions, Values, and Peak-Experiences*, Maslow exclaimed, "I want to demonstrate that spiritual values have naturalistic meaning, that they are not the exclusive possession of organized churches, that they do not need supernatural concepts to validate them, that they are well within the jurisdiction of a suitably enlarged science, and that, therefore, they are the general responsibility of all mankind."[38]

The U.S. Army leadership adopted Maslow's theory of basic and metaneeds enthusiastically after 1970. This was due, in part, because it correlated well with observable behavior among soldiers and because it was in consonance with the concept that if missions have requirements and weapons have a basic load, then soldiers must have human requirements and basic needs. In a collection of Bill Mauldin's World War II cartoons titled *Up Front*, G.I. Willie in his torn and dirty fatigues tells a medic, " Just gimme a coupla aspirin, I already got a Purple Heart."[39] Willie's basic needs, in Maslow's terms, clearly claimed priority over his esteem or recognition needs.

The October 1983 edition of Field Manual 22-100, *Military Leadership*, a standard text for thousands of the Army's leaders, incorporated Maslow's hierarchy of needs (physical, security, social) almost verbatim.[40] The manual's authors explained, "As a leader, you must understand these needs because they are powerful forces in motivating soldiers. To understand and motivate people and to develop a cohesive, disciplined, well-trained unit, you must understand human nature."[41]

However, there were three divergencies from Maslow's theory in the 1983 leadership manual. First, rather than discuss the need for esteem or recognition, which is the fourth need in Maslow's hierarchy, the leadership manual addressed "Higher

Needs," i.e., the need for religion, the need for increased competence, and the need to serve a worthwhile cause. With regard to the need for religion, the manual's writers explained that historically,

> many people not normally religious become so in time of war. The danger and chaos of war give rise to the human need to believe that a greater spiritual being is guiding one's fate for the best, regardless of whether one lives or dies. In this sense it helps soldiers to believe that they are fighting for a cause that is moral and right in the eyes of their religion. This is an important source of motivation for soldiers all over the world.[42]

Although the authors may have reflected their own beliefs accurately, Maslow would have argued that spiritual needs are universal, not dependent upon crises in war except perhaps as one of many catalysts for revealing them and not *always* leading to faith in a greater spiritual power as much as to perhaps a greater potential state of individual spirituality.

In more recent years, the Army has modified its language in describing the needs of soldiers and their families. Part of this was due to advances in medical and behavioral research, notably by the Army Research Institute for the Behavioral and Social Sciences and by the Academy of Health Sciences at Ft. Sam Houston, Texas, among others. Nevertheless, descriptively the human dimension of the soldier is still segmented into roughly the same needs Maslow propounded. The Army Well-Being Strategic Plan of 2001 produced by the Office of the Deputy Chief of Staff for Personnel defined Army well-being as "the personal—physical, material, mental, and spiritual—state of soldiers [Active, Reserve, Guard, retirees, veterans], civilians, and their families that contributes to their preparedness to perform The Army's mission."[43] The spiritual state (of well-being), according to the Army Well-Being Plan, "centers on a person's religious/philosophical needs and may provide powerful support for values, morals, strength of character, and endurance in difficult and dangerous circumstances."[44]

In summary, contemporary psychologists have challenged Maslow's hierarchy of needs, which they describe as "one size fits all," and instead point to a more complex model to explain motivation and behavior. Steven Reiss, a psychologist at Ohio State University, has identified at least fifteen fundamental motivational desires in human beings, including honor, morality, and order. Reiss does not present "spirituality" as a category of human motivation and desire, but he does say that further categories are open to scientific study.[45]

Likewise, but from a different perspective, research in the relationship between spirituality and healing has increased dramatically in the past ten years. The National Institute for Healthcare Research in Rockville, Maryland, has accumulated more than 200 studies from researchers at such prestigious institutions as Harvard, Duke, Yale, Michigan, Berkeley, Rutgers, and the University of Texas at Galveston showing that religious beliefs and practices benefit health and rates of recovery from illness in many patients.[46] According to one study by Dr. Andrew Newberg, published in 2001 under the title *Why God Won't Go Away*, research on the human brain suggests that a particular area of the brain is activated by prayer and meditation, at least among the subjects involved in the research. This led some interpreters to claim that "the human brain is wired for God."[47]

Yet none of the studies discovered to date claim that *all* people are either spiritual or religious. Levin and Maslow agreed, after sixty years of study between them, that while everyone has a range of needs, not everyone reaches an awareness of innate, or acquired, spiritual needs even though Maslow believed that all human beings are (potentially) motivated by metaneeds to some degree. [48]

What can be demonstrated, and what may be of most import for Army leadership and Army professionalism, is that the American culture, from which the military services draw support, puts a high priority on spirituality, organized religion, religious freedom, and the Constitutional right to the free exercise of religion, a right which both Congress and the Federal courts have applied to military as well as to civilian communities. *The Journal of Family Psychology* reported in 1999 that in America,

> many individuals report that religion and spirituality are integral parts of their lives. As many as 95% of American adults express a belief in God, 84% believe God can be reached through prayer, and 86% state religion is important or very important to them. Surveys also suggest religion may play a significant role in many marriages. Religiousness, as reflected by church affiliation or attendance, emerged as a correlate of higher marital satisfaction in early, classic studies on marital adjustment. More recently, greater religiousness has been tied to higher marital satisfaction and adjustment in large, nationally representative samples.[49]

The rate of attendance at religious services at least once a month among a national random sample of 1,000 families as reported by the *Journal of Family Psychology* was 37%, with 25% of the same sample reporting attendance at religious services weekly or more than once a week.[50]

Among Army soldiers in 2001 the rate of identification with one of the seven larger religious faith groups in the Army—Protestant, Catholic, Orthodox, Jewish, Muslim, Buddhist, or Hindu—was 64%, one percentage point higher than the national average.[51] Although chapel attendance figures for soldiers and family members of all faiths in the Army worldwide were not available, the U.S. Army Forces Command reported 10,563 field and chapel worship services conducted in FY 2000 for active duty soldiers.[52] In addition, FORSCOM documented 821 weddings, 611 funerals, 334 memorial services, 2,644 family skill/enhancement classes, and 1,304 separate suicide prevention classes which reached a total population of 89,979 soldiers, retirees, and family members. In a volunteer Army with 65% of its active force soldiers of all ranks married and with 52,000 physically challenged members included in the families, these services were indispensable to soldier welfare and readiness. [53]

Since church attendance by the retiree population has not been separately tabulated, the estimated church attendance figures for active duty soldiers and family members cannot be accurately determined. However, of 12,561 waiting spouses during lengthy separations due to deployments, 30.6% (3,844) reported use of worship programs and services provided by Army chaplains. [54]

Although all relevant inputs have not been considered (e.g., religious activities in the Reserve components), it seems reasonable to conclude that the active duty Army population is a microcosm of American society and culture. The majority of citizens and soldiers profess to be religious. Many more people have an interest in spirituality and religion than attend religious services, at least on a regular basis. However, during periods of prolonged stress to both individuals and families, as exemplified by deployment to a combat zone, most soldiers and spouses indicate that religion is an important support for their pre-deployment readiness, their morale, the well-being of their deploying units, the durability of their marriages, and the welfare of families back home.[55]

The Soldier's Spirit: Leveraging the Human Dimension to Build Confidence, Cohesion, and Courage

On 15 June 1941, Gen. George C. Marshall addressed the faculty and students of Trinity College in Hartford, Connecticut, a college linked to the Episcopal Church, on the subject of morale in modern war:

> I know that this association with you here this morning is good for my soul. If I were back in my office I would not have referred to my soul. Instead I should have used the word "morale" and said that this occasion increased my "morale"—in other words was of spiritual benefit to me. One of the most interesting and important phenomena of the last war was the emergence of that French word from comparative obscurity to widespread usage in all the armies of the world. Today as we strive to create a great new defensive force, we are investing the word "morale" with deeper and wider meaning. Underlying all the effort back of this essentially material and industrial effort is the realization that the primary instrument of warfare is the fighting man. We think of food in terms of morale—of clothing, of shelter, of medical care, of amusement and recreation in terms of morale. We want all of these to be available in such quantity and quality that they will be sustaining factors when it comes to a consideration of the soldier's spirit. The soldier's heart, the soldier's spirit, the soldier's soul are everything. Unless the soldier's soul sustains him, he cannot be relied on and will fail himself and his commander and his country in the end.[57]

General Marshall gave a good deal of his personal attention to supporting the soldier's morale, moral behavior, and spiritual strength, supplying more than 550 cantonment chapels and 9,111 chaplains—one for every 1,200 soldiers—to the Army and Army Air Corps.[58]

However, General Marshall recognized that the soldier's spirit—the soldier's morale—included much more and demanded much more than religious support alone. Morale is a disciplined state of mind which embraces confidence in the self and confidence in the unit. It encompasses courage, zeal, loyalty, hope, and at times grim determination to endure to the end.[59]

Morale, élan, esprit de corps,"the will to combat," and the will to win, are the human dimension's most important intangible assets. Strong morale is an emotional bonding of purpose, meaning, common values, good leadership, shared hardship, and mutual respect.[60] Of all of the factors which produce strong morale in a unit—of whatever size—leading by example and unit cohesion are frequently mentioned first.[61] Lord Moran's experiences with unit cohesion in the British regiments during World War I led him to conclude that "there was only one religion in the regular army, the regiment; it seemed to draw out of them the best that was in them."[62] Such morale, such fighting spirit, coupled with faith in their leaders, were important factors in the survival and ultimate victory of soldiers throughout military history.

However, the morale of the soldier and the esprit de corps of the unit may have a short shelf life in extended combat. Like courage, morale is an expendable commodity and needs some replenishment and support to withstand combat stress. John Keegan reflected on the experiences of British and American doctors during World War II in his book *The Face of Battle*. Of all British battle casualties during the active phase of the Battle of France in 1940, "ten to fifteen percent were psychiatric, ten to twenty percent during the first ten days of the Normandy battle and twenty percent during the two latter months, seven to ten percent in the Middle East in the middle of 1942, and eleven percent in the first two months of the Italian campaign."[63] The official American report on combat exhaustion during the same period stated:

> There is no such thing as "getting used to combat." Each moment of combat imposes a strain so great that men will break down in direct relation to the intensity and duration of their exposure [thus] psychiatric casualties are as inevitable as gunshot and shrapnel wounds in warfare. Most men were ineffective after 180 or even 140 days. The general consensus was that a man reached his peak of effectiveness in the first 90 days of combat, that after that his efficiency began to fall off, and that he became steadily less valuable thereafter until he was completely useless. The number of men on duty after 200 to 240 days of combat was small and their value to their units negligible.[64]

Not only individuals but also whole units became ineffective as a result of fatigue, stress, high casualties, poor leadership, and a loss of hope. In the Tunisian Campaign of 1942, veteran American combat troops joined newer recruits in "going to ground," "burning out," and breaking down. One 1944 report pointed out that in the North African theater nearly all men in rifle battalions not otherwise disabled ultimately became psychiatric casualties even though some of them made it as far as Cassino and Anzio.[65] Other examples of whole units becoming combat ineffective may be gleaned from the experience of some German units on the Eastern Front, American units during the Korean War, and Iraqi units during the Gulf War.[66]

What types of support did the soldiers who were able to endure find helpful in coping with the stresses of combat? John Keegan identifies four critical elements in British armies: moral purpose—believing in the "rightness" of the war; unit cohesion—formed in hard training, sports competitions, and rewards for being the "best"; selfless leadership from first-line officers; and a desire for spiritual or religious fortification before battle.[67]

William Manchester, who served as an enlisted Marine on Okinawa during the most intense fighting in the spring of 1945, wrote of his survival in his book *Goodbye Darkness*:

> You had to know that your whole generation was in this together, that no strings were being pulled for anybody. You also needed nationalism, the absolute conviction that the United States was the envy of all other nations. Today the ascent of Sugar Loaf [on Okinawa] takes a few minutes. In 1945 it took ten days and cost 7,547 Marine casualties. And beneath my feet, where mud had been deeply veined with human blood, the healing mantle of turf [I murmured a prayer: *God*] *take away this*

> *murdering hate and give us thine own eternal love.* And then, in one of those great thundering jolts in which a man's real motives are revealed to him. I understood why I jumped hospital and, in violation of orders, returned to the front and almost certain death. It was an act of love. Those men on the line were my family, my home. They were closer to me than I can say, closer than any friends had been or ever would be. They had never let me down, and I couldn't do it to them. I had to be with them, rather than let them die and me live in the knowledge that I might have saved them. Men, I now know, do not fight for flag or country, for the Marine Corps or glory or any other abstraction. They fight for one another.[68]

If morale is the human dimension's most important tangible asset, cohesion must be the most important single asset for a unit. Cohesion consists psychologically of recognition, stability, and safety.[69] Yet the coping strategies Keegan and Manchester identified, which included cohesion, did not exist as separate components. For Manchester, combat was a spiritual exercise, a willingness to sacrifice for a greater cause (moral purpose) but mostly for his fellow Marines (brotherhood). Moral purpose, selflessness, courage, and spiritual strength in Keegan's list and in Manchester's narration all contributed holistically to unit cohesion and survivability.

American surveys of other World War II combat survivors tended to center on similar coping mechanisms and their relative priorities in importance for survival of the individual. Although the methodologies involved in these surveys may be questioned, the general conclusions that spiritual strength and "not letting others down" were two of their most important motivations for endurance seem to be validated by other observers, not the least of whom were their senior officers.

In November 1945, the Research Branch in the Information and Education Division of the War Department queried a representative group of enlisted men who had returned from combat zones about their experiences in the U.S. Army during World War II. [70] There were few aspects of their experience that elicited positive responses. Most of the soldiers said they were "fed up" with the Army. When asked about coping mechanisms in combat, however, many responded that loyalty to one another and prayer for strength were important coping mechanisms in combat. [71]

In a survey of 1,433 veteran enlisted infantrymen taken in Italy in April of 1945, 84% of the privates and 88% of the noncommissioned officers said that prayer helped them more "when the going got tough" than unit cohesion, the cause they were fighting for, thoughts of finishing the job to get home again, or thoughts of hatred for the enemy.[72] Among company grade infantry officers questioned in the European and Pacific theaters in the spring of 1944, approximately 60 per cent said that prayer helped them a lot in tough circumstances.[73] In both Italy and in the Pacific, at different times, prayer as an aid to adjustment to combat generally ranked higher among enlisted men than did the other personal coping mechanisms listed in the questionnaires. While officers reported being helped by the desire not to let others down, even with them prayer ranked second. [74]

Among very senior officers who expressed religious faith, prayer seemed to be important to remind themselves and their soldiers of their dependence upon a Higher Power, to help senior leaders make decisions calmly, and to help them bear the burdens of their immense responsibilities. Lt. Gen. George Patton recognized the power of spiritual petition when he circulated 250,000 copies of a weather prayer, one for every soldier in the Third Army, during his efforts to relieve Bastogne in December of 1944.[75] President Dwight Eisenhower, in recalling his prayerful decision to launch the Normandy invasion in 1944, reflected that "prayer gives you the courage to make the decisions you must make in a crisis and then the confidence to leave the result to a Higher Power."[76] General of the Army Douglas MacArthur told the cadets at West Point in his "Duty, Honor, Country" address of May 1962:

> The soldier, above all other men, is required to practice the greatest act of religious training—sacrifice. In battle, and, in the face of danger and death, he discloses those divine attributes which his Maker gave when He created man in His own image. No physical courage and no greater strength can take the place of the divine help which alone can sustain him.[77]

In the World War II surveys of combat veterans, prayer was not of itself a sufficient indicator of religious faith; it may have been adopted as an instrument of psychological self-defense. There were no data that could prove a relationship specifically between prayer in battle and formal religion. However, the experience of combat did seem to have an effect on spiritual attitudes, for 79 per cent of combat veterans surveyed in both theaters believed that their Army experience had increased their faith in God.[78] As Lt. Gen. A.A. Vandegrift, Commandant of the United States Marine Corps, reflected on his experiences at Guadalcanal:

> The percentage of men who devoted much time to religion might not make a very impressive showing. The average marine, or soldier, or sailor, is not demonstrative about his religion, any more than he is about his patriotism. But I do sincerely believe one thing: every man on Guadalcanal came to sense a Power above himself. There was a reality there greater than any human force. It is literally true—there are no atheists in foxholes—religion is precious under fire.[79]

Thus, from the commander's point of view, the soldier's spirit, the soldier's morale, is not exactly coterminous with the soldier's personal views on, or experience with, religion. The fighting spirit of the soldier may be motivated by any emotion, idea, or complex of ideas that will inspire the soldier to accomplish the mission. These compelling drives may include personal confidence, competence, and pride in self, faith in leaders, unit bonding and cohesion, a belief in the moral necessity and rightness of the cause, a consonance between personal values and national purpose, and a belief that others are depending upon the soldier for success. As the reality of danger increases, however, and casualties pile up, religion seems to provide many soldiers a strong buttress for the spirit and will to endure.

Historically, therefore, religious support for the soldier's spirit has been an important source of strength for many in coping with difficult and dangerous situations, especially over prolonged periods of time. Religious services before battle and the presence of chaplains in the lines, at aid stations, and even in POW camps have helped thousands of soldiers face the uncertainties of war.

For example, during Operation Desert Shield, from August through December 1990, 18,474 soldiers from the XVIII Airborne Corps attended voluntary religious services. The U.S. Army Central Command (ARCENT) sponsored 7,946 religious meetings with an attendance of 341,344 soldiers. Maj. Gen. Barry McCaffrey remarked that "we had the most religious Army since the Army of Northern Virginia during the Civil War."[80]

At midnight on 17 January 1991, Gen. H. Norman Schwarzkopf held a staff meeting with 30 generals and colonels in his war room in Riyadh to read his announcement of the beginning of combat operations. In his message General Schwarzkopf reminded his staff of their purpose, their just cause, and his total confidence in them. He then asked his chaplain to offer a prayer. The chaplain reflected later that even though it was not discussed as such, the prayer for a quick and decisive victory with few casualties had a unifying, cohesive effect on the staff as they set about the business of war.[81]

In the discussion of spiritual fitness for soldiers in the Army's Health Promotion Program, the term is defined as "the development of those personal qualities needed to sustain a person in times of stress, hardship, and tragedy."[82] No matter how pluralistic the sources for spiritual fitness may be, in the estimation of many senior leaders the ability of the soldier to draw on his or her own spiritual or philosophical resources in times of stress is an undeniable component of readiness. Gen. Gordon Sullivan, former Chief of Staff of the Army, noted a relationship between courage and the spiritual fitness of soldiers in Field Manual 100-1, *The Army*, published in December 1991:

> Courage is the ability to overcome fear and carry on with the mission. Courage makes it possible for soldiers to fight and win. Courage, however, transcends the physical dimension. Moral and spiritual courage are equally important. There is an aspect of courage which comes from a deep spiritual faith which, when prevalent in an Army unit, can result in uncommon toughness and tenacity in combat. [83]

Gen. John Hendrix, a veteran of Operation Desert Storm and Commanding General of U.S. Army Forces Command, stated at a Memorial Day Prayer Breakfast at Ft. McPherson, Georgia, on 22 May 2001:

> Spirituality is an individual matter. We must not cross the line between church and state. But in general spiritual fitness is important to any organization. Spiritual fitness helps shape and mold our character. Spiritual fitness provides each of us with the personal qualities which enable us to withstand difficulties and hardship. When properly exercised, spiritual fitness enhances individual pride in our unit.[84]

Gen. George C. Marshall's comments on the subject to Army chaplains in 1944 in Washington, DC, reinforce the message: "True, physical weapons are indispensable, but in the final analysis it is the human spirit, the spiritual balance, the religious fervor, that wins the victory. It is not enough to fight. . . . It is the spirit which we bring to the fight that decides the issue. The soldier's heart, the soldier's spirit, the soldier's soul, are everything." [85]

Training Considerations: Preparing Soldiers for Future Combat

Two essential ingredients for success in combat—that is, for creating high morale, unit cohesion, bonding among soldiers, increased personal courage, spiritual strength, and determination to succeed—are inspirational leadership and tough, realistic training. [86] Many officers and noncommissioned officers in the Army, however, have never served in a combat zone. Many others have difficulty envisioning what training for future combat might mean. Therefore we shall glance briefly at the characteristics of the future battlefield and the skills soldiers must possess to prevail. Then, as part of the human dimension, we shall turn to a generalized description of the millennial generation, those Americans born in 1982 and after from whose ranks the Army will recruit its future force. Finally with regard to training, we'll examine a model battalion training program during the Gulf War, noting its holistic approach in utilizing the human dimension for success in future wars.

Battlefield Visualization and Soldier Skills

Army literature on battlefields of the future is complex and copious. For several years at the U.S. Army War College and at the U.S. Army Command and Staff College, among other institutions, numerous subject matter experts prepared briefings, training models, and articles on the Army After Next, on the digitized battlefield, and on Army transformation into a true 21st century fighting force. The purpose of these studies was to prepare the military for future wars and to tailor the reduced forces available to meet changing, possibly asymmetric, threats with a multi-dimensional National Military Strategy.

The Deputy Chief of Staff for Operations and Plans (DCSOPS) has overall responsibility for battlefield visualization. As Lt. Gen. John Miller, former deputy commander, U.S. Army Training and Doctrine Command, explained:

> Battlespace—the use of the entire battlefield and the space around it to apply combat power to overwhelm the enemy—includes not only the physical volume of breadth, depth, and height, but also the operational dimensions of time, tempo, depth, and synchronization. Commanders must integrate other service, nation, and agency assets with their own to apply their effects toward a common purpose. The digitized battle staff—a deputy commander and three planning and operations teams—is one concept to help the commander handle the current battle, the future battle, and sequels to the future battle with an information exchange system that produces virtual collocation between staff and external elements. Emerging technology includes interactive graphics, enemy and friendly force tracking, scalable map displays, three-dimensional terrain visualization, course-of-action analysis, and video-teleconferencing capabilities among other assets.[87]

At the operational and tactical levels, this meant that soldiers would have to be proficient not only with their weapon systems, but also with emerging technologies which would function in all shades of weather, terrain, and illumination. Moreover, dispersal of units, to prevent detection by an enemy with over-the-horizon targeting capabilities, would produce a force with mobile combat power and "just in time" logistics as opposed to the iron mountains of stockpiled equipment familiar on Vietnam-era firebases or on forward-deployed Desert Storm logistical bases.

The specific geography for future engagements is, of course, speculative. In the twentieth century American soldiers have fought in foreign areas on snow-bound tundra, in forests, mountains, deserts, jungles, urban areas, and on sandy beaches.

For the future, all of these settings must be considered along with the special problems of homeland defense amidst one's own citizens.

What, then, are the special skills soldiers of the future must possess? In the U.S. Army Training and Doctrine Command's Training, Leader Development and Soldiers Strategy for dealing with change, readiness, and the human dimension, some of these qualities are described:

> First, the leveraging of the human dimension is all about leading change with quality people, grounded on Army values, and inspired by an American warrior ethos. Adaptive leadership remains an essential aspect. Quality people will need to have the character and interpersonal skills to rapidly integrate individuals and groups of individuals into tailored organizations. They will need to adapt quickly to new situations, and form cohesive teams, and demonstrate competence and confidence operating in complex and ambiguous environments.[88]

In short, the Army will need not just soldiers but soldier-leaders who are committed to the professional ethic, who are talented in small-group facilitation, who are flexible and mentally agile, and who can integrate technological and interpersonal skills in the midst of uncertain and possibly chaotic combat conditions.

Concerning the human dimension, Lieutenant General Miller observed:

> Command of soldiers is, first and foremost, a human endeavor requiring the commander to be a decision-maker and leader. As is the case today, these competent commanders will establish their moral authority by tough, demanding training to standard and by the caring, holistic preparation of their subordinate leaders, soldiers, and units for mission operations. The significance of the bonds of trust and confidence between the leader and the led will grow as the potential for decentralized execution over larger battlespace increases.[89]

Needless to say, with the current shortfall in junior NCOs and company-grade officers, the Army's recruiting and leadership challenges are daunting. One would hope that there are enough time and resources to complete such complex and demanding training before combat operations on a future battlefield become necessary. In the long history of the U.S. Army, however, that has rarely been the case.

The Millennials: What Soldiers for the Future?

From approximately 210 national surveys, interviews, and studies of American young people, prolific authors Neil Howe and William Strauss have formed a description of a group they call The Millennials. These are American young people who were born in 1982 or later, or in other words were 18 or younger in the year 2000. Some characteristics of these young people may be of interest to the Army.

First, they are a large group of approximately 76 million, with 90% native born and about 10% who immigrated to the United States.[90] By the year 2002 they will outnumber the surviving Baby Boomers. They are the most diverse group ethnically in American history, with 36% nonwhite or Latinos in the 1999 youth population.[91] At least one Millennial in five has one immigrant parent, making the Millennials potentially the largest second-generation immigrant group in U.S. history. As the authors point out, their presence will contribute to the irreversible diversification of America.[92]

In terms of religious identity, approximately 20 million are Roman Catholic, which helps account for the growth of the Roman Catholic constituency in the United States from 26.6% of the U.S. Christian population in 1958 (30.6 million) to 38.3% of the U.S. Christian population in 1998 (61.2 million). That is a net increase of 30 million American Roman Catholics in forty years, making the Roman Catholic Church in America four times the size of the Southern Baptist Convention, the largest Protestant denomination.[93]

For many Millennials there is no separation of church and state in their primary education. Two million attend Catholic elementary schools and another two million attend Catholic high schools. Nine in ten private schools in the United States in 2000 had at least a nominal religious affiliation, many with their own mandatory chapel programs. Within the public schools there were no prayer clubs or circles in 1990; now, with the 1995 Federal court ruling that students had a right to organized prayer gatherings as long as they were not official school programs, there are more than 10,000 of them.[94]

Among the Millennials who are over 14, some 65% plan on attending college and 55% go to church regularly as opposed to 45% of Americans as a whole.[95] The ones in high school are bright. They have scored well in science and reading as compared with students from other industrialized countries.[96]

These figures appear even more impressive in light of the report that one-fourth to one-third of all Millennials live in single-parent families. More than half of these single parents, for whatever reason, have unmarried partners living with them.[97] It may not be surprising, then, that about 48% of Millennials have been sexually active as teenagers.[98] The *Chronicle of Higher Education* has reported that, in order to meet the psychological demands of many of these families, there has occurred an increase in "the spiritual dimensions" of social work. Edward R. Canda, a professor of social work at the University of Kansas, noted that "in a crisis or occasion of grief and loss, there is often a shaking of the foundation of one's sense of meaning, who one is, what life is about, and what reality is about. We cannot escape these questions. It would be malpractice to avoid them."[99]

Most importantly, the "war in Kosovo" is the only U.S. military action that most Millennials remember. The oldest young people in this sample were only eight to nine years old during the Gulf War, and the events surrounding the subsequent Oklahoma City Federal building bombing and the Columbine school shootings made greater impressions.[100] The vast majority of this large generation of Americans has no military experience except vicariously at the movies, and unless actively recruited will probably never serve in the armed forces.

Other sources, outside the studies surveyed by Howe and Strauss, paint a less sanguine picture. The State School Superintendent's Office for the state of Georgia reported on 5 May 2001 that of the 116,000 high school freshmen who were enrolled in 1997, only 72,000 graduated in June of 2000. This reflected a high school dropout rate of 38%.[101] Moreover, 47% of Georgia's high school seniors who graduated in the class of 1999 were unable to keep their scholarships as college sophomores because they could not maintain a B average.[102] Finally, United Way reported on 13 May 2001 that there were 230,000 troubled children under some form of care in the state of Georgia—all Millennials under the age of 18. If one assumes that Georgia, with half of its eight million people living in Atlanta, is not too different from many other states, one suspects that the rosy reports by Howe and Strauss were based on the most privileged of the Millennials.

One characteristic which has not been questioned, however, is the growing interest among older young adults in discovering their own interests, vocations, and, in some cases, spiritual insights. Many college students and young business people want to be part of an organization or movement which transcends the ordinary. The Campus Crusade for Christ, for example, has experienced an amazing growth in the past five years among college students looking for meaning in their lives. Campus Crusade has 1,000 college chapters—including one at Harvard—comprising a total of roughly 40,000 students. Donations to Campus Crusade, as reported by The *Chronicle of Higher Education*, exceeded $450 million in the year 2000. "They're bombarded and blasted with all kinds of atheistic teaching from the classroom and they need help," according to William Bright, the lay founder of the movement.[103]

In his recent book, *Capturing the Heart of Leadership*, Professor Gilbert Fairholm of Hampden-Sydney College describes a similar kind of restlessness among young workers:

> Whether we like it or not, work is becoming or has become a prime source of values in our society and our personal lives. American workers are uncomfortable, uncommitted, and adrift. They are searching for new organizational patterns and new paradigms. Integrating the many components of one's work and personal life into a comprehensive system for managing the workplace defines the holistic or spiritual approach. It provides the platform for leadership that recognizes this spiritual element in people and in all of their behavior.[104]

What Fairholm argues is that young people expect leadership to be a relationship, not just a skill or personal attribute. Leaders are leaders only so far as they develop relationships with their followers, relationships that help all concerned to achieve their spiritual as well as economic and social fulfillment.[105] This concept is not far from the Army's definition of the transformational leadership style and may be a constructive bridge in thinking about what "leadership" might mean to the next generation of American soldiers. [106]

A Model for Combat Training

In the summer of 1990, Lt. Col. Gregory Fontenot, commanding the 2nd Battalion, 34th Armor Regiment at Ft. Riley, Kansas, reflected on the training he and his soldiers had received in preparation for their wartime mission. The unit had achieved battle honors and its nickname "Dreadnought" in Vietnam. They had survived their "bloodless combat" exercises at the National Training Center and in REFORGER (Return of Forces to Germany.) But as the rumors of impending war in the Persian Gulf region grew stronger, Fontenot wondered how his current soldiers, untested in combat, would react to fear, stress, and the shock of "seeing the elephant."[108]

As a former assistant professor in the Department of History at West Point, Fontenot did his homework. He read deeply about combat, from fictional accounts such as Stephen Crane's *Red Badge of Courage* to more scholarly studies including Anthony Kellett's *Combat Motivation.* Based on his reading, training, and experience to that time, Fontenot developed the opinion that "unit cohesion stemmed from three general sources: rigorous training to high standards, credible leadership, and soldiers who believed in one another." He added:

> Patriotism and belief in the cause seemed to have little effect on units in combat, though they were important aspects of developing unit cohesion in training prior to combat. None of these conclusions is demonstrable—I came to them via subjective analysis of what I read, heard, and saw. My conclusions are decidedly not the result of objective analysis. In part they stem from my belief that not all human behavior can be reduced to objective and quantifiable data points. Some things must be felt.[109]

Fontenot also believed that the first, and last, emotion to be mastered was fear. He thought that "the stresses of combat are a euphemism for fear. I sought to develop in my soldiers an understanding of what fear would feel like and to eliminate, when possible, the uncertainty that accelerates fear."[110]

In July 1990 Fontenot and members of his color guard participated in a "welcome home" parade for Vietnam-era veterans of the 2nd Battalion, 34th Armor Regiment. It was a reminder of unit pride and tradition just before the Iraqi invasion of Kuwait.

In August it became clear that the United States would fight in the Middle East and that the 2/34 Armor would go to war. Fontenot noted that "each of us had to accept the idea of combat now—not some remote historical event but rather a 'no kidding' fact of life in the immediate present. At this stage, the need to prepare for fear in combat did not drive unit preparation generally or individual psychological preparation specifically. In August, the issue was changing our 'mind-set.' When the mind-set changed, it produced new intellectual and spiritual needs."[111]

For the next five months Fontenot worked to prepare his unit psychologically for combat. He had been working on this training mission since he took command in April of 1989; but as other units began to deploy from Ft. Riley in September, the tasks assumed a new importance and urgency. Fontenot implemented a plan to reinforce unit pride, maximize communication with every soldier in the battalion, provide accurate information, assure complete task understanding, and clarify "the arcane but important" law of war.[112]

He and his staff set up a conference room as a unit history center with slides, videotapes, and actual colors from Vietnam as the centerpiece since "to 18-year-old soldiers in 1989, World War II seemed nearly as distant as the Civil War." New soldier orientations were held in this unit shrine to heroism.

It became Fontenot's goal to meet every soldier in the battalion and to talk frequently to the entire unit from the front of his tank, with chalkboard available, in order to create clear expectations for training and personal relationships with his men. To dispel myths and misinformation about enemy capabilities, qualification gunnery exercises at Ft. Riley's Range Complex were coupled with inspections of the T-72 tank and other Soviet weapons used by the Iraqis. Fontenot required his officers to read a book each month on combat operations from the perspective of a participant and write a critical analysis for him. He personally graded the reports and led discussions on their themes.[113] In November, everyone, including Fontenot, got a "desert haircut," a skin-job which he equated with putting on his Sioux "warfeathers."

At about that time, just prior to deployment in December, Fontenot arranged for a senior chaplain, a personal acquaintance and former colleague who had taught the history of professional ethics in the Department of History at West Point, to come to Fort Riley to present a discussion of just war theory for his officers. The chaplain would himself deploy to Saudi Arabia in December, so he had more than just academic credibility with the officers. Of the chaplain's presentation, Fontenot wrote:

> As esoteric a topic as this might seem, [he] brought to it clarity, wit, and the ability to demonstrate how Thomas Aquinas' thoughts about just war affected tankers and infantrymen today. His talk melded with myriad other preparations we were making and contributed directly to our efforts to strip the coming events of their mystery. [He] showed us another side of the "elephant's" face.[114]

Actually, there were many chaplains involved in both the ethical and spiritual preparations of units for Desert Storm. Twenty of the U.S. Army's training centers had chaplain instructors who were charged strictly not to preach in the classroom, but to discuss professional values, ethics, and leadership. In field units, U.S. Army Forces Command instituted "spiritual fitness training," a command program to address, among other topics, "the full spectrum of moral concerns involving the profession of arms and the conduct of war."[115] The commander's staff officer responsible for conducting the program was the chaplain.

Part of the rationale for tasking chaplains in FORSCOM with the spiritual fitness mission was that they were deployable to combat zones, they could answer questions about morals and morale, they had connections to family support systems back home, they had legal confidentiality so that soldiers could report suspected violations of the law of war to them, as at My Lai in 1968, without fear of recrimination, and they could address the soldier's personal spiritual needs and ethical questions. Although not all chaplains were equally trained in the ethics of war, they helped religious soldiers find the bridge between their spiritual and professional values in a way no other staff officer was expected to do.

By December 1990, the 2/34th Armor was in excess of 100% of its authorized strength—12 tank platoons. The unit moved from the port in Saudi Arabia to tactical assembly area Roosevelt where the lead elements arrived on 12 January 1991. At this point the psychological preparation for combat entered another phase of realism—constant drills, alarms, and alerts for chemical attacks; rehearsals from task force to platoon level; and visits by senior officers to encourage the troops, to exchange information, and to clarify any messages. Maj. Gen. Thomas Rhame, Commander of the 1st Infantry Division, "made sure he laid hands on the troops and commanders. He was enthusiastic and confident, and he communicated that confidence as a senior leader."[116]

As the date approached when the 2/34 would be sent into combat to breach the enemy's prepared defenses, Fontenot began conducting personal chalk talks with primary groups at company and platoon level. He discussed the Clausewitzian triangle of people, state, and army, showing the soldiers how all elements were combined in support of their mission—from the citizens back home to the political leaders to their commanders on the ground. He reminded his soldiers of why the war was being fought and why their mission was important. Finally he discussed fear, how it felt, why it was natural and how it could be useful in stimulating adrenaline and directing the body's blood supply to core functions. "I concluded my discussion by promising them that we would all make mistakes," Fontenot wrote, "however I assured them that we had a technological advantage, superior training, and a good plan of operation, and that our enemies enjoyed none of these benefits."[117]

The 2/34th Armor began its attack on 18 February 1991 against an enemy outpost and fired its last shots on the 28th. The Dreadnoughts destroyed seventy armored vehicles and two dozen trucks, and captured 728 enemy soldiers while participating in night fighting with British and U.S. Air Force support.[118] Of its more than 1,000 soldiers engaged in ten days of combat with enemy tanks and artillery, the 2/34th suffered only two dead and four wounded.[119]

Upon reflection, Fontenot wrote later,

> There is no objective way to demonstrate whether this system of preparation for fear employed in the 2/34 Armor worked. The tolerance of soldiers for miserable conditions and uncertain circumstances remained very high throughout the operation and demonstrated amazing adaptability and patience on their part. There is just no way to ascertain the extent to which ritual, information, discussion, and understanding prepared us for battle. In my experience, commanders at all levels shared the conviction that they bore a responsibility to prepare troops for the stress of combat. The tools to do that included rigorous training, techniques for developing cohesion, clear communication of intent, and addressing the matter of fear forthrightly and on a personal level. [120]

Although Fontenot's account is modest to a fault, the Army was clearly impressed with his insights, not only in past conflicts but also for future integration of strategy, tactics, technology, and the human dimension of leadership. After the Gulf War, he was assigned as the Director, School of Advanced Military Studies, U.S. Army Command and General Staff College, Ft. Leavenworth, Kansas.

Conclusion

The purpose of this chapter has been to examine the literature available on the human and spiritual needs of soldiers and how they may be trained for combat in the 21st century. The working hypothesis, that all soldiers have human needs and most have spiritual needs broadly defined, seems to be supported from a wide variety of sources in a number of fields. As Karl von Clausewitz tellingly observed in his treatise *On War* : "All effects in the sphere of mind and spirit have been proven by experience: they recur constantly and must receive their due as objective factors. What value would any theory of war have that ignored them."[121] There also seems no question whether the Army Profession should continue to try to address these needs in the future as part of its leadership doctrine. Indeed, the unique aspects of the profession of arms in requiring the total commitment and unlimited liability of soldiers deployed often in difficult and dangerous situations would seem to mandate such care and concern. Moreover, the Constitution, Congress, and the American people expect and demand it.

A corollary question is to what extent the profession of arms should try to meet the spiritual needs of a military population becoming ever more ethnically, morally, and religiously diverse. At the present time the Army seems to have struck an appropriate balance between facilitating the free exercise of religion and in protecting the freedom of individual conscience for soldiers and their family members. Most religious services and ceremonies are voluntary. The Army's Chaplain Corps, in implementing commanders' religious support programs, currently represents more than 140 different denominations and faith groups in its own ranks. There is no danger that the Army will institutionalize a single religion, nor should it. There is a concern, as a shortage of young ordained clergy grows in both the Roman Catholic and Protestant communities nationally, that there will be a parallel shortage of chaplains for the Army. Part of a solution could lie in the way the Millennial generation chooses to respond to its own spiritual challenges in the 21st century.

The preparation of soldiers for future combat seems to involve more knowledge, more technological skill, and perhaps more maturity on the part of junior leaders than has been the case in the past. Yet the basic principles involved in building relationships, unit cohesion, confidence, and courage have not changed very much over the years and may not change markedly in the near future.

There are valid, practical considerations for commanders, staff officers, senior noncommissioned officers, chaplains, surgeons, psychiatrists, and other leaders in preparing soldiers for combat. Some of the more important of these, based on historical experience and analysis, may be summed up as follows:

- Soldiers and family members need to have the most accurate and most current information possible on what they may expect. The importance, necessity, and moral justification of the mission are essential elements of information for the soldiers, their families, and communities if the unqualified support of all affected parties is to be forthcoming.
- Commanders and other leaders need to spend some personal time with soldiers in their primary units to reinforce relationships, cohesion, confidence, and courage. Soldiers must know the commander's intent and their specific jobs to include how they fit into the total effort of the unit.
- Soldiers must have confidence in their leaders, training, equipment, battle plans, and ultimate chance for success. Soldiers should be encouraged to enter into dialogue with their immediate commanders on such subjects as overcoming fear and the means available for supporting one another in the chaos of battle.

- Soldiers need to have time to get their personal affairs in order. This may include time for family, and legal as well as physical, mental, and spiritual preparation.
- Rituals before deployment and before battle based on unit history, esprit de corps, and spiritual preparation are important. These should include voluntary opportunities for religious sacraments, services, or meditation.
- Soldiers need to know that their commanders, senior NCOs, chaplains, and other key personnel are present at every stage during combat operations. The soldiers' morale is strengthened if the total team is demonstrably present and involved.[122]

There is one caveat. War is not a thing which can be seen; it must be thought.[123] No one has ever seen war in all of its dimensions—physical, moral, and spiritual—because each participant sees the event from his or her own narrow, partial perspective. In the distant future, war and the professional skills needed to survive and prevail may be very different with the advent of robotics, information warfare, and even space technologies. Therefore the combat training strategies developed for the first decades of the 21st century may be of short duration, but they will also surely be important for their insights and wisdom in the evolution of future training doctrine and for appreciating the human dimension in Army professionalism.[124]

Notes

1. The views and opinions expressed in this chapter are those of the author and are not necessarily those of the Department of the Army or any other U.S. government entity.
2. S. L. A. Marshall, *Men Against Fire: The Problem of Battle Command in Future War* (New York: William Morrow Co., 1947), 208, 211.
3. Andrew Abbott, *The System of Professions: An Essay on the Division of Expert Labor* (Chicago, IL: University of Chicago Press, 1988).
4. Don M. Snider and Gale L. Watkins, "The Future of Army Professionalism: A Need for Renewal and Redefinition," *Parameters* 30, 3 (Autumn 2000): 17.
5. Richard A. Gabriel, *To Serve with Honor: A Treatise on Military Ethics and the Way of the Soldier* (Westport, CT: Greenwood Press, 1982), 5-6. Gabriel outlines many of the same handicaps during the Army's shift to the All-Volunteer Force thirty years ago.
6. Attributed to Gen. Robert E. Lee by Douglas S. Freeman as cited in Ken Burns, "The Civil War," PBS video production, 1990, part V.
7. *Sun Tzu: The Art of War,* trans. Samuel B. Griffith (London: Oxford University Press, 1963), 128.
8. FM 22-100, *Army Leadership: Be, Know, Do* (Washington, DC: Headquarters, Department of the Army, 31 August 1999), 3-3.
9. Ibid., 3-4.
10. Ibid., 3-1.
11. Ibid.
12. Ibid., 3-1 and 3-2.
13. Lord Charles Moran, *The Anatomy of Courage* (Boston, MA: Houghton Mifflin, 1967), 202.

14. Interview in Atlanta, Georgia, with James Eric Pierce, former Associate Professor of the Sociology of Religion at Pfeiffer College, on 6 April 2001.
15. Data furnished by Chaplain Michael T. Bradfield, DA Office of the Chief of Chaplains, Washington, DC, 25 April 2001. This is actually a low figure because the 64% of active duty soldiers who profess a specific faith does not count those who belong to faith groups other than the seven largest by population, i.e., Protestant, Catholic, Orthodox, Jewish, Muslim, Buddhist, and Hindu.
16. Task Force TLS, *Training, Leader Development and Soldiers Strategy* (Ft. Leavenworth, KS: U.S. Army Command and General Staff College, 1 October 1999), 75 and slides 14, 17; Lt. Col. Steven W. Shively, Project Officer, DA-ODCSPER, Directorate of Human Resources, *Draft Army Well-Being Campaign Plan*, 5 January 2001, 2.
17. Morris Janowitz, *The Professional Soldier* (New York: The Free Press, 1960), 97.
18. United States Military Academy, Office of Policy, Planning, and Analysis, "Cadet Leader Development System," Draft, July 2001, chapter 3, 3.
19. Jeff Levin, *God, Faith and Health* (New York: John Wiley, 2001), 9-10. Emphasis supplied.
20. Ibid., 10; also see David F. Swenson and Walter Lowrie, trans., *Soren Kierkegaard: Concluding Unscientific Postscript* (Princeton, NJ: Princeton University Press, 1968), 495, for Kierkegaard's condemnation of " faithless religiousness."
21. Robert D. Baird and Alfred Bloom, *Indian and Far Eastern Religious Traditions* (New York: Harper and Row, 1971), 214; and Huston Smith, *The Religions of Man* (New York: Harper and Row, 1958), 149.
22. Geoffrey Parrinder, *World Religions: From Ancient History to the Present* (New York: Facts on File Publications, 1985), 355.
23. Saeed Ahmed, "Falun Gong: Peace of Mind," *Atlanta Journal Constitution*, 19 May 2001, B-1.
24. John W. Wright, ed., *New York Times Almanac 2000* (New York: The Penguin Group, 1999), 414; Levin, 20; and Ron Feinberg, "Amen Corner," *Atlanta Journal Constitution*, 5 May 2001, B-1. This 63% may also be a low figure. The World Christian Encyclopedia published in 2001 reported 192 million Christians in the United States, 32 million more than *The New York Times Almanac* reported in 1998. The difference may be in counting numbers reported by independent congregational churches as opposed to counting only figures from major denominations. See Ron Feinberg, "Report: Christianity Still Largest Religion," *Atlanta Journal Constitution*, 20 January 2001, B-1.
25. Samuel P. Huntingdon, *The Clash of Civilizations and the Remaking of World Order* (New York: Simon and Schuster, 1996), 95-97.
26. A survey of 502 adults from across the country with a 4.5% margin of error as reported in *Atlanta Journal Constitution*, 12 May 2001, B-1.
27. Lord Moran, ix. Lord Moran must not have been as familiar with American sources as he was with British ones, for Yale University Press published William E. Hocking's *Morale and Its Enemies* in 1918.
28. See, e.g., V.V. Shelyag et al., eds., *Military Psychology: A Soviet View* (Washington, DC: Government Printing Office, 1972) ; Richard A. Gabriel, *Military Psychiatry: A Comparative Perspective* (New York: Greenwood Press, 1986); and Anthony Kellett, *Combat Motivation: The Behavior of Soldiers in Battle* (London: Kluwer-Nijhoff Publishing Co., 1982).
29. Robert W. Swezey, Andrew L. Meltzer, and Eduardo Salas, " Some Issues Involved in Motivating Teams," in *Motivation:Theory and Research* , ed. Harold F. O'Neil, Jr., and Michael Drillings (Hillsdale, NJ: Erlbaum Associates, 1994), 141; and Josh R. Gerow, *Psychology: An Introduction* (New York: Longman Publishers, 1997), 360.
30. Gardner Lindzey et al., *Psychology* (New York: Worth Publishers, 1978), 481.
31. Gerow, 360.
32. Lindzey, 481.

33. Ibid.
34. Ibid.
35. Abraham H. Maslow, "A Theory of Metamotivation," in *The Healthy Personality, eds.* Hung-Min Chiang and Abraham H. Maslow (New York: D. Van Nostrand Co., 1969), 29.
36. Ibid.
37. Ibid., 35.
38. Abraham H. Maslow, *Religions, Values, and Peak-Experiences* (New York: Penguin Books, 1970), 4.
39. Bill Mauldin, *Up Front* (New York: W.W. Norton, 2000), 133.
40 Field Manual 22-100, *Military Leadership* (Washington, DC: Headquarters, Department of the Army, 1983), 144-145.
41. Ibid., 135, 144.
42. Ibid., 145.
43. Shively, 2.
44. Ibid., 4.
45. Interview with Chaplain (Major) Daniel Wackerhagen, U.S. Army Chaplain Center and School, Ft. Jackson, SC, 15 May 2001; Steven Reiss, "Toward a Comprehensive Assessment of Fundamental Motivation," *Journal of Psychological Assessment* 10,2 (June 1998): 97-106.
46. Interview with Dr. Thomas R. Smith, Executive Director for the National Institute for Healthcare Research, Rockville, MD, 4 May 2001; see also Levin, 3-6; and Phyllis McIntosh, "Faith is Powerful Medicine," *Reader's Digest*, October 1999, 151-155.
47. Quoted from a discussion with Col. Eric B. Schoomaker, M.D., former FORSCOM Surgeon and currently Commander of the 30th Medical Brigade in Germany, 11 May 2001.
48. Levin, 9; Maslow, *The Healthy Personality*, 35.
49. Annette Mahoney et al., "Marriage and the Spiritual Realm: The Role of Proximal and Distal Religious Constructs in Marital Functioning," *Journal of Family Psychology* 13, 3 (1999): 321.
50. Ibid., 325.
51. Bradfield.
52. Chaplain (Colonel) Donald Taylor, FORSCOM Command Chaplain, Memorandum for Record, Analysis of FORSCOM Installation Activity Reports for FY 00, Office of the FORSCOM Chaplain, Ft. McPherson, GA, 2 March 2001.
53. Discussion with Dr. Bruce Bell, Army Research Institute, 11 May 2001, who kindly furnished a number of studies used in this paper.
54. U.S. Army Community and Family Support Center, *1995 Survey of Army Families III* (Alexandria, VA: Army Personnel Survey Office, 1995) 5, question 27.
55. James A. Martin et al., eds., *The Military Family: A Practice Guide for Human Service Providers* (Alexandria, VA: The Army Research Institute, 2000), 143,159.
56. Field Manual 22-100, *Army Leadership: Be, Know, Do,* 19.
57. H.A. DeWeerd, ed., *Selected Speeches and Statements of General of the Army George C. Marshall* (Washington, DC: The Infantry Journal, 1945), 121-122.
58. Ibid., 93.
59. Ibid., 123.
60. John Keegan, *The Face of Battle* (New York: Penguin Books, 1976), 274; FM 22-100, *Army Leadership, Be Know, Do*, 3-3.
61. *Sun Tzu: The Art of War*, 128.
62. Lord Moran, 184.
63. Keegan, 335.

64. Ibid., 335-36.
65. Gabriel , 39.
66. John W. Brinsfield, *Encouraging Faith, Supporting Soldiers: A History of the United States Army Chaplain Corps, 1975-1995* (Washington, DC: Headquarters, Department of the Army, Office of the Chief of Chaplains, 1997), part two, 155-156; Russell F. Weigley, *History of the United States Army* (New York: Macmillan Publishing Co., 1967) 519-20.
67. Keegan, 279, 280, 333.
68. William Manchester, *Goodbye Darkness: A Memoir of the Pacific War* (Boston, MA: Little, Brown, 1980), 391,393.
69. Jonathan Shay, "Trust: Touchstone for a Practical Military Ethos," in *Spirit, Blood, and Treasure: The American Cost of Battle in the 21st Century,* ed. Donald Vandergriff (Novato, CA: Presidio Press, 2001), E-1-2.
70. Samuel A. Stouffer et al., *The American Soldier: Combat and Its Aftermath* (Princeton, NJ: Princeton University Press, 1949), 611-13.
71. Ibid.
72. Ibid., 177.
73. Ibid., 173.
74. Kellett, 194.
75. John W. Brinsfield, "Army Values and Ethics: A Search for Consistency and Relevance," *Parameters* 28, 3 (Autumn 1998) : 79-82.
76. Ibid.
77. Ibid.
78. Kellett, 195.
79. Ellwood C. Nance, *Faith of Our Fighters* (St. Louis, MO: Bethany Press, 1944), 242.
80. John W. Brinsfield, *Encouraging Faith,* part two, 91.
81. Ibid., 123; interview with Chaplain (Colonel) David P. Peterson, Ft. McPherson, GA, 22 May 2001. Approximately an hour after General Schwarzkopf's staff meeting concluded, President George Bush, Vice President Quayle, General Powell, Secretary Cheney, and most of the President's cabinet attended a prayer service at Ft. Meyer, VA, for the same purpose. The service was also attended by the three Chiefs of Chaplains and by Dr. Billy Graham, the guest speaker. Ibid., 125.
82. Department of the Army Pamphlet 600-63-12, *Fit to Win: Spiritual Fitness* (Washington, DC: U.S. Government Printing Office, 1987), 1.
83. As cited in Brinsfield, "Army Values and Ethics: A Search for Consistency and Relevance," 82.
84. Gen. John W. Hendrix, Personal Notes, Memorial Day Prayer Breakfast Welcoming Address, Ft. McPherson, GA, 22 May 2001, 1.
85. Nance, 190-191. Chaplain Nance, a faculty member at the U.S. Army Chaplain School at Harvard University in 1944, recorded many of Gen. George C. Marshall's comments on the soldier's spirit and the type of chaplains he wanted in the Army. Nance's work was later quoted by Daniel B. Jorgensen, an Air Force chaplain, in *The Service of Chaplains to Army Air Units, 1917-1946* (Washington, DC: Office of the Chief of Air Force Chaplains, 1961), 277; and by Robert L. Gushwa, an Army chaplain, in *The Best and Worst of Times: The United States Army Chaplaincy, 1920-1945* (Washington, DC: Office of the Chief of Chaplains, 1977), 186. Evidently Chaplain (Maj. Gen.) William Arnold in Information Bulletins sent to Army chaplains world-wide as early as August 1941 also regularly quoted General Marshall. See Daniel B. Jorgensen, 146, 299.
86. Field Manual 22-100, *Army Leadership: Be Know, Do*, 20, para. 3-82. Inspirational leadership is intended to combine transformational and transactional leadership styles as indicated in FM 22-100.

87. Lt. Gen. John E. Miller and Maj. Kurt C. Reitinger, "Force XXI Battle Command," *Military Review* 75, 4, (July-August 1995): 6-9.
88. Task Force TLS, 16.
89. Miller and Reitinger, 9.
90. Neil Howe and William Strauss, *Millennials Rising: The Next Great Generation* (New York: Vintage Books, 2000), 14.
91. Ibid., 15.
92. Ibid., 16.
93. Wright, 418; Winthrop S. Hudson, *Religion in America* (New York: Charles Scribner's Sons, 1965), 354.
94. Howe and Strauss, 149, 234.
95. Ibid., 164, 234.
96. Ibid., 144, 164.
97. Chris Roberts, "Number of Single-parent Families Increases 42%," *The State* [Columbia, SC], 23 May 2001, A-1, A-12.
98. Howe and Strauss, 197.
99. D.W. Miller, "Programs in Social Work Embrace the Teaching of Spirituality," *Chronicle of Higher Education*, 18 May 2001, A-12.
100. Howe and Strauss, 19.
101. Reported by Channel 3 Television News, Atlanta, GA, 5 May 2001.
102. *Atlanta Journal Constitution*, 13 May 2001.
103. Beth McMurtrie, "Crusading for Christ, Amid Keg Parties and Secularism," *Chronicle of Higher Education,* 18 May 2001, A42.
104. Gilbert W. Fairholm, *Capturing the Heart of Leadership:Spirituality and Community in the New American Workplace* (Westport, CT: Praeger Publishers, 1997), 24-25.
105. Ibid., 1.
106. FM 22-100, *Army Leadership: Be, Know, Do*, 19, para. 3-77.
107. William Ernest Hocking, *Morale and Its Enemies* (New Haven, CT: Yale University Press, 1918), 200.
108. Col. Gregory Fontenot, "Fear God and Dreadnought: Preparing a Unit for Confronting Fear," *Military Review* 75,4 (July-August 1995): 13-24. The account of Colonel Fontenot's preparation for combat with his armor battalion is taken from this article and from his earlier one, "Fright Night: Task Force 2/34 Armor," *Military Review* 73, 1 (January 1993).
109. Fontenot, "Fear God and Dreadnought," 13.
110. Ibid., 13, 14.
111. Ibid.,14,15.
112. Ibid., 18.
113. Ibid., 16.
114. Ibid., 17.
115. U.S. Army FORSCOM Regulation 350-1, 26, para. 3-31, e.
116. Fontenot, "Fear God and Dreadnought," 19.
117. Ibid., 23.
118. Fontenot, " Fright Night: Task Force 2/34 Armor," 39.
119. Ibid.
120. Fontenot, "Fear God and Dreadnought," 24.
121. As cited in Maj. Michael D. Slotnick, "Spiritual Leadership: How Does the Spirit Move You?" Defense Technical Information Center Technical Report AD-A258 523 (Alexandria, VA: DTIC, 1992), 4.

122. A summary of findings from previous references including the U.S. Army Community and Family Support Center, the Army Research Institute, Lord Charles Moran, William E. Hocking, John Keegan, and Gregory Fontenot. For advice to chaplains see Chaplain Milton Haney's "The Duties of a Chaplain" as cited in John W. Brinsfield, "In the Pulpit and in the Trenches," *Civil War Times Illustrated*, September/October 1992, 72-73; and Field Manual 16-1, *Religious Support Doctrine: The Chaplain and the Chaplain Assistant* (Washington, DC: Headquarters, Department of the Army, November 1989), 5-2.
123. Hocking, 49.
124. For assistance in researching and writing this chapter, the author is indebted to Mrs. Teri Newsome and the librarians of the United States Army Chaplain Center and School, the Robert Woodruff Library at Emory University, and the Post Library at Ft. McPherson, GA; as well as to Chaplain Martha Hayes and Chaplain David Colwell at the Office of the Chief of Chaplains; Chaplain Daniel Wackerhagen at the U.S. Army Chaplain Center and School; Chaplain Don Taylor, the Command Chaplain, Headquarters, U.S. Army Forces Command; Chaplain Herb Kitchens, Staff Chaplain, First U.S. Army; Chaplain Kenneth Sampson, the Fifth U.S. Army Chaplain's Office; Chaplain Allen Boatright at the Installation Chaplain's Office, Ft. Stewart, GA; Chaplain Steven Colwell, the Chaplain Instructor, Ft. McCoy, WI; Chaplain Steven Paschall, Third U.S. Army; and those retired chaplains who are cited in the notes; as well as to Dr. Bruce Bell, the Army Research Institute for the Behavioral and Social Sciences; the physicians of the U.S. Army Medical Corps; Ms. Anet Springthorpe, R.N., psychiatric nurse; Dr. Thomas Smith, Director of the National Institute of Healthcare Research; Professor Don Snider of the Department of Social Sciences, USMA; Dr. Daniel Keller, Professor of Philosophy and Ethics, Alabama State University; Mr. David S. Strout of Lakeland, FL; Mrs. Marietta Branson of Atlanta; Mrs. Pat T. Johnson, Chattanooga Baptist Association; Mrs. Debra Yuhas, U.S. Army Forces Command Chaplain's Office; and Gen. John W. Hendrix, Commanding General, U.S. Army Forces Command.

20 The Role of Strategic Leaders for the Future Army Profession[1]

Gregg F. Martin and Jeffrey D. McCausland

Introduction

In the ten years following the end of the Cold War, the Army underwent a period of extreme stress and transition. The force was downsized by roughly one-third and lost the focus formerly provided by the existence of a powerful peer competitor. It was given new missions, particularly in the area of peace support operations and humanitarian assistance as well as expanded requirements to provide aid to civilian authorities for disaster relief. At the same time, the Army endeavored to maintain its preoccupation with future contingencies involving traditional war-fighting. Obviously, the attention of the Army and the nation as a whole shifted dramatically following the terrorist attacks on 11 September 2001. In the coming months and years the American Army must continue to deal with multiple tasks of enormous complexity. As a profession it must maintain its expertise and capabilities in traditional war-fighting in addition to the "other than war" missions of the past decade. The Army must also accept the reality of a new war on terrorism and asymmetric threats to the security of the nation.

But even prior to September 2001 numerous studies pointed to strain on the institution caused by the transition out of the Cold War—the explosion of new missions and operational requirements coupled with a decade-long reduction in resources. Such studies indicated that the very fabric of the institution was being torn apart. In its study of American military culture, the Center for Strategic and International Studies (CSIS) found that there exists a serious "trust gap" between junior members of the Army profession and their senior leaders.[2] According to this study, the gap was caused primarily by a disparity between the declared values of the Army's senior leaders, on one hand, and the implicit values behind their decisions and actions as perceived by their subordinates, on the other hand. Junior officers interpreted this gap as evidence that the senior leadership did not "walk their talk." Later, internal studies and surveys directed by the Army's senior leaders were conducted at the U.S. Army War College (USAWC) and the Command and General Staff College. They arrived at similar conclusions regarding the perceptions of junior officers. For example, in May 2001 the Army released the findings of its Army Training and Leader Development Panel (ATLDP) chartered by the Chief of Staff of the Army. This panel, which largely confirmed the results obtained in the earlier efforts, based its conclusions on surveys of over ten thousand officers.[3]

Given these study results, one must question how a highly skilled institution such as the U.S. Army, filled with so many intelligent, dedicated, and well-meaning people, could find itself in such an unwelcome predicament a decade after two huge successes: a triumphant resolution of the Cold War and victory in the Gulf War. And, more importantly, one must ask what can be done about it. The answer to these questions lies largely in the role of strategic leadership. Who are the Army's strategic leaders? What are their roles and responsibilities? Has the nature of strategic leadership changed, and if so, how? Have the Army's strategic leaders kept pace with rapid changes in their operating environment? A fresh look at these issues, we believe, points to the need to transform the concept of strategic leadership in the Army to one of leading a "profession" rather than leading an "organization."

Army Doctrine and Strategic Leadership

Doctrinally, the Army addresses the subject of "strategic leadership" in Chapter 7 of FM 22-100, *Army Leadership* (August 1999). Strategic leadership is described and analyzed in this manual, which offers many superb insights and historical examples. A second source on the subject is the USAWC text *Strategic Leadership Primer* (1998). This document provides a blueprint for teaching and understanding strategic leadership. It defines what strategic leadership is, explains how strategic leadership differs from lower levels of direct and organizational leadership, explores the more complex environment in which strategic leaders operate, and provides basic knowledge on vision, culture, competencies, and tasks. The primer is a useful document in that it consolidates between two covers numerous valuable insights and understandings for Army leaders that go beyond the institution's basic doctrine.

For today's Army profession, however, these two documents fall short in several ways. First, both focus almost exclusively on the Army as an "organization," and by extension on the role of Army strategic leaders as those who can lead any type of "organization." There is virtually no mention of the Army as a "profession." This omission is the source of a significant disconnect, for the Army is both a profession, as articulated in the June 2001 edition of Field Manual 1, *The Army*,[4] and an organization (in this case a bureaucracy). Professions are unique organizations and as such have unique roles for their strategic leaders. Second, both documents define strategic leaders as being located only at the Army's highest echelons.[5] We believe this interpretation is overly narrow and hence skews both the officer corps' understanding of self and its strategic leadership role within the profession. Particularly in the current setting, we believe strategic leadership of the Army profession is performable by leaders in the ranks of lieutenant colonel and above. The new security environment frequently requires officers at lieutenant colonel level and even below to apply and assess the profession's expert knowledge in situations with strategic implications; provide feedback into the profession's bank of abstract knowledge; and develop and inspire future generations of Army officers. In contrast to the reality of our present leadership environment, Army doctrine as earlier noted defines strategic leadership as applicable only to "leadership at the top of organizations," meaning from Major Commands and Department of the Army level up through the National Command Authority, and others having global, regional, national, and societal effects.[6] In terms of educating colonels at the senior service colleges for the complexities of higher-level staff and command, this rather narrow definition is perhaps understandable. However, the nature of modern professions and changes in both the external and internal environments of the Army require a broader understanding and renewal of how we think about and conceive of strategic leadership within a profession.

The USAWC also uses another foundational document titled *Strategic Art: The New Discipline for 21st Century Leaders*, by Lt. Gen. (Ret.) Richard Chilcoat, former Commandant of the USAWC and President of the National Defense University. He defines strategic art as "the skillful formulation, coordination, and application of ends (objectives), ways (courses of action), and means (supporting resources) to promote and defend national interests."[7] He then posits three roles that the "complete strategist" must perform: "the strategic leader, strategic practitioner, strategic theorist." Those who can competently integrate and combine all three roles earn the label of "master of the strategic art." Moving beyond the construct of the *Strategic Leadership Primer*, he defines the strategic leader as one who "provides vision and focus, capitalizes on command and peer leadership skills, and inspires others to act."[8] Even this comprehensive description says little, however, about leading a profession as opposed to an organization.

Lieutenant General Chilcoat does provide insight into the new phenomenon known as the "strategic corporal." He notes that the "simultaneous revolutions in military affairs, technology, and information, and a reordering of the international system, have shattered traditional boundaries." Frequently this has resulted in a merging of "the tactical, operational, and strategic levels of war into a single, integrated universe in which action at the bottom often has instant and dramatic impacts at all levels." He argues that "never [before] in history have so many strategic burdens confronted the entire chain of command, ranging from the President in the White House all the way down to the individual rifleman at a security checkpoint in Macedonia."[9] Obviously, this concept conflicts with the traditional notion that only those at the top of organizations are strategic leaders or perform actions having strategic import.

The Army Is More Than an Organization—It's a Profession

From our discussion thus far, it is clear that the Army has thought about and written a great deal on strategic leadership in and of generic organizations, but little about strategic leadership in and of a profession. This difference can best be accommodated by a more modern theory of professions recently developed by Dr. Andrew Abbott, a leading sociologist at the University of Chicago and authority on the structures and behaviors of professions.[10]

The traditional view of the Army profession as set forth in the classic treatments by Samuel Huntington (1957) and Morris Janowitz (1960) mostly emphasizes the individual role of the members of the military professions and how they were socialized into that individual role.[11] The earmarks of a profession in this traditional view are:[12]

- *Authority* delegated by society to the profession for performance of a critical social function that society cannot perform by itself.
- *Unique Expertise* in a body of knowledge that requires extensive education and training.
- *Society's Sanction* that creates the moral obligation for professional effectiveness.
- *Limited Autonomy* granted within the society's political structure.
- *Professional Culture Distinct from Society* that embodies corporateness, self-policing, and a regulative code of ethics.
- *Life-long Calling* in which intrinsic satisfactions of service strongly complement extrinsic ones.

These defining characteristics are reflected in the symbols and rituals of the profession. For example, the United States government commissions and promotes its Army officers by name both to control the membership of the profession and to certify the expertise of its members. Army officers, as moral agents of the society, are granted a limited degree of autonomy to set standards and police the profession for the good of the client society.[13]

In contrast to this traditional view, writing in the late 1980s Dr. Abbott describes professions as "somewhat exclusive groups of individuals applying somewhat abstract knowledge to particular cases."[14] He postulates that the evolution of and interrelationships among professions are determined by how well a profession controls its required knowledge and skills: "Practical skill grows out of an abstract system of knowledge, and control of the occupation lies in control of the abstractions that generate the practical techniques."[15] Abbott further argues that unless a knowledge system is governed by abstractions, it cannot redefine its problems and tasks, nor can it defend its "turf" from competitors or assume new tasks that present themselves. Abstract knowledge is what enables professions to survive in the competitive "system of professions." This knowledge system and its attendant degree of abstraction "are the ultimate currency of competition between professions."[16] Clearly one of the most vital tasks of those leading a profession at the strategic level is to tend to the dynamic nature of change that affects the particular tasks it is called upon to perform, as well as the associated knowledge base.

It is also clear that professions demand a precise system of higher education. This system must allow its professionals to master, usually in ascending stages, the appropriate body of abstract knowledge and the techniques for its application. Such an education system must also evolve as requirements of the profession grow and improved techniques are adopted. In explaining the rise and fall of various professions over time, Abbott posits that the key variable can be traced to "the power of the professions' knowledge systems, their abstracting ability to define old problems in new ways."[17] In other words, unless the strategic leaders of the profession tend to the profession's body of expert knowledge and its effective application to new situations and tasks by the members of the profession, they run the risk of competing poorly and declining in standing, or legitimacy, with their client.

Thus, an alternative to the traditional conception of professions is derived from Dr. Abbott's model. It focuses more on the professions themselves and the system of professions in which they exist and compete.[18] The defining characteristics of this concept of professions are:

- ***Unique Expertise*** based on a body of abstract knowledge that can be adapted and applied to various situations.
- ***Jurisdiction*** that defines those situations and conditions where the application of the profession's expertise is legitimate in the eyes of the client. Jurisdiction is established through effective application of the Army's expertise and negotiation between the Army profession and the civilian leaders of the society it serves.
- ***Legitimacy*** as the foundation of jurisdiction. It arises from legal, organizational, or social mechanisms that direct clients to this particular profession for the "treatment" or services that they believe the profession offers.
- ***Competition among Professions and within a System of Professions*** in which they compete for jurisdiction, legitimacy, members, and resources with other professions and nonprofessional organizations.
- ***Professional Death*** for professions that fail to compete effectively or that become overly bureaucratized. Such professions may very well "die," losing their status as a profession.

A New View of Army Strategic Leaders—Leading a 21st Century Profession

The Army's traditional view of strategic leadership of an "organization" can be profitably combined with Abbott's more recent concepts of a profession (see Figure 20-1). As explained in USAWC's *Strategic Leadership Primer*, FM 22-100, and other sources, the Army's strategic leaders must perform the traditional roles of generating a strategic vision and then leading the implementation conceived to fulfill that vision. With regard to the institution's vision, the strategic leader must provide purpose, direction, energy, motivation, inspiration, and a clear professional identity. While these elements are consistent with the traditional approach of Huntington and Janowitz, when viewed in the context of Abbott's model there are new elements to be included in the vision—the profession's expert knowledge, legitimacy, and jurisdictional competitions that ultimately determine its future.

To implement their vision, the strategic leaders of the Army profession must perform three main tasks (Figure 20-1, right side). First, they must shape the professional culture, both functional and ethical. This includes: (1) developing the Army's expert knowledge, which is the essence of any profession; (2) inculcating the ethos of "professional service to the client," which for the Army is defending American society and the nation; and (3) promoting an individual self-concept among the officer corps that, in addition to expertness at war-fighting, embraces dedicated professionalism and membership in the larger Army profession. To promote such a self-concept, the profession's strategic leaders must establish and nur-

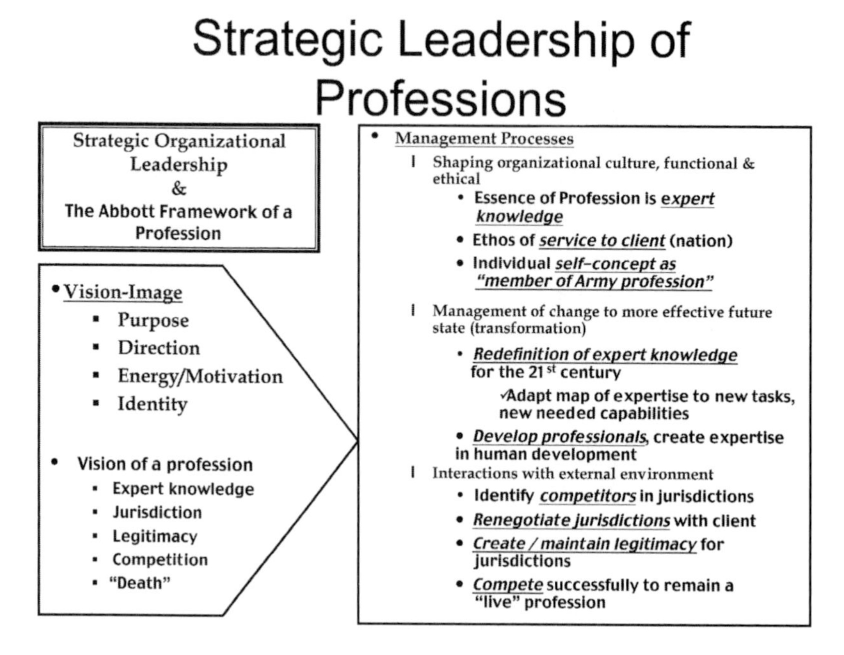

Figure 20-1. *Strategic Leadership of Professions.*[20]

ture a culture which esteems and rewards high degrees of professionalism, such as self-sacrificial service, functional competence, and a life-long commitment to professional education and development. The culture should also embody the laudable norms and values of "power-down" or decentralized leadership, initiative, and candor up the chain of command. These must be routinely practiced and rewarded throughout the profession, as opposed to being reduced to mere pious platitudes as is the case today.[19]

The second task of strategic leaders (Figure 20-1) is to bring the profession to a more effective future state. This is accomplished through the design of new policies and their implementation through the Army's management systems and its major commands. The efforts of the current Chief of Staff Gen. Eric Shinseki to transform the Army into a lighter, more strategically deployable combat force is an example of one such effort using all of the Army management and command structures. We will discuss this initiative more in the final section of this chapter.

The third task for strategic leaders (Figure 20-1) is to lead the profession's interactions with its external environment, to include all of those stimuli from American society, its political institutions, and industrial base that constantly impinge on the Army as an American institution. One very critical aspect of these interactions is the responsibility for managing the profession's competitions with other professions and vocational groups who also want to provide similar or complementary services to the client in the same context. Strategic leaders of the Army do this by recognizing the emergence of newly mandated tasks—those tasks whose time has come—and recognizing as well the identity of new competitors within the four external jurisdictions in which the Army competes (see Figure 1-1, Chapter 1). In managing those competitions, the Army's strategic leaders must constantly negotiate for legitimacy to perform the new tasks, often adapting the Army's operational capabilities to fortify their negotiations. As noted earlier, a current, positive example of this is the Army transformation program, which in terms of the Abbott construct is the Army's attempt to renegotiate legitimacy in the conventional war jurisdiction after "losing" some of that legitimacy due to perceived lack of rapid strategic mobility, particularly during the Kosovo operation of 1999.

It is in these external jurisdictions or "workplaces"—the areas in which the profession's expertise is applied—that the complexity of strategic leadership is most pronounced. If there is too much work, as some argue is the case for the Army today in the new security environment, then other professions or even non-professionals will typically move into the jurisdictions to do the required work. When this happens, boundaries between professions tend to blur or disappear.[21] This has happened throughout the last decade in the area of peace support operations and military operations other than war (MOOTW). Government agencies, nongovernmental organizations (NGOs), private contractors, other military services, and coalitions of these actors are competitively participating to varying degrees in these different operations. The aftermath of the terrorist attacks on 11 September 2001 presents another jurisdictional challenge—homeland security—that the Army must address owing to clear demands from society.

Pain relief, a branch of the medical jurisdiction, provides a good example of the unique role of strategic leaders of professions—managing competitions for the long-term legitimacy of the profession. For many years American physicians normally prescribed drugs to reduce patients' pain. In the context of the rise of "alternative medicines" in the 1980s and 1990s, patients began to demand alternative treatments for pain since drugs were not doing a sufficiently effective job. More specifically, patients went outside the medical profession to acupuncturists and herbalists. Over time the medical profession recognized this threat to their jurisdiction and decided to "compete" by bringing acupuncturists and their proven expertise in pain management formally into the medical profession, a step accomplished through licensing and regulation. The profession responded similarly in the 1990s when women began to demand the services of midwives during childbirth. Midwifery is now a licensed practice of the medical profession in many states. The opposite has occurred with laboratory technicians, who have been essentially deprofessionalized. The strategic leaders of the medical profession determined that the nature of lab technicians' work had become routinized by computer-generated analyses, thus no longer requiring the application of expert knowledge, extensive schooling, etc. that are inherent to a profession.[22]

The Army has taken actions similar to those of the medical profession throughout its own history. In its transition from the Cold War, certain tasks were jettisoned and the Army withdrew from future competition in these jurisdictions. For example, the Army no longer has the nuclear mission that began in 1956-1957 with the establishment of Pentomic divisions, which required enormous resources, manpower, and training. In addition, the associated analysis of nuclear operations, strategy, and deterrence disappeared from the Army education systems. The current Secretary of Defense, Donald Rumsfeld, has clearly articulated his desire to move the services even further in the direction of divesting themselves of non-war-fighting and irrelevant functions.[23] Still, whatever noncore missions are outsourced or privatized, care must be taken that this does not occur at the expense of the internal or external effectiveness of the Army as a profession.

While managing the profession's competition in its external jurisdictions, the Army's strategic leaders must also simultaneously attend two of its most vital internal jurisdictions (see Figure 1-1, Chapter 1)—developing and adapting the profession's expert knowledge and developing the members of the profession in order that such expert knowledge can be applied in the form of professional practice. These are essential tasks suggesting that strategic leaders must also exercise the fundamentals of face-to-face leadership with those people they touch directly, and set the conditions to assure that this is also done throughout the profession. The studies cited in the introduction to this chapter all indicate that individual professional development, mentoring of juniors by seniors, etc. are sorely deficient in the Army and are major causes of the discontent contributing to the unprecedented voluntary exodus of both junior and senior officers from today's Army.[24] Under no circumstances, however, should "strategic leadership" be construed as "strategic control." Junior officers must be given maximum flexibility to train and lead their units, make mistakes, and learn.

Like all professions, the Army must be cognizant of the needs and desires of its clients when considering its external environment. If the Army fails to provide the desired services, its clients will eventually seek alternative providers. Obviously there are not multiple armies within America competing to perform military work for society, but there are the other uniformed services, private contractors, and nongovernmental organizations (NGOs) which provide arguably satisfactory alternatives. Historically, the Executive or the Congress has arbitrated or settled jurisdictional contests between the Army and its competitors over new jurisdictions. Such outcomes, however, have at times tended to restrain the professional relevance and autonomy of the Army. In fact, when the Army has not paid close attention to the needs and desires of its client, such as during the Clinton administration, it saw the Executive become actively involved in dictating how the Army conducted itself and its professional practices, e.g., the conduct of basic training for its volunteer recruits to insure multi-gender training. The Army should resist such intrusion into its professional affairs, but its strategic leaders must recognize that they can best preclude such intrusions by close attention to their client and by their own deft management of jurisdictional competitions. Even dominant professions can be only as powerful as their clients allow them to be.[25]

When competing within a jurisdiction, a profession cannot achieve legitimacy over a new task without some potential risk of weakening its legitimacy in other jurisdictions.[26] Viewed through Abbott's theoretical lens, it is clear that the Army's reluctance to accept the non-war-fighting missions that were thrust upon it since the end of the Cold War reflects a concern that focusing on the new jurisdiction will dilute or weaken its professional expertise in what it considers its core function—high-intensity conventional warfare. The role of strategic leaders thus encompasses much more than simply managing single competitions. Balance over time across all applicable jurisdictions is the complex goal for the profession.

During periods of turmoil and intense competition, strategic leaders of some professions have followed a pattern of regression in which they withdrew to the core tasks deemed most important, or most comfortable and professionally rewarding.[27] As indicated in the preceding paragraph, throughout the last decade the Army has continued to focus most of its effort and professional expertise on the task it believes is most important, i.e., high-intensity conventional warfare. In so doing, the Army's strategic leadership has generally remained true to the belief that all other missions are "lesser included cases" that can be trained for when necessary.

We strongly believe that the fundamental purpose of the Army is to be prepared to fight and win the nation's wars (in the full context of modern 21st century warfare), and, by extension, that its professionals must be grounded first and foremost in their war-fighting skills and expertise. In addition, we believe that in the coming months and years, Army strategic leaders should reconsider the basic premise that everything below high-intensity warfare is best treated as a "lesser included case." Although the premise properly provides valuable professional and organizational focus for much of the force, thereby helping protect against the possibility of failure in high-intensity warfare, we recommend that strategic leaders manage the profession's competition within the four external jurisdictions such that the Army retains full legitimacy in each of the four jurisdictions—high-intensity conventional warfare, unconventional warfare directed against global terrorism, non-traditional MOOTW missions, and homeland security. This latter jurisdiction, homeland security, has already generated enormous new demands, particularly on reserve component forces, that are likely to continue for an extended period of time. We believe that each of these four external jurisdictions, which the client clearly wants serviced, deserve the best in terms of forces, training, and resources. This will require the Army to negotiate with its client for some combination of more resources and/or fewer missions. The point is that the nation expects full service in each of these external jurisdictions and is looking to the Army to meet those expectations. We believe our strategic leaders owe the Army's client and the profession a candid and realistic assessment of how best to address assigned missions while balancing internal jurisdictions and other vital requirements such as personnel well-being. It is critical to both the profession and the nation that an open debate take place over problematic jurisdictions and boundaries, and, by extension, forces and resources needed to meet the nation's newly enlarged strategy.

Unlike many professions—where the highest status within the profession itself is accorded to those most directly involved in the research, theorization, formulation, development, expounding, or teaching of the profession's abstract, formal knowledge system—the Army accords highest status to its practitioners.[28] In fact, the Army has too often maintained this stance even to the detriment of advancing the base of professional abstract knowledge upon which the success of its entire practice depends. Rather than holding in the highest esteem those who develop, experiment with, and then implement and teach its professional doctrines, the Army has historically tended to push them aside, neither rewarding military theorists and academics, nor recognizing their importance to the Army profession. In recent times the Army has held professional practitioners in the field Army in far higher esteem than those professionals who work exclusively, or even mostly, in the world of military doctrines and associated academic knowledge. While current operations and operators must remain in the forefront, there should be a proper balance between performing today and making ready to perform in the future. If the Army hopes to retain its brightest thinkers and harness the power of new ideas and new ways of thinking about complex security issues, it might well consider reassessing this balance.

Army Strategic Leaders...of a Profession... or of a Bureaucracy?

As the discussion thus far has shown, the changed security environment has disrupted the Army's professional equilibrium. The Army has been directed to perform a significant array of new tasks while simultaneously reducing its budget and personnel levels. Generally, this has occurred without sufficient analysis or public discussion of the costs and requirements of these expanded operations—few political elections in recent years even generated a debate on defense issues. The Army has clung to what it perceives as its core professional function—high-intensity conventional warfare—while actively engaging in a wide variety of new missions. Still, as Don Snider and others have shown, all of this has come at great cost to the profession.[29] One of those costs is increased bureaucratization.

Organization professions are those that have a dual nature as both profession and bureaucracy. In many ways the organization's day-to-day climate, and perhaps even its culture, is derivative of how well this duality is managed. As a government organization, the Army is a hierarchical bureaucracy. It has defined positions and responsibilities; extensive rules and regulations that guide members in the accomplishment of routine tasks; and a personnel system with bottom-only entry and fixed career paths. Efficiency in routine tasks, rather than effectiveness, is the primary goal of any bureaucracy, and in this regard the Army is no exception. Obviously, this principle collides with the character of the Army as a profession, in which functional effectiveness is the primary goal.

While there will always be a healthy tension between the Army's dual nature as a profession and bureaucracy, it is important for strategic leaders to strive for both efficiency and effectiveness. It is imperative, however, for the nation and the Army as a profession that the bureaucratic mandate and the professional mandate not be emphasized equally. For a military profession to exist and flourish over time, functional effectiveness must dominate the institution's culture and its behavior, both at the level of the professional ethos and at the level of the individual member of the profession. Evidence from the field, backed up by the studies cited earlier, indicates that over the past decade the Army has behaved increasingly as a bureaucracy in the pursuit of organizational efficiency. This shift has come at the expense of the Army's professional character and the pursuit of functional effectiveness. The unprecedented voluntary exodus of both junior captains and senior lieutenant colonels and colonels from the profession seems to confirm this point. How individual Army officers conceive of themselves individually, and of the Army itself—whether primarily as obedient bureaucrats striving for bureaucratic efficiency, or principally as professionals striving for functional competence and effectiveness—will determine the long-term success of the Army profession.[30]

To us this point is critical because professions offer two unique services to the American people that bureaucratic organizations do not—expert knowledge and professional behavior. Professions create and expand bodies of expert knowledge. Bureaucracies apply this knowledge, but they do not normally develop it. Secondly, professions are service-oriented, making paramount both adherence to standards of competence and loyalty to clients' needs. Professions also develop their own codes of ethics in order to prevent practitioners from taking advantage of clients. In exchange for this ethical and trustworthy behavior, society grants professions a high degree of collective autonomy.[31] In the particular case of the Army profession, it has developed and maintained a code of ethics binding members to subordinate themselves to and selflessly serve the American people, and molding individual members into coherent groups that function under morally ambiguous circumstances, to include even war. Bureaucratic controls, such as promotion and monetary rewards, have only limited ability to engender such professional behavior or to control members in such situations.[32]

In a letter from then Maj. Gen. George Patton to his classmates back in the states, Patton described the chaos and horror of his division's early battles in North Africa. He made the point that the American Army did not turn and run from the Germans—even though every shred of logic would compel one to do so—because "we were more afraid of our own consciences than we were of the Germans." Only professions—not bureaucracies—can instill that degree of discipline, which is needed for an Army to succeed in battle. In short, it is critical for the Army and for America that officers conceive of themselves and the Army primarily as professionals and a profession. Then the Army will be able to develop the military doctrine and expertise and the degree of discipline ensuring that officers and soldiers make the right choices during the heat of battle and the often complex and morally ambiguous circumstances they face during peacetime. According to the studies previously cited, much of the Army's recent thinking, culture, and behavior have become more reflective of a bureaucratic organization than of a profession.[33] To the extent that these perceptions from the field are accurate, and we think they are, then strategic leaders must rapidly take appropriate steps to reverse this trend.

Strategic Leaders for the 21st Century

> It's our duty to develop soldiers and leaders who have the skills necessary to succeed today and in the future.
>
> GEN. ERIK K. SHINSEKI, ARMY CHIEF OF STAFF[34]

Since 11 September 2001, the nation has been at war with a new foe, and the challenges confronting its Army are more complex and varied than at any time in its history. This open-ended war against terrorism occurs at a time when a profound sense of unease besets the profession, reflected, for example, in the loss of both experienced leaders and outstanding junior officers who would be the profession's future. The demands of the global war on terrorism cannot allow the profession to ignore these difficulties; rather, it should spur a greater sense of urgency in dealing with them. Several recommendations seem appropriate for strategic leaders.

First, the Army must renew its institutional identity as a profession, as opposed to merely a bureaucratic organization. The recent predominance of efficiency-focused bureaucratic thinking and behavior, at the expense of professional effectiveness, is the source of many challenges currently facing the Army. A clear and credible identity as a profession must be the foundation of any future renewal of the Army profession. All of our leadership doctrine and education should be framed within the overarching construct of the Army as a modern 21st-century profession operating within a system of competing professions.[35]

Second, the Army should expand its definition of "strategic leader" beyond the narrow circle of high-level general officers and their staffs working at the national and international levels. It should recognize who within the profession are in fact strategic leaders, and expand the definition accordingly. We believe that this concept and self-image must reach down to the level of lieutenant colonel, in particular to battalion commanders. The reason for this belief is that although their actions may be mostly at the tactical level, battalion commanders are perhaps the most important leaders *within* the profession, particularly as relates to the internal jurisdiction of developing and inspiring future members of the profession.

For not only are they the keepers and practitioners of the Army's professional expertise at the point of application, and hence responsible for assessing the state of abstract knowledge and providing appropriate feedback to the rest of the profession, they are also *the* senior leaders for the vast majority of the aspiring young professionals, i.e., the Army's lieutenants and captains. As such, they are critical actors in establishing and maintaining a climate in which young professionals may grow, develop, and be inspired to continue their service in the Army profession. Though colonels as brigade commanders as well as general officers commanding higher echelons must also set and sustain proper conditions for a positive developmental command climate within the Army's battalions, the professional role of lieutenant colonels in doing this right is even more critical. In addition to a doctrinal redefinition of the profession's strategic leaders so as to include lieutenant colonels and the grades immediately above, the Army must recognize the strategic role played by majors, captains, and lieutenants, both in terms of upward effects (i.e., the "strategic corporal") and in terms of internal effects (i.e., their obligation to develop and inspire future professionals and leaders in the profession who fall under their charge).

Third, the Army should transform strategic leader education for the 21st-century Army profession. Given the major directional shift and expansion of the Army's professional jurisdiction since the end of the Cold War, one might expect to see a corresponding shift and expansion of its professional knowledge base and in its practical application of that expert knowledge to its new missions.[36] For the most part, however, this has not yet happened. Therefore the Army must expand the base of professional knowledge imparted to its officer corps in order to bring it into closer alignment with expected performance.[37] This should be done in accordance with the Chief of Staff's intent to transform the Army and its most precious resource—people.[38] The profession must acknowledge that Army transformation includes four components: (1) equipment, (2) organization or force structure, (3) doctrine, and (4) development of leaders with the intellect to meet professional demands in the transformed force. Clearly the last two are the most difficult portions of the effort owing to their level of abstraction, but the desire by many in the Army leadership to accelerate the transformation in light of the war on terrorism must give this renewed impetus.

Fourth and finally, the Army should have an overarching intellectual and ethical framework that stresses the self-concept of officer, and leader within a profession, as opposed to the self-concept of obedient functionary within a bureaucracy. Strategic leaders must stress the obligation that *every* leader has—regardless of rank or position—to develop and inspire junior members of the profession. Every member of the profession must understand that this vital function has long-term, *strategic* consequences for the profession and its continued vitality as a servant to the nation.

Notes

1. The views and opinions expressed in this chapter are those of the authors and are not necessarily those of the Department of the Army or any other U.S. government entity.
2. Walter F. Ulmer, Jr., Joseph J. Collins, and T.O. Jacobs, *American Military Culture in the 21st Century* (Washington, DC: Center for Strategic and International Studies, February 2000).
3. See, for example, the following: Walter F. Ulmer, Jr., Joseph J. Collins, and T.O. Jacobs, *The Army Training and Leader Development Panel Officer Study Report to the Army* (Washington, DC: Department of the Army, May 2001); Leonard Wong, *Generations Apart: Xers and Boomers in the Officer Corps* (Carlisle Barracks, PA: U.S. Army War College, Strategic Studies Institute, October 2000); William M. Steele and Robert P. Walters, "Training and Developing Army Leaders," *Military Review* 81 (July-August 2001): 2-9, and "Training and Developing Leaders in a Transforming Army," *Military Review* 81 (September-October 2001): 2-11. The results of these studies were also reinforced in numerous discussions with USAWC students and faculty over the past two years.
4. FM 1, *The Army* (Washington, DC: Department of the Army, 2001).
5. FM 22-100, *Army Leadership* (Washington, DC: Department of the Army, 1999), 7-1; and Roderick R. Magee II, ed., *Strategic Leader Primer* (Carlisle Barracks, PA: U.S. Army War College, 1998), v. The full definition goes on to say, "Strategic leaders are the group that ascends to the top of the organization where indirect leadership is the norm and there is increased uncertainty and complexity."
6. See Magee, v-vi.
7. Richard A. Chilcoat, *Strategic Art: The New Discipline for 21st Century Leaders* (Carlisle Barracks, PA: U.S. Army War College, Strategic Studies Institute, 1995), iii.
8. Ibid., iii-iv.
9. This concept and the quotations in this paragraph are from Chilcoat, 2.
10. See Andrew Abbott, *The System of Professions: An Essay on the Division of Expert Labor* (Chicago, IL: University of Chicago Press, 1988).

11. See Samuel Huntington, *The Soldier and the State* (Cambridge, MA: Harvard University Press, 1959); and Morris Janowitz, *The Professional Soldier: A Social and Political Portrait* (Glencoe, NY: The Free Press, 1960).
12. This discussion of Huntington's *The Soldier and the State* and Janowitz's *The Professional Soldier* are adapted from Dr. Don Snider's lecture, briefing slides, and discussions at the U.S. Army War College, 9-10 August 2001.
13. Ibid.
14. Abbott, 318.
15. Ibid., 8.
16. Ibid., 9.
17. Ibid., 30.
18. Ibid., 86-114.
19. See for example the sources listed in n. 3. The results of these studies were also reinforced in numerous discussions with USAWC students and faculty over the past two years.
20. Figure 20-1 is adapted from Dr. Don Snider's lecture and briefing slides presented at the Army War College, 9 August 2001.
21. Abbott, 66.
22. These examples from the medical profession were gleaned from Dr. Don Snider's lecture on 9 August 2001 at the U.S. Army War College.
23. Secretary of Defense Donald H. Rumsfeld, "Bureaucracy to Battlefield" (remarks by the Secretary of Defense, The Pentagon, 10 September 2001).
24. In addition to the sources cited in n. 3 (particularly Wong, *Generations Apart*), see Thomas E. Ricks, "Younger Officers Quit Army at Faster Clip; Study Finds Little Trust in Senior Leadership," *Washington Post*, 17 April 2000, A17.
25. Abbott, 140-141.
26. Ibid., 91.
27. See Abbott for numerous examples, passim.
28. Ibid., 118-119.
29. Don Snider and Gayle Watkins, "The Future of Army Professionalism: A Need for Renewal and Redefinition," *Parameters* 30 (Autumn 2000): 8-9.
30. The discussion in this paragraph is in large part gleaned and synthesized from information presented by Dr. Don Snider in the form of a lecture, briefing slides, and discussions at the Army War College, 9-10 August 2001.
31. Taken from Lloyd J. Matthews, "Is the Military Profession Legitimate?" *ARMY*, January 1994, 22-23.
32. See n. 30.
33. Ibid.
34. Gen. Eric K. Shinseki, "CSA Keynote Address" (remarks presented at the AUSA Eisenhower Luncheon, Washington, DC, 17 October 2000).
35. See Abbott, along with Snider and Watkins.
36. Snider and Watkins, 5-6.
37. See Jeffrey D. McCausland and Gregg F. Martin, "Transforming Strategic Leader Education for the 21st Century Army," *Parameters* 31 (Autumn 2001): 17-33.
38. Gen. Eric K. Shinseki, "CSA Keynote Address."

VI

Internal Expertise and Institutional Management[1]

We continue our investigation of the Army's internal jurisdictions in this section. The expert knowledge addressed in these chapters focuses on the Army's management of its human resources and, in particular, how evolving research is (or is not) incorporated into the Army's management systems. The authors of these chapters investigated unit cohesion, diversity, well-being, physical training, and leader development.

The last three of these topics are analyzed through case studies that tell of the Army's recent professional activities regarding these important jurisdictions (Chapter 22). As we saw in Section II, outsourcing and privatization have hindered the Army's control over internal jurisdictions; and we see in Section VI that the Army's approach to research and technologies also hinders such control. Research relating to human behavior enjoys a much lower priority than other sciences with the result that the Army often relies on out-of-date research in the development of its human resource management systems.

Thus, this section's chapters highlight an important way that the Army can maintain control over its internal jurisdictions—through the development and use of current scientific research. The chapters on cohesion (Chapter 21) and diversity (Chapter 23) indicate that although contemporary expertise on human resource management exists within the civilian world, the Army is not incorporating and using it as effectively as it could. The relationship between research and management systems is further investigated in this section's three case studies mentioned above. These cases offer real-life insight into how the Army is attempting to resolve its pressing human resource concerns. In at least two of these cases, it is apparent that the Army has already started revitalizing in key areas of its professional knowledge.

Note

1. The authors express their appreciation to Capt. Elizabeth L. Tolle of the USMA Department of Social Sciences, who served as Panel Rapporteur for this section's topic during the USMA Senior Conference, June 2001.

21 Professional Closure in the Military Labor Market: A Critique of Pure Cohesion[1]

David R. Segal and Meyer Kestnbaum

Sociological research on how people come to occupy particular niches in the occupational structure has focused primarily on the ascribed and achieved individual characteristics that people bring to the labor market: race, gender, level of education, etc., and how these characteristics influence labor market behavior. In the status attainment tradition, for example, Peter Blau and Otis Dudley Duncan were concerned with "how individuals . . . locate themselves in . . . the system."[2] More recently, attention has been paid to the structural characteristics of labor markets. For example, Bradford Booth et al. have found that in labor market areas in which there is a relatively high level of military presence, unemployment among women increases, and income is depressed among those women who are employed.[3]

Far less attention has been paid to the role that the professions and other occupational groups play in limiting the access of people who have (or lack) particular characteristics to employment in those occupations. Such an approach recognizes the importance of individual attributes, but focuses less upon their influence on individual market behavior, and more upon collective action by occupational groups in restricting access to employment on the basis of such attributes. This kind of collective action allows occupational groups to reproduce themselves without significant change in their composition. However, as we shall note below, when occupational groups face external crises, the closure or exclusionary rules may change. In times of crisis, processes of reproduction are disrupted, "existing rules of the game are brought into question, and identities and interests become more ambiguous."[4]

This chapter discusses the use of the sociological concept of cohesion by the American military to continue to exclude from military service or limit the service of ascriptively-defined groups whose service has been limited or excluded in the past. That such ascriptively-based exclusion from military service operates through the legislative, executive, and judicial agencies of the state serves to legitimize closure, and makes the analysis of cohesion as a rationale for exclusion all the more important. The United States, of course, is not alone in using cohesion as a principle of professional closure in the armed forces. Similar arguments have been made by the British Ministry of Defense.[5] However, the accumulated evidence, while suggestive at an earlier point in history, does not justify this exclusion. Moreover, such closure may deprive the armed forces and the nation of the service of qualified personnel. The actions of the military in restricting access has paralleled actions by other professions. The historical trend, however, has been for ascriptive characteristics to be replaced by individual merit as criteria for admission to the professions. And in the American military, personnel shortfalls have historically led to changes in closure criteria.

The concept of cohesion also offers a point of entry into a somewhat wider set of issues concerning the armed forces as a profession. The practice of social exclusion is rooted in a body of social science research conducted during World War II that has come to be embraced as part of the "expert knowledge" of the military profession itself. More recent research, however, requires a recasting of our understanding of cohesion, highlighting two distinct challenges for the armed forces as they consider their future. First and foremost, the analysis of cohesion underscores that expert knowledge—particularly expert knowledge rooted in social scientific research—cannot be regarded by the profession as static. Rather, it is likely to change as increasingly sophisticated research methods and analytic tools produce increasingly reliable and compelling findings. Second, insofar as expert knowledge forms the foundation on which claims to jurisdiction and competition for scarce resources are undertaken by professions,[6] current social scientific challenges to the use of cohesion as a principle of closure highlight the opportunity as well as the need for the armed forces to adjust their stock of expert knowledge in order to retain their distinctiveness and robustness as a profession.

Closure Theory

Max Weber posited a general hypothesis of exclusion, or closure, whereby group boundaries are drawn, the distinctiveness of the group is affirmed, and the rewards and advantages of group membership are maximized by closing off opportunities to members of other groups.[7] Occupational exclusion or closure is a special case of this more general process. Both are predicated on the underlying tension between the social composition of the group and the maintenance of its claim to distinctiveness, revealed in the principle according to which the group limits its membership. Closure for Weber is assumed to be motivated by a desire to monopolize or at least control assets that are socially defined as valuable, including the status of the group itself. It is a zero-sum game, whereby one group gains access to rewards at the expense of another. And it is an ongoing process, with excluded groups mobilizing resources toward the goal of inclusion, and included groups mobilizing resources to protect the advantages they have gained.[8] Kim Weeden suggests that in general, closure restricts the supply of labor available in an occupation, increases demand for its product or service, consolidates an occupation's claim to monopoly over that service, and signals the quality of the service it provides, thus justifying receipt of high rewards.[9]

In the case of occupational closure, it is reasonable to expect that it will be most pronounced in those occupations that are highly rewarded in society, and are most capable of collective action. Furthermore, this tends to be self-reinforcing, insofar as successful closure secures the advantages of group membership and at the same time creates the organizational basis for mobilization around continued exclusion. It is not surprising, therefore, that most research on occupational closure has focused on the professions. Professional occupations are privileged by society, and are rewarded by high incomes, high social status, and high levels of occupational autonomy. They establish control over access to membership in the profession through establishment of educational requirements, examinations, certification, and licensure, and maintain closure through requirements for continuing certification.

Much of the research that has been done on occupational closure (and on occupations in general)[10] has in fact focused on the professions. Noel Parry and Jose Parry, for example, view closure as an element in the social mobility of the medical profession,[11] while Geoff Esland and Graeme Salaman view closure as a means of maintaining the privileged position of the professions in the class system.[12] Weeden, by contrast, assumes that the general processes through which the professions have maintained their boundaries and their privileged place in the distribution of social rewards might characterize other occupations that are organized to act collectively, with different closure strategies appearing in different areas of the occupational structure, e.g., craft occupations relying on unionization rather than the licensing and educational requirements of the professions.[13]

Weeden's research on closure is the most far-reaching in terms of the range of occupations considered. However, it explicitly excludes the armed forces.[14] This is consistent with much occupational research in the United States, which early in the twentieth century adopted the U.S. Census Bureau's convention of defining military personnel as part of the "institutionalized population" (along with people in hospitals, universities, and prisons) rather than the labor force. This convention was justified at least in part by the fact that many of our military personnel were conscripts (although we had an all-volunteer military force between the two world wars). However, in 1973, the United States abolished military conscription and elected to compete in the labor market for military personnel,[15] and by 1980 military personnel were defined as part of the labor force in the decennial census.

Sociological research on the occupational structure did not follow suit. Military personnel continued to be excluded from most research on social stratification, occupational mobility, and labor markets. As Booth points out, this is equivalent to excluding an entire industry, the labor force of which is disproportionately male, and disproportionately composed of racial and ethnic minorities.[16] This has led to misspecification of our models of occupational structure and process. This chapter seeks to include the military in sociological analysis of a general professional and occupational process.

Interestingly, the general research literature on the professions likewise tends to exclude the military, and to focus on civilian occupations such as medicine, law, and the clergy. There is a parallel literature on the military profession, rooted largely in the classic works of Samuel Huntington and Morris Janowitz.[17] This literature in fact addresses the characteristics of people in the profession.[18] However, it makes the assumption that the distribution of these characteristics in the profession is primarily the result of processes of *self*-selection, an assumption supported by recent research. In this literature, the alternative to self-selection most commonly considered is organizational socialization.[19] It does not address actions taken by the military to restrict access to the profession.

In the 1950s and 1960s, when the classic foundational works on military professionalism were written by Huntington and Janowitz, social science analysis of the professions emphasized the selfless service dimension of professionalism. In the 1970s and 1980s, however, a more critical perspective on the professions emerged, emphasizing the more self-serving elements of those occupations that were privileged by society with the recognition of professional status. Like earlier work in the sociology of professions, this critical perspective was reflected primarily in analyses of civilian professions such as medicine and the law. Analyses of military professionalism, however, did adopt this critical posture as well.[20]

One of the foci of criticism has been professional closure—the degree to which professions have excluded people on the basis of ascriptive characteristics that have not been demonstrated to be related to the performance of professional duties. This closure maximizes the rewards received by groups with professional status by closing off opportunities to other groups. A wide range of closure strategies has been developed.

Whatever the process used, closure necessarily and selectively reduces the supply of people who are available to perform the tasks required of an occupation.[21] By this means, as Weeden notes, closure indicates to the clients of the occupation that "those providing the service are well-trained and hence deserving of high rewards."[22] The tasks that the occupation performs are framed as complex and crucial to public welfare, deserving of the status and rewards with which society honors the professions. Through careful selection of members, an occupation maintains its prestige and "the consequent opportunities to enjoy honor and profit."[23] Selection can be based on the ascriptive characteristics or on the achievements of individuals who aspire to membership in the occupation.

The process of occupational closure assumes that occupational groups are capable of collective action. Thus, the ability to work cooperatively—what we will refer to below as task cohesiveness—becomes a resource in the closure process.[24] Another form of cohesion—social cohesion—may result from educational credentialing, which, as Randall Collins notes, reflects common experiences and backgrounds as much as it does knowledge and skill.[25] These commonalities of background and experience increase social cohesion, control internal fragmentation and conflict, and serve as a foundation for the mobilization of the occupational group in support of closure processes.

Processes of exclusion or closure vary among occupations and across time,[26] suggesting that two important characteristics of an occupation's closure strategies are whether or not they are enforced by the state, and whether they emphasize individual or collective rules of exclusion. Collective rules of exclusion are generally based on membership in groups defined by ascriptive or other criteria, while individualist rules are based on what individuals have (or have not) achieved. This distinction becomes especially significant when we consider the state's involvement. State-enforced exclusions from an occupation can be manifested in executive (including military), legislative, or judicial action. All have been relevant in the case of the armed forces. Closure strategies that are backed by the state may through such legal restrictions elevate public evaluations of the occupation in question and the quality of service being offered, depending in part on the legitimacy accorded exclusion of the groups in question outside of the state. The role of the state in military closure processes is particularly important because of the relationship between military service and citizenship in modern democratic states. By enforcing the inclusion (or exclusion) of classes of people regarding their eligibility for military service, the state also asserts their inclusion (or exclusion) with respect to membership in the national citizenry.[27]

Social Cohesion as a Justification for Military Closure

During the last half-century, the American military has attempted to exclude from service, or limit the military service of, three ascriptively-defined groups: African-Americans, women, and homosexuals. In all three cases, one of the major closure strategies used was to apply the concept of cohesion, and to argue (1) that in military forces, unit cohesion is essential for military effectiveness; and (2) that inclusion of previously excluded groups would adversely affect cohesion. However, there has never been a convincing empirical demonstration either of a causal relationship between cohesion and performance, or of the allegedly negative impact of increased social heterogeneity on cohesion. Moreover, when faced with shortages of personnel, the military has relaxed the constraints of closure, with no observable impact on either cohesion or performance.

The continued use of the concept of cohesion as an exclusionary device has questionable bases in the research that has been used as its foundation. The primary bases for arguing the importance of cohesion are three pieces of World War II research: Edward Shils and Morris Janowitz's study of the *Wehrmacht*;[28] surveys of American soldiers conducted by Samuel Stouffer et al.;[29] and interviews with soldiers in combat units interviewed by S. L. A. Marshall.[30] This research was seminal in the 1940s, but crude by contemporary standards. However, a romantic mythology has grown up around these studies, leading people to suspend critical judgment regarding their methods, incorrectly recall their findings, and overlook subsequent research that has suggested limits on their generalizability.

Shils and Janowitz, for example, clearly argued for the importance of primary group ties in sustaining German soldiers in combat up until 1945, when unit manning practices changed.[31] However, their data base was information collected through interrogation of German prisoners of war by American military personnel. This was not necessarily the most objective or reliable source of social science data, particularly since German soldiers could cite primary group bonding rather than Nazi party loyalty as their justification for fighting.[32] Subsequent research on the *Wehrmacht* is not all consistent with their interpretation. Martin van Creveld agrees with their emphasis on the primary group as a major basis sustaining *Wehrmacht* success.[33] Omer Bartov, however, argues that primary groups in the *Wehrmacht* were effectively destroyed as a result of heavy casualties on the Eastern front.[34] Despite this, however, German units hung together, in Bartov's view, because their commitment to an ideologically defined Nazi regime was reinforced by the brutalization of German soldiers at the hands of their own officers. R. S. Rush has more recently argued that the German Seventh Army on the Siegfried Line in 1944 was not managed according to the cohesion-building principles cited by van Creveld, and was not characterized by the cohesion that Shils and Janowitz suggest was found in the *Wehrmacht* into 1945.[35] Rather, he suggests, along the lines of Bartov, that the continued participation of German soldiers in a war that was clearly lost was based on the willingness of the German high command in 1944 to intimidate soldiers into continuing the fight through executions and threats against their families. There clearly is no consensus on the importance of cohesion in the *Wehrmacht* up to 1945. Nor for that matter is there consensus on the exclusive role of primary group ties in providing what cohesion did exist.

The surveys of soldiers after combat conducted by Stouffer et al. have been remembered as showing that cohesion (fighting for one's buddies) was the primary factor that sustained soldiers in combat. While Stouffer et al. had a minimalist measure of what we now call cohesion, this form of cohesion was *one* of the most important factors sustaining men in combat. However, it was not the most important. In the European theater, when enlisted men were asked, "Generally, from your combat experience, what was most important to you in making you want to keep going and do as well as you could?" the most frequent response, given by 39% of the soldiers, was "ending the task." By contrast, only 14% of soldiers gave responses indicating that "solidarity with the group" was important.[36] And in the Pacific theater, when enlisted infantrymen were asked about a number of things that were "helpful when the going was tough," the single most important factor was prayer or religion: 70% said that prayer helped, while 61% said that it helped "to think that you couldn't let the other men down."[37] If Stouffer's research affirmed any principle about combat, it was to provide empirical support for the proposition that there are (almost) no atheists in foxholes.

Finally, Marshall did argue on the basis of after-action interviews with infantry companies that it was a feeling of unity among soldiers that gave them the courage to fight. However, more recently questions have been raised about the quality of the data Marshall used to arrive at his findings regarding the "ratio of fire," which concluded that a low percentage of infantrymen actually fired their weapons in the direction of the enemy during combat. By questioning whether the after-action interviews really took place, several historians have cast suspicion on Marshall's other work drawn from the same purported interviews, including his account of unity and courage under fire.[38] These considerations suggest that the research conducted on the effects of cohesion among both American and German soldiers in World War II should be regarded as heuristic—a stimulus toward further investigation—not conclusive.

Such arguments concerning the performance consequences of unit cohesion comprise only one part of the expert knowledge on which professional closure in the armed forces has generally been based. The other part consists in the argument that cohesion can reasonably be expected to arise only where groups are composed of like individuals. That cohesion is purported to derive from sociodemographic homogeneity rests on two assumptions: interpersonal preferences for association form the basis of cohesive bonds in groups; and individuals prefer to associate with those like themselves, to the exclusion of others. These assumptions can be seen concretely, for example, in the notion that white soldiers would prefer not to socialize with black soldiers, men would prefer not to have women in their units, and heterosexuals do not want homosexuals in their units. Both of these assumptions have been brought into question by social scientific research, which we will examine in detail below. In the absence of a romantic mythology surrounding the relationship between sociodemographic homogeneity and cohesion, however, it is necessary first to make explicit the extent to which arguments concerning the roots of cohesion in social similarity have been reflected in survey data and explicitly articulated in policy debates.

With regard to race, Stouffer's research team asked black and white soldiers in World War II: "Do you think that white and Negro soldiers should be in separate outfits, or should they be together in the same outfits?" Eighty-four percent of white soldiers (but only 36% of black soldiers) felt that they should be in separate outfits. Percentages were similar with regard to separate post exchanges and separate service clubs.[39] The policy position reflecting these sentiments is exemplified in a statement by the General Board—the Navy's internal policy-making body: "How many white men would choose . . . that their closest associates in sleeping quarters, at mess, and in a gun crew should be of another race? . . . The General Board believes . . . that if the issue were forced, there would be a lowering of contentment, teamwork, and discipline in the service."[40] Similar sentiments were expressed in the Army.

With respect to gender, support for women in the military reflected personnel needs.[41] Americans opposed conscription of women prior to our entry in World War II, but with mobilization needs after the bombing of Pearl Harbor, a majority favored drafting them for wartime jobs, particularly when drafting women for noncombat jobs was presented as an alternative to drafting married men for the same jobs.[42] This support for women in the military did not extend to combat units. A 1974 Army survey found majorities of soldiers reporting that women would not make good front-line combat soldiers even if properly trained, and that if women were assigned to combat units, the Army would become less effective.[43] And in 1982, while a majority of Americans favored extending women's military roles to include fighter pilots, missile gunners, and serving on combat ships, there was still opposition to their inclusion in ground combat.[44] Analysis of Congressional testimony regarding appropriate roles for women in the American military between 1941 and 1985 revealed that among negative arguments appearing in the testimony, the most common was that "women would reduce effectiveness by interfering with cohesion."[45] Along similar lines, Richard Gabriel argues that "military effectiveness and cohesion are . . . the result of sociopsychological bonding—anthropologically, male bonding—among soldiers within combat groups . . . the complete integration of women in the military may have devastating consequences for the levels of cohesion and effectiveness that can be expected of integrated units."[46]

More recently, the Roper Organization, in a survey conducted for the Presidential Commission on the Assignment of Women in the Armed Forces, asked a sample of military personnel what the effects would be of assigning women to direct combat positions. Forty-one percent said that it would have a negative effect on "the cohesion or bonding that goes on among members of a military unit"; 18% said that it would have a positive effect; and 39% said that it would have neither a positive or negative effect.[47] Interestingly, among civilians who were asked the same question, 27% said that it would have a negative effect, 22% that it would have a positive effect, and 39% that it would have neither a positive or negative effect.[48]

The Presidential Commission itself recommended that women be excluded from ground combat units and positions, noting "the effects that women could have on the cohesion of ground combat units."[49] It also noted that "there are no authoritative military studies of mixed-gender ground combat cohesion, since available cohesion research has been conducted among male-only ground combat units."[50]

Unlike the issues of race and gender, where closure focused on quotas, segregation, and exclusion from certain units and specialties, the closure issue regarding sexual orientation was one of exclusion from the profession. In a series of surveys taken in 1992 and 1993, significant majorities of servicemen (and large minorities or small majorities of servicewomen) opposed allowing gay men and lesbians to serve. In surveys of almost 4,000 American military personnel, Laura Miller found that 75% of servicemen (and 43% of servicewomen) disagreed or strongly disagreed with the statement that "gays and lesbians should be allowed to enter and remain in the military."[51] In a U.S. Air Force telephone survey that asked, "How do you feel about the current policy of separating known homosexuals or discharging people who state that they are homosexuals?" 67% of servicemen and 43% of servicewomen agreed with the policy. And when the *Los Angeles Times* asked respondents, "How do you feel about lifting the ban on homosexuals in the armed forces of the United States?" 76% of servicemen and 55% of servicewomen indicated disapproval of lifting the ban. The American public at the time was more evenly divided on the issue.[52]

In testimony before both houses of Congress, numerous arguments were raised against lifting the ban, the cohesion theme figuring prominently among them. For example, William Henderson testified to the United States Senate Committee on Armed Services against allowing open homosexuals to serve in the U.S. military,[53] rooting his argument in the work of Shils and Janowitz, Stouffer et al., and Marshall, among others.[54] At the same hearings, David Marlowe argued against homosexual and lesbian integration on the basis of the same research, and more recent studies done by military psychiatrists.[55]

Personnel Crises and Changes in Closure

While closure operates to reproduce the traditional characteristics of a profession, Laura Schmidt has shown in her analysis of the health care professions that "crises are high-opportunity times because the regular process of institutional reproduction becomes disrupted."[56] Military closure rules in the United States have been no different, repeatedly shifting in the face of a "crisis that undermines existing assumptions and principles, and encourages institutional entrepreneurs to propose alternative ways of seeing things."[57]

For the American armed forces since World War II, personnel crises have been primarily responsible for the successive relaxation of historical exclusions from or limitations on military service. These crises and the revisions of closure rules they occasioned provide a kind of natural laboratory in which to assess cohesion arguments. Social scientific analysis of the post-World War II period, however, has failed to support the use of cohesion as a principle of professional closure in the armed forces on two distinct grounds. Social integration of the armed forces brought about by the demands of mobilization has not undermined military effectiveness. Nor is it the case that cohesion necessarily produced, or was even required for, high levels of performance.

The policy of racial segregation in the Army, maintained through World War II, was overtaken by the need to rapidly mobilize the relatively small cohorts born during the Great Depression for the Korean War, and to provide replacements to line infantry companies in Korea as they emerged from the training base, regardless of race.[58] The percentage of white infantrymen in all white units in Korea who expressed a preference for racial segregation had declined to 68%, and the percentage was lower still among white soldiers in racially integrated units. Only a minority of white infantrymen said that they would object strongly to being placed in an integrated unit.[59] Furthermore, unit-level bonding did not characterize rifle companies in Korea.[60] Thus, the World War II findings were not replicated in Korea. While preferences may play some role in shaping unit-level bonding, racial integration did not compromise the effectiveness of units. Recent research has found no relationship between the racial composition of Army units and soldier perceptions of unit cohesion.[61]

The gender integration of the military was advanced, in terms of numbers, under the mobilization pressure of World War II, and in terms of positions opened after 1973, because the volunteer military was not attracting sufficient male volunteers. The process of gender-integration is ongoing, with some positions still closed to women, and entry-level training still gender-segregated in the Marine Corps and in the combat branches of the Army (except for air defense artillery). While the process of social reproduction has changed, there has been no demonstration that either racial or gender integration has undercut military effectiveness.

Since homosexuals are not allowed to serve openly in the American military, there has not been any test of the impact of homosexual integration on military effectiveness. President Bill Clinton may have made a strategic error in trying to lift the ban on homosexuals serving at a time when the military was downsizing in the wake of the end of the Cold War in Europe, since there was no need for new sources of manpower. However, in two countries in which a ban on homosexuals in the military has been lifted—Canada and Israel[62]—no impact on effectiveness has been observed. Interestingly, Charles Moskos—who in the past has opposed allowing homosexuals to serve openly, who is regarded as one of the authors of the American military's "don't ask, don't tell, don't pursue" policy, and who is also an advocate of returning to military conscription—seems to have changed his view on the relationship between policies on homosexuals in the military and military accession.

Early in the debate on homosexuals in the military, Moskos suggested that dropping the gay ban would cause a decline in military recruiting, leading to a requirement to re-institute conscription.[63] The American armed forces faced a recruiting crisis in the late 1990s (with the don't ask, don't tell, don't pursue policy in place), in response to which Moskos suggested that the leaders of the gay community "support his call for a return to the draft to eradicate the ban on self-proclaimed gays serving in the armed forces."[64] His new reasoning was that with a ban on homosexuals in place, young men and women who were drafted would claim to be gay to avoid service. Faced with a consequent conscription crisis, military leaders would support lifting the ban. Having first argued that lifting the ban would require a return to conscription, he now argues that a return to conscription would require lifting the ban. The gay ban has changed from an independent to a dependent variable.

If integration has not undermined performance, neither is it clear that cohesion itself has been all that Shils, Janowitz, Stouffer, and Marshall believed it to be. Relationships based upon shared social characteristics are what have been identified in the research literature as *social cohesion*: "the nature and quality of the emotional bonds of friendship, liking, caring, and closeness among group members. A group is socially cohesive to the extent that its members like each other, prefer to spend their social time together, enjoy each other's company, and feel emotionally close to one another."[65]

These are the kinds of ties that have been referred to in analyses of the military as cohesion, solidarity, or bonding. However, high levels of this form of cohesion have not been shown to have an effect on combat performance, and may even be deleterious to performance. Paul Savage and Richard Gabriel, in their critique of the U.S. Army in Vietnam, argued that existing Army personnel policies destroyed the solidarity of the primary group as a social form.[66] However, Moskos argued that the self-interest of soldiers in Vietnam produced solidarity in combat squads,[67] and John Helmer argued that drug and alcohol subcultures among soldiers in Vietnam were reflected in primary groups that shared behavior regarding psychoactive substances.[68] Moreover, in the 1980s, when an infantry battalion that was part of an Army experiment to build unit cohesion was sent to the Sinai as part of the Multinational Force and Observers in support of the Camp David Accords, cohesion did develop at the small-unit level, but one of its manifestations was a tendency to regard senior NCOs and officers in the company and battalion chain of command as the enemy.[69]

The presence of primary groups is not simply a reflection of policy: it is an emergent group process, the results of which may run counter to organizational values and goals. What *has* been shown to be related to performance, however, is *task cohesion*: "the shared commitment among members to achieving a goal that requires the collective efforts of the group."[70]

Closure Change, Task Cohesion, and Performance

It would be politically and ethically difficult to conduct research upon the impact of racial integration on cohesion and performance in American military units today. Moreover, the absence of identified homosexuals in service makes research on the consequences of homosexual integration problematic in the extreme. Contemporary research permitting the refinement of our understanding of cohesion and the mechanisms through which it operates and is created has therefore tended to focus on gender integration. Research done in this tradition suggests that, in general, after an initial period of adaptation, gender integration enhances the performance and task cohesion of combat support and combat service support units. Research has not been conducted on gender-integrated combat units.

An analysis of cohesion and readiness in twenty-one support companies in the U.S. Army in garrison in 1988 suggested that women soldiers evaluated the task competence of women higher than did male soldiers, but that male soldiers were nonetheless willing to work with female soldiers.[71] The proportion of junior enlisted females in each company was negatively correlated with mean cohesion and readiness scores among junior enlisted males. Thus, at least for junior male soldiers in support units in the late 1980s, the presence of women initially seemed to degrade perceptions of cohesion and readiness. However, among these males, cohesion and combat readiness increased with increasing acceptance of women.

A subsequent study by the same researcher of 1,316 soldiers in garrison from thirty-four combat support and combat service support units in the summer of 1995, focusing primarily on sexual harassment, found that "gender harassment and sexual harassment as unit problems were both correlated with lower vertical cohesion, lower combat readiness, and lower acceptance of women."[72] Interestingly, horizontal cohesion, while it had been significantly related to each of such variables as combat readiness, vertical cohesion, gender harassment, and sexual harassment when examined on a paired basis, was not related to combat readiness when the effects of other variables were taken into account. The researchers noted that "there appear to be some positive changes in gender integration since our previous study."[73]

More recently, this research team looked at the relationship between gender composition and cohesion across five different research studies, including both garrison and deployment situations: the 1988 study of support units noted above; a study of about 1,000 soldiers in 48 companies deployed to the Persian Gulf for Operation Desert Storm in 1991; a survey of over 2,000 soldiers deployed to Somalia in 1993 as part of Operation Restore Hope; a study of over 1,000 soldiers deployed to Haiti in 1995 as part of Operation Restore Democracy; and the summer 1995 study discussed above.[74]

Among all these research studies, significant negative relationships between percentage of women in the unit and horizontal cohesion were revealed in only two: the Persian Gulf study and the Somalia study. These variables were not significantly related in the other three. For men only, there was a strong relationship only in the Somalia sample. This was the case among all junior enlisted personnel as well. The authors suggest that the differences in findings may be due to gender differences in approval of the missions, and differences in deployment circumstances. Most important for our purposes, the data do not suggest a robust relationship between presence of women in a military unit and low cohesion. Similarly, a RAND study asserts that:

> Perceptions about cohesion . . . tended to vary by rank more than anything else. Higher ranking men and women were more likely to report higher rates of cohesion than junior personnel. Junior personnel often gave leadership practices and guidance as an explanation for cohesion level. [Personnel] who described their units as very cohesive or loosely cohesive appeared to be personally satisfied with their situations and to believe that their units were able to meet their goals in terms of work requirements. Any divisions that may be caused by gender were minimized or invisible in these units. Gender was an issue only in units characterized as divided into conflicting groups, and then it took second place to divisions along the lines of work groups, or, within work groups, along the lines of rank.[75]

The research literature leaves one with the sense that a major determinant of the level of cohesion in units, as is the case with both organizational and personal sexual harassment, is the quality of leadership. Indeed, R. J. Deluga, in studying supervisor-subordinate relationships in the military, found trust between supervisor and subordinate to be predictive of a range of "organizational citizenship" behaviors.[76] By contrast, A. L. W. Vogelaar and H. Kuipers, studying armored infantry platoons in the Dutch army, found that leaders have an influence on behavior, but not on cohesion.[77] However, the cohesion measure used, like much cohesion research, mingled elements of *social cohesion* with elements of *task cohesion.*

Cohesion may well be a characteristic of effective military units, and may be related to performance as well. However, incorrect causal inferences have been drawn in the past regarding the nature of this relationship. B. Mullen and C. Cooper suggest that "the cohesiveness-performance effect is due primarily to commitment to task [what we refer to as *task cohesion*] rather than interpersonal attraction or group pride [what we refer to as *social cohesion*] . . . [and] the most direct effect might be from performance to cohesiveness rather than from cohesiveness to performance."[78] Similarly, E. Kier notes that "to the extent that there is a correlation [between cohesion and performance], analysts are more confident that the causal relationship leads from success to cohesion than from cohesion to success."[79] That is, when the members of a group recognize and value the ability of other members to contribute to group missions, task cohesion emerges.

A robust negative impact of gender composition on task cohesion in military units has never been demonstrated. The major influence on task cohesion seems to be successful task performance, frequently through common training. As Rush notes, "demanding and stressful training that all members of the organization endure together does more to promote primary group cohesion than any other precombat activity."[80] Neither has gender composition been shown to impact negatively on performance, whether operating indirectly (through cohesion), or as a direct influence.

Conclusion

In contrast to earlier approaches to how individuals come to occupy specific niches in the occupational structure, closure theory focuses on the collective actions taken by occupations to restrict access to their ranks. The American military, like other professions, imposes several criteria on who may enter the occupation, and the application of these criteria has been enforced by the state. Personnel are screened on the basis of educational credentials, mental aptitude, records of criminal behavior, and physical fitness. All of these have been shown to be related to performance.

Personnel have also been at times excluded and at other times restricted on the basis of race, gender, and sexual orientation. All of these criteria have been asserted to be related to unit cohesion, which in turn has been asserted to be related to performance. However, these relationships have yet to be demonstrated.

The World War II studies that have served as the basis for the argument that cohesion produces combat effectiveness are suggestive, but not conclusive. Stouffer et al. employed a minimal definition of cohesion (fighting for one's buddies may not be the same thing as unit cohesion);[81] Shils and Janowitz's[82] interrogations of German prisoners of war may be unreliable for purposes of sociological interpretation (and today such research on POWs would not be allowed by institutional review boards responsible for the use of human subjects in research); and other research on the *Wehrmacht* suggests explanations other than unit cohesion for the performance of German soldiers in the mid-1940s. Marshall's[83] interviews may not have been conducted at all. Nonetheless, these studies have been embraced by the professional military, although as early as the Korean War subsequent studies raised concerns about their validity.

Research during the Korean War failed to find unit-level bonding in rifle companies, and research during the Vietnam War found cohesion that was antithetical to effective performance. When racial and gender integration took place in the American military, changing the social reproduction of the military, they did not reduce performance (although the effect on ground combat units has not been tested). Furthermore, although the effects of sexual orientation integration have not been tested in the American military, no effects have been noted in those military forces in which such integration has occurred.

From consideration of this evidence, it is possible to make two fairly strong sets of claims organized, respectively, around cohesion as a social process, and the military as a profession. Both sets of claims consist of a pair of propositions. In the first case, these conclusions bear upon performance and the roots of social cohesion in small groups, and in the second, upon the cohesiveness of the profession of arms overall and the need of the profession to reconsider what is treated as expert knowledge. We will take these in turn.

Neither the positive association between social cohesion in small groups and military performance, nor the negative association between social integration and military effectiveness, has been strongly or uniformly supported in research. In other words, there is no clear causal link that can be demonstrated using rigorous methods between social cohesion and high levels of military performance, or between social integration and any reduction in military effectiveness. As a consequence, it is difficult to defend the use of social cohesion as a principle of professional closure, continuing to exclude from military service or to limit the service of those ascriptively defined populations whose service has been limited in the past. To do so would necessarily exclude many who are otherwise qualified and who might help meet the needs of the armed forces, especially in a time of personnel shortfalls.

If social cohesion cannot be understood to offer a justifiable basis for professional closure, nonetheless it is important in the armed forces. This is the case not least because social cohesion inheres in bonds of solidarity and can provide a supportive context in which high levels of performance may be achieved, particularly in stressful situations such as combat.[84] However, a key assumption on which arguments concerning social cohesion have been based—that people necessarily prefer to associate in small groups with those like themselves—is simply unsupportable in the face of research performed on social integration in the armed forces since World War II. Rather, as studies of gender integration have especially highlighted, it is through effective training and strong leadership that preferences for association may alter. By means of training and leadership, therefore, military personnel may not only acquire a tolerance as well as appreciation of those unlike themselves, they may also form durable affective bonds in socially heterogeneous groups.

Tolerance and appreciation of difference, furthermore, form a useful foundation for the creation of another variety of social cohesion in the military, often and inappropriately overlooked in discussion of the armed forces. As research on the *Wehrmacht* by Bartov underscores, not all social cohesion in the military is necessarily limited to small groups, nor does social cohesion necessarily inhere in bonds to particular other persons.[85] In contrast to the primary group ties examined in the work of Shils, Janowitz, Stouffer, and Marshall is the cohesion expressed in a sense of membership in a group sufficiently broad that no individual may reasonably know the actual extent of the group, let alone have strong ties to every other member. In the words of Benedict Anderson, this is the cohesion characteristic of an imagined community: *imagined* because, although the members cannot all know one another, "in the minds of each lives the image of their communion"; and *community* because, "regardless of the actual inequality . . . that may prevail in each, [it] is always conceived as a deep, horizontal comradeship."[86] This sense of imagined community is precisely what may be seen to distinguish the armed forces from the rest of society and simultaneously to bind its members together—officers and enlisted; those currently serving, those who have served, and those yet to serve. Insofar as the armed forces seek to understand themselves and to be understood by others as a single coherent profession, it is as an imagined community that their distinctiveness may be most visible and clearly asserted.

Taken together, the preceding three sets of claims draw attention to two facets of the relationship between the military and its expert knowledge as a profession. As analysis of cohesion as a principle of closure underscores, expert knowledge—especially when it is rooted in social scientific research—is highly dynamic, subject to revision and refinement over time. Insofar as the military as a profession relies upon social scientific research for the accumulation and validation of expert knowledge, significant new research findings create both opportunities and pressures to adjust. The price of failing to revise the stock of expert knowledge and of failing to alter practices in accord with new expert knowledge can be substantial—beyond merely the exclusion of otherwise qualified personnel mentioned above. In particular, the military as a profession runs the risk of undercutting its own position in competitions for jurisdictions and resources because professional competitors or the public at large perceive that the military bases its practices on invalid, inappropriate, or illegitimate "expert knowledge."

These findings underscore the task ahead for the military as a profession: a focus on effectiveness or performance—rather than supporting exclusions based on ascriptive criteria—will lead the armed forces to cultivate through leadership and training precisely those abilities that let military units perform well together, one element of which is acceptance of exactly those ascriptive differences that have historically marked exclusions from military service. By this means, task cohesion—that sense of commitment to the group's goals that resonates so strongly with the ethos of a professional force—may be most likely to emerge; and in turn, social cohesion at both the small group and "imagined community" level will also be more likely to develop.

The long-term trend has been for the professions, including the military, to increasingly eschew the use of ascriptive criteria for occupational closure, as they recognize that public reward is rooted in demonstrations of merit, not protection of privilege. However, the American military has been given more latitude than have other professions in this regard. In Europe, by contrast, the international courts have started to hold military forces to the same employment discrimination laws that are binding on other occupations. In Britain this has affected the policy of excluding homosexuals from the military, and in Germany it has affected the restriction of women to roles in military medicine and military music. In an era of globalization, and at a time when the U.S. courts have intruded increasingly on the autonomy of other professions in regulating themselves, it may be only a matter of time until the standards applied to closure by the profession of arms abroad will be reflected in the United States as well.

Notes

1. This chapter was prepared for the West Point Project on Army Professionalism. The research was supported in part by the Army Research Institute under contract DASW0100K0016. The contents of this chapter are attributable to the authors and not to the Army Research Institute, the Department of the Army, or any other federal agency.
2. Peter M. Blau and Otis Dudley Duncan, *The American Occupational Structure* (New York: Free Press, 1967), 163.
3. Bradford Booth et al., "The Impact of Military Presence in Local Labor Markets on the Employment of Women," *Gender & Society* 14 (2000): 318-32.
4. Laura Schmidt, "The Corporate Transformation of American Health Care: A Study in Institution Building" (Ph.D. diss., University of California, Berkeley, 1999), 11.
5. Christopher Dandeker and Mady W. Segal, "Gender Integration in Armed Forces: Recent Policy Developments in the United Kingdom," *Armed Forces & Society* 25 (1996): 29-47, 33.
6. Andrew Abbott, *The System of Professions: An Essay in the Division of Expert Labor* (Chicago, IL: University of Chicago Press, 1988).
7. Max Weber, *Economy and Society*, ed. Guenther Roth and Klaus Wittich (Berkeley, CA: University of California Press, 1978), 33-48, 926-55.
8. Magali S. Larson, *The Rise of Professionalism* (Berkeley: University of California Press, 1977).
9. Kim A. Weeden, "From Borders to Barriers: Strategies of Occupational Closure and the Structure of Occupational Rewards" (Ph.D. diss, Stanford University, 1999).
10. Andrew Abbott, "The Sociology of Work and Occupations," *Annual Review of Sociology* 19 (1993): 187-209.
11. Noel Parry and Jose Parry, *The Rise of the Medical Profession: A Study of Collective Social Mobility* (London: Croom Helm, 1976).
12. Geoff Esland and Graeme Salamon, eds., *The Politics of Work and Occupations* (Milton Keynes, UK: Open University Press, 1980).
13. Weeden, "From Borders to Barriers."
14. Ibid, 79.
15. David R. Segal, *Recruiting for Uncle Sam: Citizenship and Military Manpower Policy* (Lawrence: University Press of Kansas, 1989).
16. Bradford Booth, "Bringing the Soldiers Back In: An Argument for the Inclusion of Military Personnel in Labor Market Research," forthcoming; idem, "The Impact of Military Presence in Local Labor Markets on Unemployment Rates, Individual Earnings, and Returns to Education" (Ph.D. diss., University of Maryland, 2000).

17. Samuel P. Huntington, *The Soldier and the State* (Cambridge, MA: Harvard University Press, 1957); Morris Janowitz, *The Professional Soldier* (New York: Free Press, 1960).
18. David R. Segal et al., "The All-Volunteer Force in the 1970s," *Social Science Quarterly* 79 (1998): 350-411; Jerald G. Bachman et al., "Who Chooses Military Service? Correlates of Propensity and Enlistment in the United States Armed Forces," *Military Psychology* 12 (2000): 1-30.
19. Jerald G. Bachman et al., "Distinctive Military Attitudes Among U.S. Enlistees: Self-Selection versus Socialization," *Armed Forces & Society* 26 (2000): 561-85.
20. David R. Segal and Joseph J. Lengermann, "Professional and Institutional Considerations," in *Combat Effectiveness,* ed. Sam C. Sarkesian (Beverly Hills, CA: Sage, 1980), 154-184.
21. Abbott, *The System of Professions*; Eliot Friedson, *Professionalism Reborn: Theory, Prophesy, and Policy* (Cambridge: Polity Press, 1994); William J. Goode, "The Theoretical Limits of Professionalization," in *The Semi-Professions and Their Organization,* ed. Amitai Etzioni (New York: Free Press, 1969), 266-313.
22. Weeden, "From Borders to Barriers."
23. Weber, *Economy and Society,* 46.
24. Raymond Murphy, *Social Closure: The Theory of Monopolization and Exclusion* (Oxford: Clarendon Press, 1988).
25. Randall Collins, *The Credential Society: An Historical Sociology of Education and Stratification* (New York: Academic Press, 1979).
26. Randall Collins, "Changing Conceptions in the Sociology of the Professions," in *The Formation of the Professions: Knowledge, State, and Strategy*, eds. Michael Burrage and Rolf Torhstendahl (London: Sage, 1990), 11-23; Keith M. MacDonald, *The Sociology of the Professions* (London: Sage, 1995); Weber, *Economy and Society,* 45; Weeden, "From Borders to Barriers," 27-29.
27. Meyer Kestnbaum, "Citizenship and Compulsory Military Service," *Armed Forces & Society* 27 (2000): 7-36; idem, "Partisans and Patriots: National Conscription and the Reconstruction of the Modern State in France, Germany, and the United States" (Ph.D. diss. Harvard University, 1997).
28. Edward A. Shils and Morris Janowitz, "Cohesion and Disintegration in the *Wehrmacht* in World War II," *Public Opinion Quarterly* 12 (1948): 280-315.
29. Samuel Stouffer et al., *The American Soldier,* 2 vols. (Princeton, NJ: Princeton University Press, 1949).
30. S. L. A. Marshall, *Men Against Fire* (New York: Morrow, 1947).
31. Shils and Janowitz, "Cohesion and Disintegration."
32. Indeed, by contemporary standards of the use of human subjects in research, using prisoners of war as research subjects is precluded.
33. Martin van Creveld, *Fighting Power: German and U.S. Army Performance, 1939-1945* (London: Arms and Armour Press, 1983).
34. Omer Bartov, *German Troops and the Barbarization of Warfare* (New York: St. Martin's, 1996); idem, *Hitler's Army: Soldiers, Nazis, and War in the Third Reich* (New York: Oxford University Press, 1991).
35. R. S. Rush, "A Different Perspective: Cohesion, Morale, and Operational Effectiveness of the German Army, Fall 1944," *Armed Forces & Society* 25 (1999): 477-508.
36. Stouffer et al., *The American Soldier,* vol. 2, 108.
37. Ibid, 179.
38. F. Smoler, "The Secret of the Soldiers Who Didn't Shoot," *American Heritage* 40 (1989): 5, 36-45; R. J. Spiller, "S.L.A. Marshall and the Ratio of Fire," *RUSI Journal* 133 (1988): 63-71; Martin Blumenson, "Did 'SLAM' Guess at Fire Ratios? Probably," *Army* 39 (1989): 16ff.
39. Stouffer et al., *The American Soldier,* vol. 1, 568.

40. Cited in Sherie Mershon and Steven Schlossman, *Foxholes & Color Lines: Desegregating the U.S. Armed Forces* (Baltimore, MD: Johns Hopkins University Press, 1998), 50.
41. See Mady W. Segal, "Women's Military Roles Cross-Nationally: Past, Present, and Future," *Gender & Society* 9 (1995): 757-75.
42. Mady W. Segal and David R. Segal, "Social Change and the Participation of Women in the American Military," *Research in Social Movements, Conflicts and Change* 5 (1983): 235-58.
43. David R. Segal, Nora Scott Kinzer, and John C. Woelfel, "The Concept of Citizenship and Attitudes Toward Women in Combat," *Sex Roles* 3 (1977): 469-77.
44. Clyde Wilcox, "Race, Gender, and Support for Women in the Military," *Social Science Quarterly* 73 (1992): 310-23.
45. Mady W. Segal and Amanda F. Hansen, "Value Rationales in Policy Debates on Women in the Military: A Content Analysis of Congressional Testimony, 1941-1985," *Social Science Quarterly* 73 (1992): 296-309, 306.
46. Richard A. Gabriel, "Women in Combat?" *Army,* March 1980, 44.
47. Roper Organization, *Attitudes Regarding the Assignment of Women in the Armed Forces: The Military Perspective* (Conducted for the Presidential Commission on Women in the Armed Forces, September 1992).
48. Roper Organization, *Attitudes Regarding the Assignment of Women in the Armed Forces: The Public Perspective* (Conducted for the Presidential Commission on Women in the Armed Forces, August 1992).
49. Presidential Commission on the Assignment of Women in the Armed Forces, *Report to the President* (Washington, DC: U.S. Government Printing Office, 15 November, 1992), 24-25.
50. Ibid.
51. Laura L. Miller, "Fighting for a Just Cause: Soldiers' Views on Gays in the Military," in *Gays and Lesbians in the Military: Issues, Concerns, and Contrasts,* eds. Wilbur J. Scott and Sandra Carson Stanley (New York: Aldine De Gruyter, 1994), 69-85.
52. David R. Segal, Paul A. Gade, and Edgar M. Johnson, "Social Science Research on Homosexuals in the Military," in *Gays and Lesbians in the Military: Issues, Concerns, and Contrasts,* eds. Wilbur J. Scott and Sandra Carson Stanley (New York: Aldine De Gruyter, 1994), 33-51.
53. William D. Henderson, "Prepared statement by Dr. William Darryl Henderson, 'Why soldiers fight,'" in *Policy Concerning Homosexuality in the Armed Forces* (Hearings before the Committee on Armed Services, United States Senate, One hundred third Congress, Second session, 1993, S. Hrg. 103-845), 251-55.
54. Shils and Janowitz, "Cohesion and Disintegration"; Stouffer et al., *The American Soldier;* and Marshall, *Men Against Fire,* among others.
55. David H. Marlowe, "Prepared statement by Dr. David H. Marlowe, Chief, Department of Military Psychiatry, Walter Reed Army Institute of Research," in *Policy Concerning Homosexuality in the Armed Forces* (Hearings before the Committee on Armed Services, United States Senate, One hundred third Congress, Second session, 1993, S. Hrg. 103-845), 265-76.
56. Schmidt, "The Corporate Transformation of American Health Care," 11.
57. Ibid, 16.
58. David R. Segal, "Diversity in the American Military," *Sociological Forum* 14 (1999): 531-39.
59. Leo Bogart, ed., *Social Research and the Desegregation of the U.S. Army* (Chicago, IL: Markham, 1969).
60. Roger Little, "Buddy Relations and Combat Performance," in *The New Military*, ed. Morris Janowitz (New York: W.W. Norton, 1969), 195-223.
61. Guy L. Siebold and Twila J. Lindsay, "The Relation Between Demographic Descriptors and Soldier-Perceived Cohesion and Motivation," *Military Psychology* 11(1999): 109-128.

62. Franklin C. Pinch, "Canada: Managing Change with Shrinking Resources," in *The Postmodern Military: Armed Forces After the Cold War*, eds. Charles C. Moskos, John Allen Williams, and David R. Segal (New York: Oxford University Press, 2000), 156-181; Aaron Belkin and Jason McNichol, "Effects of the 1992 Lifting of Restrictions on Gay and Lesbian Service in the Canadian Forces: Appraising the Evidence" (unpublished paper, Center for the Study of Sexual Minorities in the Military, University of California, Santa Barbara, 2000); Aaron Belkin and Rhonda Evans, "Homosexuality and the Israeli Defense Forces: Did Lifting the Gay Ban Undermine Military Effectiveness?" (unpublished paper, Center for the Study of Sexual Minorities in the Military, University of California, Santa Barbara, 2000).
63. Charles C. Moskos, "From Citizens' Army to Social Laboratory," in *Gays and Lesbians in the Military: Issues, Concerns, and Contrasts,* eds. Wilbur J. Scott and Sandra Carson Stanley (New York: Aldine de Gruyter, 1994), 53-65.
64. "The Moskos Method: Do Ask and Do Tell," *Inside Washington,* 23 September, 2000.
65. R. MacCoun, "What is Known about Unit Cohesion and Military Performance?" in *Sexual Orientation and Military Personnel Policy: Options and Assessment* (Santa Monica, CA: RAND, 1993), 283-331, 330.
66. Paul Savage and Richard A. Gabriel, "Cohesion and Disintegration in the American Army," *Armed Forces & Society* 2 (1976): 340-76.
67. Charles C. Moskos, *The American Enlisted Man* (New York: Russell Sage Foundation, 1970).
68. John Helmer, *Bringing the War Home* (New York: Free Press, 1970).
69. David R. Segal and Mady W. Segal, *Peacekeepers and Their Wives* (Westport, CT: Greenwood Press, 1993), 98.
70. MacCoun, "Unit Cohesion and Military Performance," 291.
71. L. N. Rosen et al., "Cohesion and Readiness in Gender-Integrated Combat Service Support Units: The Impact of Acceptance of Women and Gender Ratio," *Armed Forces & Society* 22 (1996): 537-53, 545.
72. L.N. Rosen and L. Martin, "Sexual Harassment, Cohesion, and Combat Readiness in U.S. Army Support Units," *Armed Forces & Society* 24 (1997): 221-44, 234.
73. Ibid, 239.
74. L.N. Rosen et al., "Gender Composition and Group Cohesion in U.S. Army Units: A Comparison Across Five Studies," *Armed Forces & Society* 25 (1999): 365-86.
75. M.C. Harrell and L. L. Miller, *New Opportunities for Military Women: Effects on Readiness, Cohesion, and Morale* (Santa Monica, CA: RAND, 1997), 66.
76. R. J. Deluga, "The Relationship between Trust in the Supervisor and Subordinate Organizational Citizenship Behavior," *Military Psychology* 7 (1995): 1-16.
77. A.L.W. Vogelaar and H. Kuipers, "Reciprocal Longitudinal Relations Between Leader and Follower Effectiveness," *Military Psychology* 9 (1997):199-212.
78. B. Mullen and C. Cooper, "The Relation Between Group Cohesiveness and Performance: An Integration," *Psychological Bulletin* 115 (1994): 210-27.
79. E. Kier, "Discrimination and Military Cohesion," in *Beyond Zero Tolerance: Discrimination in Military Culture,* eds. Mary Katzenstein and Judith Reppy (Boulder, CO: Rowman & Littlefield, 1999), 25-52, 41.
80. Rush, "A Different Perspective," 502.
81. Stouffer et al., *The American Soldier.*
82. Shils and Janowitz, "Cohesion and Disintegration."
83. Marshall, *Men Against Fire.*
84. James Griffith and Mark Vaitkus, "Relating Cohesion to Stress, Strain, Disintegration, and Performance," *Military Psychology* 11 (1999): 27-55.
85. Bartov, *The Barbarization of Warfare*; idem, *Hitler's Army.*
86. Benedict Anderson, *Imagined Communities*, rev. ed. (London: Verso Press, 1991), 6-7.

22 Three Case Studies on the Army's Internal Jurisdictions

Case Study No. 1: Army Well-being[1]

Thomas A. Kolditz and Eric B. Schoomaker

Introduction

The Army has long held that care of its members is one of its primary internal jurisdictions. The importance of this jurisdiction is evident in the Army's structure that incorporates nearly all soldier support elements into the institutional Army itself. Army agencies provide all aspects of care for soldiers from housing and feeding to paying and entertaining, from health care to spiritual support. Phrases such as "mission first, people always" communicate to young leaders the Army's complex priorities. But for most of its history, taking care of soldiers meant supporting the needs of young men during the difficult preparation for and execution of wartime missions. Recently, the evolution from a draftee army to a volunteer force consisting primarily of married soldiers drastically changed what it means to care for soldiers and with it, the knowledge and skills needed to accomplish this goal.

This well-being case study describes the Army experience as it struggles to respond to changing demographics and the U. S. employment environment. Although it has taken 30 years, the Army is now regaining control over the internal jurisdiction of soldier well-being and, in doing so, is strengthening its professional expertise in managing its human resources. In addition to outlining how the Army has regained jurisdictional control, this case study elaborates the importance of such control for the Army profession. In particular, it emphasizes the value of abstract knowledge in the development of expertise and its application in jurisdictional competitions.

Like other important Army programs, Army well-being highlights the challenges of its intangible nature; defining, measuring, and directly linking well-being programs to important Army outcomes are difficult. These challenges are heightened during periods of declining resources when the pressure to publicly defend program investments increases substantially. Without a theoretical framework for linking well-being to Army outcomes, program effectiveness is very difficult to measure. As a result, efficiency becomes the default measure of a program's value as we have seen over the past two decades. Regardless of their professional understanding of program importance, in several instances Army leaders could not defend programs successfully, with the result that funding was cut and the programs went into decline.

If support for and understanding of well-being derive from a professional perspective rather than a commercial one, then its relationship with military effectiveness can be measured and its application to an evolving human resource situation can be tailored more effectively. This professional perspective can be embedded in an abstract framework, based on a theoretical understanding of soldier needs and their relationship to Army outcomes. This framework then facilitates successful competition for and defense of this internal jurisdiction.

Background

Following the end of the draft in 1973 and the subsequent establishment of the all-volunteer force, the Army's composition evolved from primarily single, male enlistees to one dominated by married soldiers of both sexes.[2] These married soldiers and their families have different needs and expectations than those of single soldiers. The U.S. employment environment of the 1980s and 1990s further complicated the effect of these demographic changes, requiring the Army to compete with the private business sector for the young, well-educated sector of the American work force. This competition focused even more attention on soldier expectations and benefits. Concurrently, financial and legislative pressures required changes in and creation of programs that helped soldiers and their families.[3]

Thus, in recent decades the Army has faced a military work force with a growing set of material and educational expectations both for initial service as well as for continued or career service. The scope and complexity of this unexpected transformation created a need for new expertise in taking care of institutional constituents—soldiers and their families. This need created a jurisdictional vacuum for the Army that served to challenge the professionals within and attract competitors from without.

The Army's initial approach to this jurisdictional void was to develop a vague yet popularized institutional commitment to Army families and their support. Collectively, the programs that grew out of these efforts were termed "Quality of Life" (QOL) programs. Unfortunately, these programs were devised incrementally without an overarching strategy or well-defined goal. The result was a proliferation of dissociated support activities that provided information piecemeal to the Pentagon bureaucracy and pulled money out to the field.

For example, in 1981 and 1982, the Army conducted a series of symposiums on family life in the military. This series was punctuated by a white paper issued by the Chief of Staff of the Army in 1983, entitled "The Army Family."[4] The year 1984 was ceremoniously proclaimed "The Year of the Family," accompanied by the issuance of the first Army Family Action Plan (AFAP) to improve Army life.[5] Successive iterations of the AFAP focused problem-solving energies and associated funds on the Army family, including single soldiers and retirees—but without an articulated intellectual framework that defined success.

On other fronts, legislative, regulatory, and internal Department of Defense (DOD) and Army management efforts were undertaken which impacted on a wide variety of QOL programs. The programs shared common themes and contextual characteristics. These include:

- Internal organizational management of the programs, especially budgeting;
- Difficulties in capturing and tracking resources expended in these programs;
- Growing expectations by soldiers and Department of the Army civilians about the benefits of these programs, and growing sensitivity to the erosion of these benefits;
- Appearance of a myriad of "champions" for these programs among the public, in Congress, in DOD and the Army; and
- Changing standards for business and service practices for each program, many of which experienced cost increases.

In the end, extensive Army QOL programs emerged but without an overarching strategy or a specific, clearly articulated vision for a suitable level of quality of life or well-being. In the absence of a coherent framework, the multitude of Army programs that affect quality of life have been neither identified nor assessed; consequently, there was no way to evaluate the effectiveness of the Army's QOL approach.

In 1984, the Chief of Staff of the Army (CSA) through the Office of the Deputy Chief of Staff for Personnel (ODCSPER) attempted to compile a record of and track resources for QOL programs.[6] However, concern that these actions would increase the programs' visibility and thus vulnerability to cost-conscious oversight agencies, as well as frustration over the apparent dilution of accountability for program management, led to suspension of this effort in 1987. In the ensuing years, budget officers took a pragmatic, yet inconsistent approach to a funding strategy. For example, written material compiled in 1990 for testimony by the DCSPER before the House Appropriations Committee stated, "Quality of Life is defined in resource management terms as those programs that directly enhance the quality of life for our soldiers and their families and exist solely for that purpose."[7] The report further outlined four principal areas included in the QOL programs—Army Community Service; Youth Programs and Child Development Programs; Morale, Welfare, and Recreation (MWR) Programs; and Community Program Construction. The author of the memorandum conceded that important issues such as pay, allowances, benefits, medical, housing, and other programs impacted upon quality of life, but were not included or tracked as components of QOL.[8] The QOL definition remained predominantly ad hoc, with management of individual components distributed among a variety of Department of the Army and DOD agencies.

Although the Army's approach to QOL was ad hoc, soldiers' expectations of high quality and accessible programs continued to grow, thus challenging the Army to efficiently plan, program, budget, and manage these programs. These expectations were further complicated by rising quality standards and attendant costs. Medical care, housing, and child care offer excellent examples of this trend. The cost of medical care has surpassed the Consumer Price Index as medicine incorporates new technologies and better trained and certified staffs.[9] Not only has the proportion of married soldiers increased but the size of the retiree population now exceeds the combined active component uniformed force, leading to a soaring demand for medical services.

Standards for Army barracks and housing and for child care services have increased, resulting in rising costs to build and execute these programs. As all of these programs have grown in use and popularity, difficulties in providing access or maintaining high standards of operation have been interpreted as an erosion of benefits by constituents. Retiree groups, elected representatives, professional associations, and others aggressively responded to this erosion with actions seeking to shore up the funding, improve staffing, and expand access to these services.

Unfortunately, since military professionals had no strategy or goals as a basis for assessing and defending QOL programs, their role in the attendant dialogue diminished and with it the ability to clearly link support to constituents with essential military outcomes such as recruiting, retention, and force effectiveness. Instead, they used commercial approaches to control costs and provide services more efficiently, and based success largely on *profitability*. This approach overlooked the strategic relationship between QOL program effectiveness and the development of the Army's human resources. For example, the Morale, Welfare, and Recreation system adopted a coding process to determine which programs would receive appropriated funds, or receive partial appropriated funds, or be self-sustaining businesses. Programs justifiable to key Congressional and DOD leaders received support, but suffered (or faced elimination) when budgets were cut. Self-sustaining businesses, however, tend to survive times of austere support, and so persist as commercially efficient enterprises, even if the articulation of how effective they are at increasing soldier quality of life or contributing to the organizational culture of the Army is unclear.[10] More importantly, not all installations had the populations to support extensive revenue generating businesses, so quality of life programs became inconsistent from installation to installation and even discontinuous over the career of a soldier. Ironically, then, commercially successful QOL programs can, in a strategic sense, erode well-being and Army culture over time by allowing commerce, rather than well-being goals or associated military outcomes, to dictate success.

The Army's Response

This was the situation that Army Chief of Staff, Gen. Eric Shinseki, found in August 1999. Rather than continue the haphazard approach to QOL, he established a study group at the U.S. Army War College (USAWC), the Army's school for strategic-level professionals, to examine QOL and to reframe these programs. General Shinseki's intent was to develop a broader, more strategically valuable concept than the QOL notion; he personally termed this concept "well-being." The guidance he gave to this study group was that well-being would apply to the totality of the Army—soldiers, civilians, family members, veterans, and retirees—and that it must link to military outcomes, specifically readiness. Eleven students at the Army War College from diverse military and civilian backgrounds were chosen by the Commandant of the War College to execute the study. Each contributor had special expertise in a quality of life area (e.g., pay, housing), special research skills, and/or experience in battalion/brigade command. The group was tasked to design and complete the study, produce a formal written report (officially titled *A Well-being Framework for the US Army: Report to the Chief of Staff, Army*), and brief the Chief of Staff on findings and recommendations by 21 January 2000. The Chief of Staff accepted the study findings and recommendations, and directed the Vice Chief (with the Deputy Chief of Staff for Personnel as lead agent) to develop a strategic plan to implement well-being Army-wide.

On 5 January 2001, the Vice Chief of Staff of the Army announced the results of this study and outlined a blueprint—the Well-being Strategic Plan—to help the Army's leaders ensure the well-being of soldiers or, in strategic terms, the human dimension of Army transformation. The USAWC report analyzed Army well-being (AWB) and the myriad programs (QOL and others) that were developed to serve the soldiers and civilians making up the internal constituency of the Army. The study group exhaustively reviewed the history of principal Army and DOD QOL programs, finding:

- No overarching institutional focus on these programs other than an undefined notion that they are to enhance "quality of life." This latter state is inferred only indirectly, principally by assessing those programs that are seen as failing to enhance QOL.
- No positive or "success" model of QOL.
- That QOL programs are "stove-piped" among several staff principals and linked loosely, rather than integrated into a system of programs.
- These stove-piped QOL programs are supplemented by major programs and processes that are not formally identified as contributing to QOL (e.g., health care, Equal Opportunity programs, the chaplaincy, and reporting on spouse employment).

Foothold for Professionals: Articulating the Strategy

Using the USAWC report as a foundation, the Army's Well-being Strategic Plan defined AWB and its essential elements. Well-being is defined as "the personal—physical, material, mental, and spiritual—state of soldiers [active, Reserve, Guard, retirees, veterans], civilians, and their families that contributes to their *preparedness to perform the Army's mission*."[11] The well-being framework charges strategic leaders with the responsibility for defining Army culture and simultaneously engaging supportive civilian communities outside our installations, posts, and stations. Although not explicitly stated, the report makes clear that ownership of and stewardship for the programs and activities impacting AWB are critical internal jurisdictions of the Army profession.

Balance between the theoretical and practical, the abstract and the concrete, is critical in the proper execution of strategic-level professional leadership and in maintaining jurisdiction over the welfare of soldiers. Running the Army as a business may appear practical at first glance, but without a theoretical framework business proceeds with technical, not professional, expertise. Expert action absent an underlying abstract foundation creates "craft knowledge," which lacks the special legitimacy of a profession.[12] Craft knowledge is concrete and, although effective in the specific situation, is inflexible and cannot be applied to new domains and situations. In contrast, it is abstract knowledge that is most adaptable to evolving environments and demands. Evolving environments, problems, and knowledge provide opportunities for professions to expand their jurisdiction, assuming their expertise can be applied to these new situations. Similarly, organizations and nonprofessional occupations may also compete in such jurisdictions, seeking to routinize the knowledge and skills required.

This is where AWB falls short; it is an evolving enterprise that, in the past, has been controlled by the Army but without using abstract knowledge. As such, the craft knowledge that the Army developed during the conscription era has not been adaptable to the present population of soldiers and families. In fact, until recently the Army did not recognize well-being as an essential internal jurisdiction that it needed to control. Therefore, the Army's lack of appropriate knowledge, either abstract or concrete, and its disinclination to dominate the jurisdiction of well-being opened it to external competition.

Current Army leaders must build a comprehensive, coherent, and integrated system of well-being programs that attends to the needs of both the profession and its members. Programs have been clustered in the Army's Well-being Strategic Plan in four tiers (serve, live, connect, grow) that relate particularly to the following four functional roles, respectively: fundamental, essential, defining, and enhancing. **Fundamental** features of Army values, leadership, doctrine, and the like meet members' aspirations to **serve** while establishing the organizing principles for the Army's work. **Essential** programs meet individual aspirations to **live**, while satisfying the Army's requirement to attract high-quality soldiers. **Defining** programs meet individual desires to **connect**, i.e., to be a part of a team, and they assist the Army in building such a team. **Enhancing** programs provide opportunities for personnel to **grow**. All four needs—serve, live, connect, grow—are important for both individuals and an Army that relies on high-performing, self-reliant soldiers who want to "be all they can be." As the Army's transformation progresses, AWB programs will ideally be arrayed according to a rational, functional scheme that links an individual's personal (or "occupational") goals in seeking military service with the Army's (or "institutional") requirement to foster a military culture aimed at achieving its principal outcomes. This composite framework of individual aspirations, fundamental Army attributes, well-being programs, and linkage to the civilian community defines the institutional strength of the Army (see Figure 22-1).

It is on the basis of the Army's institutional strength that principal military outcomes—effectiveness, recruitment, retention, job performance, and the like—are pursued and eventually realized. In the context of this discussion of Army professionalism, professional aspirants must view these outcomes as necessary (but not sufficient) not only for the short-term effectiveness of the standing force, but also for something much more important: the maintenance of a military profession.

The AWB framework is based on a hierarchical developmental model derived from C. P. Alderfer's (1969) ERG theory.[13] This framework, when combined with fundamental Army elements such as leadership, organizational structure, doctrine, values, materiel, and the like, constitutes the institutional strength of the Army.

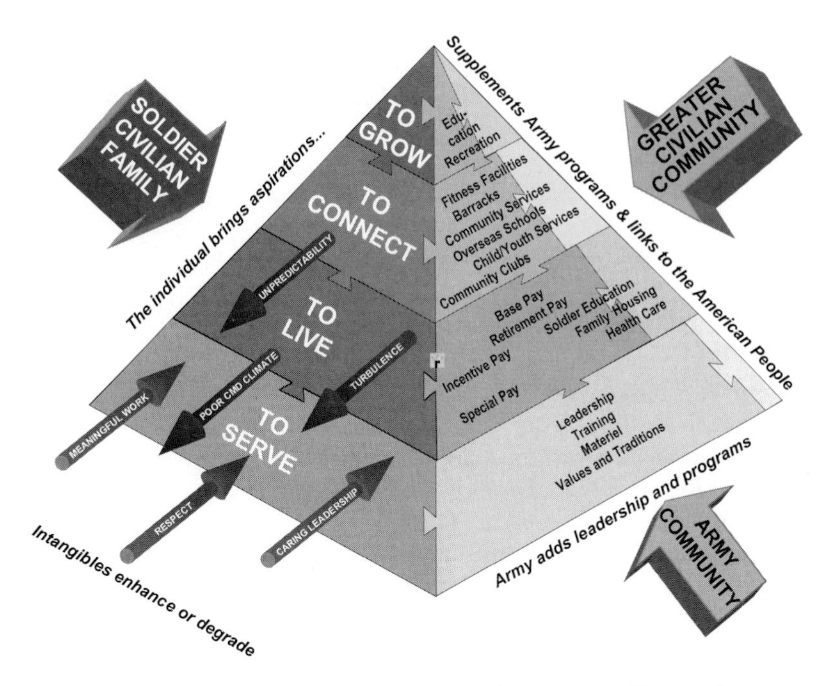

Figure 22-1. *Framework for Army Well-Being and Institutional Strength.*

The institutional strength of the Army ultimately ensures the accomplishment of critical military outcomes, such as recruitment and retention of the force, job performance, and preparedness to accomplish the Army's war-fighting missions. Representing the profession's central values, institutional strength is measured by the degree of success in these desired military outcomes, rather than by measuring values attitudinally.[14]

The limits of jurisdiction are therefore:

- Ownership of and stewardship for the elements comprising the **fundamental** function (e.g., values); and oversight and effective execution—with very heavy DOD and private sector partnering—of **essential** functional elements (e.g., pay and allowances, healthcare, housing);
- Cultivation of **defining** functional programs (e.g., barracks, fitness centers, family member education overseas) with selective partnering with non-Army civilian professionals so as to foster an appropriate image (e.g., education partnerships with local schools), an accession stream of qualified recruits (e.g., education partnerships; fitness and sports sponsorship, like coaching off-post, the Army 10-miler, etc.), and maintenance of a unique culture that both accommodates and perpetuates the characteristics of the Army as a profession;
- Private sector involvement in providing access to high-quality **enhancing** functions (e.g., outdoor recreation, CONUS family member education).

Analysis of these jurisdictional boundaries will result in a consolidated set of recommendations to recast the well-being framework within the wider context of military professional responsibility. Charles Cotton here points to such a wider context:

> Military culture must be managed through systematic institution building by military and civilian leaders who are conscious of their corporate responsibility, and this need takes on critical significance in periods of rapid change and growing ambiguity regarding the goals of the military.[15]

Taking care of the constituents of the Army, like other areas of professional leadership, is both science and art. The scientific aspect of leading the Army toward an optimal state of well-being is a process of balancing resources against requirements in ways that ensure that individuals may serve in reasonable comfort, with a reasonable rate of personal growth and satisfaction. The artful aspect of the process is to ensure that such efforts result in a close-knit Army community committed to the institution and aligned with its values—the aforementioned state of "institutional strength." Research linking well-being to military outcomes, especially field research, leads us to the conclusion that institutional strength results in desired military outcomes such as success in recruiting and retention, and in overall Army effectiveness in deploying to, fighting, and winning the nation's wars.[16]

The initial USAWC report—and the Army's Well-being Strategic Plan derived from it—sought to determine what constitutes the critical features of a system of programs that support the well-being of the totality of the Army. This is a diverse group comprised of all active and reserve component soldiers and their families, retirees and their families, veterans and their families, and DA civilians and their families. The end result was a framework for the human dimension of Army transformation,[17] the process by which the Army Chief of Staff seeks to engender meaningful change in the force. This framework also sets the stage for renewed control over this internal professional jurisdiction, Army well-being.

The Value of the Well-Being Framework

One of the key benefits of the USAWC report is that it offers a conceptual framework, a tool for encouraging and organizing thought. In this case, the framework organizes the Army's abstract knowledge pertaining to human development and management. It incorporates theoretical elements into the dialogue, thus simultaneously expanding and focusing the Army's goals beyond the simple commercial models available in the market. Furthermore, the framework establishes a systematic means of analyzing Army constituents' well-being.[18] Such analysis provides support for and informs senior leader decisions. Yet, the framework is simple enough to be understood by everyone in the Army who chooses to or must implement its principal features.

Prior to the establishment of the well-being framework, the direct link between AWB and military outcomes—outcomes central to its purpose—was supported by neither theory nor research. Studies correlating well-being programs with military outcomes, such as deployability levels, reveal only weak relationships.[19] However, such research is based on very limited definitions of well-being programs, focusing primarily on morale and welfare, and overlooks a basic rule of soldiering in a volunteer Army—personal responsibilities and needs are subordinated when duty calls. It is a soldier's job to ensure that personal issues do not influence his ability to deploy. It is not surprising, then, that the limited elements of programs intended to promote safe, comfortable, and satisfying lives do not reveal a strong direct relationship to soldiers' desire and willingness to perform their military duties.

Alternatively, the well-being framework forms the basis for a shared understanding of well-being programs and the functions they perform for the Army. It establishes a language for professional dialogue about this area of expertise; the value of a program or policy is not reflected merely in its compensation value to individuals, and its value should not be measured in purely commercial terms. A professional approach to well-being conceptually links Army leaders' efforts at promoting institutional strength to soldiers' needs and aspirations. In their purest form, well-being efforts are directed at the most personal needs and aspirations of individuals in the Army. Thus, a framework is more useful than readiness indices in understanding the state of well-being, and more predictive of many desirable organizational goals, such as retention of a high-quality force. A benefit to the Army as an institution must also be shown relative to advantages to the individual, so that the expense of such programs and policies can be justified to the Army leadership and the Congress.

One critical element of professional expertise is the ability to measure the benefit of well-being programs and policies to both the individual and the Army. From the standpoint of the military profession, AWB is a systems approach. One of the primary advantages of a systems approach is that it incorporates feedback loops—measurements that point to continuous, automatic adjustments for maximizing effectiveness. Without ascertainable results, there is less and less need to prefer a professional versus commercial approach and thus a weaker and weaker professional hold on the problem area—in this case the role of professional soldiers in taking care of the force.[20] If links between the well-being systems and military outcomes are measured on a periodic basis, they serve to update the framework over time. This "successive approximations" approach is clearly preferred as compared to estimating future conditions and making decisions on possibilities, rather than present reality. In nonprofessional commercial approaches, such metrics are limited to measures of popularity, efficiency, and profit margin.

An abstract framework gives professionals the ability to make strategic decisions about the absolute and relative value of well-being programs, and about the effectiveness of policies. All programs are not of equal value to the soldiers they serve nor do all programs and policies serve equivalent functions for the Army as an institution. Without organizing principles, the only measure of a program's value is the amount of resources committed to it; success is measured as increases in budgets for the programs. This is the current method of assessing QOL program performance.

Professional abstractions are neither pliant nor accommodating. For example, the framework will not necessarily support the retention of all the QOL programs, policies, and practices currently in use in the Army. A good framework reflects reality, and reality might be that some programs and policies are superfluous, outdated, or dysfunctional. Neither will the framework provide an automatic or immediately compelling rationale for additional funding or quantify specific dollar amounts corresponding to specific programs. As a theoretical abstraction, AWB brings a ruthless logic to program decision-making, as contrasted with the fuzzier, open-ended "quality of life" approach.

How This Competition for Jurisdiction Will End

The outcome of strategic competition for the jurisdiction of well-being of soldiers is not assured. In his book *The System of Professions*, Andrew Abbott offers a range of possible jurisdiction outcomes or jurisdictional settlements.[21] Of these, the ideal resolution for the Army would be for the profession to dominate in both the intellectual and executive senses while permitting others to operate in limited fashion in support of the abstract professional principles controlled entirely by Army professionals.

This limited settlement, entailing various forms of subordination of the Army's competitors, seems the only outcome palatable to the military profession at this time, and only if senior military leaders are in the superordinate role. The challenge is that without the abstraction provided by AWB (or some other equally clear and comprehensive vision), subordination can reverse. Those with some other model, or perhaps just craft skills, may come to dominate the profession. Once the profession and its attendant abstractions are subordinated, jurisdictional competitors use decision-making based on their own models, be they commercial, economic, or political, to call the shots with regard to the well-being of soldiers. If that were to happen, the military as a profession would be, both by definition and in practice, subordinated to others in its responsibility for its internal constituents—soldiers—and their well-being. In this way, AWB, as an abstract representation of both individual and organizational goals, would be hostage to the dynamic tension between an efficiency-seeking bureaucracy and an effectiveness-seeking profession.

The AWB framework is only one case of professional abstraction that competes for jurisdiction among those inside and outside the military. The acquisition of materiel, the recruitment and selection of soldiers, and the development of national security strategy are examples of processes guided by abstraction and potentially vulnerable to jurisdictional competition among professionals engaged in their own forms of expert work. Thus the relationship between the AWB framework and military professionalism can be viewed as a case study illustrating the levels of abstraction and expertise demanded of the truly professional officer across multiple areas of responsibility.

Conclusions

Addressing the well-being of soldiers, civilians, retirees, veterans, and their families is a responsibility inherent to the Army profession. Maintaining control over this jurisdiction demands an abstract framework in which military professionals can perform expert work. Well-being is unusual because jurisdiction over it is internal to the Army, yet is subject to competition from both within and outside the organization. The AWB framework is the only abstraction currently formulated that directs expert work toward achieving specific organizational functions—providing essentials, defining the culture, and enhancing the lives of people in the Army and those who are members of the greater Army community.

Although there are no other comprehensive frameworks, Army well-being competes to subordinate other approaches undertaken to support soldiers. It explicitly seeks to protect and permit the strategic management of Army culture by Army professionals. It can reveal that bureaucratic efficiencies are false economies and at the same time articulate the limits of the Army's support responsibility to individuals. In examining its place in the system of professions and its role in officership and in the military as a profession, the importance of the abstract well-being framework becomes most evident. Jurisdictional contests are won by professions that successfully articulate compelling abstractions to clients, and serve those clients with expert work. The well-being of the force is a jurisdiction that must always be secured by the Army profession.

Case No. 2: Physical Readiness Training and Assessment

Maureen K. LeBoeuf and Whitfield B. East

> Military Leaders have always recognized that the effectiveness of fighting men depends to a large degree upon their physical condition. War places a premium upon the strength, stamina, agility, and coordination of the soldier because victory and his life are so often dependent upon them. Warfare is a grueling ordeal for soldiers and makes many severe physical demands upon them. To march long distances with full pack, weapons, and ammunition through rugged country and to fight effectively upon arriving at the area of combat; to drive fast-moving tanks and motor vehicles over rough terrain; to make assaults and to run and crawl for long distances, to jump in and out of foxholes, craters, and trenches, and over obstacles; to lift and carry heavy objects; to keep going for many hours without sleep or rest—all these activities of warfare and many others require superbly conditioned troops.[22]

The purpose of this case study is to examine the impact of the Army's current practices involving physical readiness training and assessment (PRT/A) on the profession and on the effectiveness of those practices. Specifically, we will look at today's Army Physical Fitness Test (APFT), its standards, the existence of gender and age norms, and current practices in Army units for administering and using the results of the test. Three issues have surfaced during our examination of this issue: (1) whether the APFT is currently reflective of the expert body of knowledge in measuring baseline physical fitness; (2) who within the Army profession has been/is/should be responsible for maintaining the Army's expert body of knowledge related to physical readiness training and assessment; and (3) whether the Army now has the best set of tests in place to assess the physical readiness of its soldiers and units to accomplish their assigned missions. We will make recommendations in each area at the end of the case.

A Short History of the APFT

During World War II as the ranks of the Army swelled into the millions, the mobility, endurance, and strength of many recruits were substandard. As a result, the Army introduced a systematic physical development program during the recruit's Combat Basic Training (CBT) course. To assess the benefits of these specific programs, formal fitness testing for men was first introduced into the Army in 1942 with the Army Ground Forces Test.[23] Despite the introduction of physical fitness tests, however, the Army recognized that it was significantly behind the power curve of expert knowledge in human physical development. In an attempt to expand the expert body of knowledge, the Physical Training School was organized at Fort Bragg, North Carolina, in 1946. The first task of the school's staff was to significantly revise Field Manual 21-20, *Physical Training*, and to modify what was now called the Physical Efficiency Test. Only seven years later, despite the costly lesson in physical and military readiness learned in Korea through the tragedy of Task Force Smith, the Physical Training School was closed due to insufficient funding. Thus, many of the gains in Army physical readiness were lost.

Such an on-again, off-again approach to PRT/A is seen throughout Army history. By 1980 a group of civilian and military physical fitness experts was asked, once again, to develop a new physical fitness test for the Army. The group's task was to design a physical training test that (1) assessed a baseline level of physical fitness, and (2) required no equipment.[24] The result, finalized in 1984, was the three-event Army Physical Readiness Test (APRT), which was to be administered semiannually. The events included push-ups, sit-ups, and a two-mile run. A scoring standard was established for each age/gender group. The APRT score was determined by converting raw scores to a 100-point scale for each event. The maximum score a soldier could earn on the APRT was thus 300 points, or 100 points per event. To pass the APRT, soldiers were required to attain at least 60 points on each of the three test events for a minimum score of 180. Soldiers were to wear fatigue trousers, t-shirt or fatigue shirt, and combat boots while being tested. The Army Physical Fitness Badge (APFB) was also established as an incentive to soldiers to score well on the test. In the mid-1980s, the test was renamed the Army Physical Fitness Test (APFT), and is still called that today.

Physical fitness standards remained essentially unchanged during the next decade. In the early 1990s, Gen. Frederick Franks, the Commanding General of the Training and Doctrine Command (TRADOC), commissioned an extensive study to review the scoring standards. The focus of the study was threefold: (1) ensure the APFT measures baseline Army physical fitness; (2) provide scientific review of the APFT; and (3) seek gender equity. Participating agencies included the United States Army Research

Institute of Environmental Medicine (USARIEM), Army Research Institute (ARI), the Office of the Surgeon General, and the United States Army Physical Fitness School (USAPFS). Researchers analyzed the physical fitness levels of 2,588 active U.S. Army soldiers as demonstrated by scores on the APFT administered between September 1994 and March 1995. As a result of this study, APFT scoring standards were lowered for soldiers 18-22 years old and raised for older soldiers.

As can be inferred from these very brief highlights, the evolution of Army physical readiness training and testing reflects a complex history marked by changing standards and norms, and by periodic swings from training/testing for unit missions to training/testing for individual physical fitness. We will now turn our attention to the issue of gender standards for physical readiness.

Gender Issues (Standards?) and Physical Readiness

The inclusion of women in the U.S. Army over the past hundred years has been met with general resistance by a large segment of male soldiers, society, and elected officials. "I just don't think that the hand that rocks the cradle ought to be shooting the heads off the enemy."[25] This comment was made by Representative Roscoe Bartlett, Member of the House Armed Services Committee, during an interview about segregation of the sexes during basic training. For whatever religious or social reasons, there have also been and continue to be male soldiers at every level of the enlisted and officer corps who object to women in the Army. It is our contention that the disparity between male and female scoring standards on the APFT has served to exacerbate this tension and strain the professional fabric of the Army.[26] There are two interesting points related to this issue. First, it is male officers who have almost exclusively established gender standards on the APFT. Second, since the initial APFT in 1984, scoring standards for women have increased significantly, and in the case of the sit-up event, women now have the same standards as the men.

Interestingly, during this period, the Army, which generally prides itself on attacking difficult internal issues head-on, has not resolved the festering problems of women serving in combat MOSs and gender specific physical performance standards. In stark contrast, during this same period many civil organizations such as police, fire, and emergency rescue have confronted and resolved the gender performance issue.

Scoring Standards for Required Physical Performance for Civilian Organizations

With the passage of the Civil Rights Acts of 1964, the United States has moved towards leveling the playing field in the employment arena. Earlier federal legislation pertained mostly to race-related civil liberties. However, in follow-on legislation such as the Age Discrimination in Employment Act (ADEA) in 1967, Americans with Disabilities Act (ADA) in 1990, and the rewrite of the Civil Rights Acts of 1991, the federal government has afforded significant levels of protection from employment bias in the areas of gender, age, and handicapping conditions. There are, however, instances where physical performance requirements are required to successfully perform a job or task. In these cases agencies/businesses have struggled to establish reasonable scoring standards that can effectively discriminate between qualified and unqualified candidates. Up to the early 1990s, many employers utilized physical performance standards based upon some notion of the physiological characteristics of the applicants, to include form, shape, size, etc. These characteristics generally pertained to differences predicated by gender.[27]

Although there will always be individual female performers who out-perform their male counterparts, the literature is conclusive with regard to strength and endurance performance by gender. A simple review of current track and field records provides evidence of a gender gap favoring males among elite performers (see Table 22-1).

The Cooper Institute for Aerobics Research notes that if physical performance/fitness can be established to be job-related, "then fitness tests, standards, and programs can discriminate against anybody. . . , regardless of age, gender, race, or handicap conditioning."[28]

Men		Women	
Event	**Record**	**Event**	**Record**
100	9.79	100	10.49
400	43.18	400	48.83
10,000	27:20.56	10,000	31:19.9
Pole Vault	19' 9"	Pole Vault	15' 1.5"
High Jump	7' 10.5"	High Jump	6' 8"
Shot Put	75' 10.25"	Shot Put	66' 2.5"
Discus	237' 4"	Discus	216' 10"
20K Walk	1:22:17	20K Walk	1:31:51

Table 22-1. *Current U.S. Track and Field Records.*

According to the applicable legislation, standards should be job-related and based on the principle of "same job = same standard."[29] In the past, norm-referenced standards based upon age and gender were used to evaluate performance because there were relatively little data to support criterion-referenced (go/no go) standards for performance evaluation. Many criterion-referenced standards, such as the 15 pull-up requirement for Ranger School, have been established rather arbitrarily and have little scientific support. But it is also clear that the legislation of the 1990s has made mandatory the principle that standards are to be job-related, a practice the Army has yet to fully implement. As noted in the literature from the Cooper Institute, "with a required emphasis on job relatedness . . . , age and gender-based norms (as mandatory standards) are currently not recommended."[30] Two problems emerge: first, rather than establish valid criterion-referenced performance standards for entry into Army branches and MOSs, Congress has used a philosophical argument to prohibit women from selecting certain combat branches and MOSs; second, the Army continues to use norm-referenced gender-based standards for Army physical fitness testing. As a result of such inconsistencies, few are satisfied with current policy. It is our view that if a soldier meets the performance criteria for a branch or MOS, gender is irrelevant.

Over the past ten years most civilian organizations have adopted a go/no go physical performance standard for admission to and continuation in the profession. Emergency services, fire, and police departments are examples of professional organizations that require members to take physical aptitude and/or fitness tests. For example, the police departments of El Paso, Las Vegas, and New York City, as well as the fire departments of Albuquerque, Houston, and New York City, all have a *common* physical performance standard for their recruits, whether male or female. Minimum performance standards have been established as part of the hiring and retention process.

Relative to Army APFT performance standards, the performance standards for civil organizations are generally low and there is a great variance in physical requirements. For example, the El Paso Police Department has a four-event test: sit-and-reach, push-ups, sit-ups, and 1.5-mile run. Criterion-referenced standards are as follows: sit-and-reach, 16.4"; push-ups in one minute, 18; bent-leg sit-ups in one minute, 27, and 1.5-mile run, 15:20 minutes.[31] The Las Vegas Metropolitan Police Department has a three-event fitness test that requires 15 sit-ups (one minute); 18 push-ups (no time limit), and 1.5-mile run (17:17 minutes).[32] The Albuquerque Fire Department fitness test is conducted over a two-day period. On the first day the candidates take a 1.5-mile run (minimum 12:10 minutes). On the second day the candidates take part in five job-specific events that focus on muscular strength, muscular endurance, and flexibility. These events include a high-rise stair climb, carrying two 50-foot sections of high-rise hose (total weight = 42 pounds) to a fifth floor landing; the hose hoist, which requires the pulling of a rolled 50-foot section of fire hose weighing 45 pounds to the fifth floor landing; forcible entry which requires driving a steel I-beam five feet using a 9-pound hammer; the hose advance, which requires dragging a hose 75 feet; and a victim rescue requiring the candidate to hold and move a 175-pound dummy 100 feet.[33] The Houston Fire Department test focuses on fitness and job-specific events that measure balance, coordination, strength, endurance, and cardio-vascular fitness. The test includes balance beam walk, ladder extension, stair climb, equipment hoist, portable equipment carry, rescue attempt, and 1.5-mile run.[34]

In spite of an all-out recruiting effort to attract women to join the ranks of the New York City firefighters, women represented only 2.6 percent of the 17,000 applicants who took the written exam. Of the 354 women who passed the written exam, only a third showed up to take the physical fitness test. Of the 117 women who took the physical fitness test, only 11 passed. Of the 11,000 firefighters in New York City only 36 are women.[35]

Current Soldier Attitudes Toward the APFT

Over the past twenty years the APFT has evolved into an organizational icon for the Army. The perceptions, attitudes, and behaviors of officers and soldiers towards the APFT should be of concern to those worried about the future of the Army profession. We believe there are four primary issues relative to the APFT: (1) perceptions of APFT scoring standards, including norm-referenced standards based on age and gender; (2) APFT test items; (3) APFT testing protocol; and (4) the status of physical readiness/ training. To explore these issues, we developed a "U.S. Army Physical Readiness Questionnaire," and administered it to approximately 2,000 active duty personnel. The questionnaire was developed to assess the attitudes and opinions of current active duty soldiers towards unit combat readiness, individual physical readiness, and the APFT, including its scoring standards.

Forty-five closed-ended questions were based on a Likert-scale response format that required choice among five responses ranging from strongly agree to strongly disagree. Following established developmental and sampling protocols, the questionnaire was first field tested, then refined, and subsequently administered electronically to approximately 800 officer and senior enlisted personnel at the United States Military Academy located at West Point, New York, and to approximately 1,200 officer and senior enlisted personnel of the 10th Mountain Division located at Fort Drum, New York.

Demographics of Respondents

Interest in the questionnaire among the targeted samples was intense. The Internet response site was left open for approximately seven days for each response group. By the end of the open response period 1,557 soldiers had responded. The response rate was in excess of 75%, an unusually high response for such surveys.[36]

Of the 1,557 respondents, 926 (59.47%) also responded with written responses to the open-ended question (How would you change the APFT: the test items, test administration, physical training?). There were 97.5 printed pages of open responses with a word count of 64,939. This level of response again signals a very significant interest in the APFT/physical readiness issue. For analysis purposes the open-ended responses were factored into five categories: (1) modify APFT test items; (2) no change; (3) issues with APFT scoring standards; (4) issues with APFT testing protocol; and (5) issues with physical training. The number and percentage of categorized responses are presented in Table 22-2. Some respondents wrote extensive open-ended comments suggesting changes in more than one category. Since each comment was recorded as a separate response, the number of responses, 1,295, was greater than the total number of subjects, 926.

Modify APFT Test Items	585	45%
Issues with APFT Scoring Standards	311	24%
Issues with Physical Training	242	19%
Issues with APFT Testing Protocol	110	8%
No Change In APFT Test Items	47	4%
Total	1,295	100%

Table 22-2. *Item Response to Open-ended Survey Question.*

Issues with APFT Test Items

Respondents were asked to select from a menu listing items they would include in a revised APFT, given that the purpose of the test is to measure general physical fitness as it relates to physical readiness. The top ten responses by category are shown in Table 22-3.

Items	Selection Rate
Push-ups	77.10%
Body Composition	52.20%
2 Mile Run	46.10%
Crunches	42.10%
Sit-ups	38.80%
Obstacle Course	32.90%
Pull-ups (over)	30.20%
Chin-ups (under)	26.70%
5-mile Ruck March	24.90%
10-mile Ruck March	21.10%

Table 22-3. *Top 10 Items Selected for a Revised APFT.*

Regarding the open-ended question (Table 22-2), modifications in the APFT test items stimulated a great deal of interest. Of 1,295 total responses, 585 (45%) of the respondents suggested changes in the APFT test items. Approximately 65% of the 585 respondents recommended in written comments that the sit-up be deleted from the APFT, with most suggesting that it be replaced by abdominal "crunches." The second and third most common suggestions, respectively, were to extend the run distance to three miles and add pull-ups. Based upon the open-ended responses, the most strongly supported revised APFT would contain push-ups, three-mile run, crunches, and pull-ups. Respondents also recommended that some type of standardized obstacle course, a 5-10 mile ruck march, and the shuttle run be added to the APFT.

Respondents were quite forthright about the sit-up as currently conducted, mostly with regard to injury. Typical responses were: "After 15 years of sit-ups, I am now recovering (six months later) from a herniated disk operation. After consistently maxing the APFT, I now focus on passing the event without causing further physical damage to myself"; "Get rid of injury-producing sit-ups"; "Replace sit-ups with a crunch exercise."

The second most common open response relative to the current APFT test items dealt with the 2-mile run. Respondents were almost equally divided in their opinion of the appropriate length of the run, with half wanting it shortened, half wanting it lengthened. Regardless of their response, respondents were unanimous in their belief that performance on the 2-mile run was not a valid indicator of mission capability. Many respondents expressed the belief that running endurance did not equate to battlefield endurance. The following quotations are typical of those respondents who want to shorten or eliminate the 2-mile run:

> Eliminate 2-mile run—God forbid I meet an enemy face to face—I have no intention of running. The enemy dies for his country or I die for mine—the two sides of the coin in war. No holds barred.

> The two-mile run may be a basic measure of fitness, but not the kind of endurance necessary in combat. I have known soldiers who could run like the wind in Nikes and shorts, but put a full rucksack on their back, and an M-16 in their hands, and they couldn't hang more that a few miles.

> The Army has become obsessed with pop trends in body fat and appearance, and this is turning us into an Army of jogger types, rather than ass-kicker types.

Then there were the responses that suggested the run be lengthened. These respondents were concerned that two miles were not enough to evoke an aerobic response.[37] The following responses are typical:

> The current run distance is so short that many individuals with a very poor level of cardiorespiratory endurance are able to gut out the test. Even one more additional mile of length would provide a much better fitness indicator.

> A four-mile run provides a better indicator of overall cardiovascular endurance.... The end result is that soldiers now fail to train for the 20-minute minimum aerobic threshold.

> For stamina, I think that a longer duration cardiovascular fitness based test is needed. Two miles may be nice, but I've never run two miles in the field at a six-minute pace. I'd prefer to see a fitness measurement that stressed whether or not I could rely on a soldier after an extended period of strenuous physical activity.

Perceptions of APFT Scoring Standards

The purpose of this line of questions was to determine the effects of APFT scoring standards on the Army profession, particularly on the degree to which soldiers believe that Army preparedness is being maintained and the Army is truly ready for war. Army doctrine generally holds that physical readiness of soldiers is vital to success in war. Moreover, there is an ethical dimension involved, for if Army leaders fail to optimize the physical fitness of soldiers and their units to face the rigors of war, they jeopardize the lives of those soldiers and the nation's security. Thus if soldiers perceive that Army leaders are not scrupulous in assessing and otherwise maintaining the combat readiness of units and individuals, including their physical readiness, morale generally declines and the fitness of the Army profession itself is brought into question.

Question 28 of our Physical Readiness Questionnaire (Currently [are there] significant morale issues in the Army?) was designed to shed light on the morale aspect of the soldier's organizational climate. In response, 74% agreed or strongly agreed (A/SA) that there are significant morale issues in the Army (Table 22-4). With regard to age and gender scoring standards, 72% of respondents A/SA that scoring standards based on age were appropriate (Q 19). Fifty-nine percent of respondents disagreed or strongly disagreed (D/SD) that scoring standards based on gender are unethical (Q 31); however, there was a mixed response to the question that asked should women have a lower APFT scoring standard than men (Q 18).

Question	SD/D	Neutral	A/SA
6. Physical readiness should reflect Mission Essential Task List (METL) needs without regard for age and gender.	28%	14%	59%
8. Mission accomplishment should take precedence over gender equity.	7%	14%	79%
17. The Army should lower APFT scoring standards to meet the declining physical fitness level of today's society.	90%	6%	4%
18. Women should have lower APFT scoring standards than men.	46%	20%	34%
19. A 40-year-old soldier should have lower APFT scoring standards than a 20-year-old soldier.	16%	12%	72%
20. APFT scoring standards motivate soldiers to improve physical performance.	29%	19%	52%
23. The Army should have different APFT scoring standards for soldiers in combat arms, combat support, and combat service support units.	71%	10%	19%
24. All MOS classifications should have the same APFT scoring standards.	20%	10%	69%
26. A soldier's APFT score should be a part of his/her efficiency report (NCOER/OER).	21%	12%	67%
27. Providing promotion points to enlisted soldiers based upon their APFT score effectively motivates soldiers to improve performance.	9%	13%	78%
28. Currently there are significant morale issues in the Army.	12%	14%	74%
29. Lowering APFT scoring standards would cause lower Army morale.	25%	22%	53%
30. Soldiers should not be separated from the Army for failure to meet minimum APFT standards.	77%	10%	13%
31. APFT scoring standards based on gender are unethical.	59%	21%	20%
38. US Army Reserve and National Guard soldiers should have the same fitness standards as the active Army.	8%	9%	83%

Table 22-4. *Questions Related to APFT Scoring Standards.*

There was strong consensus that physical fitness standards should remain challenging, with 90% D/SD that Army standards should be lowered (Q 17). Moreover, 53% A/SA that lowering APFT standards would hurt Army morale (Q 29). Respondents were strongly negative toward having different scoring standards across branches of service (71% D/SD with Q 23), and strongly positive toward having identical fitness standards among Active Army, Army Reserve, and Army National Guard with 83% A/SA (Q 38). They were nearly as supportive of common APFT scoring standards across Military Occupation Specialties (MOS) with 69% A/SA (Q 24). Only a little over half of the respondents (53%) A/SA that the APFT scoring standards motivate soldiers to improve their individual physical performance (Q 20).

There is, however, a strong perception among officers and NCOs that the APFT plays a vital role. Sixty-seven percent of the respondents believe APFT scores should be a part of Officer Evaluation Reports (OER) and Non-Commissioned Officer Evaluation Reports (NCOER). With regard to minimum APFT standards, 77%

believe failure to meet them should be grounds for separation (Q 30); and 78% A/SA that providing promotion points to enlisted soldiers based upon APFT performance motivates soldiers to improve performance (Q 27).

On the open-ended question (Table 22-2), 24% of the written comments were directed at the APFT scoring standards. The responses focused on three general issues: age/gender standards, recent changes in scoring standards, and the validity of height/weight standards. Although 59% of respondents A/SA that scoring standards based on gender are ethical (Question 31, Table 22-4), 46% D/SD that women should have lower fitness requirements (Question 18, Table 22-4). We interpret this to mean that the issue of gender norming is not considered an issue of professional ethics per se, but is just bad professional practice. In the open comments the majority of respondents stated that scoring standards by gender should be removed or significantly modified to eliminate disparities in performance standards between males and females. Conversely, nearly all of the respondents stated that scoring standards should be adjusted by age. Typical comments related to gender norms are:

> I think the APFT Maxes are challenging but the Minimums unfairly set women up for failure. Women, especially older ones, can pass an APFT run but, running at that same pace, will fall out of a unit 9-minute-per-mile run. Then they get publicly humiliated in the dud formation for fall-outs at the end of the run, loudly harangued by a 1SG or CSM for being piss poor in leadership and fitness and a number of other dimensions. We have to figure out what minimum running pace we really want everyone, men or women, to be able to meet.

> Raise the female standards to eliminate inequality and animosity. (Same work, same pay.) If a female is expected to do less PT, then how will they be able to keep up in combat (or regular duty for that matter) when they will be needed the most.

> I would keep the age-based PT standards, but I am still split on the gender-based standards. I have seen women who can do just as much if not more push-ups and sit-ups and run faster then most of the males in the Army. Gender is an excuse to not have to work as hard.

> There should be a baseline standard that all soldiers meet regardless of MOS, age, or gender. Combat does not discriminate based on age or gender. All soldiers should meet the baseline standard. Then standards for excellence can be based on other factors (age, gender) if you like.

The last area of concern with scoring standards pertains to the height/weight tape test. Currently, in accordance with Army Regulation 600-9, soldiers are required to have their height and weight screened semiannually. If their weight exceeds the recommended weight for their height, their girth measurement is then taken at different sites on the body. If it is determined that a soldier has excessive body fat, then he/she is placed in a weight reduction program. Soldiers who do not make satisfactory progress can be recommended for separation from the Army. Typical comments were as follows:

> Put a standard on it for infantrymen (maybe engineers), other combat arms, combat support, and CSS and include the electronic body-fat meter. Current height/weight tables are ridiculous. I have been taped for 10 years at 5-10 pounds "overweight" and each time my body fat is 17% (on a very bad day, after Christmas Leave) or less. I am certainly not overweight.

> I think there is an issue with AR 600-9. I have seen many soldiers who score 270-280 . . . during the APFT but are getting taped and being put on overweight programs due to an obvious genetic inheritance. The genetic issue is one that needs to be addressed. I realize this is hard to do but there are soldiers who are in great shape and are not getting any smaller regardless of the programs or classes we send them to.

APFT Testing Protocol

One closed-ended question on the Physical Readiness Questionnaire pertained to FM 21-20 testing protocol. Q 25 sought to elicit the degree of responder concurrence with the following statement: "Units in which I have served strictly enforced FM 21-20 APFT testing protocols." Responses revealed that 12% D/SD, 10% were Neutral, and 78% A/SA. However, many soldiers used the open-ended question (8% of the written responses) to express significant concerns about testing protocol. Most of the testing protocol comments focused on two issues: inter-rater reliability and command influence on scoring. Typical responses on these issues were:

> Despite demonstrations, push-ups are not graded to standard. Various sorts of sub-standard push-ups are accepted during most tests.

> Have different units administer test and don't identify soldiers by rank until test completed to avoid bias and intimidation of administrators because of rank.

> The administration of the APFT is often not to standard. This is normally done at a local level and the desire to achieve competitive scores has an influence.

We conclude from responses that, while in a definite minority, many soldiers do have strongly felt concerns with the Army's testing protocols for the APFT, in particular with the inequity with which it is administered across Army units.

Relationships between Physical Readiness, Physical Training, and the APFT

One of the objectives of our Physical Readiness Questionnaire was to measure soldier perceptions of the relationships among physical readiness, physical readiness training, and the APFT. The questions addressing these relationships are shown in Table 22-5.

To us the data clearly support our hypothesis that the purpose and function of the APFT have come into serious question over the past decade or so. The APFT was developed as a diagnostic tool for commanders to use in assessing the physical fitness of their soldiers. But only 44% of respondents A/SA that the APFT is a good measure of physical fitness, and only 35% A/SA that the APFT was a good measure of physical readiness.

During a recent survey at the Army War College, former battalion commanders expressed concerns about the spotlight that has fallen on the APFT.[38] These commanders reported that they were being held accountable on Quarterly Training Briefs for poor unit APFT scores. This increased level of visibility and pressure has accorded the

Question	SD/D	Neutral	A/SA
1. Physical readiness is an important part of combat readiness.	2%	2%	96%
3. Physical readiness is dependent upon a unit's Mission Essential Task List.	36%	14%	50%
4. A soldier's physical readiness should be measured by his/her ability to perform critical MOS-related tasks and unit mission-essential tasks.	28%	12%	59%
7. APFT scores are a reliable indicator of physical readiness.	45%	19%	35%
9. In my unit improving APFT performance was rarely the focus of our physical training plan.	60%	14%	26%
10. Most soldiers enjoy physical training as presently conducted.	41%	26%	33%
11. Most unit commanders place too much emphasis on physical training.	65%	22%	13%
12. Performance on the APFT does not directly relate to performance on mission essential tasks.	24%	14%	62%
16. The APFT is a good measure of physical fitness.	36%	19%	44%

Table 22-5. *Questions Related to Physical Readiness, Physical Training, and APFT.*

APFT a level of significance far beyond the original intent. Since APFT scores have made their way onto Quarterly Training Briefs, unit commanders believe they must commit specific training time to the three-event test to maintain parity with other commanders. While 65% of respondents D/SD that most unit commanders place too much emphasis on physical training (Q 11), 60% A/SA that improving APFT performance was often a focus of the physical training plan (Q 9). Our interpretation of this series of questions is that units do not spend an excessive amount of time on physical training as compared to other training; however, they do spend a disproportionate amount of physical training time preparing for the 3-event APFT. Further, a majority of Army soldiers surveyed agreed that performance on the APFT does not indicate readiness to perform functions on the Mission Essential Task List (Q 12), and prefer instead that assessment of physical readiness be measured directly against those tasks (Q 4). Moreover, there is little enjoyment in physical readiness training as currently conducted (Q 10).

Thus, we conclude that the APFT does not inspire confidence in a great many Army soldiers. The implementation and application of APFT training have deviated significantly from its original purpose. The following comments responding to the open-ended question adequately summarize the general attitude concerning the APFT relative to physical training and readiness:

> The problem with the current APFT is that there is way too much focus on it—even in infantry units getting ready for real-world missions it is very difficult to break units away from this inappropriate focus. I believe the reasons for this are: (1) APFT score tied to promotions for NCOs . . . ; (2) Our leadership doctrine—we as leaders, at least some of us (me included), cannot perform to the level of 22 year olds on the vast majority of MOS-specific tasks in the infantry, but we generally can hold our own executing "corporate fitness" tasks like running 2 miles or doing sit-ups. In my estimation, we are holding our soldiers back because we are insecure about being able to hang—with

> good reason, because we've institutionalized the expectation that we must hang; (3) Most leaders, MAJ through GO, who were raised in our current system still don't "get it" and lend credence to the unit APFT averages on quarterly training briefs, which according to both current and emerging doctrine, really don't mean much.
>
> The only thing a good APFT score (300+) tells me is that the individual puts time and effort into his personal program. It DOES NOT reflect that individual's technical/ tactical capabilities, and should not carry the same weight. . . . I've witnessed PT "studs" who couldn't lead a Boy Scout troop in the pledge of allegiance let alone a Platoon into combat. I personally witnessed a 340 (extended scale) senior NCO who wasn't worth a damn on the third day of a JRTC rotation due to a lack of mental toughness. Meanwhile, I watched NCOs/soldiers with 230-270 averages display exactly what the Army Values mandate. Yet on the NCO [evaluation report] the first example receives an excellent for "maxing" the APFT. There's something wrong there.
>
> Units spend too much time focusing on the three measures of fitness used in the test, and not enough time on being physically healthy. We need a broader based test that incorporates more events checking the overall strength, agility, and endurance of our soldiers.

The Way Ahead: Reclaiming the APFT

Is the professional effectiveness of today's Army still largely dependent upon the physical fitness of its members, or does this criterion no longer apply to today's "high tech" military organizations? Some would argue that the Army's dominance in information technology, advanced target acquisition capabilities, increased weapon stand-off ranges, and smart weapons negate the necessity for soldiers to possess a high level of physical fitness. We believe emphatically that physical fitness remains important. However, the priority and importance accorded to physical fitness in today's Army do not reflect such a view.

Technological advancements in equipment and weapon systems have naturally increased the complexity of employment techniques, use, and maintenance. As we continue to seek equipment and weapon systems that can perform with increased capabilities, should we not also expect the same from the soldiers who operate them? Once again, we believe the answer should be a resounding, Yes. But is the Army's current physical readiness training and testing program utilizing advancements in science and technology to ensure that our soldiers gain and maintain optimal physical performance levels? Our own experiences as well as an alarming number of our survey responses would indicate the answer is No.

Department of Defense Directive 1308.1, issued on 20 July 1995, states:

> It is DoD policy that physical fitness is essential to combat readiness and is an important part of the general health and well-being for armed forces personnel. Individual service members must possess the cardiorespiratory endurance, muscular strength and endurance, and whole body flexibility to successfully perform in accordance with their service-specific mission and military specialty. Those qualities, as well as balance, agility, and explosive power, together with levels of body composition, form the basis of the DoD Physical Fitness and Body Fat Program.[39]

Over the last twenty years the U. S. Army has assessed the physical fitness levels of individual soldiers based on results of the APFT. The APFT is nothing more than a three-event physical performance test used to measure cardiorespiratory and upper and lower body muscular endurance. The APFT was designed as an assessment tool to establish a baseline level of fitness for all soldiers regardless of MOS or duty. Current Army Regulation 350-41, *Training in Units* (1993), states: "The APFT is the commander's tool for measuring minimum physical fitness. Temporary training periods solely devoted toward meeting APFT requirements are discouraged."[40] Yet, in violation of stated Army policy, the APFT has become a primary focus of their unit's physical training program according to a majority of survey respondents.

It appears that the Army has a twofold problem with the current Army physical training program. First, the APFT, which was designed to measure minimal baseline levels of individual fitness, is now being used as a comprehensive assessment of the physical readiness of soldiers and Army units to go to war and physically accomplish their METL functions. Second, the Army is not adhering to the doctrine found in AR 350-41, which clearly states that "as a first priority, commanders will conduct physical fitness programs that enhance soldiers' abilities to complete critical soldier or leader tasks that support the unit's mission essential task list (METL)."[41] Thus the Army is training incorrectly, and it is assessing that training incorrectly. The practical reality is that the Army has inadvertently developed a culture that values success on the APFT above physical capability to perform mission-essential tasks. Army policy to provide promotion points to enlisted soldiers based on their APFT score, commanders requiring subordinates to provide unit APFT averages during quarterly training briefs, and the lack of unit leaders properly trained in the principles of exercise science and human movement have all played a role in causing these policies and priorities. The bottom line is that too many units are spending valuable training time focused on the APFT, which by itself does little to enhance physical combat readiness—the physical ability of the unit's soldiers collectively to do their job.

A more comprehensive approach to Army physical readiness training/assessment (PRT/A) is necessary to restore a training focus on mission-essential tasks and critical soldier skills. The first step in the genesis of a new physical development paradigm is for the Army profession to renew the body of expert knowledge for developing soldiers physically.

Andrew Abbott states that true professions develop, maintain, and control the body of expert knowledge that defines their expertise and enables their professional practice in behalf of the client.[42] In order for the Army to maintain its status as a profession, it must gain firm control of the body of expert knowledge concerning PRT/A. There are two significant constraints in accomplishing this goal. First, there is no single organization within the Army responsible for the development and application of such expert knowledge. Second, there are not sufficient personnel cross-trained in military operations and the discipline of exercise science and human performance to support the development of an expert body of knowledge. Over the past fifty years the Army has spent billions of dollars in research, development, and procurement of hardware and software systems to achieve technological superiority in warfare. However, during those same fifty years the Army has tolerated a bare subsistence budget for personnel and R&D in the sciences of human performance and training for individual soldiers and units.

Organizations such as the Natick Soldier Center (NSC), the U.S. Army Center for Health Promotion and Preventive Medicine, and the U.S. Army Physical Fitness School (USAPFS) have attempted to keep pace with advances in the area of human performance. These organizations are currently tasked with the development and application of the Army's expert knowledge in physical development. We believe the Army's inability to stay current in this vital area of expert knowledge is primarily due to an ambiguous allocation of responsibilities among these organizations. The mission of the NSC is to "maximize the individual war-fighter's survivability, sustainability, mobility, combat effectiveness, quality of life; [that is, to] treat the war-fighter as a system."[43] Enlarging on this idea, the NSC states that:

> We are deeply committed to making our soldiers, and other members of the American Armed Forces, the best equipped, best clothed, best fed, and best protected in the world. . . . As the most important weapon on the digitized battlefield of the future, we owe war-fighters nothing less than our best products, technologies, and customer service.[44]

Although the NSC conducts research on the physiological and biomechanical aspects of human performance, its primary mission is in human factors engineering (HFE). HFE applies the principles of human physiology and biomechanics to the design of war-fighter technology with the intention of maximizing sustainability by increasing mobility, survivability, and combat effectiveness. The NCS is well funded, and its cutting-edge "research and development efforts culminate in combat items that save lives, protect soldiers against battlefield threats and environmental hazards, and steadily elevate their quality of life.[45]

The mission of the USAPFS, on the other hand, is "to develop and field physical fitness doctrine, performance standards, and training programs for leaders and soldiers."[46] The primary objective of the USAPFS, sometimes known as the Soldier Fitness School, is to develop the Army's physical training doctrine. However, the Army's physical training doctrine must be based on an empirical application of the principles of exercise science and human performance. In the execution of its mission the USAPFS has never been funded or staffed to effectively study the physical performance needs and capabilities of soldiers. The development and maintenance of an expert body of knowledge are predicated upon this type of empirical scientific investigation. The USAPFS currently has only one doctoral-trained exercise physiologist on its staff and no laboratory or research facilities. In recent years, the USAPFS commandants have been field soldiers with no background in human performance or exercise science. Without sufficient numbers of properly trained and educated military and civilian professionals, there is no chance of developing truly adequate expert knowledge. As currently configured, the USAPFS has little opportunity to fulfill its organizational responsibilities to develop PRT/A doctrines and training programs for the Army based upon current knowledge in exercise testing and prescription.

As a result, the USAPFS and the Army are mired in a dated paradigm of physical training that has evolved through anecdotal record and trial and error. Two examples from recent doctrine are illustrative of this problem. In the Index/Preface to the recent revision of FM 21-20 (now FM 3-25-20) by USAPFS personnel, it is stated: "Army manuals have undergone cycles of change after major periods of armed conflict, reflecting the lessons learned from the soldiers who endured the rigors of combat. Our objective is to write a manual . . . rooted in time-tested theories and principles."[47] While it is true that historical lessons are valuable in the overall development of doctrine for PRT/A, it is also true that intuition and anecdotal record should not be the sole foundation. The expert knowledge for war-fighter fitness should be based upon replicable results from scientific investigation. A second example of reliance on a dated paradigm is this statement: "If soldiers are to be brought up to the desired standard of physical readiness, a well-conceived plan of mass military physical training must be an integral part of every unit training program."[48] However, in an era of the digitized and decentralized battlefield, the mass athletic model appears as arcane as the musket. There are certainly many other training paradigms that could more effectively be used to physically develop soldiers. Elite units like Special Forces and Rangers often conduct physical readiness training (PRT) at the platoon, squad, or even team level. In response to our Physical Readiness Questionnaire, several officers stated that their unit's PRT program often interferes with their personal training regimen. The presumption that "mass military physical training must be an integral part of every unit training program" is highly problematic in developing training doctrine for soldiers who will fight using the next-generation individual soldier systems.[49]

Mired in traditional PRT paradigms, the USAPFS has found it difficult to keep pace with current trends in physical training/fitness. In sum, the three primary reasons are: (1) significant lack of resources resulting in limited applied scientific research; (2) lack of personnel with expert understanding of the exercise sciences; and (3) lack of synergy between all Army organizations that can assist in the accomplishment of the profession's need for sound doctrine in this area. As a consequence, the expert body of knowledge for developing and maintaining a physically fit Army has fallen significantly behind civilian sport/athletic/exercise models of physical development.

By equating training to education, the Army has fallen victim to the same siren song that has duped many other organizations: we are trained, therefore we are educated. Even officers who have personally achieved a high level of physical fitness usually have little understanding of the scientific principles of exercise frequency, intensity, duration, and recovery. There is a well-known phrase found in FM 21-100, *Army Leadership*, that prescribes the Army's leadership ideal: "Be, Know, Do." With regard to physical development and training, the Army has the Be and Do part down pretty well. However, Being and Doing do not often result in Knowing. Being and Doing should be derived from Knowing. In the athletics/sports world there is an extensive body of expert knowledge relating to exercise adherence, periodization, specificity, cross training, and recovery. Our concern remains: who in the Army is systematically studying, developing, and applying these scientific principles as relating to the fitness of our soldiers and to the accomplishment of the METL functions of our units? To our knowledge, the answer currently is no one. Why do we continue to train soldiers physically pretty much the same way soldiers were trained seventy-five years ago, considering that since that time the science of physical training has advanced light-years? The answer: probably because Army leaders have been wedded to the doctrine from which they themselves emerged. "Tradition-bound and rigidly hierarchical in popular imagination, military organizations are famous for their resistance to change."[50]

Recommendations

With regard to physical readiness training and assessment, the Army has clearly failed to maintain parity with the discipline-based body of knowledge in exercise and sport science. In an attempt to reestablish a professional body of knowledge in Army physical readiness training/assessment (PRT/A), we have developed three recommendations. First, the Army should create an Army Combat Fitness Center with the mission to conduct scientific research into the application of the theories of exercise and human performance to mission-essential tasks and the physical combat readiness of Army soldiers. The Center would have a blend of military and civilian researchers with appropriate education and training to nurture and enrich the body of knowledge for physically developing soldiers. The military researchers would comprise a corps of officers who can understand and apply such scientific principles of exercise and human performance to mission-essential tasks. The physical location of an Army Combat Fitness Center is not important; however, it must be funded and staffed to insure that the expert body of knowledge on Army physical readiness training is cutting-edge.

Our second recommendation is to eliminate the APFT in its current form and function and then replace it. If the Army is to move into the 21st century, the APFT's death grip on physical fitness training within Army units, as well as its influence on organizational climates, must be broken. For too long individual fitness and unit combat readiness have been misunderstood within the Army. Physical fitness is the general capacity of an individual to adapt and respond favorably to physical effort. Physical combat readiness is defined as the level of physical fitness, motor fitness, and movement skills necessary to execute mission-essential tasks. Yes, it is imperative that soldiers develop and maintain a baseline level of physical fitness in order that they might achieve, collectively within units, an adequate level of combat readiness; but these are two very different goals.

We must change the paradigm for physical readiness training/assessment and replace the APFT with a new baseline Army Soldier Fitness Test (ASFT). Similar to fitness tests developed for police, fire, and emergency rescue organizations, the ASFT should be a go/no go test designed to determine whether the soldier is fit for military service, i.e., fit to wear the uniform of the United States Army. As a starting point, we recommend that four components of physical fitness be measured: shoulder-girdle endurance, abdominal endurance, cardiorespiratory endurance, and lower back/hamstring flexibility. Lower back and hamstring flexibility are critical to many aspects of physical training as well as a generally healthy lifestyle. We also believe it is time for the Army to reassess the no-equipment paradigm for fitness testing. An Army that can afford to digitize the battlefield can also afford the modern equipment necessary to physically develop and test the soldiers it places on that battlefield.

Since the ASFT would fulfill the role of a health- and performance-related fitness test, we further recommend a single criterion-referenced standard for each event, one standard for all regardless of age and gender. Although there should be a comprehensive study to determine the test items and corresponding criterion-referenced standards, we recommend for initial consideration the following four exercises, corresponding respectively to the four components of physical fitness listed in the previous paragraph: abdominal crunches (50 repetitions), push-ups (40 repetitions), four-mile run (36:00 minutes), and sit-and-reach (15 inches). In conjunction with a reliable and valid measure of body composition, the ASFT would provide an excellent baseline measure of a soldier's physical fitness. Moreover, in order to maintain the motivational component now missing in unit climates, we further recommend that the Army Physical Fitness Badge be retained with an appropriate criterion-referenced standard.

Once the Army Soldier Fitness Test is established, our third recommendation would come into play: specifically, each branch in the Army would be given sway to develop what might be called Branch-Specific Fitness Tests (BSFT), that is, rigorous, combat-focused physical performance tests. Such tests should be developed from the best research available and strongly focused on combat tasks, both individual and unit. As mentioned earlier, research into successful practices of other professions such as firefighters would be a good place to start. Moreover, we recommend that when introducing the BSFT, the Army drop its practice of gender discrimination in branch or MOS selection. When interpreting current civil rights and labor laws, it is clearly acceptable to discriminate against any protected class when a specific physical requirement is necessary to successfully perform a job or task. But under that same body of laws, it is illegal to arbitrarily discriminate on the basis of age or gender. Following appropriate test development protocols as recommended by the Cooper Institute for Aerobic Fitness in their interpretation of such law, we believe the metric for inclusion in the profession of arms and the various branches should also be performance-based.

We are in a new century, and the Army of the 21st century is transforming into a lighter and more lethal force.

> Strategically responsive Army forces will move wherever needed, and be capable of transitioning quickly from one type of operation to another. Forces must be capable of shifting from engagement to deterrence to war to postwar reconstruction—seamlessly. As missions change from promoting peace to deterring war, or from resolving conflict to war itself, operations become more complex.[51]

This mission, outlined in the new FM 1, *The Army* (2001), clearly describes an Army that must be flexible and adapt quickly to changing situations. In order to effectively execute these missions to the highest standard, soldiers must be physically fit, not just to max a three-event or a six-event physical fitness test, but fit enough to endure the sustained physical rigors of combat. This transformation can be accomplished only by renewing the profession's expertise on individual soldier physical readiness and on creating and maintaining physically ready Army units. Our analysis indicates this transformation will require that strategic leaders of the U. S. Army resource an Army Combat Fitness Center that can maintain professional expertise on physically developing soldiers for success in combat; implement a single-standard baseline physical fitness test; and develop and implement challenging branch-specific physical fitness tests. The time is right to make the U. S. Army truly *An Army of One!*[52]

Case No. 3: The 2000 Army Training and Leader Development Panel

Joe LeBoeuf[53]

Introduction

> The unprecedented display of prowess and efficiency in Desert Storm revealed a military unlike any the nation had ever seen before. From disarray in the aftermath of Vietnam to the "Be All You Can Be" campaign in the 1980s, the US Armed Services did no less than reinvent themselves with stunning results.
>
> James Kitfield[54]

As Samuel Huntington reminded us over forty years ago in *The Soldier and the State*, our military must reflect both its functional imperatives—its professional essence—and and at the same time recognize and adapt to emerging societal forces at any given time.[55] An adaptive capacity, which is illustrated by the epigraph at the beginning of this case study, is a key aspect of organizational learning; moreover, it is an essential quality, if not an organizational imperative, in the emerging nature of successful modern professions. Learning organizations will be the successful organizations of the future.[56] Professional organizations must be learning organizations, focused on constant renewal of their expertise and the knowledge that underlies it, which "are the coins of the professional realm."[57]

As described by James Kitfield, the Army demonstrated impressive organizational learning and professional orientation in the face of rapidly changing environmental demands by reinventing itself (to include its expert knowledge). This reinvention culminated with stunning success during Operation Desert Storm in 1991. Since Desert Storm, the Army has come under criticism for losing its professional focus, particularly in its inability to articulate and apply its expertise, a systemic problem pointed out by several authors in the present anthology. These authors go on to suggest that the Army, as a profession, has become increasing more bureaucratic and occupational in nature.[58]

Such trends have led the Army to question its professional self and even its own legitimacy in an emerging professional world characterized by jurisdictional competition and fluidity. We have witnessed a significant junior officer exodus from the Army, a perceived lack of trust between junior and senior officers, stifling micromanagement, and a perceived lack of reciprocal commitment from the Army to its officers, noncommissioned officers, soldiers, and their families commensurate with their dedication and sacrifices. To its credit however, the Army hierarchy is demonstrating the type of strategic leadership associated with a learning organization, a key characteristic of a profession. Within it internal jurisdiction, the Army's strategic leaders have begun to display an encouraging measure of introspection, studying the

Army profession to insure it is able to meet the demands of the society its serves through renewing and updating its expertise. The ultimate goal is to retain its professional legitimacy in applying its military expertise in a rapidly changing geo-political world.[59]

The Army Training and Leader Development Panel (ATLDP), conducted from June 2000 to January 2001, offers an impressive example of the Army's potential for professional renewal, for revitalizing its professional self in response to the problems noted. In their article, "The Future of Army Professionalism," Don Snider and Gayle Watkins argue that if the Army is going to remain a viable profession, the profession must change itself to meet emerging demands and expanding requirements. This change must be led by the Army's strategic leaders and implemented by and through its officer corps.[60] The ATLDP was created to study the officer corps as a prelude to a process of strategic analysis and communication across multiple levels within the Army profession. This would be the critical first step in the Army's professional renewal.

Purpose

This case will describe the ATLDP in its role as a rigorous self-evaluation leading to reprofessionalization. It will explain why and how the ATLDP was created, and the manner in which it was conducted. It will set forth the key panel findings, including the need to link training and leader development inextricably. The panel's conclusions and recommendations to the Army's strategic leaders will be presented along with comment on the strategic implications of the panel process, particularly as it relates to the reprofessionalization of the Army and its key leadership: officers, noncommissioned officers, warrant officers, and Department of the Army civilians. Finally, although the ATLDP is a very positive case of the introspective process characteristic of a learning organization, some cautions will be registered toward the end of keeping the reprofessionalization efforts on track and consistent with Army transformation.

Background

During the 1990s, the Army experienced a high degree of turbulence and uncertainty as it initiated a transformation to equip it to meet changing operational demands. A high operational pace, with resources and force structure significantly reduced from its status during Desert Storm, added to the turbulence. Resulting conditions included a significant junior officer retention issue, manifest mistrust between junior and senior officers, and significant differences in perceptions among the ranks concerning commitment and the Army's service ethic. Additionally, the Chief of Staff of the Army (CSA) believed there had developed a significant disconnect between training and leader development that if not corrected would make it difficult, if not impossible, to effectively transform the Army.

With these and other issues in mind, the CSA commissioned the ATLDP in June 2000, with the Commander of Training and Doctrine Command (TRADOC) tasked as executive agent, and the Commander of the Combined Arms Center (CAC) as the study director, to review and assess the status of training and leader development in the Army based on the following mission directive:

> Focusing initially on commissioned officers, convene an Army panel to review, assess, and provide recommendations for the development and training of our 21st Century leaders. Identify Objective Force leader characteristics and skills, and the training requirements to support the Objective Force. Ascertain whether the current training construct—institutional training, unit training, and self-development programs—meets future needs. Assess the viability of the existing leader development model to meet mission requirements in full-spectrum operations. Based upon our review, make recommendations to improve Army training and leader development programs to meet the Army's transformational objectives, and address factors directly affecting retention of key officer grades. Develop an action plan to implement and monitor execution of these recommendations. Prepare to conduct similar studies and analysis for the Warrant Officer and NCO corps following completion of the Commissioned Officer portion of the assessment.[61]

Based upon subsequent mission analysis by the panel members, the notions of the service ethic and Army culture were added as significant topics in the study effort.

Study Process

To execute this mission for the commissioned officer portion of the study, the ATLDP was task-organized into four study groups, each focused on a different set of specified tasks that emerged from the TRADOC and CAC commanders' analysis of the CSA's directive. These study groups were as follows: (1) the Army Service Study Group, tasked to examine Army culture, leader development, service ethic, and accession and retention issues; (2) the Army Unit Study Group, tasked to examine training at home station, the Combat Training Centers, the Systems Approach to Training (SAT), and training doctrine; (3) the Army Institution Study Group, tasked to examine the Officer Education System (OES), particularly at the Army's "school houses"; and (4) the Army Self-Development Study group, tasked to examine self-development and mentoring in the Army. Each of these study groups was led by an Army colonel with extensive subject-matter expertise in the given study area, and each group was staffed with company and field grade officers and senior noncommissioned officers from units throughout the Army who served as subject-matter experts.

To enhance the reliability and validity of the study process, three other small organizations were created. First, an integration team was established, with its core of members from the Center of Army Leadership (CAL), the Center for Army Lessons Learned (CALL), and the Army Research Institute (ARI) field research cell. This integration team provided analytical, planning, and logistical support for all four study groups to include the worldwide data-gathering effort. Secondly, an Executive Panel was established, staffed with senior officers, senior noncommissioned officers, civilian subject-matter experts, and academic scholars to provide the study groups and the study director and staff with advice and direction. Finally, a Red Team was created with membership similar to that of the Executive panel and headed by a promotable colonel. The Red Team served as the study's "conscience." Its task was providing real-time critical reviews, insights, and general azimuth checks in the ATLDP's investigative process and subsequent findings and recommendations.

Overall, this unique methodology represented a significant investment in intellectual and experiential capital to insure that the structure, process, and content of the study would be reliable, valid (inclusive of wide-ranging disciplines), and in keeping with the CSA's intent. The study methodology, a *disciplined, iterative process* with a strong feedback mechanism, provided credible data and analysis.[62]

This methodology was developed by CALL and was included in the lessons learned database as the model for how information is gathered from the field, analyzed, and categorized. This methodology is built around four activities: *Organize* (educate and analyze), *Assess* (evaluate), *Visualize* (conclude), and *Describe* (recommend). Issues were determined from a detailed mission analysis conducted by each study group, from which a survey instrument and individual and focus group interview protocols were developed. Four travel teams were constituted that included representation from each study group. These four travel teams were assigned to visit various posts, camps, and stations across the depth and breadth of the Army, gathering data to be used in the study.[63]

The findings and conclusions of the ATLDP covered over twenty-five major operational areas in the Army. The findings were organized into several major categories: Army culture, the Officer Education System (OES), Army training and the Systems Approach to Training (SAT), and the Army training and leader development linkage. The following subsections will summarize the key findings and conclusions of the panel across these major categories.

The Army Culture

Culture is the accepted understanding of what is right and true about any organization. In essence, it is the members' agreement on how things get done. It is the "amalgam of values, customs, traditions, and their philosophical underpinnings that, over time, has created a shared professional ethos."[64] From culture springs a common framework for the members of the military describing the behavioral expectations within an organization. Although Army culture is not always well defined and is often misunderstood, it has an enormous impact on leader behavior and the attitudes of followers in interpreting that behavior. It significantly influences everything that occurs in an organization.[65]

The culture of the military is the set of beliefs through which the accepted behavioral norms of the profession are prescribed and maintained. Army culture is the underlying foundation for all aspects of the Army's training and leader development programs. It is pervasive and must be attended to, otherwise it will evolve and drift away from the proven norms, and will ultimately reinforce leader behaviors that are unwanted and inimical to the Army's training and developmental needs.

The Army's cultural condition is most directly assessed at any point in time through the measurement of organizational climate. Climate is a reflection of how organizational members feel about various aspects of their organization, particularly how well or poorly the organization is living up to its cultural norms. Climate considerations involve perceptions by organizational members across a number of dimensions to include work load; job performance expectations; fairness of reward, punishment, and administrative systems; flow of communications; and the example set by all organizational leaders.[66] These dimensions, as well as many others, were directly assessed by the ATLDP. Over the long term, changes in climate significantly influence the organization's underlying culture.

Based on climate considerations, the ATLDP assessed Army culture as "out of balance" and failing to provide the conditions necessary to preserve the Army's professional standing. Army culture has in places breached an acceptable "band of tolerance,"[67] thereby compromising unit readiness and leader growth, and reducing individuals' overall satisfaction with and motivation for service.[68] In other words, the professed principles of officers do not always coincide with their actual practices. Army culture must reflect a set of conditions that embody a mutually supportive and trustworthy relationship between individual professionals within the organization and the Army as a profession. A military culture that is out of balance can be seen in a number of ways, but most importantly through an excessively stressful operational pace or tempo, lack of reciprocal commitment, a misunderstood notion of service ethic, unfulfilled junior officer expectations, and micromanagement by superiors.

Operational pace (OPACE) is the flow of activity that pervades all aspects of the professional and personal lives of officers and their families. Since the end of the Cold War and the Army's success in Desert Storm, the Army has endured a reduction in force structure by nearly 40%, but has been subjected to an increase in OPACE by nearly 300%. This excessive OPACE results in junior officers not having sufficient time to conduct their own unit training and insufficient time to assess and retrain at all echelons. The lack of "white space"[69] on training calendars is systemic, created in large measure to satisfy multiple, non-mission-essential task list training requirements generated by strategic leaders at MACOMs and higher headquarters. Examples of these training distracters include non-mission-related taskings, borrowed military manpower (BMM), and support for non-training-related events, particularly garrison and community support missions.[70] Army units face relentless training and operational requirements and literally race from one event to the next driven by a culturally reinforced "can do" attitude to all assigned tasks. "No" is not widely used in the Army lexicon, particularly by higher-level commanders. A "can do" approach, plus unwillingness to say No to discretionary requirements, places an unsustainable demand on the commitment, service ethic, and good will of junior Army leaders and their families.[71]

Commitment is a military imperative, one of the most important aspects of military operational readiness, both in peacetime and in war. It has often been said that commitment is the sine qua non of effectiveness in the volunteer military. Charles Moskos, a leading military sociologist, maintains that commitment is an essential component of motivation for effective performance in a military environment.[72] Few civilian occupations require the same high levels of commitment as do the military services, particularly from the officer corps. The Army is a greedy institution demanding a level of professional commitment from its officers that conflicts with the familial and emotional commitment demanded by spouses, many with careers, and families. Meanwhile, the professional commitment of Army officers is not perceived by these officers as being matched and reciprocated by the institutional Army in terms of support, rewards, and understanding, particularly in terms of service ethic.[73]

A consistent finding in the ATLDP study, for example, was that many junior officers are leaving the Army because they believe there is a lack of a return commitment by the Army to them and their families. The Army demands a commitment that includes unlimited liability and the potential sacrifice of life, and in turn it is duty-bound to demonstrate a commensurate sense of responsibility for the welfare of the professional officer and his family.

The panel also found that senior and junior officers differ significantly in their understanding of the notion of commitment and the service ethic, creating a huge gap in perceptions and expectations from which a significant trust issue arises.[74] The service ethic is the commitment by military professionals to serve the nation, the Army, its soldiers, and their families above self. It is a self-concept of "servant" with a moral obligation to serve the society externally and the members of the professions internally. Such service is manifested as a duty established in the officer's Oath of Office, which requires the officer to "support and defend the Constitution of the United States against all enemies, foreign and domestic." This oath binds the officer to the profession, but when fulfilled also "strengthens the claim of the military profession on the affections and support of the American people," thereby enhancing the legitimacy of the Army profession.[75]

A strong sense of service and commitment to the nation and the Army permeated the study findings. Officers value operational and educational experiences, and understand the benefit of leadership opportunities when they are available. But the Army's service ethic and concept of officership are not well understood across the officer corps, nor clearly defined in our doctrine. Such lack of mutual understanding fuels the misperceptions and mistrust between senior and junior officers. The study's data showed that trust is broken between senior and junior officers due to a different understanding of the service ethic. Specifically, the junior officers wonder whether the senior officers truly place their commitment to the nation, the Army, its soldiers, and their families above their own narrow career interests. This concern is compounded by the lack of effective senior-subordinate contacts and lack of confidence in the officer evaluation report (OER) and system.[76]

Expectations of junior officers are not being met, particularly as to opportunities to serve as small unit leaders, master their craft, and develop the warrior ethos. The flame of the warrior ethos and confidence in the Army's war-fighting foundation is quickly extinguished when operational expectations and leader development opportunities for junior officers are not met. Junior officers' experience at platoon and company level branch-qualifying jobs are regarded as too brief; young officers perceive a lack of freedom to fail and learn.[77] The flame of the martial ethos is further stifled by reduced Mission Essential Task List (METL) training opportunities as a result of significant resource reductions, non-mission-related requirements, and a high operational pace that have generated a leadership climate characterized by micromanagement.[78]

Junior officers, because of a significant captain shortage, are frequently asked to do jobs they are not prepared for, and are being micromanaged by senior officers pressed by high operational pace to get jobs done without latitude for mistakes or failure.[79] The Army Personnel Management System primarily focuses on assigning a "face to a space," without an apparent concern for leader development needs of the officer. Young officers come into the Army expecting a deliberate process of professional development to occur, and are disappointed when it does not. The Army does not emphasize the importance of self-development, mentoring, and counseling. In these areas, Army doctrine is vague and inadequate, but more importantly the profession's strategic leaders do not demonstrate a good understanding of or support for the professional development process. The study results indicate that rather than a profession consciously developing its junior members, it is more often the case of a bureaucracy simply injecting warm bodies into required slots.[80]

The Officer Education System (OES)

The Army's OES is one of the most comprehensive in the world and has been a hallmark of the profession. But the OES needs to be redesigned to meet the educational and training needs of the 21st century officer-soldier. It is in need of reform from pre-commissioning level and continuing throughout an officer's professional service. OES must transit from a Cold War focus to one that prepares officers for a new operational environment characterized by regional threats, full-spectrum operations, and information age technology. In addition to conveying the profession's expert knowledge, it must develop in officers the key competencies of self-awareness and adaptability, and instill a commitment to lifelong learning.

The OES must also use information technology to provide just-in-time learning capable of meeting rapidly changing educational and training needs. The OES must link schools horizontally and vertically in a manner that synchronizes and integrates educational and operational experiences of officers. Additionally officers must be trained to a set of common standards, *those that certify professionals to the nation*, consistent with the emerging principles of officership. Current initial-entry education and training for officers vary significantly among sixteen basic branch courses. Furthermore, the Army, through its school selection process, deliberately excludes 50% of its majors from the command and staff level schooling they need to be effective in a rapidly changing Army. The panel found this practice detrimental to officer development and retention.

Army Training and the Systems Approach to Training (SAT)

The panel concluded that the Army is training hard in the field, but is not training to doctrinal standards. A significant number of non-mission-related taskings, an excessive operational pace coupled with an untempered "can do" attitude, and a shortage of training resources make it difficult to effectively execute home station training consistent with established doctrinal standards. This lack of effective training is having a significant impact on the leader development of junior officers who have difficulty in learning what "right looks like" and as a result do not master sound principles and practices of effective training. Additionally, many of the Army's training manuals need to be rewritten to meet the rapidly changing training demands of Army transformation. Training aids, devices, simulators, and simulations (TADSS) are outdated and do not adequately meet the demands required of realistic training. Many units in the Army have weapons and command and control systems lacking the appropriate TADSS.

The Combat Training Centers (CTCs) remain the crown jewels of the Army's training system and have enabled the Army to achieve great operational success and a world-class reputation. The CTCs provide the most realistic tactical leader development experience, short of actual combat, of any training operations. But the CTCs are getting old and need to be revitalized through significant recapitalization and modernization. Modernization of the CTCs must be combined with increased resourcing of home station training in order to maximize the training and leader development experiences that occur at the CTCs.

Finally, the systems approach to training (SAT) was found to be fundamentally sound, but not being executed in the manner intended by its designers. Due to a significant reduction in training and development resources, force reductions, and reduced manning levels within TRADOC, the Army is considerably short in up-to-date training and educational products, the foundation for standards-based training and leader development. In other cases, Army training standards are outdated or have not been adequately defined or described.

Army Training and Leader Development Linkage

Training and leader development, the cornerstones of an effective Army, are not thoroughly linked and integrated within the Army culture. Part of the problem is that responsibility for training and leader development program management is split between two separate sections of the Department of the Army General Staff. More important, however, is that leader development is not adequately understood as a process that includes more than just operational experiences. Rightly understood, it also includes continuous learning, regular performance feedback through reflection and mentoring, and sufficient time in developmental positions.[81] In the field, training and leader development are too often seen as competing requirements, not as linked imperatives specifically articulated in the Army's definition of effective leadership. Leader development falls "below the line" of attention when commanders are faced with unrelenting operational imperatives.

Leader development is not well resourced, and competes with training and officer personnel management systems that are. As noted earlier, leader development is driven to a great degree by a bureaucratic Officer Personnel Management System that seeks simply to assign bodies rather than function as a true human resource management system driven by the need for a profession to develop leaders. The Army's strategic leadership must focus leader development on those emerging competencies that reflect changes in the operational environment and Objective Force requirements. The aim should be to produce leaders committed to life-long learning and who are self-aware and adaptable.

The Army's personnel management system is focused on assignments to positions that serve as promotion and command qualification gates rather than opportunities to complete significant developmental experiences based on articulated standards. This process results in unmet expectations on the part of many junior officers, who recognize they are not receiving required developmental experiences and opportunities and thus are not being qualified for subsequent command positions. The process can also result in excessive micromanagement. The resulting dissatisfaction is, the panel concluded, a significant source of the junior officer retention problem.

Finally, the Army's leader development model (consisting of three separate pillars of institutional development, operational experience, and self-development) is out of date, failing to meet the needs of Army transformation. It is a Cold War, industrial-age construct resting on standards that no longer exist, and it is not well linked to today's and emerging training requirements. The Army's current culture reinforces officer performance at the expense of learning and leader development. The Army has yet to fully embrace the definition of leadership in Field Manual 22-100, *Military Leadership*: "Leadership is influencing people—by providing purpose, direction and motivation—while *operating* to accomplish the mission and *improving* the organization."[82] Observe that leadership by the Army's seniors is more than just operating, it is also making the organization better. The panel concluded that the Army now is a "doing" organization rather than a "learning" organization. Attempts at learning to improve run into direct conflict with a high operational pace and the high priority of accomplishing associated tasks.

Recommendations

Based on its findings and conclusions, the ATLDP developed a set of eighty-four recommendations across the four major categories of assessment for implementation by the Army's strategic leaders. The recommendations are designed to revitalize our expertise in those areas so critical to the Army's internal jurisdiction. Major recommendations in each category are set forth in the four subsections that follow.

Culture

- The strategic leadership of the Army must reinforce and sustain a military culture that promotes the evolution of professional expertise, but that also emphasizes "timeless values as a stable foundation for service members in the face of profound change."[83]
- The profession must better define and teach an Army service ethic and the concept of officership throughout the OES from pre-commissioning through senior service colleges attendance. Particular emphasis must be placed on the company grade educational and developmental experiences.
- Strategic leaders must demonstrate a commitment to Army families commensurate with the level of commitment they require of service members. A comprehensive review of all Army management systems must be conducted to

insure they demonstrate that the Army leadership is committed to its soldiers and families.

- Strategic leaders must reestablish disciplined restraint in the training management process, eliminating training not directly related to unit mission readiness.
- Considerable work must be conducted to establish centralized control over internal and external taskers to insure they are valid and consistent with unit capabilities.
- Fundamentally, strategic leaders must reduce operational pace across all operating systems and increase predictability, balance, and time for the well-being of all Army members and their families.
- To retain officers, the Army must protect junior officers' developmental experiences and meet established expectations. Assignments must be based on a more comprehensive understanding of development that requires adequate time in jobs; officers must receive the time and resources to meet associated developmental standards.
- All majors should receive Intermediate Level Education (ILE),[84] thereby eliminating it as a selection hurdle. Our Army needs well-educated field graders.
- The Army's strategic leadership must revise its career management guidance as set forth in DA PAM 600-3 to reflect a focus on quality operational and educational experiences that are truly developmental in design as opposed to a series of promotion and command qualification wickets. Additionally, the personnel system must shelter junior officers, particularly lieutenants, from premature post-developmental assignments. They must be allowed to remain as small unit leaders until they have developed appropriately based on an established standard. These standards must be qualitatively established to insure effective branch qualification at company and field grade level, qualification based on operational experience and not just time in a given position.
- Additionally, strategic leaders must revitalize our counseling and mentoring process, and be accountable for insuring that officers are counseled, mentored, and developed consistent with their development needs.
- Army strategic leaders must be more actively committed to leader development and to the effective building of our future replacements.

Officer Education System (OES)

- Quality of the OES experience from the Officer Basic Course (OBC) through senior service college must be improved to meet the demands of the 21st century and its focus on full-spectrum operations.
- A major transformation of education and training from OBC through ILE must be designed and implemented. This revised OES must provide officers with realistic, challenging, and relevant training and education so that they can better lead immediately upon arriving in operational assignments.
- The redesigned OES must facilitate the rapid socialization of incoming officer cohorts, who must develop a sound concept of the profession and officership and what it means to serve as an officer in the Army before they are proficient and knowledgeable in their branch-related expertise.

- The Army strategic leadership must also change the manner in which they recruit, educate, and assign faculty to our key schools. Only the best-qualified, most experienced officers should be teachers, mentors, and role models in our service schools. This should be a most valued and career-enhancing assignment.
- The Army's strategic leaders must establish a comprehensive OES accreditation process to insure that the highest of academic standards are established and met for the faculty, curricula, facilities, and students.

Training and Systems Approach to Training

- Although our fundamental training doctrine is sound, the Army's strategic leaders must rapidly adapt this doctrine to reflect emerging operational considerations, and institutionalize a more responsive process to keep the doctrine current and disciplined in its application.
- The Army's strategic leaders must orchestrate the rewriting of our training doctrine to adapt to full-spectrum operations and Army transformation.
- The Army's strategic leaders must preserve the Combat Training Centers (CTCs) as our premier training and leader development experience, but allocate the resources for significant modernization to provide full-spectrum, multi-echelon combined arms training and leader development experiences.
- Additionally, the Army's strategic leaders must redesign the Systems Approach to Training (SAT) development and support structure so that it better leverages the subject matter expertise in the CTCs for training and leader development and doctrinal writing. This process needs to be more attentive to leveraging retired officer/senior NCO expertise through the use of Title 10 positions for the development of training publications.

Training and Leader Development

- The disconnect between training and leader development exists both structurally and functionally. The Army's strategic leadership must establish a single Army Staff (ARSTAF) component for training and leader development, thus structurally linking training and leader development policy and resourcing. The Army's doctrine must be rewritten to reflect this proponency change.
- The Army's strategic leaders must create the conditions to foster lifelong learning on the part of our leaders through a balanced approach embracing educational and operational experiences, self-development, and just-in-time learning through distance education.
- The Army's strategic leaders must develop, build, and fund a Warrior Development Center as a single source for all developmental needs.
- Finally, the Army's strategic leaders must doctrinally rethink the current training and leader development model, and redesign it to reflect the emerging emphasis on the structural and functional linkages.

Implementation Issues and Considerations

The strategic leaders of the Army have taken a significant step in the direction of professional renewal through the creation of the ATLDP and its study process. While the findings of the officers' study have been vetted and prioritized for implementation by the Chief of Staff of the Army (CSA), the work on professional renewal continues. The CSA has initiated follow-on studies of the noncommissioned officer corps, warrant officers, and Department of the Army civilians, which will consume a substantial two plus year effort devoted to introspection, organizational change, and professional revitalization. These actions are clear examples of attempts to control and discipline the Army bureaucracy so as to support the renewal of the Army profession by enhancing the development of its expert knowledge and imbuing its members with that knowledge covering the internal jurisdiction of the profession. But these initiatives should not stop there. They must be reflected in a long-term perspective based on a solid understanding of organizational culture and an institutionalized commitment to organizational learning.

Several other studies of the Army, the Army professionalism study in the early 1970s and the Sullivan study in the 1980s in particular, drew many of the same conclusions about issues facing the Army.[85] Unfortunately, 20-30 years later, many of these issues still remain and once again are negatively impacting the profession. While there are several explanations that could be adduced to explain this recurrence, the most compelling and relevant to the future involve organizational culture and change. The key issue is that previous findings were not implemented based on a sound understanding of how militaries change and were not ultimately made an enduring feature of the Army's professional culture.

John Kotter, a leading authority on organizational change, suggests that transformational efforts fail because of the inability of the strategic leaders to ultimately embed changes in an organization's culture.[86] As mentioned earlier in this case study, culture matters and powerfully influences behavior, and it is very difficult to change.

> Culture matters because it is a powerful, latent, and often unconscious set of forces that determine both our individual and collective behavior, ways of perceiving, thought patterns, and values. Organizational culture in particular matters because cultural elements determine strategy, goals, and modes of operating. If we want to make organizations more efficient and effective, then we must understand the role culture plays in organizational life.[87]

When changes made through transformational efforts (such as the ATLDP) are not compatible with an organization's culture, they will always be subject to regression back to former modes of behavior.[88] Obviously, ideas and initiatives can be enormously useful, but the Army has failed to institutionalize a process to monitor the evolution of the Army culture to insure that changes are embedded and reinforced over time. The ability to understand the strengths and limitations of the Army's culture and to condition the culture to readily adapt to rapidly changing circumstances is the ultimate challenge of the profession's strategic leaders.

The Army's strategic leaders must also reinforce their own commitment to promote the growth of a true professional, learning organization. A learning organization is characterized by systems thinking, personal mastery (including self-awareness and adaptability), shared vision, changing mental models (how things are thought about), and team learning.[89] The Army's strategic leaders have moved a good distance along these dimensions. They have declared in favor of the proposition that the success of Army transformation is dependent on the mastery of three enduring competencies: self-awareness, adaptability, and lifelong learning. They have created a compelling vision of transformation, and all are working hard on changing the thinking of a classic bureaucracy, institutionally resistant to change. The Army had always has a team perspective, building successful organizations through collective action. But to be effective as a profession, our strategic leadership must become more adept at integrating the three competencies named above into a coherent body of theory and practice that is passed on to succeeding generations of officers.

The Army remains largely a "doing" organization, and has not yet made the shift to a learning organization. To become a true learning organization, the Army must continue to demonstrate the capacity as a profession to be self-regulating, bringing practice into alignment with doctrine within it internal jurisdiction as recommended in the ATLDP. The Army cannot train its way out of these problems, but must include a substantial educational and developmental component for all of the profession's members. Actions must be top-down, with strategic leaders creating conditions for change, and bottom-up, with junior officers educated, trained, and developed in a manner more consistent with the demands of the profession and Army transformation.

Notes

1. The views and opinions expressed in this study are those of the authors and are not necessarily those of the Department of the Army or any other U. S. government entity.
2. In 1969 44% of the Army was married, with the figure rising to 55% in 1989 and 63% in 1999. This trend is attributable largely to increases in the proportion of married enlisted soldiers. Greg Baxter, "Demographics of Well-being" (briefing commentary, U.S. Army War College, Carlisle Barracks, PA, 12 November 1999).
3. Examples are: (1) PL 93-112, Section 504, Rehabilitation Act of 1973, protects the rights of and prohibits discrimination against disabled people. This law resulted in DODD 1342.17, *Family* Policy, that directs the establishment of coordinated multi-agency services for families with special needs, i.e., the Exceptional Family Member Program (EFMP); (2) PL 104-1006, Military Child Care Act of 1989; (3) PL 101-189, Defense Authorization Act of 1989, Section 661, Military Relocation Assistance (requiring DOD to provide relocation assistance services).
4. U.S. Army Chief of Staff, Department of the Army Pamphlet 608-41, *The White Paper* (Washington, DC: Department of the Army, 1999).
5. Further details on Army Family Support Programs and the AFAP, with references to specific documents and legislative highlights, is provided in Appendix F of *A Well-being Framework for the US Army: Report to the Chief of Staff, Army*, US Army War College Well-being Committee, 21 January 2000.
6. Details of the history of program compilation and tracking of budgets for this period were provided in a series of personal interviews. Interviews by authors with Karin McArdle, Budget Officer, Deputy Chief of Staff, Personnel, 13 and 22 December 1999, Pentagon, Washington, D.C.
7. Congress, House of Representatives, Committee on Appropriations, *Department of Defense Appropriations for 1990*, 101st Cong., 1st sess., 7 March 1989.

8. Despite this definition and outline for component programs, the DCSPER, in fact, tracked programs for selective medical services for Real Property Maintenance, Army (RPMA), and for motor pool hardstand construction in the FY90-94 Program Objective Memorandum.
9. Haynes Johnson and David S. Broder, *The System—The American Way of Politics at the Breaking Point* (Boston, MA: Little, Brown and Company, 1996), 61-62.
10. *A Well-being Framework*, Appendix H-3.
11. Department of the Army, Office of the Vice Chief of Staff, Army, Memorandum dated 5 January 2001, SUBJECT: Well-being Strategic Plan, 2.
12. Andrew Abbott, *The System of Professions: An Essay on the Division of Expert Labor* (Chicago, IL: University of Chicago Press, 1988), 103
13. C. P. Alderfer, "An Empirical Test of a New Theory of Human Needs," *Organizational Behavior and Human Performance* 4 (1969), 142-175.
14. Abbott, 54.
15. Charles A. Cotton, "The Institutional Organization Model and the Military," in *The Military: More Than Just A Job?* eds. Charles C. Moskos and Frank R. Wood (New York: Pergamon Press, 1988), 46.
16. *A Well-being Framework*.
17. Information on Army transformation is available from numerous sources; see the Army Transformation Homepage, *http://www.army.mil/usa/Cover%20Sheet.htm,* accessed 1 June 2001.
18. Borrowing from Abbott: "Academic knowledge legitimizes professional work by clarifying its foundations and tracing them to major cultural values." Abbott, 54
19. Michael J. Stehlik, "Systematic Planning for Raising the Future Force," in *Population Diversity and the U.S. Army*, eds. Lloyd J. Matthews and Tinaz Pavri (Carlisle Barracks, PA: U.S. Army War College, Strategic Studies Institute, 1999), 69-82.
20. Abbott, 46.
21. Abbott, 69-79.
22. Headquarters, Department of the Army, FM 21-20, *Physical Readiness Training* (Washington, DC: U.S. Government Printing Office, 1950).
23. J.P. Ladd, "U.S. Army Physical Fitness Testing: Past, Present and Future" (Unpublished thesis, location unknown, March 1971).
24. Louis F. Tomasi et al., "Age and Gender Performance on the U.S. Army Physical Fitness Test (APFT) 1995," a study conducted for the Chief of Staff of the Army, available from *http://www.benning.army.mil/usapfs/Research/1995APFTUpdateSurvey.html;* Internet; accessed 27 October 2000.
25. Vince Crawley, "Conservatives Push Return to Same-sex Training," *Army Times*, 6 August 2001, 39.
26. A key finding in the 1997 Secretary of the Army's Senior Review Panel on Sexual Harassment was that gender discrimination is more common than sexual harassment.
27. Cooper Institute for Aerobics Research, *Common Questions Regarding: Physical Fitness Tests, Standards, and Programs in Law Enforcement* (Dallas, TX: January 1999), booklet on-line available from *http://www.cooperinst.org/lawenf.asp,* accessed 20 April 2001.
28. Ibid.
29. Ibid.
30. Ibid.
31. El Paso Police Department, *Career Opportunities*, (El Paso, TX), booklet on-line; available from *http://rgfn.epcc.edu/users/eppd/recruit1.htm,* accessed 23 April 2001.
32. Las Vegas Metropolitan Police Department, *Police Recruitment,* (Las Vegas, NV), booklet on-line; available from *http://www.lvmpd.com/recruit/police/study/htm,* accessed 23 April 2001.

33. Albuquerque Fire Department—Ph-ysical Fitness Test Home Page (Albuquerque, NM), booklet on-line; available from *http://www.cabq.gov/fire/fit.html,* accessed on 19 April 2001.
34. Houston Fire Department Recruiting Division Home Page (Houston, TX), booklet on-line; available from *http://www.hfd.ci.houston.tx.us/Recruiting/,* accessed on 23 April 2001.
35. Kevin Flynn, "Despite Recruiting, Few Women Do Well in Firefighter Tests," *New York Times*, 3 February 2000, B1.
36. There was self-selection on the part of the respondents; however, it is obvious by the response rate that interest in the area of physical fitness and testing is high. The baseline demographics of the sample are: males = 89.9%, females = 10.1%; officers = 63%, enlisted = 37%; median age = 35 years; median rank (O) = Major; median rank (E) = SFC; commissioning source: USMA = 36.1%, ROTC = 47.2%, Other = 16.7%; respondents by branch (major categories): 21.5% Infantry, 8.6% Field Artillery, 7.9% Medical, 7.6% Aviation, 6.2% Adjutant General, 5.8% Engineers, 5.5% Signal, 5.4% Ordnance; 48.4% currently serving in a TOE unit; median education, 4-year college degree; median reported APFT, 270-300.
37. In a laboratory environment cardiorespiratory efficiency is measured via a procedure called maximum oxygen uptake (VO_2 max). VO_2 max is a measure of the body's ability to do aerobic work, based on oxygen consumption and utilization. Rates are measured in milliliters of oxygen consumed per kilogram of body weight per minute (ml/kg/min). Depending upon age, or physiological maturity, there is a positive correlation between VO_2 max and length of run; generally the longer the distance, the higher the correlation with VO_2 max. Shaver (1975) studied the relationships between VO_2 max and length of run for 30 college-aged men. He concluded that relationship increased with distance (1M - r_{xy}= -.43; 2M – r_{xy}= -.76; 3M – r_{xy}= -.82).
38. During the past four years, members of the Department of Physical Education faculty at West Point have conducted focus interviews with groups of former battalion commanders who were attending the Army War College. In 2000 a theme emerging during these focus groups was the importance the APFT has taken on in their units. APFT scores are reported as a part of Quarterly Training Briefs.
39. Department of Defense Directive 1308.1, 20 July 1995: available from *http://www.dtic.mil/whs/directives/corres/pdf/d13071072095/d13081p.pdf,* accessed on 13 November 2001.
40. AR 350-41, *Training in Units*, 19 March 1993, 16.
41. Ibid, 17.
42. Andrew Abbott, *The System of Professions: An Essay on the Division of Expert Labor* (Chicago, IL: University of Chicago Press, 1988).
43. Natick Home Page: available from *http://www.natick.army.mil/soldier/index.htm,* accessed on 25 June 2001.
44. Ibid.
45. Ibid.
46. U.S. Army Physical Fitness School Mission available from *http://www-benning.army.mil//usapfs/AboutUs/index.htm,* accessed on 25 June 2001.
47. U.S. Army Physical Fitness Doctrine: available from *http://www-benning.army.mil/usapfs/doctrine/FM3-25-20/index.htm,* accessed on 25 June 2001.
48. Ibid.
49. Ibid.
50. Kim Field and John Nagl, "Combat Roles for Women: A Modest Proposal," *Parameters*, 81 (Summer 2001): 74-88.
51. Headquarters, Department of the Army, FM 1, *The Army*: available from *http://www.adtdl.army.mil/cgi-bin/atdl.dll/fm/1/ch1.htm3-6,* accessed on 14 June 2001.
52. The views and opinions expressed in this case study are those of the authors and are not necessarily those of the Department of the Army or any other U.S. government entity.

53. Col. Joe LeBoeuf served as a study group team chief (Army Service Study Group) for the officer's study, and a member of the executive panel for follow-on meetings and briefings through the summer of 2001. Additionally, he is currently serving as a member of the executive panel and the culture study group for the NCO study, which followed the officer study. He shared, with many others, in the writing of the final report of the officer's study, and is thoroughly versed in study findings, conclusions, and recommendations. The views and opinions expressed in this case study are those of the author and are not necessarily those of the Department of the Army or any other U.S. government entity.
54. James Kitfield, *Prodigal Soldiers: How the Generation of Officers Born of Vietnam Revolutionized the American Style of War* (New York: Simon and Schuster, 1995). This quotation comes from the summary on the dust jacket. The book is a detailed history of the transformation in the armed services from Vietnam through the successes in Desert Storm, as seen through the eyes of some of its most well-known flag officers.
55. Samuel P. Huntington, The *Soldier and the State* (Cambridge, MA: Harvard University Press, 1957).
56. Peter M. Senge, *The Art and Practice of the Learning Organization* (New York: Doubleday, 1990), 3-11.
57. Don M. Snider and Gayle L. Watkins, "The Future of the Army Profession," *Assembly*, November/December 2001, 46.
58. Charles C. Moskos and Frank Woods, eds., *The Military: More Than Just a Job?* (Washington, DC: Pergamon-Brassey's, 1988). This edited volume is devoted to adducing evidence that military institutions, to include the U.S. military, are moving away from a professional format to one more resembling a civilian occupation.
59. In the present anthology, Andrew Abbott (Chapter 24) and James Burk (Chapter 2) provide more detailed arguments and discussion concerning the evolving nature of the professional practice and the impact this evolution is having on the U.S. military as a professional organization.
60. Don M. Snider and Gayle L. Watkins, "The Future of Army Professionalism: A Need for Renewal and Redefinition," *Parameters* 30 (Autumn 2000): 20.
61. *Army Training and Leader Development Panel Report (Officers): Final Report* (Ft. Leavenworth, KS: Combined Arms Center, 2001), Appendix A A-1. The quotation is taken from the CSA's directive for the Commanding General, TRADOC to establish the ATLDP charter.
62. William M. Steele and Robert P. Walters, "Training and Developing Leaders in a Transforming Army," *Military Review* (accessed at *http://www-cgsc.army.mil/milrev/English/SeptOct01/steele.htm*, September/October 2001), 9. Lt. Gen. Mike Steele, as the ATLDP director, built the study group structure and process, integrating the expertise of the Army Research Institute (ARI) along with a civilian contracted analysis firm to insure an acceptable level of reliability and validity for the study's findings.
63. The study groups' data-gathering effort lasted 40-45 days, and resulted in over 13,500 contacts with Army leaders and their spouses (the largest survey of its type of the officer corps in the history of the U.S. Army). A majority of those surveyed and interviewed were company and field grade officers through the rank of major. Individual interviews focused on commanders at brigade level and above. The data-gathering effort of this study was also the largest effort of its type ever conducted. It represented a broad sample representative of the Army across all categories to include gender, ethnicity, rank, and arm of service. Additionally, the research effort allowed for the travel teams to get a "sense of the issues" directly from officers and senior noncommissioned officers in the field.
64. Center for Strategic and International Studies (CSIS), *Report: American Military Culture in the 21st Century* (Accessed at *http://www/csis.org/pubs/am21/exec.htm,* February 2000), 4.
65. "Culture matters because it is a powerful, latent, and often unconscious set of forces that determine both our individual and collective behavior, ways of perceiving, thought patterns, and values. Organizational culture in particular matters because cultural elements determine strategy, goals, and modes of operating." From Edgar H. Schein, *The Corporate Culture: Survival Guide* (San Francisco, CA: Jossey-Bass Publishers, 1999), 14.

66. CSIS, 4.
67. The "band of tolerance" is a term coined during the discussion of the culture issues. The Army states that it must always operate within a band of excellence to insure our units are always prepared to fight our nation's wars. The "band of tolerance," as a similar metaphor, describes the range of cultural imbalance that is acceptable to Army officers before we are outside the band of excellence. Currently, the Army culture has moved outside the band of tolerance. Additionally, the cultural imbalance refers to the notion that Army beliefs (what it says it values) and its practice (what it actually does) are in conflict and out of balance. The Army wants a certain type of behavior among its officer members, but is organizationally reinforcing many of the opposite behaviors. See the Steele and Walters article cited in note 10 for a detailed discussion of the "band of tolerance" and its relationship to the cultural imbalance as identified by the ATLDP.
68. ATLDP: Final Report (Officers), ES-8.
69. "White space" was an indictment of gated training strategies wherein everything a unit had to do for a CTC rotation and operational deployment was laid out 6-9 months beforehand, so that units simply executed what higher headquarters had determined were the training requirements for that "event" and thus locked into the training calendar.
70. ATLDP: Final Report (Officers), 2-44.
71. The Army has a "can do" culture that is both a blessing and a curse. It is a blessing in that it describes an organizational culture and individual behavior that are willing to take on any and all tasks assigned, and get the job done. It is a curse in that this is so embedded in its culture that the Army and its leaders will not say No because it would seem unprofessional; thus a high OPACE, and an over-worked professional force.
72. Charles C. Moskos, "The All-Volunteer Military: Calling, Profession, or Occupation?" *Parameters* 7, no.1 (1977): 2-9.
73. Lewis A. Coser, *Greedy Institutions: Patterns of Undivided Commitment* (New York: Free Press, 1974). The Army is often characterized as a "greedy institution." Coser characterizes greedy institutions as those seeking exclusive and undivided loyalty from their members. "[They]. . . are characterized by the fact that they exercise pressures on component individuals to weaken their ties, or not to form any ties, with other institutions or persons that might make claims with their own demands" (p. 11). The military is probably one of the most "greedy" of institutions, demanding unlimited liability and commitment (often referred as 24 hours/7 days a week, or 24/7) from its members, particularly its officer corps. The family is also a "greedy" institution competing for the commitment of service members, particularly as families move away from a traditional structure with a dramatic increase in dual earning/career couples. More and more dual earner and dual career couples are present in the military and placing increasing demands on military members' time and attention. This tension has caused a shifting pattern of commitments and is increasing the need for the military to be more sensitive to and understanding of family considerations. These create a new demand on the military's strategic leadership to insure a reciprocal commitment of the military to their members.
74. In his monograph, *Generations Apart: Xers and Boomers in the Officer Corps*, published by the Strategic Studies Institute of the Army War College (2000), Dr. Lenny Wong presents evidence of significant differences in perceptions and attitudes between our junior and more senior officers. The major findings of Dr. Wong's study were that "today's senior officers do not understand today's junior officers or their perspectives. Senior officers think they understand the world of lieutenants and captains, but many junior officers and others are convinced that they do not" (p. 3). Junior officers are leaving the Army at alarming rates, almost 4% higher over the last 2 years, and "many are blaming their departure on senior officer lack of understanding" (p. 5).
75. United States Military Academy, *Cadet Leader Development System* (CLDS/Draft) (West Point, NY: USMA, 2001), 6.
76. ALTDP: Final Report (Officers), 2-2.
77. "Warrior ethos" is characterized as an attitudinal orientation that all soldiers must share, not just those soldiers in the combat arms. It is a belief that compels soldiers to never accept defeat, and to drive on to mission accomplishment, no matter what the cost, no matter how long it takes.

78. ATLDP: Final Report (Officers).
79. The Army is experiencing a shortage of captains based on what is required and a higher-than-envisioned attrition rate, which has caused several things to happen. The Army has over-accessed lieutenants to make up for this officer shortage, so there are more lieutenants than available platoon leader positions. So, as senior officers try to get all lieutenants some time with troops as a platoon leader, their time in this key developmental position is cut way short, and is not what these lieutenants expected nor what they were "promised" by the Army: time to learn their basic craft and be warriors at platoon level. They are quickly cycled through these jobs (often no more than 6 months) and then placed in captain's slots on battalion and brigade staffs, jobs they are not trained nor experienced enough to do. Because of the high OPACE, they are micromanaged by senior officers in these positions, and not developed. They quickly get frustrated, and many decide to leave the Army, making the problem worse.
80. DA Pamphlet 600-3, *Commissioned Officers Development and Career Management* (1997), states that the central engine driving the officer personnel system is Army requirements (the Army's needs and not the preferences or concerns of the officer based on developmental needs). The pamphlet does not address what developmental experiences or operational assignments an officer should have before moving to the next. More than 70% of the officers in the field felt that they did not spend enough time in jobs to properly develop, and over 50% stated they were not getting the developmental experience they thought they needed to be effective (ATLDP, 2-16). Officers want the assignment process to be more balanced in terms of developmental needs and preferences. However, it is perceived as simply "a face in a place" mentality.
81. Cynthia D. McCauley, Russ S. Moxley, and Ellen Van, eds., *Center for Creative Leadership Handbook of Leadership Development* (San Francisco, CA: Jossey-Bass, 1998), 1-25. Their theory of leadership development, the "cutting-edge" theory to date in the leader development literature, is presented in this section of the handbook.
82. FM 22-100, *Military Leadership* (Washington, DC: Headquarters, Department of the Army, 1999), 1-4.
83. CSIS, 49.
84. Intermediate Level Education (ILE) is a new label for command and staff schooling. Within the next several years, the Army will no longer select officers for ILE, but provide this type of in-residence education for all majors consistent with rating comments from the field. An understanding of the importance of this type of education to the leader development of our officer corps and the long-term health of the Army is finally emerging. The educational experience will not be the same for each branch, but tailored to meet the needs of the arm of service.
85. See ATLDP: Final Report (Officers), Appendix F, 11, for a listing of the remarkable similarities between the major findings of these two studies.
86. John P. Kotter, "Why Transformation Efforts Fail," *Harvard Business Review* (March-April 1995), 61.
87. Edgar H. Schein, *The Corporate Culture Survival Guide* (San Francisco, CA: Jossey-Bass, 1999), 14.
88. John P. Kotter, *Leading Change* (Boston, MA: Harvard Business School Press, 1996), 148.
89. Senge, 6-12. Senge discusses these 5 dimensions of a learning organization. He defines systems thinking as the Fifth Discipline, the most important of the dimensions due to the integrative nature of the dimensions. This is particularly important in understanding the systemic nature of cultural change. For a straightforward discussion of changing organizational culture and making it stick, see Donald M. Bradshaw, "Changing the Organizational Culture," in *Building and Maintaining Healthy Organizations: The Key to Future Success*, ed. Lloyd J. Matthews (Carlisle Barracks, PA: Department of Command, Leadership, and Management, U.S. Army War College, January 2001), 41-68.

23 Professional Leadership and Diversity in the Army[1]

Mady Wechsler Segal and Chris Bourg

Introduction

In their essay calling for a renewal and redefinition of Army professionalism, Don Snider and Gayle Watkins highlight three indicators of the Army's deteriorating relations with its client, American society.[2] They assert that recruiting shortfalls, a widening "gap" between the attitudes and perspectives of the military and American society (something that is subject to debate), and the adverse public reaction to various well-publicized scandals involving the unethical behavior of some Army leaders reflect increasing societal dissatisfaction with the Army. These indicators make it clear that leadership and personnel issues shape the public's perception of the Army and its members. For the Army profession, organizational effectiveness and public support for, and satisfaction with, the profession are interrelated. An effective military can expect to receive a high level of support from a satisfied client. By the same token, a military with a positive, mutually supportive relationship with society will be more effective. A volunteer military divergent from its own populace in a democratic society will face continued problems of recruitment, retention, and legitimacy. Military effectiveness is well served by an Army supported by its wider society.

While many factors influence public perception of and support for the Army, we concentrate on issues of personnel diversity. Diversity refers to the degree to which members of a group or organization differ in terms of both social identities and individual characteristics. We argue that the military's relationship with society and its effectiveness are enhanced by a commitment to the successful leveraging of diversity among Army leaders at all levels. "Leveraging of diversity" or "capitalizing on diversity" means turning diversity into an advantage by using it to enhance performance and social legitimacy (which we discuss further below) . We provide a general assessment of the current state of diversity in the Army, followed by recommendations designed to improve the manner in which the Army capitalizes on diversity.

Diversity and Professional Effectiveness

As a profession, the Army competes within a system of professions for members, resources, and jurisdiction.[3] An important element of this competition involves professional claims within the court of public opinion to the social and cultural authority over the right to perform the work it wishes as well as the right to decide how the work is performed and by whom.[4] Snider and Watkins note that the Army must adapt to massive changes in the nature of its work and "is missing (and thereby losing) competitions with other professions and organizations at the boundaries of its expertise."[5] In addition, the Army must adapt to significant changes in the composition and attitudes of American society.

As recent U.S. census figures show, the American population is now more diverse than ever. This is especially true in the labor force, where the influx of women and racial minorities represents one of the most profound changes in the American workforce in recent years.[6] These less traditional sources of labor will soon come to constitute the majority of workers. By 2025, the labor force is expected to be 48% women and 36% minority.[7] In addition, there is increasing diversity among the college and college-bound population. Recent research indicates that the military increasingly competes with colleges, in addition to the labor force, in recruiting enlisted personnel.[8] The military also draws its officer corps from college graduates. The increasing diversity of both the labor force and the college-bound population means that for the Army to meet its recruitment needs, it must appeal to members of this new majority. In other words, in order to compete successfully within the system of professions, the Army must follow the lead of other professions and adapt its personnel policies to an increasingly diverse population.

Adapting its personnel policies and organizational culture to embrace and leverage diversity more fully will also help to close part of the supposed civil-military attitude "gap." As one of the few organizations in the United States with explicit legal restrictions on the employment of women and openly gay men and lesbians, the military's culture and policies are increasingly at odds with dominant public attitudes favoring equality. For example, research indicates that American high school students have become increasingly more egalitarian in their attitudes towards women's roles, with significant majorities of both males and females expressing support for gender equality in the workplace.[9] The attitudes of high school students are especially relevant to the future military, because today's high school students represent the Army's major recruiting pool, its source of future officers, and future civilian leaders, policymakers, and voters. The disconnect between military policy and prevailing public attitudes contributes to a civil-military "gap," hindering the Army's ability to compete successfully with other professions both for members and for public support. As part of a commitment to a new professionalism, the Army should adapt its culture and policies to reflect more closely the egalitarian attitudes of its client. In other words, the Army must adopt and enforce policies and practices that reduce bias and discrimination and which contribute to the successful management of a diverse workforce.

Diversity is directly related to professional effectiveness as well. Arguments of degradation of military effectiveness have been used in the past to exclude members of some groups, but such arguments derive from preconceived attitudes (i.e., prejudice) and are not supported by the accumulated scientific evidence. For example, while some argue that homogeneous groups are more cohesive and therefore more effective, there is little scientific support for that assertion. The scientific evidence linking cohesion with performance is mixed and inconclusive, as is the evidence that cohesion is lower in groups composed of diverse individuals.[10]

Traditional definitions of cohesion are not specific. The general definition is that group cohesion is the social glue that results from all the forces that keep group members attached to the group.[11] It is a group property. Traditional concepts of cohesion emphasized peer relationships. Relationships with those in authority were considered part of leadership, not cohesion. Attempts to measure cohesion have produced multiple definitions and have led to distinctions among different components or types of cohesion. For example, the term "horizontal cohesion" has been used to refer to the peer bonding of early conceptualizations, while "vertical cohesion" is used to refer to bonds between leaders and followers.[12]

The latest theoretical and methodological advances make a distinction between "task cohesion" and "social cohesion."[13] Task cohesion is the extent to which group members are able to work together to accomplish shared goals. This interdependence is the sort that Emile Durkheim conceptualized as organic solidarity based on a division of labor.[14] (He saw mechanical solidarity, based on similarity, as less functional.) Task cohesion includes the members' respect for the abilities of their fellow group members. For combat situations, it translates into the trust that group members have in each other, including faith that the group can do its job and thereby protect its members from harm. Task cohesion can be horizontal or vertical. The latter is the respect and confidence that unit members have in their leaders' competence. Social cohesion is a more affective dimension and includes the degree to which members like each other as individuals and want to spend time with them off duty. Vertical social cohesion would include the extent to which unit members believe that their leaders care about them.

Why is unit cohesion important for the military? The common wisdom is that units with higher cohesion are more effective, especially in combat. The accumulated evidence shows that there is sometimes a relationship between cohesion and group effectiveness, but there are three very important qualifiers to this relationship. First, the direction of causality is not established. Some evidence indicates that causality works in the direction opposite to what is usually assumed, i.e., that it is group success that produces cohesion success.[15] Second, the evidence for a relationship between cohesion and group performance shows that it is task cohesion, not social cohesion, that is related to success.[16] Indeed, high social cohesion sometimes negatively affects performance.[17] Third, there is evidence that vertical cohesion, or what we prefer to call effective leadership, affects both horizontal cohesion and performance.[18] Groups in which members have confidence that their leaders are competent and care about what happens to them are more likely to be successful in various ways. Good leaders by definition organize task activities within the unit in ways that foster task effectiveness, respect, and caring among group members.

Thus, even if performance is enhanced by cohesion (and the evidence is not clear on this), it is likely to be task cohesion, not social cohesion, that provides the positve effects. There is no evidence showing that diversity of race, gender, or sexual orientation interferes with task cohesion.

With the increasing variety of missions and tasks within the Army, the organization and units within it are most effective when they are composed of people with different strengths. Task cohesion and performance are based on a division of labor and the diverse capabilities of individuals within the group. Readiness and mission accomplishment are enhanced when there are people with diverse characteristics, including abilities, skills, and problem-solving styles. Diversity in these attributes is more likely when the group is composed of people from diverse social identity groups.

Retention of qualified and trained personnel has been an important problem for all the services. To retain people, the armed forces must treat them in such a way that they are satisfied with their lives in the service. If individuals perceive that the Army is not a place where they are treated well, then they "vote with their feet." On the positive side, when people are treated with respect within an organization, they tend to develop loyalty and commitment to the organization.

We argue that by adopting policies and leader practices designed to manage and leverage diversity successfully, the Army will improve not only the effectiveness of the organization, but also its relations with American society. This, in turn, will enhance the Army's ability to compete successfully within the system of professions. As the Army engages in the process of renegotiating its status and position as a profession, it must attend to an increasingly diverse and egalitarian-minded public. The effectiveness of the future Army will be enhanced by policies and practices that contribute to a public perception of the Army as a profession which successfully adapts to and turns to advantage the talents, skills, perspectives, and abilities of a diverse workforce and a diverse public client.

In applying the concepts of "managing" and "leveraging" diversity to the Navy, George Thomas provides the following definitions (which are equally applicable to the Army):

> To manage diversity is to *lead* in a manner that maximizes the ability of personnel to contribute to the Navy's missions. . . . *Leveraging diversity* is the linkage between diversity characteristics and force readiness. People feel that their differences make up an essential part of their worth and they feel most valued when they believe they are seen in their fullest dimensions, both as individuals and as members of their own group(s). To reach its fullest potential the Navy must capitalize on socially relevant differences and tap into the strength of all personnel, including those regarded as different[19].

Managing diversity should not to be viewed as Equal Opportunity programs or Affirmative Action. It is "*not* a set of programs that are intended to improve the positions of women and minorities. . . . It is concerned with "organization culture and leadership" rather than compliance.[20] We view the successful leveraging of diversity as being composed of two interrelated elements: (1) the representation of diverse groups throughout an organization, and (2) the treatment of members of diverse groups by the organization and its leaders. In the next section, we provide a brief assessment of the Army's managing and leveraging of diversity.

Assessment of the Army's Leveraging of Diversity

In our assessment of the Army's treatment of diversity we concentrate on issues of the social identities of race, gender, and sexuality. This is part of a broader context where diversity includes differences among individuals in such characteristics as mental abilities, socio-economic status, region of origin, and parental status. We cover both representation and treatment.

In many ways, the Army is widely considered to have achieved enviable results when it has diversified by integrating previously excluded groups. In particular, the Army was ahead of other social institutions in racial integration. As of 5 September 2000, 20.6% of Army officers and 44.6% of enlisted personnel are members of racial/ethnic minorities; 11.3% of officers and 29.2% of enlisted personnel are black.[21] Some analysts hold the Army up as an example in race relations for the rest of the society to follow.[22]

While the Army's success in representation of diverse racial and ethnic groups is to be applauded, there remain areas of concern regarding racial diversity with the profession, both in representation and treatment. In the area of representation, people of color have higher concentrations among enlisted personnel than among officers. However, representation among officers has been rising over the past two decades. Analysis by rank shows that the representation of black officers (both men and women) is higher at the major and captain levels than at the more senior ranks. For enlisted personnel, the representation of blacks (both men and women) rises within grade cohorts going from E-1 to E-6, then declines going from E-7 to E-9. For Hispanic enlisted men, representation declines within grade cohorts going from E-1 to E-5, rises going from E-6 to E-7, and then declines going from E-8 to E-9. Representation of Hispanic enlisted women declines within grade cohorts going from E-1 to E-6, rises going to E-7, declines going to E-8, and then rises going to E-9.[23]

Of greater concern are issues of treatment. For example, results of personnel surveys indicate that many African-American soldiers are dissatisfied with the racial climate.[24] Even high-ranking black officers report experiences with racial discrimination.[25] Despite these shortcomings, the Army still appears to have a more positive race relations climate than most civilian institutions.

The Army has also made considerable progress in gender integration in the last several decades. Women's representation in the armed forces has increased substantially over the past thirty years, from approximately 2% in the early 1970s to about 15% of active duty personnel today.[26] Among enlisted personnel in the Army, women constitute 15.5 % of active duty personnel, 11.6% of the Army National Guard, and 24.9% of the Reserve.[27] Women's representation in the Army's commissioned officer corps is similar to that of the enlisted forces. Women are 15.4% of active duty officers, 10.2% of the Army National Guard, and 25.7% of the Reserve.[28] However, women continue to be significantly underrepresented in the Army relative to their representation in the civilian workforce. While we would not expect that this representation to be equal, it could be higher than it is, even with current exclusions. Moreover, women are still excluded from 8% of the military occupational specialties in the Army, which constitute 30% of all active duty positions.[29]

Analysis of women's representation by rank shows generally that, among both enlisted personnel and officers, the higher the rank, the smaller the percentage of women. Women constitute 16.4% of grades E-1 to E-3, 17.5% of E-4, 15.0% of E-5, 12.2% of E-6, 11.7% of E-7, 10.4% of E-8, and 6.6% of E-9.[30] For officers, women's representation is 19.6% of second lieutenants, 19.4% of first lieutenants, 15.5% of captains, 13.4% of majors, 12.2% of lieutenant colonels, 8.6% of colonels, 4.7% of brigadier generals, 2.1% of major generals, and 2.3% of lieutenant generals.[31] While it takes time for women to reach the higher ranks—and women's representation at the lower ranks has been increasing over the past 20 years—women are still underrepresented at the higher ranks compared to their percentages of the earlier entry cohorts.

However, statistics show that women's representation at the higher ranks has been increasing over the years. For example, women's representation among colonels (O-6) has increased from 2.4% in 1988, to 4.0% in 1992, to 6.2% in 1996, to 8.1% in 2000.[32] Similarly, in grades E-7 to E-9, women's representation has grown from 3.9% in 1988, to 7.1% in 1992, to 10.1% in 1996, to 11.0% in 2000.[33] Furthermore, percentages of female lieutenant colonels in command assignments have increased over the past few years.[34] In 1995, 10% of female lieutenant colonels (and 13% of male lieutenant colonels) were in command assignments. The figures for 1996, 1997, and 1998 were similar: 11% of women (14% of men), 13% of women (14% of men), 14% of women (14% of men), respectively. For 1999, the situation changed dramatically for both male and female lieutenant colonels, with 23% of women and 22% of men in command assignments. These figures bode well for women's eligibility for promotion to colonel (O-6).

There is greater representation of African-Americans among military women than their percentage of the population and even greater than the percentage of military men who are African-American. In the Army, 47% of enlisted women and 22% of women officers are black. An extraordinary 62% of enlisted women are members of racial/ethnic minorities.[35] This shows the Army's attraction for women of color, who face the greatest obstacles to economic advancement in the civilian sector.[36] For example, the national median income of black women is only 85% of white women's—and that percentage has *decreased* since 1975.[37] Nationally, Hispanic women's median income is only 76% of white women's. Black women's median income is 83% of black men's, while Hispanic women's income is 90% of Hispanic men's.[38]

Evidence shows that there is much room for improvement in the treatment of women in the Army (as well as the other services). Sexual harassment is a common occurrence, especially crude and offensive behavior and unwanted sexual attention.[39] Lower-ranking women are most likely to be victims of harassment.

Even more troubling is the high frequency of gender harassment, including statements and behavior by men to women indicating continued resistance to accepting women into the organization as legitimate participants worthy of respect as soldiers.[40] Gender harassment takes various forms, including resistance to women's authority, constant scrutiny of women, passing untrue rumors and gossip about women, sabotaging women's work, and making indirect threats.[41] Research on harassment shows that women are more likely to view sexist behavior (such as treating women differently, making offensive sexist remarks, and putting women down because of their sex) as having a more negative effect on them than other forms of sexual harassment. When asked in a sexual harassment survey about the situation involving certain behaviors in the previous 12 months that had the most effect on them, 35% of respondents cited an instance of sexist behavior.[42] This was more than any other category of behaviors. However, women were less likely to label this category of behavior as "sexual harassment." This shows that the Army needs to pay as much or more attention to gender harassment as to sexual harassment. Gender harassment may negatively affect morale, which in turn may interfere with effectiveness.

Resistance to accepting military women is sometimes voiced in terms of inequitable treatment or lowering of standards. Military men cite violations of principles of justice and of readiness goals to justify their opposition to women. This is especially apparent in complaints about the use of gender norms for physical fitness tests and the perception that the scoring is unfair.[43] But the evidence shows that the purpose of these tests is widely misunderstood by military personnel. Experts, both in and out of the military, say that these are intended as measures of physical fitness and health, not job performance. Gender norming is required for a valid measure of physical health. Interestingly, there are few complaints about the age norming of the tests. Even the politically diverse Congressional Commission on Military Training and Gender Related Issues (1999) unanimously recommended that military personnel be educated about the real purpose of the tests.

Carol Cohn's analysis shows that the protests against the gender norms on the fitness test standards show "strong feelings of loss and anger about changes in the way the organization is gendered"[44] and demonstrate men's antipathy toward women in the military. Cohn maintains that even if the test requirements were exactly the same, men would find some other "focal point" for their dissatisfaction.

Changes in civilian society in the social construction of gender have been affecting the armed forces. There has been a transformation of values, norms, and beliefs about gender that has affected every society institution (educational, political, legal, family, economic, etc.). As one of the most predominantly male institutions—and a gender-defining one—the military has been one of the last social organizations to gender integrate and adapt to this changing construction of gender. [45] Forces of resistance to change are evident in sexual harassment and gender harassment in the military. Resistance to gender integration in the armed forces can also be seen in conservative stances with regard to military gender integration by some members of Congress and by political pressure groups, despite considerable evidence of public support for greater gender integration in the military.[46]

With regard to sexuality, the military has officially excluded homosexuals from service since World War II. Gay men and lesbians have, of course, served in the military throughout its history. Enforcement of bans on the service of homosexuals has generally been lax during times of heightened manpower needs, such as wartime.[47] In 1993 the U.S. Congress codified a revised anti-gay policy. This policy moved the military in the direction of seemingly greater tolerance for the presence of gay and lesbian service members by dropping the statement that homosexuality, per se, is incompatible with military service and by prohibiting asking recruits and others about their sexual orientation. However, the "Policy Concerning Homosexuality in the Armed Forces,"[48] colloquially known as "Don't Ask, Don't Tell, Don't Pursue," still requires gay and lesbian service members to maintain a level of secrecy about their sexuality and personal relationships not required of heterosexual service members.[49]

The secrecy requirement of current military policy makes estimating the percentage of service members who are gay or lesbian even more difficult than the already contested and difficult task of estimating the proportion of homosexuals in the larger population. We do know, however, that in 1998, 1,149 service members were discharged from the military for violating the "Don't Ask, Don't Tell" policy. This figure translates to a rate of three to four discharges per day, representing a significant increase in the number of service members discharged for homosexuality since the new policy was instituted in 1993. Many more thousands of gay men and lesbians are currently serving.[50]

The military's "Don't Ask, Don't Tell" policy has been criticized on grounds of both its failure to be justifiable on the basis of evidence and its lack of proper implementation. While Army leaders are not responsible for establishing policies regarding the service of openly gay men and lesbians, they are responsible for ensuring adherence and proper implementation of the current policy at all levels.

A recent Inspector General's survey of 75,000 service members found that 80% of service members had heard anti-gay remarks during the past year, and that 37% had witnessed or experienced targeted incidents of anti-gay harassment.[51] Discipline, morale, and cohesion are compromised when harassment and derogatory remarks directed at any group of people are tolerated. Moreover, the presence of harassment and maltreatment of any group, including homosexuals, violates the trust placed in military leaders both by the American public and by those who serve.

On a positive note, the Army is recognized as providing "the best illustration of trying to do 'what's right' in its anti-gay harassment prevention efforts."[52] In the past year, the Army led the other services in providing training to soldiers on preventing anti-gay harassment and in upholding the investigative limits of the current policy.[53] Continued leadership emphasis on training and accountability is necessary for the Army to ensure that all soldiers are treated with respect, dignity, and honor. When soldiers are assured of treatment that is consistent with the Army's core values, effectiveness and readiness are enhanced, as is the Army's public image and relationship with society.

What policies and leader practices will build on and improve the Army's situation with regard to acceptance and leveraging of diverse personnel? We deal with this in the next section.

How Does Leadership Relate to Diversity? Promoting Effective Leadership

The degree to which the organization accomplishes successful integration of previously excluded groups is a function of leadership commitment to that integration at all levels. Considerable social science research indicates the kinds of policies and practices that are likely to minimize bias and discrimination, and to promote the successful leveraging of a diverse workforce. Recent scholarly work on employment discrimination suggests that accountability in decision-making, equal resource policies, open information about pay, and the construction of heterogeneous, cooperative, and interdependent workgroups are key factors which appear to minimize discrimination and bias.[54] The fact that the Army currently employs many of these practices and policies accounts for much of the Army's success in the arena of racial integration.[55] In addition, there is social science research to guide successful policy and individual leader behavior.

What policies and practices (including the behavior of individual leaders) will improve the Army's ability to recruit, retain, manage, and leverage a diverse workforce successfully? Social science evidence points to at least three areas for improvement: (1) decision-making accountability, (2) recognition of effects of discriminatory policies, and (3) leader behaviors and conditions that foster respect for diversity.

Decision-making Accountability

Social psychological research has consistently and conclusively shown that stereotypes and biases invariably influence our perceptions and evaluations of others.[56] It shows that it is extremely difficult to get people to attend to individuating information rather than stereotypes in assessing others. For example, even when given information that men and women in the target population were distributed equally across college majors, survey subjects continued to rely more on gender stereotypes than on information about individual interests in predicting whether a target individual was an engineering or a nursing major.[57]

However, the biasing effects of stereotypes on evaluative judgments have been shown to be greatly reduced when decision-makers know that they will be held accountable for the criteria they use for decision-making.[58] While current Army policies and practices emphasize accountability in many career-relevant decisions such as officer promotions and senior school selections, many career-relevant decisions are made by decision-makers who are not currently held accountable for their decisions. For example, junior officers are usually assigned to career-enhancing jobs such as company command by battalion commanders who are not currently held formally accountable for those decisions. Given white male predominance among key Army decision-makers, unconscious in-group preferences and reliance on gender and racial biases and stereotypes are likely to lead to systematic discrimination in selection for key jobs at the junior officer level. This may be one cause of the lower representation noted earlier of women and racial and ethnic minorities in the higher ranks, including among officers.

Research shows, however, that the effects of these tendencies are reduced when decision-makers know they will be held accountable for their decisions. We recommend that the Army develop programs that hold decision-makers at *all* levels responsible for ensuring that their decisions are not tainted by in-group preferences. Decision-makers should also be held accountable for the outcomes of their decisions in terms of representation of women and minorities selected for career-enhancing jobs, training, and school assignments. This accountability need not be implemented through rigid, restrictive, and bureaucratic policies. For example, decision-making accountability can be accomplished through educating leaders on the effects of unconscious stereotyping and bias, and through the establishment of decision-making procedures such as selection boards for all levels of career-relevant decisions. Including others in decisions such as company command selection will not only provide accountability but will also increase the effectiveness of the decision-making process by allowing input from other informed leaders.

Recognition of Effects of Discriminatory Policies

Research shows that when valued rewards are distributed among people working together in a goal-oriented context, individuals will implicitly assume that those with greater rewards are more competent than those with less.[59] The exclusion of women from units and military occupational specialties most closely associated with the Army's core combat mission systematically denies women access to organizationally valued positions. One expected consequence of this is that members of the Army come to believe that women are less valued and less competent members of the organization. In this way, the Army's combat exclusion policies not only contribute to the growing civil-military values "gap," but also contribute to gender integration difficulties within the Army.

In a similar way, military policy regarding homosexuality systematically denies gay and lesbian service members access to organizationally valued resources. For example, under current policy gay and lesbian service members are denied support for their personal and family relationships. Research has consistently shown that policies and practices that are perceived as supportive of military members' personal and family relationships have positive effects on job satisfaction and retention.[60]

The military's combat exclusion policy and anti-gay policy both systematically discriminate against specific groups of service members by denying them access to valued resources. Moreover, an expected consequence of these policies is that they encourage or reinforce expectations of lower competence and worth for members of the excluded groups. When policies treat certain groups in a discriminatory manner, organizational participants are likely to treat members of those groups in a discriminatory manner as well. For these reasons, the successful leveraging of diversity is hindered by the military's exclusionary policies regarding the full participation of gay men and all women.

What can Army leaders do to counter the effects of policies that require differential treatment of some service members? First, Army leaders need to be educated on the potential negative effects of discriminatory policies on the attitudes and perceptions of all service members. In order to mitigate the effects of such policies, Army leaders must aggressively seek to eliminate formal statements and informal banter within their units that reflect a devaluing of the contributions of any group of soldiers. Army leaders should consistently emphasize the value of all soldiers, regardless of social characteristics such as race, gender, and sexuality. In addition, Army leaders must communicate to their subordinates that the Army needs and values the contributions of all military specialties and units, not just those with combat designations. This will reduce the negative effects of the military's discriminatory policies on both Army morale and readiness and on public perceptions of the Army.

Individual Leader Behaviors

At the interpersonal level, leaders' behavior has been demonstrated to have strong effects on the treatment of diversity characteristics within military units. Leaders serve as role models for personnel in their units: military personnel often model their behavior toward others on the basis of the behavior of their leaders. Further, the degree to which leaders enforce nondiscriminatory behavioral guidelines affects the likelihood of such behavior occurring and recurring.

We know a great deal about the conditions that affect the success of managing diversity in groups. Much of the early theory and research was based on the "contact hypothesis," originally developed with regard to racial relations and adapted to integration of other previously excluded groups.[61] Research demonstrates that the process of integrating members of previously excluded groups into organizations and groups within organizations generally does not proceed smoothly. Problems are encountered when integrating the military and other social institutions. Integration problems tend to occur with various social characteristics, including race, ethnicity, and gender.

Early phases of integration are often characterized by negative attitudes toward the members of the newly admitted group. Such negative attitudes are usually accompanied by negative behaviors involving discrimination against the new group. Among such negative behaviors are social isolation and harassment of the new members. Breaking with tradition is hard and there are always sources of resistance to change in any institution. Resistance to change is great when there is long experience with, and/or identification with, the old ways of doing things. Part of men's resistance to women in previously all-male roles, especially those that have served as rites of passage to manhood, is the difficulty in proving masculinity if the challenge can also be met by a woman. Military men may also retain definitions of their roles as masculine by accepting individual women as exceptions—by disconnecting perceptions of an individual woman's success from their conceptions of "women" in general.

But research also shows us the conditions that tend to foster more effective integration.[62] More positive attitudes tend to develop when:

- interaction is sufficiently close and sustained so that the members of the majority group have the opportunity to get to know the individual members of the minority group well;
- the minority group members are of at least equal status to the majority group members;
- the minority group members constitute more than a small (token) minority of the work group;
- there are commonly shared goals;
- the situation is one that fosters cooperation rather than competition among members of the group;
- the social norms support equality and integration;
- those in positions of authority support the integration.

This last condition is very important. It is also amenable to control in the military services: integration of diverse members proceeds most smoothly and with the fewest problems if leaders are committed to making integration work and if they communicate that commitment. The greater the degree of public commitment expressed by leaders at each organizational level, the more successful will be the leveraging of diversity and the more effective the military units will be.

In addition to voiced commitment to diversity, leaders are role models for their subordinates. Soldiers observe their leaders' behaviors, and those actions often speak louder than their words. When leaders show in their behaviors on a day-to-day basis that they respect soldiers with diverse characteristics, this value is transmitted to their troops (both officers and enlisted personnel). These behaviors include the quality of their interactions with service members with diverse characteristics as well as what they say about members of identifiable groups. For example, the positive effects of anti-sexual harassment workshops or leaders' statements about commitment to diversity are negated if the same leaders treat women or members of other diverse groups with disrespect or tell sexist or racist jokes (the effects are negative whether or not members of the derogated groups are present).

Another area in which the Army has room for improvement involves creating situations which foster cooperation among diverse members of groups rather than competition. Current Army policy and culture encourage individual competition in many settings, despite official endorsement of cooperation. Emphasis on individual rankings at Officer Basic and Advanced Courses and at the United States Military Academy are examples of settings in which diverse individuals are placed in direct competition with one another. Social psychological research shows that competition highlights differences and encourages stereotyping.[63] We recommend that the Army eliminate unnecessary competition in its professional training environments, and develop training programs that truly require cooperation among diverse individuals.

We further recommend that leaders at all levels be educated about the way their behaviors affect respect for diversity among their subordinates (and their peers). Of course, we expect that they will want to act in ways that have positive effects on leveraging and managing diversity. We also recommend that leaders be held accountable for the degree to which their behavior contributes to respect for diversity.

Conclusions

We have analyzed the ways in which diversity contributes to the Army's core value of respect, to military professionalism, and to military effectiveness. Our assessment of Army diversity shows increasing representation in the recent past of members of racial and ethnic minorities—both men and women. Women's representation has increased, especially among women of color. Women are more concentrated in the lower ranks, but their representation in the higher ranks has been increasing. While people of color perceive a need for improvement in their treatment, the Army's racial climate appears better than that of many civilian institutions. Treatment of women has been improving, but still requires attention to eliminating sexual harassment and gender harassment. Gay men and lesbians are precluded from serving openly, but private sexual orientation is by policy not to be the basis of attention or harassment; appropriate education about policy is needed at all levels, as is enforcement of anti-harassment policy. Much social science knowledge exists that can be applied to assisting Army leaders in improving the leveraging of diversity to enhance personnel retention, mission readiness, and military effectiveness. Improvement will come from accountability in decision-making, recognizing effects of differential treatment, promoting equitable treatment, educating leaders in the effects of their behaviors, providing them with positive models, and holding them accountable for their behaviors with regard to diverse personnel. Implementation of these recommendations will enhance the Army's core values and its mission effectiveness.

Notes

1. Support for writing this chapter was provided by the U.S. Military Academy and by the U.S. Army Research Institute for the Behavioral and Social Sciences under Contract No. DASW 0100K0016. The views expressed in this chapter, however, are the authors' and not necessarily those of the contracting agencies or any other government entity. We appreciate the very able research assistance and advice provided by Darlene M. Iskra, and we thank Don M. Snider and Gayle L. Watkins for helpful comments on drafts of this chapter.
2. Don M. Snider and Gayle L. Watkins, "The Future of Army Professionalism: A Need for Renewal and Redefinition," *Parameters* 30 (Autumn 2000): 5-20.
3. Andrew Abbott, *The System of Professions: An Essay on the Division of Expert Labor* (Chicago, IL: University of Chicago Press, 1988); Snider and Watkins.
4. Abbott.
5. Snider and Watkins, 6
6. Howard N. Fullerton, Jr., "Labor Force Projections to 2008: Steady Growth and Changing Composition" *Monthly Labor Review* (November 1999): 19-32.
7. Howard N. Fullerton, Jr., "Labor Force Participation: 75 Years of Change, 1950-98 and 1998-2025," *Monthly Labor Review* (December 1999): 3-12.
8. Chris Bourg, "Trends in Intentions to Enlist and Attend College," in *Recruiting College-Bound Youth into the Military: Current Practices and Future Policy Options*, MR-1093, eds. M. Rebecca Kilburn and Beth J. Asch (Santa Monica, CA: RAND, forthcoming).
9. Shelley J. Correll and Chris Bourg, "Trends in Gender Role Attitudes, 1976-1997: The Continued Myth of Separate Worlds" (paper presented at the annual meeting of the American Sociological Association, 1999).
10. Elizabeth Kier, "Homosexuals in the U.S. Military: Open Integration and Combat Effectiveness," *International Security* 23 (Fall 1998): 5-39; David R. Segal and Meyer Kestnbaum, "Closure in the Military Labor Market: A Critique of Pure Cohesion," chap. 21 of the present book.
11. Dorwin Cartwright, "The Nature of Group Cohesiveness," in *Group Dynamics: Research and Theory*, 3d ed., eds. Dorwin Cartwright and Alvin Zander (New York: Harper & Row, 1968), 91-109; Leon Festinger, Stanley Schachter, and Kurt Back, "Operation of Group Standards," in *Group Dynamics: Research and Theory*, 3d ed., eds. Dorwin Cartwright and Alvin Zander (New York: Harper & Row, 1968), 152-64; Stanley Schachter, "Deviation, Rejection, and Communication," in *Group Dynamics: Research and Theory*, 3d ed., eds. Dorwin Cartwright and Alvin Zander (New York: Harper & Row, 1968), 165-81.
12. Robert MacCoun, "What is Known about Unit Cohesion and Military Performance," in *Sexual Orientation and U.S. Military Personnel Policy: Options and Assessment* (Santa Monica, CA: RAND, 1993), 283-331.
13. Albert V. Carron and Lawrence R. Brawley, "Cohesion: Conceptual and Measurement Issues," *Small Group Research* 31, no. 1 (February 2000): 89-106.
14. Emile Durkheim, *The Division of Labor in Society* (New York: Free Press, 1893; reprint 1964, 1997).

15. Brian Mullen and Carolyn Copper, "The Relation Between Group Cohesiveness and Performance: An Integration," *Psychological Bulletin* 115 (March 1994): 210-27.
16. Ibid.
17. Carol Burke, "Pernicious Cohesion," in *It's Our Military, Too! Women and the U.S. Military*, ed. Judith H. Stiehm (Philadelphia, PA: Temple University Press, 1996), 205-91; Robert MacCoun and Donna Winslow, *The Canadian Airborne Regiment in Somalia: A Socio-Cultural Inquiry* (Ottawa: Canadian Government Publishing, 1997); Donna Winslow, "Rites of Passage and Group Bonding in the Canadian Airborne," *Armed Forces and Society* 25, no. 3 (Spring 1999): 429-57.
18. Margaret Harrell and Laura Miller, *Opportunities for Military Women: Effects Upon Readiness, Cohesion and Morale* (Santa Monica, CA: RAND, 1997); Nora Kinzer Stewart, *Mates & Muchachos: Unit Cohesion in the Falklands/Malvinas War* (Washington, DC: Brassey's, 1991).
19. George W. Thomas, *Managing Diversity in the 21st Century Navy*, vol. 1, *Introduction* (Monterey, CA: Naval Postgraduate School, Center for Diversity Analysis, 2000), 8.
20. Ibid, 10.
21. *Defense Almanac 2000*, "Minorities in Uniform" [journal on-line] available from *http://www.defenselink.mil/pubs/almanac/people/minorities.html*; Internet; accessed 26 March 2001.
22. Charles C. Moskos and John Sibley Butler, *All That We Can Be* (New York: Basic Books, 1996).
23. ODCSPER, *DCSPER 441*. [instruction on-line] (Washington, DC: Office of the Deputy Chief of Staff for Personnel, U.S. Army, 1997); available from *http:www.odcsper.army.mil/Directorates/hr/demographics/DCSPER-441.txt;* Internet.
24. Brenda L. Moore and Schuyler C. Webb, "Perceptions of Equal Opportunity Among Women and Minority Army Personnel," *Sociological Inquiry* 70 (Spring 2000): 215-39; Jacquelyn Scarville et al., *Armed Forces Equal Opportunity Survey*, DMDC Report No. 97-027 (Arlington, VA: Defense Manpower Data Center, 1997).
25. Craig Thomas Johnson, United States Army Officer Professional Development: Black Officers' Perspectives (Carlisle Barracks, PA: U.S. Army War College, 1997).
26. Lory Manning and Vanessa R. Wight, Women in the Military: Where They Stand, 3d ed. (Washington, DC: Women's Research and Education Institute, 2000).
27. Margaret C. Flott, Women in the U.S. Army (Washington, DC: Department of the Army, Office of the Deputy Chief of Staff for Personnel, 2001). These figures are for 4 December 2000.
28. Ibid.
29. Ibid.
30. Ibid.
31. Ibid.
32. Manning and Wight, 17.
33. Ibid.
34. Margaret C. Flott, *Lieutenant Colonel Command* [data presented to DACOWITS spring conference, 18-22 April, 2001] (Washington, DC: Department of the Army, 2001).
35. Manning and Wight.
36. Francine D. Blau, Marianne A. Ferber, and Anne E. Winkler, *The Economics of Women, Men and Work*, 3d ed. (Upper Saddle River, NJ: Prentice Hall, 1998); Barbara Reskin and Irene Padavic, *Women and Men at Work* (Thousand Oaks, CA: Pine Forge Press, 1994); Paula S. Rothenberg, ed., *Race, Gender & Class in the United States: An Integrated Study*, 3d ed. (New York: St. Martin's Press, 1995); Daphne Spain and Suzanne M. Bianchi, *Balancing Act: Motherhood, Marriage and Employment Among American Women* (New York: Russell Sage Foundation, 1996).
37. Blau, Ferber, and Winkler, 139. The 85% figure is for 1995.

38. Ibid. These figures are for 1995.
39. Lisa D. Bastian, Anita R. Lancaster, and Heidi E. Reyst, *Department of Defense 1995 Sexual Harassment Survey*, DMDC Report No. 96-014 (Arlington, VA: Defense Manpower Data Center, 1996.); Department of the Army, *The Secretary of the Army's Senior Review Panel Report on Sexual Harassment*, 2 vols. (Washington, DC: Department of the Army, 1997); Juanita M. Firestone and Richard J. Harris, "Sexual Harassment in the U.S. Military: Individualized and Environmental Contexts," *Armed Forces and Society* 21 (Fall 1994): 25-43; David E. Rohall, Maria Bina Palmisano, and Mady Wechsler Segal, *Gender and Power Issues Affecting Perceptions of Sexually Harassing Behaviors among Military Personnel* (College Park: University of Maryland, 2001).
40. Rohall, Palmisano, and Segal.
41. Laura L. Miller, "Not Just Weapons of the Weak: Gender Harassment as a Form of Protest for Army Men," *Social Psychology Quarterly* 60 (March 1997): 32-51.
42. Rohall, Palmisano, and Segal.
43. Carol Cohn, "How Can She Claim Equal Rights When She Doesn't Have to Do as Many Push-Ups as I Do? The Framing of Men's Opposition to Women's Equality in the Military," *Men and Masculinities* 3 (October 2000): 131-51; *Congressional Commission on Military Training and Gender-Related Issues, Final Report, July 1999,* vol. II, *Transcripts and Legal Consultants & Reports* [book on-line]; available from *http://www.house.gov/hasc/reports/miscmaterials.html;* Internet.
44. Cohn, 131.
45. Chris Bourg and Mady W. Segal, "Gender, Sexuality, and the Military," in *Gender Mosaics: Social Perspectives: (Original Readings)*, ed. Dana Vannoy (Los Angeles, CA: Roxbury Publishing Company, 2001).
46. See James A. Davis, Jennifer Lauby, and Paul B. Sheatsley, *Americans View the Military: Public Opinion in 1982,* NORC Report 131 (Chicago, IL: National Opinion Research Center, University of Chicago, 1983.); Roper Organization, *Attitudes Regarding the Assignment of Women in the Armed Forces: The Public Perspective* (survey conducted for The Presidential Commission on the Assignment of Women in the Armed Forces, 1992).
47. Bourg and Segal, "Gender, Sexuality, and the Military."
48. *U. S. Code*, vol. 4, sec. 654.
49. Bourg and Segal, "Gender, Sexuality, and the Military."
50. Ibid.
51. Stacy L. Sobel, Jeffrey M. Cleghorn, and C. Dixon Osburn, "Conduct Unbecoming: The Seventh Annual Report on 'Don't Ask, Don't Tell, Don't Pursue, Don't Harass'" (Servicemembers Legal Defense Network, 2001).
52. Ibid, 64.
53. Ibid.
54. William T. Bielby, "Minimizing Workplace Gender and Racial Bias," *Contemporary Sociology* 29, no. 2 (March 2000): 120-129; Barbara Reskin, "The Proximate Causes of Employment Discrimination," *Contemporary Sociology* 29, no. 1 (January 2000): 319-28; Cecilia Ridgeway and Shelley J. Correll, "The End(s) of Gender: Limiting Inequality through Interaction," *Contemporary Sociology* 29 (2000): 110-120.
55. Charles C. Moskos and John Sibley Butler, *All That We Can Be.*
56. Galen V. Bodenhausen, C. Neil Macrae, and Jennifer Garst, "Stereotypes in Thought and Deed: Social Cognition Origins of Intergroup Discrimination," in *Intergroup Cognition and Intergroup Behavior*, eds. Constantine Sedikides, John Schopler, and Chester A Insko (Mahwah, NJ: Lawrence Erlbaum Associates, 1998), 311-35.
57. Thomas E. Nelson, Michelle Acker, and Melvin Manis, "Irrepressible Stereotypes," *Journal of Experimental Social Psychology* 32 (January 1996): 13-38.

58. Barbara Reskin, Gerald R. Salanick, and Jeffrey Pfeffer, "Uncertainty, Secrecy, and the Choice of Similar Others," *Social Psychology* 41, no.3 (September 1978): 246-55; Phillip E. Tetlock, "The Impact of Accountability on Judgement and Choice: Toward a Social Contingency Model," *Advances in Experimental Social Psychology* 25 (1992): 331-76; Phillip E. Tetlock and Jennifer S. Lerner, "The Social Contingency Model: Identifying Empirical and Normative Boundary Conditions on the Error-and-Bias Portrait of Human Nature," in *Dual Process Theories in Social Psychology,* eds. Shelly Chaiken and Yaacov Trope (New York: Guilford Press, 1999), 571-85.
59. Wendy J. Harrod, "Expectations from Unequal Rewards," *Social Psychology Quarterly* 43, no. 1 (March 1980): 126-130; Ridgeway and Correll; Penni Stewart and James Moore, "Wage Disparities and Performance Expectation," *Social Psychology Quarterly* 55, no. 1 (March 1992): 78-85.
60. Mary C. Bourg and Mady W. Segal, "The Impact of Family Supportive Policies and Practices on Organizational Commitment to the Army," *Armed Forces & Society* 25 (Summer 1999): 633-52; Rose M. Etheridge, *Family Factors Affecting Retention: A Review of the Literature,* Research Report 1511 (Alexandria, VA: U.S. Army Research Institute for the Behavioral and Social Sciences, 1989); Dennis K. Orthner, *Family Impacts on the Retention of Military Personnel,* Research Report 1556 (Alexandria, VA: U.S. Army Research Institute for the Behavioral and Social Sciences, 1990.); Mady Wechsler Segal and Jessie J. Harris, *What We Know About Army Families,* Special Report 21 (Alexandria, VA: Army Research Institute for the Behavioral and Social Sciences, 1993).
61. Gordon W. Allport, *The Nature of Prejudice* (Cambridge, MA: Addison-Wesley, 1954); Morton Deutsch and Mary Evans Collins, *Interracial Housing: A Psychological Evaluation of a Social Experiment* (Minneapolis: Univ. of Minnesota Press, 1951); Theodore M. Newcomb, "Autistic Hostility and Social Reality," *Human Relations* 1 (1947): 69-86; Samuel A. Stouffer et al., *The American Soldier: Adjustment During Army Life,* vol. I (Princeton: NJ: Princeton University Press, 1949); Robin M. Williams, *The Reduction of Intergroup Tension* (New York: Social Science Research Council, 1947).
62. Allport; Deutsch and Collins; Norman Miller and Marilynn B. Brewer, *Groups in Contact: The Psychology of Desegregation* (Orlando, FL: Academic Press, 1984); W. T. Moxley, "Leadership Considerations and Lessons Learned in a Mixed Gender Environment," *Minerva* 17, no. 3 and 4, (Fall/Winter 1999): 58-67; Newcomb; Thomas F. Pettigrew, "Intergroup Contact Theory," *Annual Review of Psychology* 49 (1998): 65-85; Stouffer et al.; Williams.
63. Susan T. Fiske, "Stereotyping, Prejudice, and Discrimination," in *Handbook of Social Psychology*, eds. D. F. Gilbert, S. T. Fiske, and G. Lindzey (New York: McGraw Hill, 1998), 357-411.

VII

Conclusions

The final chapters conclude this anthology by offering two views of the research presented herein. The first is an insightful look at the authors' efforts by Andrew Abbott, author of *The System of Professions* (Chapter 24). In addition to a critique, Professor Abbott's chapter offers the Army's leadership broad contextual food for thought concerning efforts to strengthen the profession's competitive position through a revitalization of its members' career patterns and development. He also amends his own turf-war theory of professions vis-à-vis the Army by acknowledging that the Army's enmeshment in a vast web of interactive security bureaucracies and constituencies greatly complicates its efforts to compete successfully for the jurisdictions it determines upon.

The book concludes with Chapter 25 by Don Snider and Gayle Watkins, the directors of the "Future of the Army Profession" study. Here they present nine key conclusions garnered from the study, as well as some policy recommendations for improving the Army profession's competitive position as it moves into the 21st century.

24 The Army and the Theory of Professions

Andrew Abbott

It is common today to say that the U.S. Army stands at a crossroads in its history. The thirty years since the end of the conscription force have seen at least two dramatic transformations, and we seem to face yet another now. To reimagine the military, writers have turned in a variety of directions. One such turn has been towards the literature on the sociology of professions. Perhaps conceptual tools that have proved helpful in understanding the professions more generally can be applied usefully to the Army case in particular.

In this chapter, I consider recent work invoking professions literature to analyze the military's current situation, in particular, as treated by the chapters in this anthology. My chapter has three parts. I begin within an analysis of the assumptions that seem to underlie the earlier chapters, assumptions about the uniqueness of the Army experience, both with respect to its own history and with respect to the experience of other professions. I then turn to the Army's dual nature as simultaneously an organization and a profession, focusing in particular on the career mechanisms that try to reconcile the conflicting demands involved. I close with a brief discussion of the ways in which the chapter authors' reference to the professions literature must of necessity be limited.

My chapter's chief substantive conclusion is that the current moment is much less unique than we think. First, the changes happening to the military are widespread not only in the other professions but in society at large. Second, the security situation is by no means unprecedented in world history. Third, the "vocational crisis" is one the Army shares with other "organization professions." What seems most important is the maladjustment between the officer career structure and the actual current necessities of professional work. All these substantive results funnel into a theoretical conclusion, which is that the Army case sets important boundaries around the utility of the "turf-war" model, a model I once thought to be fully descriptive of professional life.

Let me start with the three major assumptions that undergird the chapters gathered here: that trends in current Army organization are radical and unique, that the current ambiguities in the Army's goals are without precedent, and that the Army faces a possibly unique crisis of vocation.

It is a basic assumption of many of these chapters that there is a radical and quite unusual shift in the Army as an organized system for production—in its case the production of destruction. Without judging the correctness and incorrectness of these empirical assertions, let me just list them. Some chapters see a decline in officer commitment, evidenced by retention problems. Some chapters see a blurring of lines of command that follows in part from technological developments and in part from the rise of joint operations. Some focus on increasing roles for women or other underrepresented groups. A number speak of a decline in public respect for the Army. Some see an increasing permeability of the Army, evidenced by outsourcing and civilianization of functions. Some see a different kind of sharing, an increased emphasis on joint operations with other American military services, with other American quasi-military forces like the drug or immigration enforcement agencies, and indeed with the military forces of other nations.[1]

This set of changes—perceived by many as some kind of general trend and, indeed, as in some way radical and unprecedented—provides the problematic aspect of many of these chapters. Now the problem is not that these changes aren't radical and unusual. Indeed, the Army has not seen such things before. The problem is rather that they are in fact changes characteristic of the entire American production system. Outsourcing and subcontracting are now widely characteristic of heavy industries like automobile manufacture and light industries like textiles, not to mention the high-tech world where they are completely dominant. Joint operations between firms are today almost more the rule than the exception, whether conducted through contracting agreements or managed through merger. Collaboration in production across national borders is as common as airplanes and cell phones. As for the change of missions, American production—indeed all first-world production—is characterized today by a ceaseless changing of aims of production. Many large firms, particularly in the technology area, are now organized largely on a project basis, mobilizing whole divisions around a continuing sequence of new product roll-outs. Missions change constantly.[2] As for the blurring of lines of authority, what else is matrix management than precisely this change from the clear and stable hierarchical lines of earlier firms? Throughout the advanced sector of the economy, hierarchical organization of production is fast disappearing. Bosses are bosses only for personnel purposes; their workers report to various others for their actual assignments. As for women, the concern with women's place in traditionally male preserves is of course very general, the worries about women in high-technology occupations in particular exactly paralleling the Army's concerns about women in combat. Finally, the worries about a perceived loss of status by the Army are exactly paralleled by a strong decline in general attitudes towards business in America, attitudes that fell sharply in the 1970s and have not recovered since.[3]

A summary example captures this insight clearly. AT&T prior to 1984 was a giant firm, with a clear mission and stable lines of authority, producing everything necessary to telephony from chips and switches to wire and telephones, doing all of its own service and sales, conducting almost no joint ventures with any other firm. Today, the entire area of communications has fragmented. AT&T's successor firms shift their missions continually. Both manufacturing and service in telecommunications have been broken down into mind-bogglingly complex webs of subcontracting. Even the big successor firms like Lucent Technologies are internally fragmented and project organized, with complex matrix reporting lines that have many workers reporting dotted-line to three or four others—superiors or in some cases peers—for their various projects.

In short, the sea change in the Army is not a sea change in the Army alone but a sea change in something much larger. These changes are a society-wide phenomenon. Undoubtedly, they have come to the Army in part because the Army borrows business principles both directly and indirectly. But the basic point is that the phenomenon of fragmenting production is not "something that is happening to the Army" but "something that is happening to production and services in advanced capitalism." Generally, we must not mistake the true "level" of the event that is occurring.

Having pointed out this out, however, I should at once note that by seeing the changes in military "production" as part of this larger change, we see at once what is actually unique about the Army case. In the first place, the Army does not seem to have moved towards the very short-term outcome judgments that are characteristic of American production under the governance of the stock market. One might possibly see evidence of short-run judgment in the short-circuiting of mature military judgment by excessively intimate news media coverage, but that is nothing like the almost absolute tyranny that quarterly results now have in American production and the almost complete lack of long-term planning in many firms. Nor does the Army show the overtly anti-worker sentiment characteristic of many firms since the Reagan years. Indeed, one of the peculiarities of the Army seems to be that it continues to treat its workers in the manner characteristic of welfare capitalism long after the American commercial world has greatly curtailed such programs as health insurance, child care, disability coverage, pensions, housing, and so on. Another way of putting this is to say that the Army has remained leery of acquiring and governing its workforce by pure market principles, even while its actual modes of production have paralleled trends in market production throughout the manufacturing and service economy.

The second assumption supporting and sustaining many of these chapters is the idea that the security environment of today is unprecedented; that not only the organization, but much more important the goals of the Army are shifting in directions never before seen. We are concerned here with the long list that begins with unconventional warfare and continues through the various activities covered by the phrase "operations other than war": counterdrug activities, immigration supervision, peacekeeping, humanitarian missions, civil affairs, psychological operations, stability operations, domestic support, and so on. The list is a long one and the activities constant; Charles Moskos and others count 54 military operations by Western nations in the eight years after Desert Storm. We are seeing a truly extraordinary change.[4]

But this new security milieu is by no means unprecedented. The British military in the nineteenth century faced more or less the same environment. Like America today, Britain was then hegemonic over much of the world. Her economy spread to the antipodes, and her missionaries not only accompanied but often preceded the traders who spread her economic empire. Like the American government today, the nineteenth-century British government was very often wary of overseas entanglements and was usually dragged into them by other interests, often economic but almost equally often humanitarian.[5]

The security situation, too, was closely similar. The fundamental security of the home islands of the British empire was guaranteed by British control of the sea, a deterrence unchallenged even by the emergence of ironclads until Admiral Sir John Fisher astonished the world by launching HMS *Dreadnought* in 1906, outmoding his own deterrent overnight.

In the shadow of this oceanic deterrent, the sixty-four years from Victoria's accession in 1837 to her death in 1901 saw only two substantial British wars—the

In short, the revolutionary new security situation of today is not so revolutionary. We have seen before the situation of a more or less single-great-power world in which that power's basic interest was to stabilize existing international political and economic arrangements because it did best under those particular arrangements.[11] Under such security conditions, we see lots of small unconventional conflicts punctuated by occasional major conventional ones. And in the past, the "imperial" armies involved have not been conspicuously well-prepared for those large-scale conventional engagements (e.g., the Crimean and Boer Wars). This is the nugget of historical truth behind the fears of those who worry about an excessive emphasis on unconventional warfare.

As with the assumption regarding production fragmentation, reducing the halo of uniqueness around the Army's current missions makes much clearer what is in fact really unusual about those missions. First, internal domestic policing, which was one of the basic functions of the British military in the nineteenth century, is not part of the picture today. This is true even though the American Army was itself used extensively in the nineteenth and early twentieth centuries for labor suppression, ultimately playing an important role in the fall of organized labor in America.[12] But today the labor movement, which provided most of the internal threats during the period from 1880 to 1945, is virtually dead. The most heavily unionized sector is now white-collar government employment, where the typical employee is hardly the embittered young male of the turn-of-the-century labor movement. Race riots have been the most common internal disorders of recent years, but they have been handled with relative ease by National Guard units. One can envision larger racial conflicts in the United States, but they seem very unlikely at the present juncture. So a first real difference between the current security world and the earlier example is that internal domestic policing is unimportant for today's active Army, though obviously under certain scenarios of terrorism or illegal immigration this might however change.

Second, comparison with the past reveals that America's involvement in the rest of the world is less isolated than was that of Britain in the nineteenth century. We are not imperialists protecting our turf against other imperialists. Rather, we and the others are mostly in this together. Unlike Victorian Britain, we are bound to other developed nations by extraordinarily powerful links. Not only are there much stronger political and cultural links, there are also stronger financial ties. Nearly half of the U.S. national debt is owed to those abroad. About one tenth of private U.S. capitalization is in foreign hands. This interlock implies a continuation or increase of coalition-based military actions, just as it also defines our potential major opponents as being those outside this inner circle of co-investment.[13]

Third, there has been a fundamental change in the temporal geography of the system. When English soldiers fell in botched wars in Afghanistan or Matabeleland, the news might take some time to travel back to England. Today we hear—and see—at once. More important, where for the British only the Irish could bring terrorism to the home islands, in a world integrated by international tourism, employment, and migration, terrorism is a much more present problem. And geographic integration also lies behind the twenty-first-century versions of the drug and illegal migration problems. Put another way, new forms of contiguity make new forms of quasi-military problems much more

likely, just as a highly technologized society creates new points of vulnerability to unusual forms of force.

I have so far challenged the assumptions linking these chapters to the analysis of organizations and to the analysis of the security jurisdiction. I want now to question a third assumption, that reaching into the literature on professions. My aim here is not the link between these chapters and my own work; I leave that for later. I am interested, rather, in the abiding assumption that Army service is a special kind of calling.

There are two versions of this assumption in these chapters. One makes the idea of a special calling or legitimacy the mark of all true professions. This view derives from the professions literature of the early twentieth century, when the things we now call professions were busy turning themselves from occupations indistinguishable from the rest of the labor force into occupations set apart from the common herd. In that context, the claim of special calling was a performative, that is, the group became a profession by saying it was and then doing something special to live up to that claim. Of course it helped if your occupation was old and of high status in the early modern world and, above all, if the members of your occupation were men. But the problem with the all-professions-are-callings view, which sustains a number of these chapters, is that accepting it ignores our obligation to explain the many expertise-based occupations that aren't marked—either in their own eyes or certainly in the public's—by any profound sense of special calling. There are many such occupations—librarians, engineers, and so on—and they cannot be ignored by anyone interested in understanding the vagaries of organized expertise in the modern world.

So that line of argument is indefensible. What about the other one, the narrower view that the military itself has some special calling? Surely this view is central to most of these chapters. It is also central to Army thinking about recruitment and retention, to the notion of the Army as somehow held to higher standards than its civilian masters, to the obsessional worry about civil-military relations, all of which themes surface repeatedly here.[14]

Now this notion of a special calling is indeed characteristic of some occupations, usually professions, or, more properly, of reforming subgroups within those occupations. William Osler the physician, Louis Brandeis the judge, Frederick Haynes Newell the engineer—all of these had their almost sacerdotal dreams of a zealously dedicated professionalism. Yet one striking phenomenon characteristic of nearly all of these would-be professions as callings is that associationalism or corporateness, which we might take as a strong indicator of "calling," is in retreat in all of them. The percentage of professionals who are members of the national professional associations is falling steadily. The lawyers turned downward in 1980 after almost a century of growth, the dentists after 1975, the architects in 1970, the physicians and the electrical engineers in 1960. Robert Putnam—who noted these trends—sees them as part of a general decline in associational life in America, the professional version of "bowling alone." And he would no doubt see the current crisis in Army professionalism as yet another part of this broader phenomenon.[15]

But Putnam is wrong. The timing of the Army's vocational decline doesn't fit with his model, which is basically about long-term cohort decline. For Putnam, the big causal variable is the supposed failure of the baby boom generation—those born

between 1946 and 1965—to raise its children on the values that had been taught to it by its own depression- and war-tested parents. If that were the mechanism, the Army decline would not have been as sudden as many seem to think; the quick falloff from the strong Army of the early 1990s would have been impossible. Moreover, much of the steady falloff in associational membership in the other self-styled professions as callings is clearly a result of proliferation; as the denominators have gotten bigger, these occupations have moved beyond the size where a personal sense of belonging can matter. There are now nearly a million lawyers in America, about seven-hundred thousand doctors, a roughly equal number of electrical engineers, and even a quarter of a million archictects. These cannot be tightly bound groups as they could be in 1950 when each numbered around a hundred and fifty thousand.[16]

The Army in peacetime is of course smaller—less than a hundred thousand officers—still within the size that used to produce relatively cohesive professionalism in other fields. So size is unlikely to be the major issue in vocational decline in the Army's case. Moreover, the military idea of calling has always been much more resonant than that of the lawyers or doctors: not just a medical or law school, but an academy on a hill; not just a graduation ceremony but a legally binding oath; not a choice of careers, but orders to be obeyed. Such structuring reaches beyond the military only in France, where all the higher civil services are organized into the *grands corps de l'etat*: selected as young people, taught in special schools, organized into paramilitary structures, where junior members' careers are directed by senior members of the corps.

What is striking about the French case is that some French are today complaining of precisely the same demographic crisis in the *grands corps* that many analysts are describing in the U. S. Army. Resignation of junior corps officers to join private firms—often firms closely related to those regulated or otherwise overseen by those very officers—has reached such epidemic proportions that new laws have been created to regulate the traffic, which can be two-way under certain conditions and which has the rather romantic French name of *pantouflage* (literally, walking in one's slippers). Thus, the Army is not the only strongly vocational profession whose high calling seems to be fading.[17]

Closer to home, however, is an even better example, the Roman Catholic clergy. The parallel between the Army officer corps and the clergy is very close. Like officers, priests have suffered a sudden change in mission, occasioned by the promulgations of Vatican II. Gone is the priest whose only mission was to illustrate the life of Christ in the parish by performing the central sacramental acts of Catholicism. In his place came the urban priests and the social activist priests and even the priests who practiced other professions during the week and celebrated mass on Sundays. By the 1990s, sociologist Scott Appleby was describing the typical parish priest as an orchestra leader: a man charged with overseeing dozens of activities but expert in few of them, a workaday manager who happened to celebrate mass.[18]

This change of mission was connected, as in the Army case, to broader changes in the mode of production in American Catholic parishes. As in the Army, there was massive civilianization and outsourcing. The parochial schools hired more and more secular teachers as the population of sisters fell into rapid decline. The laity invaded the liturgy and became so responsible for those parts of the service that

precede the actual Eucharist that some spoke of "protestantization." The liturgy itself was restructured along lines suggested by the ecclesiastical equivalent of "coalition warfare"—ecumenicalism. Blurring of lines of authority well describes the new structure of parishes and dioceses. And as in the Army, the question of women pervaded the church's heartland jurisdiction, both in their character as potential celebrants of the mass and as potential wives of celibate priests. And finally, the Army and the clergy remain the lone professions that still start all their leaders' careers with extended periods performing the very basic functions, something that has long since disappeared in the commercial sector, where leaders have for many decades come from sales or finance, not from operations.

I need not remind readers that this set of changes exactly parallels the changes so often attributed to today's Army. Moreover, the changes are associated directly with a new precariousness of the priesthood's missions, just as the Army's heartland of battlefield combat is threatened by the gossamer violence of unconventional warfare. What is deeply different about the two cases is that the priests' own heartland—the Eucharist itself—faces no makeover of its own. There is no new technology—no technological Revolution in Military Affairs—that can replace the celebration of mass. There may be new words, even new language, perhaps even new types of people. But the act of Eucharist is, in Catholic eyes, as unchanging as the God who commanded it two thousand years ago.

In terms of calling, then, the Army is located within several concentric sets of occupations, ranging from the usual professions with their relatively weak vocations to the Catholic clergy where "vocation" is a subject of burning interest. Like many of the so-called higher professions, it has endured a decline in associational and vocational life, as I have noted. And although there is a smaller number of occupations—both here and abroad—that share the Army's intensity of calling and commitment, among them too we see a general decline, not confined to the U. S. Army. Vocationalism, or calling, is in general retreat.

Let me summarize. I have attacked three problematic assumptions: that the change in mode of production is specific to the Army, that the current security environment is uniquely new, and that the Army is seeing a unique decline in calling. First, I have shown that the changes in the Army's modes of production are broadly similar to larger trends in the commercial sector, with the two exceptions that the Army is not under such short-term surveillance as are other producers and that it has, unlike them, retained a welfare capitalist model for its workforce. Second, I have shown that the security environment is more or less the standard imperial security environment, with three differences: internal domestic policing has temporarily atrophied as an Army jurisdiction, new financial and political connections dictate new types of coalitions and opponents, and tightened geographic and temporal contiguities imply new forms of security threat. I have shown third that the Army's worries about calling are not unique, even if we set very high standards for the level of calling expected.[19]

We should now note a contradiction that becomes obvious once we see the nonuniqueness of the Army's experience. The Army is changing like most other producing organizations on the one hand, and the Army is changing like most other

professions on the other. Since those trends are moving in unrelated directions, the result has been to put a well-nigh unbearable strain on the Army's longstanding attempt to be an organization and a profession at the same time.

It is not then surprising that this strain is, underneath the surface, the common theme of the chapters. Indeed, the common theme that brings all these many topics into sharp coherence is the concept of career. Many chapters have focused on calling, and the problem of calling stands at the center of concerns about the Army career. Other chapters have focused on jurisdictions and their control, and much of the worry about jurisdictions concerns whether the full spectrum of work areas from traditional combat to MOOTW can be embraced in a single career. And even the many chapters touching on problems of politics concern, perhaps more than anything else, the use of political knowledge and the exercise of political judgment and values at different stages of the officer's career.

That career is central should hardly surprise us. The idea of career—that is, the idea of a single occupational skill or identity characterizing individuals for their entire working lives—is probably the most central single constituent of the idea of profession as it emerged in the nineteenth century. But more important in the present case is the fact that the development of formally patterned careers is the Army's chief mechanism for reconciling the demands of its dual nature as an organization and as a profession.

Let me then address the question of career directly. Can we imagine a notion of career that works in the current Army context? My earlier arguments help us little here. American production may have experienced the same general shifts as has the Army. But it has dealt with their impact on individuals' lives by simply ignoring the fact that the new high-turnover, project-oriented workplace is a horrible place to work for anybody with more than minimal personal obligations. Because most career advancement takes place by geographic and interfirm mobility, youth and portability become the central, sometimes the only, criteria of success. Single-firm careers have become considerably less common.[20] In fact, only a small proportion of the labor force spends an occupational lifetime in a single occupational identity. Even in the high-status professions like law and medicine, many leave active practice to become administrators, consultants, professors, trainers, executives, business people, and so on.[21]

In short, the experience, and to some extent even the ideal, of career is disappearing from the labor force. Military officers, clergy, and professors are among the few holdouts. Nor is the clergy experience much help. Clergy careers have been a lot like military ones in terms of moving from one set of obligations to another, but many fewer people move up the hierarchy. Nor is there an equivalent in the clergy setting to the military's enlisted ranks.

For a partially successful example of career structuring under contemporary conditions, we might turn to the French administrative elite mentioned earlier. The members of the *grands corps* are an extreme elite; they constitute about two percent of college-trained civil servants in France, ten thousand out of half a million people. Selected very young, they are bound to state service for no less than ten years. Their careers are complex, typically including stays in various ministries, public enterprises, and, often, private industry as well. In some corps, like the Council of State, most members work within the administrative activities of the corps itself. In others—like

the corps of financial inspectors—members typically work in other ministries, while retaining their identities as part of the corps. Note that the extreme centralization of French society means that the vast majority of this employment is in the vicinity of Paris; geographic mobility is not involved, as it would be in the United States.[22]

The example of the *grands corps* suggests that one way of preserving professional Army careers under current conditions is to create such a super-elite corps, with members meticulously selected at the outset of their careers for prestigious, immensely desirable career tracks. There is a strong suggestion in a number of these chapters that Army careers in general—and these elite careers in particular—should include experience in a very broad set of positions: not only Army combat commands, but joint commands, staff positions in the Army and joint services, and, indeed, positions throughout the other security bureaucracies and perhaps beyond. To be sure, bifurcating the officer corps into the elite and the nonelite is not a particularly American thing to do, and by resting heavily on early identification it risks missing late developers. But it does begin to address the crucial question of demography, to which I would now like to turn.

Essentially, the structure of careers as well as the rank and age structure of the Army is dictated by the entrance and exit rates at various ranks. Entrance is determined by recruitment and promotion policies and by externally imposed constraints on the sizes of certain rank cohorts. Exit is determined by resignation and retirement and the policies affecting them. Resignation in turn reflects internal forces arising in the demographic structure; if there is no possibility of promotion or other interesting career developments, exit rates rise. It is a fairly straightforward exercise—presumably it has been done—to model the various Armies possible under various officer population parameters. And, more important, it is possible to reason backwards from the desired career structure to the demographic parameters necessary to create the possibility for it. Of course one has to factor in the possibility of significant surrounding events—wars—that might suddenly result in increases at various levels. And one can, as in the French corps, create a certain flexibility in the system by allowing extensive detachment to other related services. But the basic fact is that there is no way to have an Army in which most officers can aspire to a certain model career without managing the formal population and rank structure of the officer corps very carefully. Moreover, if as many of the chapters here imply, the Army in one fashion or another continues to straddle the jurisdictional divides between traditional battlefield combat, unconventional warfare, and MOOTW, the career structure has to be designed to knit these various jurisdictions together.

There are a number of traditional models for this knitting. In medicine, all doctors have a common beginning, but then specialize fairly rapidly. In law, a similar specialization sets in, but only after an abstract mastery of the "principles of legal reasoning." Neither of these professions has a routine system of midcareer education, although the doctors have various kinds of short brush-up courses. Some areas in academics have midcareer conference-based education—the Gordon conferences, for example—but again these are not really like the military system. Indeed, no other American profession really has such a system, deliberately designed to advance practitioners to a new level. It exists in business in the guise of executive MBA programs, but its not really clear whether those exist primarily as a mode of training or as a simple selling of credentials.

A more common system for knitting jurisdictions together—the one characteristic of most well-designed bureaucratic systems of employment—is circulation of fast-track employees through complex rotations in various parts of the overall organizational or professional operation. This is the pattern characteristic of most modern militaries and of the French administrative elite. The real question is how much to try to knit together. Just traditional combat and staff positions? That plus some unconventional experience? All of that plus some MOOTW? Where does one stop?

It helps to realize that there is a certain pattern to these jurisdictions. We think of the traditional Army jurisdiction as battlefield combat, the use of force pushed to a logical conclusion. But the nature of force is not a constant. In a time of hegemony, a time in which predominant force as customarily defined is very much at the disposal of one power or set of powers, those who oppose this power will generally seek to redefine the nature of force. Under these conditions, we often see terrorism, which is just such a redefinition. Terrorism is the hegemon's denomination of those types of deadly force over which it cannot exercise monopoly control or in which it does not choose to exercise absolute superiority. In current conditions, for example, terrorism usually involves violating the convention that separates combatants and noncombatants; and, of course, that's the whole point of force—it violates conventions. Thus, we consider bombing of noncombatant buildings, poisoning of water supplies, hostage-taking, and assassination of civilians to be terrorism. But from the point of view of those who do them, they are simply Clausewitzian force, policy by other means. Since they are the only force-based weapons available to those over whom we exercise crushing conventional superiority, they become the inevitable weapons of the weak.[23] In this obvious sense, crushing conventional superiority automatically, logically, breeds exercises of force to which it is potentially vulnerable or even powerless. The use or potential use of force is thus curiously recursive. We have an ultimate deterrent, nuclear weapons, so opponents use levels of force against us that don't justify our invoking them—namely, conventional warfare. We then build conventional forces that can address nearly any challenge at this next lower level, so our opponents seek to develop types of force that evade that conventional superiority. This is unconventional warfare, to which our ancestors had recourse during the revolution against the British with their relatively strong conventional forces, and to which our own enemies today have recourse for the same reasons. And indeed if we were to develop a masterfully powerful anti-unconventional-warfare capability, that in turn would lead our opponents to develop yet other counter.

The fundamental question is whether these various recursively organized jurisdictions can or should be experienced within the same U.S. Army careers. There is some logic for thinking that they should, at least for some subgroups of officers, because the excessive use of force at any one of these levels leads to expansion of force at the next one below, unless there is utter destruction of the enemy. A serious grasp of how to prevent this move to the next level of force is obviously important to proper exercise of force at any given level.

The topic of jurisdiction brings me to a final issue. It remains to consider the implications of the Army case for the system-of-professions model that I argued 13 years ago in my book *The System of Professions: An Essay on the Division of*

Expert Labor.[24] I argued that the history of professions is to be understood as a continual clash over areas of work, over what I called jurisdiction. Professions tried to claim jurisdiction over this or that form of work by inventing some abstract knowledge to accomplish the work and then presenting that claim in front of various audiences: in the workplace, in public opinion, before the organs of the state.

I assumed in that book that these various audiences of professional claims were just "there," fixed in place. For example, I thought physicians figured out some abstract knowledge governing bodily ailments, then went about convincing the state that they ought to be given exclusive legal control of doing that work. But I never asked why the political system should care about physicians and their licensing. Yet no politician would push licensing unless it did something for him, in his home world, that is, in the world of politics. In that sense, licensing had to work both as a claim for the physicians (against their competitors such as the homeopaths, the botanical doctors, and so on), but also as some kind of political fodder for people in the government, perhaps for administrative types who were looking for something new to administer and thought licensing would be a good object for that. Licensing had to work in two systems, both the system of professions and the system of politics.

Thus, I had originally seen the system of professions as a dynamic ecology surrounded by relatively static environments. But in my new view of "linked ecologies," I am insisting that at the margins, things move forward only for both ecologies (professions and politics), and only because they can come together over some policy that works well in two systems.[25] This immediately poses the empirical question of when it is useful to think of the social world as ecologies with static environments (as I did in the original book) and when as linked ecologies coming together at the edges (as I do now).

But the case of the Army pushes me to see that an institution can be so enmeshed in so many different "ecologies" that thinking about it chiefly as a profession involved in jurisdictional adventures (over who controls traditional combat, unconventional warfare, and MOOTW) may really capture only a tiny portion of the systems of environing relations that govern it. It is for this reason that the great books about the bureaucratic politics of the military—Morton Halperin's *Bureaucratic Politics and Foreign Policy* (1974), for example—spend most of their time elaborating an enormous cast of characters—the branches of the service, the civilian and military bureaucracies, the National Security Council, the Central Intelligence Agency, the congressmen with their worries about base closure and military business, the manufacturers with their desire for military profits—and then tracing various particular stories about these characters through to their conclusions.[26] Such books do not see any particular system of ecological relations or competitions that governs actors in the security field. They see simply a mass of countervailing complexities. And they don't even touch the other constituencies mentioned in many of the chapters of the present book—veterans, civilian employees, subcontractors, and various political interest groups.

This then is the payoff of the Army case for the theory of professions. The dominant "turf-war" theory of professions presupposes an isolation of the competitive system of professions that is probably not correct even in the general case, but that

is certainly false in the Army one. The Army case thus poses a major theoretical question of more general import: when and why does it make sense to see interprofessional competition as isolated from other forms of occupational determination, in contrast with the Army's manifest enmeshment? Relatedly, how does one begin to think systematically about systems that are as densely populated with actors as is the system that surrounds the Army? These questions must be addressed in depth before the sociology of professions can present an effective analysis of the Army.

Notes

1. These assumptions are by no means limited to the chapters in this volume. For another example, see Charles C. Moskos's conception of the "postmodern military": "Towards a Post-modern Military," in *The Postmodern Military*, eds. Charles C. Moskos, J.A. Williams, and David R. Segal (New York: Oxford University Press, 2000).
2. Indeed, one could argue that the original models for this mode of production are actually military ones, arising in the project-like conception of operations in the Second World War and civilianized afterwards in the concept of operations research (OR). On the history of OR, see H.J. Miser, "The History, Nature and Use of Operations Research," in *Handbook of Operations Research*, eds. J.J. Modern and S.E. El.maghraby (New York: Van Nostrand, 1978); and M. Fortun and S.S. Shweber, "Scientists and the Legacy of World War II," *Social Studies of Science* 23 (1993): 595-642.
3. On attitudes to business, see W.G. Mayer, *The Changing American Mind* (Ann Arbor, MI: University of Michigan Press, 1993), 98-100.
4. For this list, see Moskos, Williams, and Segal, appendix.
5. In this section, I have relied heavily on G. Harries-Jenkins, *The Army in Victorian Society* (London: Routledge Kegan Paul, 1977).
6. The figure of 72 campaigns is from Harries-Jenkins, 35. For a general review of the bewildering variety of campaigns, see chapter 6.
7. For the Calwell quotation, see Harries-Jenkins, 184; the Thomas Adams book was published by Frank Cass (Portland, OR).
8. Harries-Jenkins, 186-187.
9. On the French army's hyper-professionalism, see W. Sermon, *Les officiers Français dans la nation, 1848-1914* (Paris: Aubier Montaigne, 1982). On the current U.S. Army, see Moskos, 14-31; and S.C. Sarkesian and R.E. Connor, *The U.S. Military Profession into the Twenty-First Century* (Portland, OR: Frank Cass, 1999).
10. For a general history, see B. Porter, *The Lion's Share* (London: Longmans, 1975).
11. Of course, the details of the arrangements preferred by the great power have changed. A century and a half ago, those arrangements meant free trade, Western notions of property, and protection for British (and other Western) citizens from the "traditional" rule of local authorities. Today, these earlier conditions have been supplemented by an insistence on political forms that can be at least loosely labeled democratic, privatization of religion, and a much greater openness to and security for Western investment. These added conditions of course flow from the possibilities opened by the end of the Cold War.
12. To be sure, the Army generally carried out anti-labor policies that originated with civilians, as sometimes did pro-labor policies. When Franklin Delano Roosevelt used the Army, both during the great labor troubles of the 1930s and during the Second World War, he aimed to stabilize situations to allow arbitration actions, which themselves generally favored labor. On this extensive and quite well-studied history, see J.M. Cooper, *The Army and Civil Disorder*

(Westport, CT: Greenwood, 1980); J.M. Jensen, *Army Surveillance in America, 1775-1980* (New Haven, CT: Yale University Press, 1991); C.D. Laurie, *The Role of Federal Military Forces in Domestic Disorders, 1877-1945* (Washington, DC: Center for Military History, 1997); and S. Yellen, *American Labor Struggles* (New York: S.A.Russell, 1936).

13. These figures are much greater than the equivalent figures in 1900, when the hegemon (Britain) owned the rest of the world, rather than the other way around. On a market basis the U. S. international investment position (including governmental and private debts and assets) is about negative 1.5 trillion dollars. These figures come from the *Statistical Abstract of the United States* for the year 2000. For the total U.S. international investment position by type of holding, see Table 1310, p. 782. For current (annual) figures see Tables 829 and 830, p. 524. Foreign-owned assets in the United States in 1998 totaled about 7.5 trillion dollars, including about 2.5 trillion in U.S. securities, at a time when the total U.S. debt was about 5.5 trillion (Table 542, p. 346). Total private capitalization was about 20 trillion in 1998 (Table 827, p. 523), of which about 2 trillion was in foreign hands. (I am disregarding nearly 1 trillion in "foreign official assets" in the United States, about half of it being Treasury securities.) The United Kingdom is still the largest single foreign investor in the United States.
14. It is interesting to note that Parsons theorized the doctor-patient relationship in the same kind of "sacral" framework that many people use for the issue of civil-military relations. See T. Parsons, *The Social System* (New York: The Free Press, 1951), chap. 10.
15. These figures come from R.D. Putnam, *Bowling Alone* (New York: Simon and Schuster, 2000), 444.
16. For current figures, see *Statistical Abstract 2000*, Table 669, 416-18. Historical figures can be found in the *Historical Statistics of the United States* (Washington, DC: Government Printing Office, 1976), vol. 1, 140-145.
17. The current status of the *grand corps* is well chronicled in the work of Jean-Michel Eymeri. See his *La fabrique des enorques* (Paris: Economica, 2001); and and J. M. Eymeri and J. Pavillard, *Les administrateurs civils* (Paris: Documentation Française, 1997).
18. R.S. Appleby, "Present to the People of God," in *Transforming Parish Ministry*, eds. J.P. Dolan et al. (New York: Crossroad, 1990), 1-107.
19. Space constraints prevent me from speaking to the issue of bureaucratization and micromanagement, issues in which the Army's worries are close to those of the physicians under the new regime produced by the dominance of medical insurers.
20. It should thus not surprise us that many pensions now become partially available at age 55 and that rates of labor force participation are declining at all ages over 55. The rest of the labor force is in fact coming around to the Army's "thirty years and out." For labor force participation rates, see P.J. Purcell, "Older Workers," *Monthly Labor Review* 123, no. 10 (2000): 19-30.
21. M.D. Evans and E.O. Laumann, "Professional Commitment," *Research in Social Stratification and Mobility* 2 (1983): 3-40.
22. Eymeri, see also A. Claisse et al., *An Introduction to French Public Administration* (Paris: International Institute of Public Administration, 1993); M.-C. Kessler, *Les grands corps de l'etat* (Paris: Presses de la Foundation Nationale des Sciences Politiques, 1986); and J.-L. Quermonne, *L'appareil administratif de l'etat* (Paris: Sueil, 1991).
23. This phrase is taken from J.C. Scott, *Weapons of the Weak* (New Haven, CT: Yale University Press, 1985).
24. Andrew Abbott, *The System of Professions: An Essay on the Division of Expert Labor* (Chicago, IL: University of Chicago Press, 1988).
25. Andrew Abbott, "Ecologies Liees," in *Les professions et leur analyses sociologiques*, ed. P.M. Menger (Paris: INED, forthcoming 2002).
26. Morton H. Halperin, *Bureaucratic Politics and Foreign Policy* (Washington, DC: The Brookings Institution, 1974).

25 Project Conclusions[1]

Gayle L. Watkins and Don M. Snider

Intention of the Research: A Review

As discussed in Chapter 1, our intentions in this research project were essentially twofold: (1) to apply to the Army the newer theories about professions, particularly that of the "competitive systems of professions" theory developed by Andrew Abbott, and (2) in so doing to establish a broad research foundation concerning the Army profession's expert knowledge and expertise that serves to describe accurately the current state of the profession. The subsequent chapters of the book addressed both intentions in considerable detail. These chapters represent some of the best contemporary research and thought available on the Army as a profession. As a body, they summarize much of what we know about the historical evolution and current state of the Army profession.

But since they also suggest a number of critical crossroads the Army is facing as it moves into the 21st century, one task remains, that of looking at the assembled chapters whole and drawing out the major conclusions emerging from the aggregated research. We have attempted to do this in a manner that is useful for civilian and uniformed policy-makers, those leading the Army's transformation efforts. Recalling Figure 1-1 in Chapter 1, which displays the jurisdictions of the Army profession, both internal and external, we need to be very clear about the framework used by the researchers, that of a profession competing within a system of professions. There are many "good-news" stories concerning today's Army, particularly if it is viewed as just another producing organization. But when viewed as a competitive profession in the midst of ongoing 21st century competitions, we found significant challenges, particularly in the two internal jurisdictions of expert knowledge and the development of professionals. Thus what we present here are the broad, compelling insights that became apparent to us as we worked with the other researchers—nine conclusions that best define the path for renewing the Army profession as part of the ongoing transformation.

Conclusion 1—*At present, the Army's bureaucratic nature outweighs and compromises its professional nature. This is true in practice, but, of greater importance, it is regarded as true in the minds of the officer corps. Officers do not share a common understanding of the Army profession, and many of them accept the pervasiveness of bureaucratic norms and behavior as natural and appropriate.* This is the core conclusion of the project, underlying everything that follows. As explained in

Chapter 1, throughout its history the Army has had to balance its two natures: the Army is at once a government bureaucracy and a military profession. Although this dual nature results in tension and stress for the Army, both aspects are necessary. However, if the U.S. Army is to provide the military capabilities expected of and needed by a 21st century military force, its professional nature must dominate.[2] Our conclusion is that it does not; and in fact the current ascendancy of the institution's bureaucratic nature is undermining professional identity and performance.

The research presented in this volume validates the Army's own Training and Leader Development Panel finding that the Army's bureaucratic nature dominates the perceptions of the officer corps. Our research indicates that the officer corps shares basic Army values, but members do not have a shared understanding of the Army as a profession; they have little of a profession's common language, conceptions, or identity.[3] Many see themselves as employees—as mere occupational timeservers—rather than professionals and members of a life-long calling. This inclination is further highlighted by their uncritical acceptance of bureaucratic norms and behavior, which have been reinforced internally by pervasive, bureaucratic Army management systems and externally by a decade of force structure cuts and under-resourcing. Increasingly centralized management systems have shifted decision-making power upward while eroding the flexibility and authority of responsible professionals on the scene. The bureaucratizing effect of this movement has been compounded by a barren fiscal environment that pushes leaders toward efficiency measurements and "doing more with less." As a result, today's Army is more bureaucracy than profession—more preoccupied with efficiency than effectiveness—and is perceived as such by its members, including the officer corps.

Unless its members hold common basic conceptions of the profession, the Army faces a near impossible task in trying to rebalance its dual nature in favor of the profession. Establishing a new equilibrium is of overriding importance, for if a profession's members do not recognize its essential character, then its long-term survival in the intense jurisdictional competitions of the 21st century seems in jeopardy.

To tip the balance back toward the preeminence of its professional nature, to restore the Army profession as a proud calling for experts in national defense, the Army's strategic leaders must cry halt to the present trend, and then initiate the process of rebuilding a professional culture. In other words, the Army's strategic leaders are going to have to discover the time to become serious about professional renewal. In the broadest terms, they must first work to reverse prevailing conceptions, that is, the leader attitudes, language, and behavior demonstrated throughout the Army, and then they must begin work to reform the Army's management systems. Key decisions must be approached from the perspective of a profession first, rather than a government bureaucracy or a business venture. The most important of these key decisions is addressed in our next conclusion.

Conclusion 2—*The Army needs to redraw the map of its expert knowledge and then inform and reform its educational and developmental systems accordingly, resolving any debate over the appropriate expertise of America's Army.* The fact that expert knowledge can be embodied in people for professional practice in new and unfore-

seen circumstances is the hallmark of professions. Professions establish their position in society by adapting, extending, developing, and refining their expert knowledge. Furthermore, professions negotiate with their client as to where and when this knowledge will be applied. The scope and boundaries of a profession's expertise, as well as those of related professions, make up the "map" of its expert knowledge. This map then provides a framework for professional education and development and human resource management, as well as other institutional systems.

Since the end of the Cold War, the Army's external environment has changed dramatically. Adapting to these changes requires significant shifts in the map of the profession's expert knowledge. Although the Army has debated these changes, it has not conclusively determined the new scope and boundaries; it has not redrawn the map of its expert knowledge. As a result, its organizational systems do not support the current map, and may be even more maladapted for the future map. Although the Army is performing an increasing number of so-called "nontraditional" missions (a misnomer if there ever was one), its primary organizational systems—i.e., structuring, manning, equipping, training the force, and assessing its readiness—remain focused primarily on fighting conventional wars. This irresolution has resulted in professional education and development systems that are not clearly linked to the Army's evolving expert knowledge. Unit officers are learning skills in the schoolhouse they are not using in their deployments, and are failing to learn skills in the schoolhouse that are critically necessary for them to succeed in present deployment missions. Unit training by template in just-in-time preparation for deployments, which is highly centralized in organization and design, has filled this void, but at an immense price in loss of junior leader initiative and satisfaction.

This loose coupling between the Army education/developmental systems and current mission requirements has been further weakened by the Army's diminishing professional claim on the development of its members and doctrine. As contractors take on ever-greater roles—in professional education, ranging from Reserve Officer Training Corps instruction, to CGSC curriculum development, to general officer executive education; in analysis and writing of doctrine; in staff functions on the General Staff—the Army profession is abdicating education of and control over these critical elements of its professional life. Unlike businesses, for which the education and development of employees is typically a noncore function, the education and development of the members of the Army profession are its core, as are the creation and development of the doctrinal principles by which professional work will be undertaken.

The chapters in this volume assist in identifying the Army profession's expert knowledge by presenting a wide range of subject matter areas that have not recently been well explicated, including soldier spirituality, civil-military relations, strategic leadership of a profession, unit cohesion, physical development, and soldier well-being. In addition to identification, the prioritizing and bounding of the expertise that falls within the Army's professional realm are fundamental to mapping its expert knowledge. For example, taking care of soldiers has become an end in itself for many Army officers, an end that is no longer linked in their minds to the accomplishment of the mission. This decoupling has effectively moved this aspect of an Army professional's work into the realm of social work. If that priority reflects the

intent of the profession's strategic leaders, then drawing on the well-explicated discipline of social work would aid Army officers in this work. If, however, the Army does not intend officers to be spending such a high proportion of their time and energy on social work, then clearly prioritizing and bounding this expertise are required, as is the subsequent realignment of institutional management systems to maintain such priorities and boundaries.

Conclusion 3—*Since people are the Army's most important resource, understanding them should be an uppermost component of its expert knowledge, and Army management systems should be aligned to reflect that knowledge and its priority.* Knowledge pertaining to people and human development is the foundation of Army expertise. Yet if resources are the measure, then the Army's desire and ability to understand people fall well behind its desire to understand hardware and software. Technological research dominates funding, billets, and priorities while the human domains—behavioral science, social science, and physical development research programs—are under-funded and out of date. In the last decade, the Army has slashed funding for research in the human disciplines of the social and behavioral sciences. Rather than pursuing cutting-edge knowledge in these areas, the Army bases much of its "expert" knowledge of human behavior, specifically human behavior during war, on research done over fifty years ago. These earlier findings did not benefit from the increasingly sophisticated research methods and analytic tools that have produced over recent decades more reliable and compelling findings about such subjects as unit cohesion.

It is true that the Army has devoted significant contemporary research to another discipline pertaining to people—physical readiness. The present Army Physical Fitness Test (APFT) is the result, and it offers the Army a means of measuring baseline individual fitness. However, the Army has not educated its members as to the APFT's purpose, use, and limitations. As a result, the APFT's role has been distorted so that it now dominates unit fitness programs, and is often poorly correlated with units' physical readiness to perform their mission-essential tasks.

The Army has not well integrated extensive study of human behavior from such disciplines as psychology, sociology, and political science into the career-long professional education system. These disciplines have particular relevance to the Army's internal jurisdiction because decisions regarding personnel management are grounded in our strategic leaders' understanding of people. Therefore, since the Army's understanding of human resources has fallen behind cutting-edge scientific knowledge, its management systems have not evolved in tandem with the scientific understanding of human behavior. Instead, many of the Army's recent personnel management policies have derived from business practices, which are more appropriate for profit-driven firms with employees. Professionals, however, expect their profession to provide support and systems that differ from those of other work organizations. Today, the Army's retention problems and high levels of dissatisfaction are evidence that its management systems, such as soldier acquisition, evaluations, assignments, career development, housing, and community development, are not meeting the needs of the profession, its members, and their families.

Conclusion 4—*The most important management concept needing alignment with the demands of the Army profession is career progression.* The Army has historically reconciled its dual nature as both bureaucracy and profession through its system of career management, wherein a unique combination of centralized bureaucracy and individual development balanced the institution's conflicting natures. However, the evolution of Army personnel management practices over recent decades has shifted the balance away from individual development and toward a lock-step, centralized system that requires all officers to follow specific timelines and fill certain positions if they are to succeed. If they do not, most officers have no choice but to leave the Army. The result is reduced variety in the careers of successful officers and a strong reluctance to pursue opportunities outside of the mainstream; officer development is not tailored to either the officer or the Army but to the promotion and the assignment systems.

Sadly, this rigid system has evolved at an inopportune time for the Army; rather than officers with essentially identical career patterns, this is a time when the Army needs educational, skill, and experiential diversity in its pool of members. To its credit, the Army moved slightly in this direction with the adoption of OPMS XXI, offering a limited degree of specialization. However, the rigidity of officer career timelines actually tightened under the new system. Unfortunately, officers themselves fight against the Army's specialization needs because the promotion, assignment, and educational systems do not in their view support it. Few officers are willing to sacrifice their career (and with it their family's welfare) to pursue a career track outside the norm.

Officially, an Army career is still based on the traditional but increasingly obsolescent idea of a single occupational skill or identity characterizing individuals for a working lifetime. Though the system remains intact, however, the profession's younger members, like their peers across America, are seeking alternatives to the traditional concept of a 20- or 30-year career with a single company or institution. Unlike their predecessors, these young people see jobs and positions as opportunities to gain the individual expertise and developmental experiences needed for personal growth and long-term upward mobility. Short employment contracts and increased inter-job movement pervade the motives of the generations filling the Army's junior ranks. By and large they enter the Army neutral in their attitude toward a lifetime career. Far from winning them over to the idea of a long-term commitment, however, the dominance of the Army's bureaucratic nature in its personnel management systems is accomplishing the exact opposite. Young officers become cynical about "check-the-block" career management and tend to see the Army as just another transient job.

To accommodate this generational shift within the demands of a military profession, the Army must accept radical change in its concept of "career." The new conception will likely require diluting the centralized authority of the "face-to-space" personnel management system, allowing senior officers in the field more flexibility to develop individually the professionals in their units, truly complementing the institutional and self-development components of the Army's professional development system. It would encourage officers to seek experiences outside of the Army organization and joint force structure, providing knowledge and expertise

that comes from working closely with other government agencies, international organizations, educational institutions, and even businesses. It also requires a review of, and perhaps consolidation within, the Army's commissioned and noncommissioned rank structures and progressions, a review that should seek to identify the singular needs of the Army profession, those that are quite different from the other two military professions. Further, the Army's sole reliance on an internal labor market may no longer be appropriate; creating the opportunity for lateral entries and exits with later returns may enable the development of specialists to the degree needed by a 21st century Army. Lastly, this revised system would most likely require reversing the trend of recent decades for higher-level defense bureaucracies to homogenize personnel policies across America's three military professions.

Conclusion 5—*Military character and the professional ethic are the foundation for the trust the American people place in their military and the foundation for the trust Army officers place in their profession.* No matter how extensively the Army must transform nor how much of the profession must be redefined, we conclude that one foundation must remain unchanged: by the nature of the Army profession, only individuals of firm moral character can discharge adequately their professional obligations to the nation and to the soldiers they are called to lead—good officers, good professionals are first of all men and women of good character.

The Army profession, because of its responsibility for wielding deadly force to defend the nation and its Constitution, has developed over two centuries an ethic that provides objective norms and standards for the behavior of the profession and its members. This ethic is cooperative and cohesive in spirit, selfless but meritocratic, and fundamentally anti-individualistic and anti-careerist. It requires that members transcend the norms of the pack, particularly when under chaotic, stressful, and fearful situations such as combat.

Although the jurisdictions within which the Army is currently operating present new and different challenges for the Army, the need for foundational professional ethics remains. This requirement stems both from the American people and from Army professionals themselves. One of the strongest findings in our research was the importance that Army officers correctly place on the profession's ethics, particularly the duty concept and the moral courage of Army leaders. Their commitment to a culture of candor—clear truth-telling by all leaders at all levels at all times—is universally seen as an absolute requirement.

Conclusion 6—*The "trust gap" between junior and senior Army officers, the junior and senior members of the profession, has reached dangerously dysfunctional levels.* This gap, which tears at the sinews of vertical cohesion in the chain of command, is both a cause and effect of the Army profession's inability to inspire and retain officers, company grade and field grade. In one sense, this conclusion is not new—the Army has long had, and likely always will have, some degree of misunderstanding or miscommunication between junior and senior officers.[4] This is the unavoidable result of perspectives informed by different levels of responsibility and years of experience within the profession.

However, today's gap, now documented by numerous investigations, has widened to a degree that threatens the Army's core functionality, as evidenced most prominently by the continued exodus of well-qualified officers into civilian life.[5] Perhaps more than civilian occupations, trust in the military goes to the heart of the profession's ethic and therefore to its effectiveness on the battlefield. Unless commanders establish a culture of trust within Army units, soldiers will not feel free to tell the truth, and without transparent honesty in interpersonal relations and official reporting systems, effectiveness suffers. This downward spiral induces micromanagement on the part of leaders and risk-averse responses on the part of followers.

To resolve this issue, a task the Army has already undertaken, leaders at all levels must urgently restore climates of candor that permeate communications, interpersonal relations, and the immensely bureaucratic management systems of the Army. Only professionals who trust the integrity, competence, and intentions of their leaders will choose to remain in the Army profession.

Conclusion 7—*The Army faces increasing jurisdictional competitions with new competitors. Thus its jurisdictional boundaries must be constantly negotiated and clarified by officers comfortable at the bargaining table and skilled in dealing with professional colleagues on matters touching the profession's civil-military and political-military boundaries.* The Army's competitors have multiplied well beyond the group comprising its traditional rivals—the maritime and aerospace military professions. Today the Army faces professional pressure from other professions, occupations, businesses, governmental agencies, nongovernmental agencies, and international organizations. One cause of this increased competition is the Army's expansion into arenas across the spectrum of conflict since the end of the Cold War—drug wars, border patrols, homeland defense, disaster assistance, and so on. While the Army has expanded its missions into new jurisdictions, other organizations have pursued work in the Army's traditional fields. Today we have private corporations providing logistical support during Army operations, teaching Reserve Officer Training, and writing Army doctrine, while the U.S. Marines are executing inland expeditionary missions traditionally assigned to the Army.

This increase in competition facing the Army is no different than that confronting other professions in the western world as they compete for legitimacy. If there is a difference, it is that the Army appears unaware that it is in a fight to defend its current jurisdictions and to determine other jurisdictions in which it should become legitimate. Throughout this jurisdictional battlefield, the Army is proceeding too often in an analytically superficial and haphazard manner. In fact, the boundaries of the profession's jurisdiction are unclear to those outside the profession, confusing them as to the optimum allocation of forces and responsibility for undertaking military operations, but, even more disconcerting, the boundaries are unclear to members of the Army profession itself, thus weakening their professional identity and commitment. This jurisdictional ambiguity makes it appropriate for the nation's civilian leadership to ask the Army to undertake most any task, and it makes it difficult for the Army to advise otherwise. Since effective advice-giving by the Army's strategic leaders increasingly necessitates expertise in negotiating at these

critical political interfaces, the profession needs a larger cohort of leaders at all levels well developed in the skills and norms of such civil-military and political-military relations.

As a corollary to this conclusion, we note that the boundaries *within* the Army profession itself have also become unclear. As the Army outsources more and more inherently professional tasks to private corporations that employ retired military members, it is acknowledging the continued professional affiliation of these former members. However, the nature of their affiliation with the Army is unclear. Although often committed and loyal, these former professionals no longer have the legal obligations to the nation and the profession they once served. Now they have strong ties to for-profit commercial firms with goals quite different from and often incongruent with the Army's. Thus it is with some peril to its own legitimacy that the Army treats these retired professionals as an extension of itself but without imposing the bonds of professional membership—certification, licensing, and legal obligation.

Conclusion 8—*Contrary to the preoccupations of current Army doctrine, renewing the Army as a profession will require strategic leaders to focus rather on (1) the profession's expert knowledge, (2) the development of professionals (future and present), and (3) the Army's jurisdictional competitions.* Owing to its unique characteristics as a profession, strategic leadership of the Army differs from that of other types of organizations, such as a government bureaucracy or a commercial business. Professions are built upon expert knowledge and expertise, and the embodiment of that knowledge in future and existing professionals. Thus, leaders of a profession must focus on updating and redefining expert knowledge through research, investigation, and analysis. They adapt the profession's capability to apply that expert knowledge to present and future situations; and they concentrate the profession's efforts and resources on imparting this expertise to its members, and inculcating in them a professional ethic.

Strategic leaders of the Army must therefore focus its research on people, their behavior, and development; on new forms of conflict, violence, and killing; on culture, world politics, and changing social organizations; as well as on the more familiar experimentation in military doctrines and applications of technology. A sharper research focus will also be required at the next stage, i.e., in discovering how best to develop this knowledge in professionals, the experts in these domains, throughout their Army careers. Interestingly enough, these are the very core activities of the profession that the Army is increasingly outsourcing to private corporations.

Professions also engage in competition and negotiation for legitimate jurisdictions within which to apply their expertise. Therefore, the profession's strategic leaders' task is to identify its competitors, carefully analyze the competition, renegotiate jurisdictions, and create and maintain professional legitimacy to ensure that the profession can attract new members and compete successfully to remain a viable profession. Since jurisdictions and legitimacy are strongly influenced if not determined from outside the profession, these activities require strategic leaders with high degrees of familiarity with external constituencies, such as the press, the Executive, the Congress, international agencies, and the American people. Such literacy

appears best developed over many assignments and years of interaction with these constituents. Thus assignments outside of the Army, perhaps in other government agencies or international organizations or even civilian corporations, are essential components of the career of a successful military leader. Yet, as we noted in Conclusion 4, the prevailing Army career pattern today discourages sufficient numbers of officers from venturing out of the mainstream and into these less touted but strategically relevant billets.

Conclusion 9—*Though daunting, renewing the Army profession and redressing the balance between bureaucracy and profession are not impossible tasks; they have been done before and quite successfully so.* As a profession, the Army has accomplished a similar revitalization in its recent past, from roughly 1972 to 1985. During that renewal, a handful of extraordinary strategic leaders redefined the profession's jurisdiction by focusing on the Soviet threat in the European theater. They reestablished the Army's identity as a profession that could "fight outnumbered and win." The ensuing war-fighting and training revolution produced a transformed Army profession of immense effectiveness, as exemplified in the Gulf War. Today, the Army faces a more complex international environment, though with far less intellectual consensus among its leaders. Yet, the experience of the 1970s and 1980s indicates that the Army profession's renewal can, and must, be an essential element of the on-going transformation.

A critical part of such renewal will be negotiating with external constituencies to achieve needed changes in legislation and executive policy and in political guidance to the profession's strategic leaders. Although any significant change in the Army can occur only after the nation's political leaders have given the profession's strategic leaders such guidance and autonomy, ultimately it will be Army professionals who redefine and renew their profession. This is consistent with other professions where the most successful transformational actions have come from within. Led by the Chief of Staff of the Army, it will be the Army profession that examines itself, decides on the path to pursue, and implements a plan to reach this goal.

How Can I Be a Professional if There Is No Profession?

During the field research reported in Chapter 5 of the present volume, a sincere mid-level female officer, questioned on officer attitudes and reasons for the exodus of officers from the service, spoke these poignant words:

> For the past few years, I've heard generals telling us that junior officers are uncommitted and aren't as professional as earlier generations. They claim that is why we are getting out. But I keep asking myself, "How can I be a professional without a profession?" I don't think the Army is acting much like a profession these days, so I think we are even.

These sentiments, echoed in one guise or another by scores of her contemporaries, were a major stimulus behind the "Future of the Army Profession" study project.

If we have accomplished what we set out to do, then this book will have addressed the concerns expressed above. The book will thus be a beginning rather

Cohesion, and Leadership. Recent publications include the chapter "Hardiness as a Resiliency Resource under High Stress Conditions" in *Post-Traumatic Psychological Stress: Individual, Group, and Organizational Perspectives on Resilience and Growth* (in press); and "Factors Influencing Small Unit Cohesion in Norwegian Navy Officer Cadets" in *Military Psychology* (in press).

Lance Betros is Academy Professor in the Department of History at the U.S. Military Academy. He is a graduate of the Academy and holds M.A. and Ph.D. degrees from the University of North Carolina in Chapel Hill. Colonel Betros has served in command and staff assignments in the United States and Europe. His current research interests are civil-military relations and the development of officer professionalism in the early twentieth century. His most recent publication is "Political Partisanship and the Military Ethic in America" in *Armed Forces and Society* (2000).

James A. Blackwell, a graduate of the U.S. Military Academy, is Corporate Vice President of Science Applications International Corporation, where he is the Director of the Strategic Assessment Center. The Center conducts studies and analyses on the long-range future of warfare. He was the principal investigator for the Director, Net Assessment, Office of the Secretary of Defense, in producing the study titled "The Strategic Meeting Engagement: Experimentation and Transforming the U.S. Military." He was also the author of *On Brave Old Army Team: The Cheating Scandal That Rocked the Nation, West Point 1951* (Presidio Press, 1996). Dr. Blackwell's research interests are the operational level of war, doctrine, technology, and the defense industrial base.

Chris Bourg is a Ph.D. candidate in the Department of Sociology at Stanford University. In her 10 years as an Army officer, she occupied various positions in Europe before serving as Assistant Professor at the U.S. Military Academy. Her research interests include social psychology, social inequality, and military and political sociology. Her most recent publications are "Trends in Intentions to Enlist and Attend College" in *Recruiting College-Bound Youth into the Military: Current Practices and Future Policy Options* (RAND MR-1093, forthcoming); and "Gender, Sexuality, and the Military" (coauthor) in *General Mosaics: Social Perspectives* (2001). Her dissertation, titled "Gender Mistakes and Inequality," examines the link between sex categorization and inequality.

John W. Brinsfield is Deputy Command Chaplain at U.S. Army Forces Command, Ft. McPherson, Georgia. From 1980 to 1984 he served as an assistant professor in the Department of History at West Point. Subsequent assignments included teaching assignments at the U.S. Army Chaplain School and in the Department of Command, Leadership, and Management of the U.S. Army War College. His most recent publications include *Encouraging Faith, Supporting Soldiers: A History of the U.S. Army Chaplain Corps, 1975-1995* (1997); and *Civil War Chaplains*, written with Dr. James Robertson and due to be released in the fall of 2002. Chaplain (Colonel) Brinsfield is currently assigned by the Chief of Chaplains, as an additional

duty, to interview 50 chaplains and assistants who responded to the 11 September 2001 terrorist attacks on the World Trade Center and the Pentagon. He will edit an anthology of their memories of ministry in these crucibles of fear and faith.

James Burk is a professor of sociology at Texas A&M University. He previously taught at McGill University and the University of Chicago, where he earned his master's and Ph.D. degrees. He has served as editor of the journal, *Armed Forces and Society*, and presently is Chair-Elect of the Peace, War, and Social Conflict Section of the American Sociological Association. He also sits on the executive council of the Inter-University Seminar on Armed Forces and Society, and is a member of the board of editors of *Armed Forces and Society*. A student of political sociology, he has published numerous articles on problems of democratic renewal with an emphasis on civil-military relations and political culture. His latest work is "The Military's Presence in American Society, 1950-2000," in *The Military Profession and American Society in Transition to the 21st Century*, eds. Peter D. Feaver, Richard H. Kohn, and Lindsay Cohn (MIT Press, in press).

Randi C. Cohen, Ph.D., is the founder of Beetrix LLC, a research and consulting firm in Boston. She works with diverse clients, from Fortune 500 companies to non-profit organizations, helping them gain fuller insights into their constituents. Employing advanced qualitative methodologies and sociological frameworks, her studies have ranged from evaluations of financial services to assessments of military actions. Dr. Cohen has been a visiting scholar at the School of Management of Boston College, and has taught marketing in Boston University's MBA program.

Joseph J. Collins, is currently the Deputy Assistant Secretary of Defense for Peacekeeping and Humanitarian Affairs. Dr. Collins rose to the rank of colonel in an Army career that was equally divided among infantry assignments in the United States, Germany, and the Republic of Korea; teaching assignments in the Department of Social Sciences at West Point; and duty as a strategic analyst in the Pentagon. From 1998 to 2001, he worked at the Center for Strategic and International Studies, where he supervised studies on economic sanctions, homeland defense, and American military culture. His chapter in the present book, coauthored with T. O. Jacobs, stems from his work at the Center. He holds a Ph.D. in international relations from Columbia University and is a member of the Council on Foreign Relations.

Martin L. Cook, Ph.D., is Elihu Root Professor of Military Studies and Professor of Ethics, Department of Command, Leadership, and Management, U.S. Army War College. He previously was a tenured member of the faculty at Santa Clara University, and has taught at the U.S. Air Force Academy and The College of William and Mary, among other institutions. His most recent publications include "Immaculate War: Constraints on Humanitarian Intervention" in *Ethics and International Affairs* (2000); "Why Serve the State? Moral Foundations of Military Officership" in *The Leader's Imperative: Ethics, Integrity, and Responsibility*, ed. J.

Carl Ficarrotta (Purdue University Press, 2001); and "Two Roads Diverged, and We Took the One Less Traveled: Just Recourse to War and the Kosovo Intervention" in *Kosovo: Contending Voices on the Balkan Intervention*, ed. William Buckley (Eerdmans, 2000).

Whitfield B. East is Director of Instruction in the Department of Physical Education at the U.S. Military Academy. He received a master's degree in physical education from the University of North Carolina and the Ed. D. at the University of Georgia, specializing in mathematical techniques for estimating motor performance. Prior to arriving at West Point, he was a professor and department chairman in physical education at East Tennessee State University. He has lectured and published widely in the field of human physical fitness, including most recently coauthorship of the article "Relationships of Fitness with Gross Motor Skills for Five- to Six-Year-Old Children," appearing in *Perceptual and Motor Skills* (1999).

George B. Forsythe, Ph.D., is Professor and Vice Dean for Education at the U.S. Military Academy. A career Army officer with over 31 years on active duty, Colonel Forsythe has served in infantry units in Berlin and Korea and has been a member of the USMA faculty for two decades. His scholarly interests include leader development, education in the professions, and program assessment. Recent publications include *Practical Intelligence in Everyday Life* (coauthor), published in 2000 by Cambridge University Press; and "Experience, Knowledge, and Military Leadership" (coauthor) in *Tacit Knowledge in Professional Practice*, eds. Robert Sternberg and Joseph Horvath, published in 1999 by Lawrence Erlbaum Associates.

Frederick M. Franks, Jr., General, U.S. Army Retired, occupies the Simon Chair, Center for the Professional Military Ethic, at the U.S. Military Academy, and works with the Army as a senior observer in the Battle Command Training Program for senior commanders and staffs. He also teaches senior level battle command in seminars at military schools in the United States and United Kingdom. He is a graduate of the U.S. Military Academy, holds two master's degrees from Columbia University, and is a graduate of the National War College. He served in the Vietnam War in 1969-70, where he was twice wounded, and later commanded VII Corps during Operations Desert Shield/Desert Storm in the Gulf War. He completed 35-plus years of military service in 1995 as commander of the U.S. Army Training and Doctrine Command. General Franks has published in magazines and journals on battle command and most recently collaborated with Tom Clancy on the book, *Into the Storm: A Study in Command*, published in 1997.

T. Owen Jacobs occupies the Leo Cherne Distinguished Visiting Professor Chair at the Industrial College of the Armed Forces, National Defense University. Before coming to the Industrial College in 1995, he was with the U.S. Army Research Institute where he directed research focusing on strategic leadership. He received B.S. and M.A. degrees from Vanderbilt University (1950, 1954) and the Ph.D. in psychology from the University of Pittsburgh (1956).

Douglas V. Johnson II is a research professor at the U.S. Army War College's Strategic Studies Institute where he has served since 1985. A 1963 graduate of the U.S. Military Academy, he served two tours each in Germany and Vietnam, and took his master's degree in history at the University of Michigan and his Ph.D. in history at Temple University. He has taught history of the military art at West Point and was one of the founding faculty of the School of Advanced Military Studies at Ft. Leavenworth. He has taught the USAWC Army After Next (now Transforming the Army) Special Program seminar for the past six years. During his sixteen years at the War College, he has supervised the revision of Field Manual 100-1 (now FM 1), *The Army*, in 1986, 1991, 1994, and 2001. His chapter in the present book, coauthored with Leonard Wong, stems from his long association with that field manual and his history background. His fields of study are military history (specializing in the AEF in WW I), future studies, civil-military relations, national security strategy, and technology.

Meyer Kestnbaum is Assistant Professor of Sociology and Research Associate of the Center for Research on Military Organization at the University of Maryland, College Park. He received his A.B., M.A., and Ph.D. degrees from Harvard University. His recent publications include "Citizen-Soldiers, National Service, and the Mass Army: The Birth of Conscription in Revolutionary Europe and North America" in *Comparative Social Research* (in press); "Are Post-Cold War Militaries Postmodern?" (coauthor) in *Armed Forces and Society* (2001); and "Citizenship and Compulsory Military Service: The Revolutionary Origins of Conscription in the United States" in *Armed Forces and Society* (2000). His current research examines the construction and transformation of relations between the state and its citizens during war, with particular emphasis on regimes of military service and the dynamics of voluntarism and compulsion.

Thomas A. Kolditz, Ph.D., is Head of the Department of Behavioral Sciences and Leadership and Director of the Tactical Officer Education Program at the U.S. Military Academy. Colonel Kolditz's Army service included command of a nuclear-capable artillery battery in Germany during the Cold War and of a cannon battalion in support of multinational forces in the Demilitarized Zone of the Republic of Korea. A social psychologist, he has researched and coauthored several strategic analyses for the Army, including the Senior Review Panel Report on Sexual Harassment (1997) and the Well-being Framework for the U.S. Army (2000). His current research interest is the formation of professional dyads. His most recent work (coauthored), "Feedback Seeking Behavior and the Development of Mentor-Protégé Relationships," was presented at the Inter-University Seminar on Armed Forces and Society (2002).

Joe LeBoeuf, Ph.D., is Professor and Director, Organizational Studies and Military Leadership, in the Department of Behavioral Sciences and Leadership at West Point. A graduate of the U.S. Military Academy with over thirty years of uniformed service, Colonel LeBoeuf served in a variety of command and staff positions as an

Army combat engineer before returning to West Point as a member of the permanent faculty in 1994. His research interests include the effective teaching of leadership, leader development, and organizational culture. His most current research and study activity was with the Army Training and Leader Development Panel (May 2000), conducted for the Army Chief of Staff and focusing on Army culture, leader development, and the service ethic.

Maureen K. LeBoeuf assumed the duties of Professor and Head of the Department of Physical Education at the U.S. Military Academy in 1997. Her previous assignments in the Department of Physical Education included instructor and administrative officer. In 1991 she was selected to be an Academy Professor and was Director of Instruction from 1994 to 1996. Colonel LeBoeuf entered the U.S. Army in 1976 and has held various command and aviation staff assignments in the United States and Germany. Her current position carries the unique title *Master of the Sword*. She is the first woman department head at the U.S. Military Academy since it was founded in 1802. Colonel LeBoeuf holds an Ed.D. degree in curriculum and instruction from the University of Georgia.

William J. Lennox, Jr., is the 56th Superintendent of the U.S. Military Academy. He is a graduate of the Academy and holds M.A. and Ph.D. degrees from Princeton University. He was a White House Fellow in 1986-88 and a National Security Fellow at Harvard University in 1990-91. Lt. Gen. Lennox's past positions have included service as an instructor in the West Point Department of English, commander of a field artillery battalion, commander of divisional artillery, Deputy Commander of the Field Artillery School, Special Assistant to the Secretary of the Army, Chief of Staff of III Corps, Deputy Commanding General of Eighth Army in Korea, and Chief of Army Legislative Liaison.

Philip Lewis is a professor of psychology at Auburn University where he directs the graduate program in Industrial and Organizational Psychology and teaches undergraduate and graduate courses on personality, adolescent and adult development, and managerial development. His research on the conceptual development of leaders has been supported by the U.S. Army Research Institute for the Behavioral and Social Sciences and the National Research Council. In 1993-94 he was the Distinguished Visiting Professor at West Point in the Department of Behavioral Sciences and Leadership. While at West Point he helped initiate a longitudinal study of the conceptual development of cadets in the class of 1998.

Gregg F. Martin teaches at the U.S. Army War College in the Department of Command, Leadership, and Management, where he is Director of Leadership and Command Instruction. In addition to serving in a variety of command and staff assignments throughout the Army, he taught in the Department of Social Sciences at the U.S. Military Academy. A graduate of West Point, he holds advanced degrees from both the Army and Naval War Colleges, as well as from Massachusetts Institute of Technology, where he earned a Ph.D. His latest publication is the article

"Transforming Strategic Leader Education for the 21st Century Army" (coauthor), in *Parameters* (Autumn 2001). He is slated to command the 130th Engineer Brigade in Europe upon rotation from his present assignment.

Lloyd J. Matthews is an independent contractor performing writing and editorial projects principally for the U.S. Army War College. He received the B.S. degree from the U.S. Military Academy, M.A. from Harvard University, and Ph.D. from the University of Virginia, and is a graduate of the Army War College. His military assignments included command at platoon, company, and battalion levels; advisory duty in Vietnam; editorship of *Parameters*, the Army War College quarterly; and the associate deanship of the U.S. Military Academy. Following retirement from the Army, he served as a project manager in Saudi Arabia and Turkey. Colonel Matthews has published some 100 articles, features, reviews, monographs, and editions on professional topics, including the article "Is the Military Profession Legitimate?" in the January 1994 issue of *ARMY*. His most recent publications are the entry on military ideals in *The Oxford Companion to American Military History* (1999) and an edition titled *Building and Maintaining Healthy Organizations*, published by the Army War College's Department of Command, Leadership, and Management (2001).

John Mark Mattox is a nuclear policy planner on the staff of the Supreme Headquarters, Allied Powers Europe, the military headquarters of the North Atlantic Treaty Organization. He holds a B.A. degree from Brigham Young University, a Master of Military Arts and Science degree from the United States Army Command and General Staff College, and M.A. and Ph.D. degrees from Indiana University. Lieutenant Colonel Mattox earlier served in such positions as acting operations officer (S-3) of an air assault field artillery battalion of the 101st Airborne Division during the 1991 Gulf War, commander of an air assault field artillery battery, and assistant professor of English at the U.S. Military Academy. He has published articles on theoretical linguistics, military tactics, and ethics. His recent articles include "Henry V: Shakespeare's Just Warrior" in *War, Literature, and the Arts* (2000); "From Commander's Mind to Steel on Target (coauthor) in *Infantry* (1999); and "Fifth-Century Advice for 21st-Century Military Leaders" in *Military Review* (1998).

Jeffrey D. McCausland holds the Class of 1961 Chair of Leadership at the U.S. Naval Academy. He is a retired U.S. Army colonel, having culminated his military career as Dean of Academics, U.S. Army War College. He is a graduate of the U.S. Military Academy and holds master's and Ph.D. degrees from the Fletcher School of Law and Diplomacy, Tufts University. He has served in a variety of command and staff positions in both the United States and Europe, including the position of Director for Defense Policy and Arms Control on the National Security Council Staff. He has also worked on the Treaty on Conventional Armed Forces in Europe (CFE) as a member of the Office of the Deputy Chief of Staff for Operations, U.S. Army Staff. Following this assignment he assumed command of a field artillery battalion stationed in Europe and deployed his unit to Saudi Arabia for Operations Desert Shield/Storm in 1990-91.

He has published broadly on leadership, military affairs, European security issues, the Gulf War, and arms control. He served as a member of the Army Chief of Staff's Blue Ribbon Panel on Training and Leader Development in 2000. Dr. McCausland has provided advice and assistance to the Air Force Chief of Staff's Developing Aerospace Leaders initiative. He has been a visiting fellow at the Center for International Affairs, Harvard University; Stiftung Wissenshaft und Politik, Ebenhausen, Germany; George C. Marshall Center for European Security Studies, Garmisch, Germany; and the International Institute for Strategic Studies, London.

Thomas L. McNaugher is Acting Director of the Arroyo Center, the Army's federally-funded research institute within the RAND Corporation. Before going to RAND in 1995 he was a Senior Fellow in the Foreign Policy Studies Program at the Brookings Institution, specializing in U.S. military strategy and politics. His books include *The M16 Controversies: Military Organizations and Weapons Acquisition* (Praeger, 1984); *Arms and Oil: U.S. Military Strategy and the Persian Gulf* (Brookings, 1985); and *New Weapons, Old Politics: America's Military Procurement Muddle* (Brookings, 1989). Dr. McNaugher received a B.S. degree from the U.S. Military Academy as well as M.P.A. and Ph.D. degrees from Harvard University.

Suzanne C. Nielsen is a graduate of the U.S. Military Academy and is currently assigned to the Academy's Department of Social Sciences, where she is the course director of the core course in international relations and teaches an elective titled "The Strategy and Politics of Cyberwarfare." Prior to her assignment to West Point, she served in military intelligence units in Germany and the United States. As the company commander of an intelligence company, she deployed her company twice from Fort Bragg, North Carolina, to Europe to support peace enforcement operations in Bosnia. Major Nielsen is currently a doctoral student in the Department of Government, Harvard University. Her research pertains to civil-military relations and the use of force.

Peter J. Roman is Associate Professor of Political Science at Duquesne University. He is author of *Eisenhower and the Missile Gap* (Cornell University Press); "Ike's Hair Trigger: U.S. Nuclear Predelegation, 1953-1960" in *Security Studies*; "The Joint Chiefs of Staff: From Service Parochialism to Jointness" in *Political Science Quarterly* (coauthored with David Tarr); and "Is There A Gap between Civilian and Military Leaders? If So, Does It Matter?" in *The Civil-Military Gap and American National Security in the 21st Century,* eds. Peter D. Feaver and Richard H. Kohn (MIT Press, forthcoming). He is currently writing a book with David Tarr on the role of the Joint Chiefs of Staff in national security policy-making.

Eric B. Schoomaker is Commander of the 30th Medical Brigade, V Corps, headquartered in Heidelberg, Germany. He was commissioned as a second lieutenant from the ROTC program of the University of Michigan, from which he received a B.S. degree. Colonel Schoomaker later received his medical degree from the University's Medical School and a Ph.D. in human genetics from the University's

Rackham School of Graduate Studies, and graduated from the U.S. Army War College. In previous assignments, he served as Commander of the Evans Army Community Hospital at Fort Carson, Colorado, and as Command Surgeon for U.S. Army Forces Command.

David R. Segal, Ph.D., is Distinguished Scholar-Teacher, Professor of Sociology, Affiliate Professor of Government and Politics and of Public Affairs, and Director of the Center for Research on Military Organization at the University of Maryland. He is also president of the Inter-University Seminar on Armed Forces and Society. Professor Segal's recent publications include *The Postmodern Military* (co-editor) (Oxford University Press, 2000); "Postmodernity and the Modern Military" (coauthor) in *Armed Forces and Society*, (2001); "Downsizing the Russian Army" (coauthor) in *Journal of Political and Military Sociology* (2001); "Is a Peacekeeping Culture Emerging among American Infantry in the Sinai MFO?" in *Journal of Contemporary Ethnography* (2001); and "Attitudes of Entry-level Personnel: Pro-military and Politically Mainstreamed" (coauthor) in Peter D. Feaver and Richard H. Kohn, eds., *Soldiers and Civilians* (MIT Press, 2001). His current research focus is youth attitudes toward the military and changing military organization and missions.

Mady Wechsler Segal, Ph.D., is Professor of Sociology at the University of Maryland, Distinguished Scholar-Teacher, faculty affiliate of the Women's Studies Program and the Center for International and Security Studies at Maryland (CISSM), and Associate Director of the Center for Research on Military Organization. She is Vice President of the Research Committee on Armed Forces and Conflict Resolution of the International Sociological Association. Her recent publications include "Gender and the Military" in Janet Saltzman Chafetz, ed., *Handbook of the Sociology of Gender* (Kluwer, 1999); "The Impact of Military Presence in Local Labor Markets on the Employment of Women" (coauthor) in *Gender & Society* (2000); and "Gender, Sexuality, and the Military" (coauthor) in Dana Vannoy, ed., *Gender Mosaics: Social Perspectives* (2001). Her current research interest is military women and military families.

Anna Simons is Associate Professor of Defense Analysis and a member of the Special Operations Academic Group at the Naval Postgraduate School. Previously, she was an associate professor of anthropology at UCLA. She has conducted ethnographic fieldwork both in Africa and with U.S. Army Special Forces, and is author of *Networks of Dissolution: Somalia Undone* (Westview Press, 1995) as well as *The Company They Keep: Life Inside the U.S. Army Special Forces* (The Free Press, 1997). Her research and publications continue to focus on the bonds that tie people together as well as those that drive people apart.

Don M. Snider, Ph.D., is Professor of Political Science, U.S. Military Academy, appointed in 1998. As a career Army officer his service ranged from rifle company command in Vietnam to the staff of the National Security Council in the White

House. As an academic, he has served some nine years in the Social Sciences Department at West Point, including three years occupying the Olin Chair as Distinguished Professor of National Security Studies. His most recent publications are "America's Post-Modern Military" in *World Policy Journal* (Spring 2000); "The Future of Army Professionalism: A Need for Renewal and Redefinition" (coauthor) in *Parameters* (Autumn 2000); and *Professionalism, the Military Ethic, and Officership in the 21st Century* (coauthor), USAWC Strategic Studies Institute, 2000).

Scott A. Snook is Academy Professor in the Department of Behavioral Sciences and Leadership at the U.S. Military Academy where he teaches courses in systems leadership and leading organizations through change. He also teaches on leading complex organizations in the Harvard Business School's executive education programs and works with Professor John Kotter as a member of Harvard's emerging Global Leadership Initiative. He is currently on sabbatical, serving as the Superintendent's strategic planner in the Academy's Office of Policy, Planning, and Analysis. He has served in various command and staff positions for the past 20 years, including participation in operations in Grenada in 1983. He has an MBA from the Harvard Business School and earned his Ph.D. from Harvard University in organizational behavior. His book *Friendly Fire* was published by Princeton University Press in 2000. He has also coauthored a book that explores the role of "common sense" in leadership titled *Practical Intelligence in Everyday Life* (Cambridge University Press, 2000).

Elizabeth A. Stanley-Mitchell is Associate Director of the Center for Peace and Security Studies and Associate Director of the Security Studies Program in the Edmund A. Walsh School of Foreign Service at Georgetown University. A Ph.D. candidate in government at Harvard University, she is currently finishing her dissertation, which deals with the domestic politics of war termination. Until 1996, Ms. Stanley-Mitchell served as a U.S. Army captain in military intelligence, with postings in Korea, Germany, Italy, Macedonia, and Bosnia. Her most recent publication is "Technology's Double-Edged Sword: The Case of U.S. Army Digitization" in *Defense Analysis* (Winter 2001). She holds an MBA from MIT's Sloan School of Management and a B.A. in Soviet and East European Studies from Yale University. Ms. Stanley-Mitchell has taught at MIT, Harvard, and Georgetown.

David W. Tarr is Emeritus Professor of Political Science at the University of Wisconsin-Madison. A former chairman of the Political Science Department, his teaching and research focus on U.S. national security. Currently he is coauthoring a book with Peter J. Roman on the role of the Joint Chiefs of Staff and other senior military leaders in the policy-making process. Recent publications include two coauthored articles: "Military Professionalism and Policy-Making: Is There a Gap at the Top? If So, Does It Matter?" in Peter D. Feaver and Richard H. Kohn, eds., *Soldiers and Civilians: The Civil-Military Gap and American National Security* (MIT Press, 2001); and "The Joint Chiefs of Staff: From Service Parochialism to Jointness" in *Political Science Quarterly* (Spring 1998).

James H. Toner is Professor of International Relations and Military Ethics at the U.S. Air War College, Air University, Maxwell AFB, Alabama, where he teaches in the Leadership and Ethics Department. He has taught at the University of Notre Dame, Norwich University, Auburn University, and Auburn University-Montgomery. Professor Toner took his B.A. degree in history from St. Anselm College, M.A. in government from The College of William and Mary, and Ph.D. in government and international studies from the University of Notre Dame. He was on active duty from 1968 to 1972, attaining the rank of captain in the Army. He received a diploma from the Air War College in 1992. Professor Toner has published widely in such journals as *Air University Review*, *Armed Forces and Society*, *ARMY*, *Joint Force Quarterly*, *Marine Corps Gazette*, *Military Review*, *Naval War College Review*, *Parameters*, *Review of Politics*, and *Social Science*. He is author of *The American Military Ethic: A Meditation* (Praeger, 1991); *The Sword and the Cross: Reflections on Command and Conscience* (Praeger, 1992); *True Faith and Allegiance: The Burden of Military Ethics* (University Press of Kentucky, 1995); and *Morals Under the Gun: The Cardinal Virtues, Military Ethics, and American Society* (University Press of Kentucky, 2000).

Marybeth Peterson Ulrich is Associate Professor of Government in the Department of National Security and Strategy at the U.S. Army War College. Dr. Ulrich is a graduate of the U.S. Air Force Academy and earned her Ph.D. in political science from the University of Illinois. She served 15 years on active duty as an Air Force officer, which included tours as a navigator on KC-135Q refueling tankers and as an instructor in the Department of Political Science at the Air Force Academy. She is currently serving in the USAF Reserve. She has written extensively in the field of strategic studies with special emphasis on national security democratization issues in postcommunist Europe, European security, and civil-military relations. Her book, *Democratizing Communist Militaries: The Cases of the Czech and Russian Armed Forces*, was published by the University of Michigan Press in February 2000.

Gayle L. Watkins, Ph.D., is the founder of Clove Brook Enterprises, a research-consulting firm in Cold Spring, New York. During her 23-year Army career, she served in logistics positions in the United States and Europe before joining the faculty of the U.S. Military Academy. While at West Point, she directed the Leadership and General Management programs and conducted studies in personnel issues for the Army. Her research has ranged from studies of social identity to qualitative assessments of military actions. Most recently, she coauthored "The Future of Army Professionalism: A Need for Renewal and Redefinition" in *Parameters* (Autumn 2000).

Leonard Wong is Associate Research Professor of Military Strategy (Human and Organizational Dimensions) in the U.S. Army War College's Strategic Studies Institute. He served in the Army for over twenty years, including service as a leadership instructor at the U.S. Military Academy. During assignments on the Army Staff in the Pentagon, he was an analyst in the Program Analysis and Evaluation

Directorate and later in the Office of the Deputy Chief of Staff for Personnel, and Director of the Office of Economic and Manpower Analysis. He has authored several articles, chapters, and papers on organizational issues in the Army such as downsizing, leadership, and junior officer retention. His monograph *Generations Apart: Xers and Boomers in the Officers Corps* was published by the Strategic Studies Institute in 2000. He is a registered Professional Engineer and holds a B.S. from the U.S. Military Academy plus M.S.B.A. and Ph.D. degrees from Texas Tech University.